Piezoelectric and Acoustic Materials for Transducer Applications

Ahmad Safari • E. Koray Akdoğan
Editors

Piezoelectric and Acoustic Materials for Transducer Applications

Springer

Editors
Ahmad Safari
Department of Materials Science
and Engineering
The Glenn Howatt Electronic Ceramics
Laboratory
Rutgers University
Piscataway, NJ 08854

E. Koray Akdoğan
Department of Materials Science
and Engineering
The Glenn Howatt Electronic Ceramics
Laboratory
Rutgers University
Piscataway, NJ 08854

Consulting Editor
D.R. Vij

ISBN: 978-0-387-76538-9 e-ISBN: 978-0-387-76540-2

Library of Congress Control Number: 2008927196

Printed on acid-free paper

springer.com

Preface

The idea of creating a book of this genre came with the realization that there were no books in the open literature that combined the materials aspect of piezoelectric and acoustic materials together with the principles of transducer design and recent advances in piezoelectric transducers, as well as their application. By combining topics such as Fundamentals of Piezoelectricity (Part I), Piezoelectric and Acoustic Materials for Transducer Technology (Part II), Transducer Design Principles (Part III), and Piezoelectric Transducer Fabrication Methods (Part IV), our purpose was to provide a comprehensive and self-consistent volume whereby the aforementioned lack of a reference book in the open literature can be remedied.

This book is comprised of four complementary sections. In Part I, a concise treatment of piezoelectric phenomena in solids is presented. Chapter 1 by Akdoğan and Safari establishes the solid-state thermodynamic foundation for ferroelectricity as all piezoelectrics used in today's transducer technology are ferroelectric. It also delineates the origin of very important concepts such as spontaneous polarization, hysteresis loops, and piezostrain coefficients, which are needed to describe the macroscopic behavior of piezoelectrics. The chapter by Kholkin, Pertsev, and Goltsev (Chap. 2) deals with the symmetry aspects of the piezoelectric effect in various materials (single crystals, ceramics, and thin films). It defines the third-rank tensor of piezoelectric coefficients in reference to the fundamentals of crystallography, and then discusses the orientation dependence of the longitudinal piezoelectric response in ferroelectric single crystals. Also, a concise discussion on the effective piezoelectric constants of polydomain crystals, ceramics, and thin films and their dependence on crystal symmetry is provided. The domain-wall contribution to the piezoelectric properties of ferroelectric ceramics and thin films is also given. Finally, the crystallographic principles of piezomagnetic, magnetoelectric, and multiferroic materials are presented. In Chap. 3, Trolier-McKinstry presents a detailed description of the crystallochemical principles of ferroelectricity and piezoelectricity in solids. The discussion covers perovskites, tolerance factors, domains and domain walls, the lead zirconate titanate (PZT) system, bismuth-layer structure, $LiNbO_3$, tungsten bronze structure, and SbSI, which is the point of departure in the development of piezoelectric composites. The emphasis is on the atomic arrangements leading to polarization and its consequences on material properties.

In Part II, three chapters are devoted to the most prominent piezoelectric and acoustic material systems currently either in use or in development for transducer applications. Damjanovic (Chap. 4) gives a detailed account on lead-based ferroelectrics/piezoelectrics, including $PbTiO_3$, PZT, relaxor ferroelectrics, and single crystals. Therein, one could find an in-depth discussion of the morphotrophic phase boundary in the PZT system as well as the anisotropy in piezoelectric response in $PbTiO_3$. Systems with compositional modifications are treated as well, where hard and soft ferroelectrics are introduced. Kosec et al. (Chap. 5) gives a full overview of the (K, Na)NbO_3 system, which is the most important lead-free ferroelectric system whose compositional modifications have great promise for applications in the future. The emphasis is on the processing of ceramics such as $KNbO_3$, $NaNbO_3$, (K, Na)NbO_3, and the recently discovered KNN–LT–LS ternary system, thereby effectively establishing processing-property relations. Also, a wealth of material properties appertaining to the (K, Na)NbO_3 and related systems are included therein. Chap. 6 by Takenaka is on bismuth-based ferroelectric ceramics, which is another lead-free system of utmost importance. The major focus there is on the $Bi_4Ti_3O_{12}$ and its compositional modifications. Most importantly, Takenaka provides a very thorough treatment of grain orientation (texturing) of such lead-free ceramics, and discusses its consequences on piezoelectric properties. Cheng et al. (Chap. 7) presents a very detailed overview of the advances in electromechanically active polymers in the context of mechatronics and artificial muscle. After a concise review of the appertaining phenomenology on electromechanical behavior in solids, he and his colleagues present systems such as PVDF, P(VDF–TrFE) copolymer, and terpolymers. A very comprehensive discussion on phase transitions, structure, and electromechanical response is provided along with sets of material properties. In Chap. 8, Yamashita et al. introduce the recent advances in acoustic lens materials and provide a systematic account on recent advances in silicon-based technologies. Therein, the evolution of acoustic properties of silicon lenses are discussed from the processing vantage point, and the most prominent dopants for composite-making are identified. Part II ends with another chapter on acoustic lens materials by Kondo, which summarizes the use of carbon fiber composites and how a wide range of acoustic impedances can be fashioned for transducer applications.

Part III is devoted to transducer design principles. First, Lethiecq et al. (Chap. 10) discuss the design principles governing medical ultrasound transducers. The performance metrics of such transducers are introduced, operation principles of single element and transducer arrays are elaborated on, and the underlying acoustic principles are discussed. In Chap. 11, Tressler discusses the design principles of transducers for sonar applications. Various in-air and under-water calibration techniques are first presented, followed by various projector designs (ring, bender bar, flexural disks, flextensional, tonpilz) are introduced and appertaining design methodologies are elaborated on. Also included in that chapter are hydrophone design principles, including cylinder and sphere configurations. This section ends with Chap. 12, where Hladky-Hennion provides a comprehensive overview of finite element modeling (FEM) principles as applied to piezoelelectric transducers. First, the mathematical foundations are provided through constitutive equations, the variational principle, and appertaining functionals. Then, the application of the FEM method is demonstrated. Analysis methods such as static, harmonic, modal, and transient are given. The chapter concludes with examples on the analysis of cymbal transducers and cymbal arrays, and 1–3 composite transducers.

Part IV is on transducer fabrication methods. Therefore, it is inherently tied into applications as well since fabrication methods tend to be application specific. In Chap. 13, Schoenecker addresses the status of piezoelectric fiber composite fabrication, and focuses on three topics: the preparation of sol–gel-derived PZT fiber/polymer composites, the soft-mold method with high achievement potential for preparing tailor-made composites and understanding the structure–property relationships, and the preparation of powder suspension-derived PZT fiber/polymer composites as the technologically advanced and commercialized process. Therein, it is iterated that the use of piezoceramic fibers allows for the fabrication of high-quality fiber composites, surpassing performance metrics that cannot be achieved by the conventional dice and fill technique. Beige and Steinhausen (Chap. 14) discuss different types of composition gradient systems for bending actuators. The combination of hard and soft piezoelectric ceramics and electrostrictive and electroconductive materials are introduced. Processing strategies are summarized and a mathematical models for modeling of poling in such combined systems are presented. The results of theoretical analysis are compared with experimental data for lead-free systems based on barium titanate. In Chap. 15, Smay et al. present the advances made in the use of robocasting solid freeform fabrication (SFF) technique based on the direct writing of highly concentrated colloidal gels. They show that robocasting offers facile assembly of complex three-dimensional geometries and a broad pallet of ceramic, metallic, and polymeric materials from which devices can be developed. Examples of PZT skeletons for direct use or to create epoxy-filled composites suitable for hydrostatic piezoelectric sensors are given. PZT composites of (3–3, 3–2, 3–1) connectivity are demonstrated. It is shown that the figure of merit ($d_h g_h$) increases by up to 60-fold compared with bulk PZT in such composites. In Chap. 16, Jantunen et al. present a comprehensive overview of the general properties and requirements of piezoelectric micropositioners. Special attention is paid to stiffness related to other actuator properties, with general rules and examples. Also the control and sensor techniques required to avoid or minimize the nonlinearities of the piezoelectric device are discussed in detail. And it is shown that in micropositioning a profound overall know–how about material properties, actuator design, and control, sensor and driving techniques are required in addition to an in-depth knowledge of application requirements. Finally, some commercial applications utilizing piezoelectric micropositioners are given as examples. Doğan and Uzgur (Chap. 17) present a broad review of the application of piezoelectric transducers. Piezoelectric actuators are compared with magnetically active and thermally active actuators, and then piezoelectric actuators and their design and fabrication, especially traditional piezoelectric transducers with newly designed flextensional transducers, are compared.

Finally, application-related issues are discussed on Chap. 18 by Shashank et al. provides strategies for the selection of the piezoelectric transducers based on the frequency and amplitude of the mechanical stress in the context of energy harvesting. The figure of merit for the material selection is shown to be directly proportional to the product $(d \times g)$. The criterion for maximization of the product is discussed in depth, and results are reported on various devices utilizing piezoelectric bimorph transducers. Or and Chan (Chap. 19) present recent advances in piezocomposite ultrasonic transducers for high-frequency wire bonding of semiconductor packages. The principles of wire-bonding and appertaining challenges are summarized. The use of 1–3 piezocomposite rings in such applications is shown, and the electromechanical characteristics are presented. The chapter also includes a section on the evaluation of wire-bonding performance utilizing such transducers. Bassiri-Gharb (Chap. 20) presents the use of piezoelectrics in MEMS applications. In that chapter, the properties of AlN, ZnO, PZT, and PMN–PT films are reviewed, and the effect of substrate clamping is briefly discussed. Then fabrication and integration issues are addressed. Various applications of piezoelectrics, including AFM probe tips, RF switches, micromirrors, micropumps, and microvalves are given. In Chap. 21, Shung et al. discuss high-frequency ultrasonic transducers and arrays. After a thorough overview of the state of the art, $LiNbO_3$ single-crystal transducers, and PMN–PT single-crystal needle transducers for Doppler flow measurements are presented. A multitude of transducer designs are shown (annular arrays, linear arrays, and their derivatives), and underlying processing issues are elaborated on. The last chapter of the section and the book is on micromachining of piezoelectric transducers by Pappalardo et al. (Chap. 22). The basic principles, the fabrication processes, and some modeling approaches of novel micromachined ultrasonic transducers (MUTs) are described. It is shown that these transducers utilize the flextensional vibration of an array of micro-membranes called cMUT (capacitive MUT). It is also shown that good echographic images of internal organs in the human body have been obtained, demonstrating the possibilities of this technology to be utilized in commercial 1D and 2D probes for medical applications.

In conclusion, we thank all the contributing authors for their great zeal and industry in composing their respective chapters. Were it not for their willingness to participate in this exciting project, none that was accomplished could have been possible. A project of this magnitude cannot see the light of day without a constant source of encouragement and support. To that end, we express our gratefulness to our families, friends, and colleagues. Last but not the least, we thank Springer for giving us great flexibility in regard to the size of this volume.

Piscataway, New Jersey

A. Safari
E.K. Akdoğan

Contents

Preface v

Part I Fundamentals of Piezoelectricity

1 **Thermodynamics of Ferroelectricity** 3
E. Koray Akdoğan and Ahmad Safari

2 **Piezoelectricity and Crystal Symmetry** 17
A.L. Kholkin, N.A. Pertsev, and A.V. Goltsev

3 **Crystal Chemistry of Piezoelectric Materials** 39
Susan Trolier-McKinstry

Part II Piezoelectric and Acoustic Materials for Transducer Technology

4 **Lead-Based Piezoelectric Materials** 59
Dragan Damjanovic

5 **KNN-Based Piezoelectric Ceramics** 81
Marija Kosec, Barbara Malič, Andreja Benčan, and Tadej Rojac

6 **Bismuth-based Piezoelectric Ceramics** 103
Tadashi Takenaka

7 **Electropolymers for Mechatronics and Artificial Muscles** 131
Zhongyang Cheng, Qiming Zhang, Ji Su, and Mario El Tahchi

8 **Low-Attenuation Acoustic Silicone Lens for Medical Ultrasonic Array Probes** 161
Y. Yamashita, Y. Hosono, and K. Itsumi

9 **Carbon-Fiber Composite Materials for Medical Transducers** 179
Toshio Kondo

Part III Transducer Design and Principles

10 Piezoelectric Transducer Design for Medical Diagnosis and NDE ... 191
Marc Lethiecq, Franck Levassort, Dominique Certon, and Louis Pascal Tran-Huu-Hue

11 Piezoelectric Transducer Designs for Sonar Applications 217
James F. Tressler

12 Finite Element Analysis of Piezoelectric Transducers 241
Anne-Christine Hladky-Hennion and Bertrand Dubus

Part IV Piezoelectric Transducer Fabrication Methods

13 Piezoelectric Fiber Composite Fabrication 261
Andreas Schönecker

14 Composition Gradient Actuators 289
Ralf Steinhausen and Horst Beige

15 Robocasting of Three-Dimensional Piezoelectric Structures 305
James E. Smay, Bruce Tuttle, and Joseph Cesarano III

16 Micropositioning 319
J. Juuti, M. Leinonen, and H. Jantunen

17 Piezoelectric Actuator Designs 341
Aydin Doğan and Erman Uzgur

18 Piezoelectric Energy Harvesting using Bulk Transducers 373
Shashank Priya, Rachit Taneja, Robert Myers, and Rashed Islam

19 Piezocomposite Ultrasonic Transducers for High-Frequency Wire Bonding of Semiconductor Packages 389
Siu Wing Or and Helen Lai Wa Chan

20 Piezoelectric MEMS: Materials and Devices 413
Nazanin Bassiri-Gharb

21 High-Frequency Ultrasonic Transducers and Arrays 431
K. Kirk Shung, Jonathan M. Cannata, and Qifa Zhou

22 Micromachined Ultrasonic Transducers 453
Massimo Pappalardo, Giosue Caliano, Alessandro S. Savoia, and Alessandro Caronti

Index 479

Contributors

E.K. Akdoğan
The Glenn Howatt Electronic Ceramics Laboratory, Department of Materials Science and Engineering, Rutgers University, Piscataway, NJ 08854, USA

N. Bassiri-Gharb
George W. Woodruff School of Mechanical Engineering,
Georgia Institute of Technology, Atlanta, GA 30332-0405, USA

H. Beige
Institute of Physics, Martin-Luther-Universität Halle-Wittenberg,
Friedemann-Bach-Platz 6, 06108 Halle, Germany

A. Benčan
Jožef Stefan Institute, Jamova 39, 1000 Ljubljana, Slovenia

G. Caliano
Aculab – Department of Electronics, University "Roma Tre", Roma, Italy

J.M. Cannata
NIH Resource on Medical Ultrasonic Transducer Technology,
Department of Biomedical Engineering, University of Southern California,
Los Angeles, CA 90089-1111, USA

A. Caronti
Aculab – Department of Electronics, University "Roma Tre", Roma, Italy

J. Ceserano III
Sandia National Laboratories, Albuquerque, NM 87185, USA

H.L.W. Chan
Department of Applied Physics, The Hong Kong Polytechnic University,
Hung Hom, Kowloon, Hong Kong

Z. Cheng
Materials Research and Education Center, 275 Wilmore Lab, Auburn University, Auburn, AL 36849-5341, USA

D. Certon
Ultrasonics Group (LUSSI), Francois Rabelais University, Tours, France

D. Damjanovic
Ceramics Laboratory, Swiss Federal Institute of Technology–EPFL, Lausanne, Switzerland

A. Doğan
Department of Materials Science & Engineering, Anadolu University, Eskisehir, Turkey

B. Dubus
ISEN Department, IEMN, UMR CNRS 8520, 59046 Lille, France

A.V. Goltsev
Department of Physics, University of Aveiro, 3810-193 Aveiro, Portugal

A.-C. Hladky-Hennion
IEMN, ISEN Department, UMR CNRS 8520, 59046 Lille, France

Y. Hosono
Corporate Research & Development Center, Toshiba Corporation, 1 Komukai-Toshiba-cho, Saiwai-ku, Kawasaki 212-8582, Japan

R. Islam
Materials Science and Engineering, University of Texas Arlington, TX 76019, USA

K. Itsumi
Corporate Research & Development Center, Toshiba Corporation, 1 Komukai-Toshiba-cho, Saiwai-ku, Kawasaki, 212-8582, Japan

H. Jantunen
Microelectronics and Materials Physics Laboratories, EMPART Research Group of Infotech Oulu, University of Oulu, P.O. Box 4500, 90014 Oulu, Finland

J. Juuti
Microelectronics and Materials Physics Laboratories, EMPART Research Group of Infotech Oulu, University of Oulu, P.O. Box 4500, 90014 Oulu, Finland

A.L. Kholkin
Center for Research in Ceramics and Composite Materials (CICECO) & Department of Ceramics and Glass Engineering, University of Aveiro, 3810-193 Aveiro, Portugal
and
A. F. Ioffe Physico-Technical Institute of the Russian Academy of Sciences, 194021 St. Petersburg, Russia

T. Kondo
Faculty of Engineering, Tokushima Bunri University, Sanuki-shi, 1314, Kagawa 769-2101, Japan

M. Kosec
Jožef Stefan Institute, Jamova 39, 1000 Ljubljana, Slovenia

M. Leinonen
Microelectronics and Materials Physics Laboratories, EMPART Research Group of Infotech Oulu, University of Oulu, P.O. Box 4500, 90014 Oulu, Finland

M. Lethiecq
Ultrasonics Group (LUSSI), Francois Rabelais University, Tours, France

F. Levassort
Ultrasonics Group (LUSSI), Francois Rabelais University, Tours, France

B. Malič
Jožef Stefan Institute, Jamova 39, 1000 Ljubljana, Slovenia

R. Myers
Materials Science and Engineering, University of Texas Arlington, TX 76019, USA

S.W. Or
Department of Applied Physics, The Hong Kong Polytechnic University, Hung Hom, Kowloon, Hong Kong

M. Pappalardo
Aculab – Department of Electronics, University "Roma Tre", Roma, Italy

N.A. Pertsev
A.F. Ioffe Physico-Technical Institute of the Russian Academy of Sciences, 194021 St. Petersburg, Russia

S. Priya
Materials Science and Engineering, Virginia Tech, Blacksburg, VA 24061, USA

T. Rojac
Jožef Stefan Institute, Jamova 39, 1000 Ljubljana, Slovenia

A. Safari
The Glenn Howatt Electronic Ceramics Laboratory, Department of Materials Science and Engineering, Rutgers University, Piscataway, NJ 08854, USA

A.S. Savoia
Aculab – Department of Electronics, University "Roma Tre", Roma, Italy

A. Schönecker
Fraunhofer Institute for Ceramic Technologies and Systems, 01277 Dresden, Winterbergstrasse 28, Germany

K.K. Shung
NIH Resource on Medical Ultrasonic Transducer Technology, Department of Biomedical Engineering, University of Southern California, Los Angeles, CA 90089-1111, USA

J.E. Smay
School of Chemical Engineering, Oklahoma State University, Stillwater, OK 74078, USA

R. Steinhausen
Institute of Physics, Martin-Luther-Universität Halle-Wittenberg, Friedemann-Bach-Platz 6, 06108 Halle, Germany

J. Su
Advanced Materials and Processing Branch, National Aeronautics and Space Administration, Hampton, VA 23681, USA

M. El Tahchi
Applied Physics Laboratory, Physics Department, Faculty of Sciences II, Lebanese University, Fanar Campus, BP 90656, Jdeidet, Lebanon

T. Takenaka
Department of Electrical Engineering (EE), Faculty of Science and Technology, Tokyo University of Science, Yamazaki 2641, Noda, Chiba-ken 278-8510, Japan

R. Taneja
Materials Science and Engineering, University of Texas Arlington, TX 76019, USA

L.P. Tran-Huu-Hue
Ultrasonics Group (LUSSI), Francois Rabelais University, Tours, France

J. Tressler
Naval Research Laboratory, Washington, DC 20375, USA

S. Trolier-McKinstry
Materials Science and Engineering Department & Materials Research Institute, The Pennsylvania State University, University Park, PA 16802, USA

B. Tuttle
Sandia National Laboratories, Albuquerque, NM 87185, USA

E. Uzgur
Department of Materials Science & Engineering, Ondokuz Mayis University, Samsun, Turkey

Y. Yamashita
Toshiba Research Consulting Corporation, Kawasaki 212-8582, Japan

Q. Zhang
187 Materials Research Laboratory, The Pennsylvania State University, University Park, 16802 PA, USA

Q.F. Zhou
NIH Resource on Medical Ultrasonic Transducer Technology, Department of Biomedical Engineering, University of Southern California, Los Angeles, CA 90089-1111, USA

Part I
Fundamentals of Piezoelectricity

Chapter 1
Thermodynamics of Ferroelectricity

E. Koray Akdoğan* and Ahmad Safari

1.1 Introduction

This chapter is about the macroscopic theory of ferroelectrics, which is based on the Landau theory of phase transformations. It was first proposed by Landau in the 1930s to study a wide range of complex problems in solid-state phase transformations in general, and ordering phenomena in metallic alloys in particular (Landau and Lifshitz 1980). It was further developed and brought to the realm of ferroelectrics by Devonshire (1949, 1951). It has since then proven to be a very versatile tool in analyzing ferroelectric phenomena, and has become a de facto tool of analysis among ferroelectricians.

At the heart of the Landau theory of phase transformations is the concept of order parameter, which was first proposed in the context of order–disorder transformation involving a change in the crystal symmetry. In such systems, the material of interest transforms from a high-symmetry disordered phase to a low-symmetry ordered phase (de Fontaine 1979; Ziman 1979). The so-called broken symmetry of the crystal due to ordering is represented by an order parameter. Landau has shown that the (Helmholtz) free energy of an order–disorder transformation can be expressed very simply as a Taylor series expansion of the order parameter describing the degree of order (or disorder). Generally, the order parameter can be any variable of the system that appears at the phase transition point. In the case of ferroelectrics, it is the spontaneous polarization. Originally, the Taylor expansion of the order parameter was intended to be limited to temperatures close to the phase transformation temperature, as any mathematical series has a radius of convergence. However, quite oddly, it is a well-established fact that the free energy functional provides a very good approximation even at temperatures far below the transition temperature once higher-order expansion terms are included.

E.K. Akdoğan
The Glenn Howatt Electronic Ceramics Laboratory, Department of Materials Science and Engineering, Rutgers University, Piscataway, NJ 08854, USA

A. Safari, E.K. Akdoğan (eds.) *Piezoelectric and Acoustic Materials for Transducer Applications.*

It should be remembered that the original Landau theory was constructed for second-order phase transformations, which are also known as continuous phase transitions as there is no discontinuity in the entropy at the transition point. Moreover, the heat capacity and the elastic moduli display a so-called λ-type anomaly in the neighborhood of the phase transformation temperature (Landau and Lifshitz 1980). As most ferroelectric phase transitions are weakly first order, the Landau formalism has successfully been extended to the study of ferroelectrics, for which the entropy is discontinuous (Devonshire 1949, 1951).

The literature on the Landau formalism is well-matured and extensive. Classical texts covering wide range of topics include Landau and Lifshitz's *Statistical Physics* (1980) and Salje's *Phase Transformations in Ferroelastic and Co-Elastic Crystals* (1990). Landau theory has been used quite successfully in describing phase transformations in a variety of materials systems, including ferromagnetic, martensitic (ferroelastic), and ferroelectric transitions with appropriate order parameters. For instance, proper ferroelectric, ferromagnetic, and ferroelastic phase transformations can be described via the Landau potential, with the polarization P_i, magnetization M_i, or the self-strain as the order parameter; respectively (Wadhawan 2000).

From a historical perspective, it should be pointed out that the extension of Landau theory to ferroelectric phase transformations, characterized by the noncentrosymmetric displacements of atoms in the ferroelectric phase, was accomplished by Ginzburg (Ginzburg 1945), which precedes Devonshire's work on ferroelectrics (Devonshire 1949, 1951, 1954). Ginzburg and Levanyuk incorporated the effect of spatially inhomogeneous thermal fluctuations of polarization near the phase transformation (Levanyuk 1959; Ginzburg 1957). They introduced what is now known as correlation length, which defines the region confining polarization fluctuations. Such further developments have become the so-called Landau–Ginzburg–Devonshire (LGD) thermodynamic theory of ferroelectric phase transformations.

The LGD phenomenology enables one to study a variety of physical properties of ferroelectrics in a rather straightforward fashion without referring to all the degrees of freedom. Excellent treatment of LGD phenomenological theory can be found in the seminal articles by Devonshire (1949, 1951), in the textbooks by Strukov and Levanyuk (1998), Fatuzzo and Merz (1967), Jona and Shirane (1962), and Lines and Glass (1977), and in the articles by Cross and coworkers (Haun et al. 1987, 1989a–e) and Damjanovic (1998).

This chapter was written to provide an uninitiated reader an initial exposure to ferroelectric materials and their applications, thereby providing them the fundamentals for further study. In addition, this chapter also complements the ensuing two chapters of this volume, where the concepts introduced herein will be further developed from the perspective of crystallography, crystal chemistry, and structure–property relations.

1.2 Landau–Ginzburg–Devonshire Formalism

1.2.1 Crystallographic Considerations

The evolution of spontaneous polarization and thus ferroelectricity in homogeneous nonlinear dielectric media is attributed to spontaneous displacement of atoms relative to each other below a certain temperature, creating permanent dipole moments, as will be discussed in detail in Chaps. 2 and 3. The density of such dipole moments is termed polarization, which is a vector. It is the most fundamental parameter on which all the magnitudes of intrinsic (monodomain) dielectric, pyroelectric, ferroelectric, and piezoelectric properties depend. The orientation of this vector is, in turn, dictated by the point group of the polar phase, which forms upon the paraelectric–ferroelectric phase transition.

Most ferroelectric materials that are of practical interest have a perovskite structure with the chemical formula ABO_3. Some typical examples are $BaTiO_3$ (BTO), $PbTiO_3$ (PTO), $(Ba,Sr)TiO_3$ (BST), and lead-based solid solutions such as $Pb(Zr,Ti)O_3$ (PZT). Figure 1.1a shows the perovskite structure of PTO above the transformation temperature, where the TiO_6 octahedra are linked in a regular cubic array forming the high-symmetry nonpolar $m3m$ prototype for many ferroelectric materials. The six-fold coordinated site in the center of the octahedron is filled by a small, highly-charged Ti^{4+} (or other transition metal oxide), and the larger 12-fold coordinated "interstitial" site between octahedral carries typically a larger Pb^{2+}. For instance, the spontaneous polarization in PTO arises from the spontaneous displacement of Ti^{4+} and O^{2-} ions relative to Pb^{2+} ions, thereby creating a noncentrosymmetric lattice. Figure 1.1b depicts an example of such a shift in the tetragonal phase, resulting in a net polarization along the [001] (c-axis) of the tetragonal unit cell. In general, the axis of polarization may be parallel to the unit cell edge, the face diagonal, and the body diagonal as in the case of tetragonal

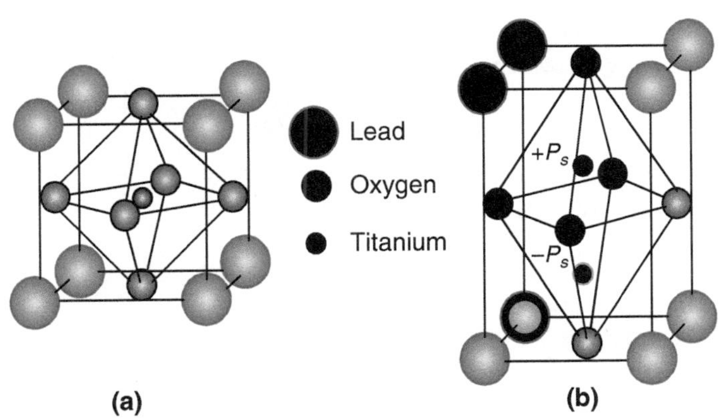

Fig. 1.1 (**a**) The perovskite crystal structure of $PbTiO_3$ and (**b**) off-center shift of titanium ion along $\langle 001 \rangle$ below the ferroelectric phase transformation temperature in the tetragonal phase

(4*mm*), orthorhombic (*mm*2), and rhombohedral (3*m*) point groups, respectively. In lower symmetry point groups (monoclinic and triclinic polar point groups), the orientation of the polarization vector is not trivial (see Chap. 3 for more details).

The polarization is accompanied by the so-called eigenstrain or spontaneous strain, which is known as the secondary order parameter in ferroelectrics. The said strain is due to coupling of the spontaneous polarization vector with the strain tensor via electrostriction – the most fundamental electromechanical coupling mechanism in all materials. In the case considered herein, a spontaneous polarization along the cube edge results in an elongation along the direction of polarization and a corresponding contraction along directions perpendicular to it, reducing the symmetry to tetragonal Laue class 4/*mmm*. This is not a polar crystal class. The polar point group 4*mm* forms because of the off-center displacement of the Ti^{4+} ion, which breaks the mirror plane. Following the same line of reasoning, it can be stated that polarization along the face diagonal reduces the symmetry to orthorhombic (*mm*2) and polarization along the body diagonal reduces it to rhombohedral (3*m*), as in the BTO system as shown in Fig. 1.2. The high-symmetry, nonpolar cubic phase transforms sequentially to low-symmetry ferroelectric phases, with decreasing temperature.

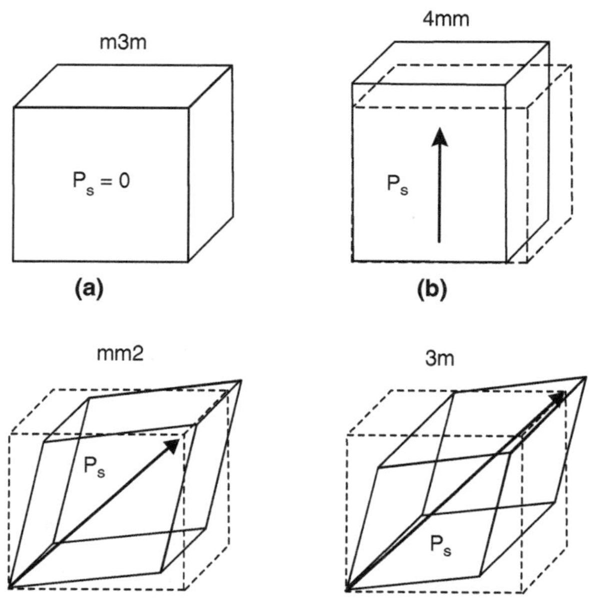

Fig. 1.2 The symmetry change and orientation of the spontaneous polarization direction for (**a**) cubic (*m*3*m*), (**b**) tetragonal (4*mm*), (**c**) orthorhombic (*mm*2), and (**d**) rhombohedral (3*m*) phases in $BaTiO_3$ for (**a**) $T > 132°C$, (**b**) $5 < T < 132°C$, (**c**) $-90°C < T < 5°C$, and (**d**) $T < -90°C$

1.2.2 Thermodynamic Formalism

Consider a simple ferroelectric transformation where, upon cooling, the paraelectric (nonpolar) cubic *m*3*m* phase transforms to a ferroelectric tetragonal 4*mm* phase. As seen in Fig. 1.2b, the spontaneous polarization and the accompanying lattice distortion is along six-fold crystallographically degenerate $\langle 001 \rangle$ directions. For stress-free single crystals, the (Helmholtz) free energy density can be described by an LGD potential. In constructing such a free energy functional, a Taylor series representation is utilized and the free energy density is expanded into the powers of polarization. Only the even powers are retained in the said series as the free energy functional has to be an even function, i.e. $\Delta A(-P) = \Delta A(P)$. Since the example we deal here is a uniaxial ferroelectric (tetragonal), only one order parameter is needed. Then, the free energy functional takes the following simple form:

$$\Delta A_{\mathrm{L}}(P,T) = \frac{1}{2}\alpha P^2 + \frac{1}{4}\beta P^4 + \frac{1}{6}\gamma P^6 + \cdots, \tag{1.1}$$

where P is the polarization (which is the order parameter of the phase transformation), and α, β, and γ are the so-called expansion (or the dielectric stiffness) coefficients, which are Taylor series as a function (or any other intensive thermodynamic variable) of temperature themselves. Traditionally, only the α coefficient is taken as temperature-dependent, which is the first nonvanishing term of its Taylor expansion in temperature. It takes the following form, which one immediately identifies as the reciprocal Curie-Weiss law:

$$\alpha = \frac{T - T_{\mathrm{C}}}{\varepsilon_0 C}, \tag{1.2}$$

where T_{C} and C are the Curie-Weiss temperature and constant, respectively, and ε_0 is the permittivity of free space. The other two expansion coefficients are usually taken to be independent of temperature.

The Landau potential should exhibit only one minima at $P = 0$ above T_{C} corresponding to the centrosymmetric cubic nonpolar phase. Below T_{C}, the free energy curve should have two minima corresponding to two identical ferroelectric states but with opposite orientation of the polarization direction along the (001) orientation (P_{S} and $-P_{\mathrm{S}}$). The variation of the free energy given in (1.1) as a functional of temperature and polarization is illustrated in Fig. 1.3 for a second-order phase transition, while Fig. 1.4 illustrates the same for a first-order transition. If $\beta < 0$, the phase transition from the paraelectric state is of first order, which reveals itself as a discontinuity in P_{S}, and lattice parameters at T_{C}, as well as thermal hysteresis in the same parameters around T_{C}. The same transition is said to be second order if $\beta > 0$, which results in a gradual variation in P_{S}, and lattice parameters below T_{C} with no thermal hysteretic. Most importantly, both paraelectric and ferroelectric phases co-exist at the transition temperature; there is latent heat associated with the transition, and the atomic displacements in the unit cell are finite. On the other hand, no phase coexistence is possible in a second-order phase transition; there is no latent heat (although there is a discontinuity in heat capacity), and the atomic displacements in the unit cell can –in principle – be infinitesimally small. The γ is positive for stability reasons.

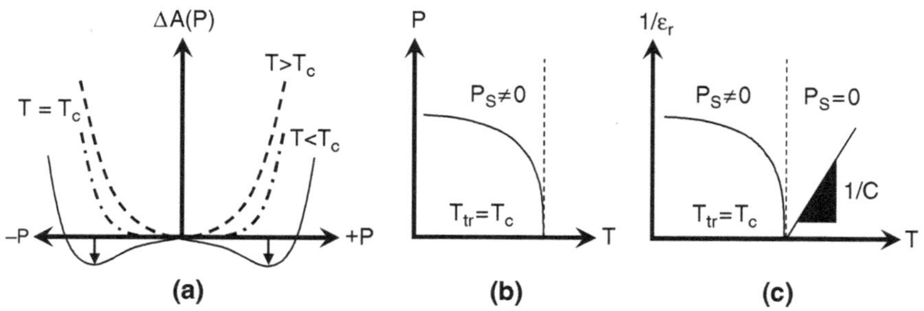

Fig. 1.3 (**a**) Variation of free energy density with temperature and polarization. Arrows in wells indicate spontaneous polarization states, (**b**) Variation of spontaneous polarization with temperature, and (**c**) Variation of reciprocal dielectric constant with temperature for a second-order ferroelectric phase transition

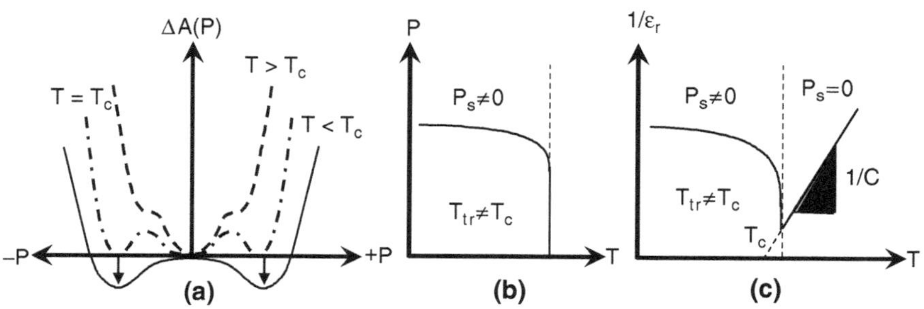

Fig. 1.4 (**a**) Variation of free energy density with temperature and polarization. Arrows in wells indicate spontaneous polarization states, (**b**) Variation of spontaneous polarization with temperature, and (**c**) Variation of reciprocal dielectric constant with temperature for a first-order ferroelectric phase transition

The spontaneous polarization P_S in the tetragonal phase can be obtained from the condition for thermodynamic equilibrium; i.e. $(\partial \Delta A/\partial P) = 0$ such that:

$$P_S^2(T) = \frac{-\beta + \sqrt{(\beta^2 - 4\alpha\gamma)}}{2\gamma}. \tag{1.3}$$

The spontaneous polarization (1.3) results in spontaneous strains, which are the so-called structural component of the phase transformation. However, it should be noted that self-strains are not true strains but rather a geometric description of the lattice distortion induced by the displacement of atoms in the unit cell upon the phase transition. The components of the spontaneous strain tensor for point group 4*mm*, for example, are defined as:

$$x_1^s = x_2^s = \frac{a_t - a_c}{a_c} = Q_{12}P_S^2, \quad x_3^s = \frac{c_t - a_c}{a_c} = Q_{11}P_S^2, \tag{1.4}$$

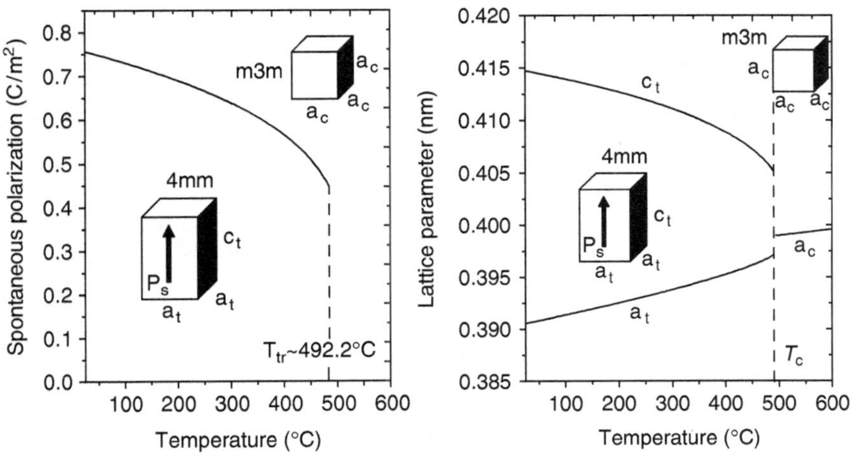

Fig. 1.5 The variation of the spontaneous polarization (*left*) and lattice parameters (*right*) with temperature in $PbTiO_3$ (adapted from Haun et al. 1987)

where a_t and c_t are the lattice parameters in the tetragonal ferroelectric state and a_c is the lattice parameter of the cubic phase extrapolated to the temperature of interest for $T < T_C$. Q_{ij}s are the fourth-rank electrostriction coefficients (in Voigt notation), coupling the spontaneous polarization with the spontaneous strains.

In Fig. 1.5 the variation of the spontaneous polarization and the lattice parameters of PTO as a function of temperature are depicted; respectively (Haun et al. 1987). The said variations were obtained using experimentally determined expansion parameters in conjunction with the formalism presented herein as reported by Haun et al. (1987) in their phenomenological theory. It should be borne in mind that the macroscopic description of the ferroelectric state obtained from such a thermodynamic formalism applies to monodomain single crystals. Multidomain ferroelectrics and polycrystalline ceramics require more sophisticated theoretical treatments as discussed to some extent in Chap. 2.

1.2.3 Further Development of the Thermodynamic Formalism: Elastic and Electrostatic Free Energy Functionals

Up to this point, we have taken into account the temperature dependence of the free energy associated with the ferroelectric phase transition only. To provide a macroscopic description of a ferroelectric's piezoelectric properties, the effect of applied stresses has to be incorporated. Likewise, one must incorporate an electrostatic energy term into the free energy functional to account for the dielectric properties of a given ferroelectric. In what follows, such further developments of the thermodynamic formalism will be presented.

The energy density due to applied elastic stresses imposed on the prototypical cubic crystal is given by:

$$\Delta A(\sigma) = \tfrac{1}{2}S_{11}(\sigma_1^2+\sigma_2^2+\sigma_3^2) + S_{12}(\sigma_1\sigma_2+\sigma_1\sigma_3+\sigma_2\sigma_3) + \tfrac{1}{2}S_{44}(\sigma_4^2+\sigma_5^2+\sigma_6^2), \tag{1.5}$$

where S_{ij} are the elastic compliances at constant polarization and σ_i are components of the applied elastic stress tensor. The free energy expression in (1.5) describes solely the change in free energy of the crystal due to an applied mechanical perturbation. However, one needs to remember that the applied stress is coupled with the polarization because of the electrostrictive effect. In other words, the spontaneous polarization induces a spontaneous strain, following a second-order variation in the form of $x_i = Q_{ij}P_s^2$, to which there corresponds a strain energy density as well. Then the free energy contribution can be expressed as:

$$\Delta A(\sigma : P) = x_1^0(\sigma_1+\sigma_2) + x_3^0\sigma_3 = Q_{12}P_s^2(\sigma_1+\sigma_2) + Q_{11}\sigma_3 P_s^2. \tag{1.6}$$

The stored electrostatic energy density contribution due to an applied field is simply given by

$$\Delta A(P) = E \cdot P, \tag{1.7}$$

where E is the electric field along the polarization direction.

Having developed pertinent free energy functionals for all important phenomena relevant to a ferroelectric crystal, the overall free energy functional is simply obtained by scalar addition of each expression owing to the fact that energy is additive:

$$\Delta G(P,T,\sigma_i,E) = \Delta A(P,T) - \Delta A(\sigma) - \Delta A(\sigma : P) - \Delta A(E). \tag{1.8}$$

By inserting all individual free energy functionals, the following explicit form of (1.8) is obtained:

$$\begin{aligned}\Delta G(P,T,\sigma_i,E) = {} & \tfrac{1}{2}\alpha P^2 + \tfrac{1}{4}\beta P^4 + \tfrac{1}{6}\gamma P^6 - \big(\tfrac{1}{2}S_{11}(\sigma_1^2+\sigma_2^2+\sigma_3^2) \\ & + S_{12}(\sigma_1\sigma_2+\sigma_1\sigma_3+\sigma_2\sigma_3) + \tfrac{1}{2}S_{44}(\sigma_4^2+\sigma_5^2+\sigma_6^2)\big) \\ & - \big(Q_{12}P_s^2(\sigma_1+\sigma_2) + Q_{11}\sigma_3 P_s^2\big) - E \cdot P.\end{aligned} \tag{1.9}$$

It should be noted that the overall free energy functional (1.9) is the one to use (called the elastic Gibbs function) in describing ferroelectrics under various circumstances as the Helmholtz free energy functional (1.1) describes the internal energy of the system in the absence of external perturbations. Once the elastic Gibbs function is in hand, virtually all material properties can be obtained by evaluating the first and second derivatives. For example, the dielectric susceptibility along the polarization axis at zero stress is given by $(\partial^2 G/\partial P^2)^{-1}$ such that:

$$\left(\frac{\partial^2 \Delta G}{\partial P^2}\right) = \left(\frac{\partial E}{\partial P}\right) = \chi_3 = (\alpha + 3\beta P^2 + 5\gamma P^4)^{-1}, \tag{1.10}$$

where the relative dielectric constant along the same axis is defined as:

$$\varepsilon_r \varepsilon_0 = 1 + \chi_3 \quad \text{or} \quad \varepsilon_r \cong \chi_3/\varepsilon_0 \tag{1.11}$$

because $\chi_3 \gg 1$.

The piezoelectric charge coefficient, which is very important in the context transducer applications, describing induced strain due to an applied electric field along the (001) direction of a tetragonal ferroelectric crystal is defined as:

$$d_{33} = \chi_3(\partial^2 G/\partial P \partial \sigma_3) \tag{1.12}$$

from which the explicit expression for d_{33} can be found as:

$$d_{33} = \varepsilon_0 \chi_3 Q_{11} P_s. \tag{1.13}$$

It is peculiar to observe that piezoelectricity is nothing but electrostriction that is biased by spontaneous polarization based on (1.13).

Up to this point, the thermodynamic development has been based on a single-order parameter, which is sufficient to take into account for the cubic to tetragonal phase transitions in ferroelectric. For phases of reduced symmetry, as exemplified by the orthorhombic and rhombohedral phases in BTO, the elastic Gibbs functional becomes mathematically more demanding but the approach remains the same. Specifically, in the case of the orthorhombic phase, the atomic shifts and the corresponding polarization is along the $\langle 110 \rangle$ directions, which requires one to use two order parameters, i.e. P_x and P_y. For the rhombohedral phase, the said shifts occur along $\langle 111 \rangle$, which requires one to use three order parameters, i.e. P_x, P_y, and P_z. For even further reduced symmetries such as the monoclinic and triclinic, Taylor expansion beyond the sixth order would be needed, making the manipulation of tensor quantities very tedious (Devonshire 1949, 1951). Additional terms may arise if there are structural variations due to the tilting or rotation of the oxygen octahedron as it is the case for the "incipient" ferroelectric $SrTiO_3$ (Slonczewski and Thomas 1970; Pertsev et al. 2000). In such cases, two coupled sets of order parameters, one set describing displacements and the other set describing rotations, may have to be employed (Salje 1990).

1.3 Ferroelectric Hysteresis

Perhaps the most important property of ferroelectrics is the reversal of the polarization vector upon to the application of a physically realizable electric field. Such a reversal is called switching, and refers to the change in orientation of the polarization vector from one crystallographically equivalent polar direction to another.

These crystallographically equivalent directions are, in turn, called domain states or orientational variants. From a thermodynamics point of view, the said domain states refer to energetically equal states on the free energy density hypersurface, which differ in orientation only. In common ferroelectric parlance, one would encounter the term *domain switching* as well. However, what is switched is actually the polarization vector. In each domain, the polarization vectors are aligned parallel and change orientation in a cooperative manner.

The aforementioned switching property of ferroelectrics is the very reason why they are the material of choice in application since a multidomain polycrystalline ferroelectric can be made piezoelectric upon the application of a strong enough electric field in a process called poling as discussed in the ensuing chapters of this book.

1.3.1 Definition of Terms

As pointed out earlier, the polarization (or the relative displacements of ions) in a tetragonal ferroelectric can be along the positive or negative direction of the $\langle 001 \rangle$ family of direction (see Fig. 1.1 for the [001] direction). The equilibrium condition $\partial \Delta G / \partial P = 0$ at $P = P_S$ and $P = -P_S$ is, however, identical as changing the orientation of the polarization vector should not change the free energy of the system, i.e. $\Delta G(P) = \Delta G(-P)$. Therefore, one has to conclude that these orientation variants should be equally probable, resulting in regions (or domains) with opposite direction of polarization, yielding zero net polarization in a so-called "virgin" ferroelectric – a ferroelectric to which no electric field is ever applied.

If a uniform electric field is applied along the positive [001] direction, the volume fraction of the domains that are favorably oriented with respect to the field increases will increase via domain wall motion, giving rise to a measurable net polarization. The polarization behavior is nonlinear as shown in Fig. 1.6. The extent of the reorientation of the domains along the applied field direction is governed by the point group of the crystal in question. Once all the domains have "switched," further increase in the applied field results in a proportional increase in the polarization until a saturation value is reached. Upon the removal of the applied field, the ferroelectric does not revert to its initial state although some domains do switch back. The value of the polarization at zero field after poling is called the remnant polarization (P_R). The subsequent application of a uniform field along the negative z-axis increases the volume fraction of domains with $P = -P_S$. At a critical field $-E_C$, called the coercive field, the volume fractions of the domains become equal and the net polarization is zero. Again, an increase in the negative field eventually yields a single-domain state with the polarization pointing in the negative z-direction. Upon removing the negative field, the polarization of the ferroelectric reverts to $-P_R$. This process results in a hysteresis corresponding to an energy loss. This energy is essentially spent to nucleate and grow domains with the opposite polarization.

1.3.2 Theoretical Analysis of Hysteresis Phenomena

Consider an unclamped ferroelectric crystal of point group $4mm$ to which is applied a uniform electric field along the [001] direction in an infinitesimally slow fashion. Let us further assume that the said experiment is carried far away from the phase transition so that Landau formalism can be employed, with no need to consider thermal fluctuations in the polarization. The appertaining free energy functional one has to use is:

$$\Delta G(P,T,E) = \frac{1}{2}\alpha P^2 + \frac{1}{4}\beta P^4 + \frac{1}{6}\gamma P^6 - EP. \tag{1.14}$$

By taking the derivative of (1.14) with respect to polarization and equating it to zero, the so-called dielectric equation of state is obtained as follows:

$$E = \alpha P + \beta P^3 + \gamma P^5. \tag{1.15}$$

The said equation of state defines the equilibrium polarization of a ferroelectric in the presence of an applied electric field. The variation of polarization with applied field far below the transition temperature is depicted in Figure 1.6. In the region A–B, the field-induced response is monotonic, followed by a region of instability region B–C as the second derivative of the Gibbs function is negative. This means that at B, defined by a critical field $-E_C$, the orientation of the polarization vector along the positive $\langle 001 \rangle$ direction becomes unstable. The same is true for point C, where the polarization vector along the negative $\langle 001 \rangle$ axis becomes unstable at E_C. The analytical expression given in (1.15) traces the path A–B–C–D. However, the B–C portion of the trace represents metastable states in the sense indicated above, and cannot be observed experimentally. What is observed in an experiment is, however, the trace A–B–B′–C–C′, which forms the so-called hysteresis loop of the ferroelectric.

From the hysteresis loop, one obtains very useful information: From the intersection of the polarization axis at $E = 0$, the remnant polarization is obtained. In the case of a single crystal, if the measurement axis if parallel to the polar axis, the remnant polarization is indeed the spontaneous polarization itself. Otherwise, the

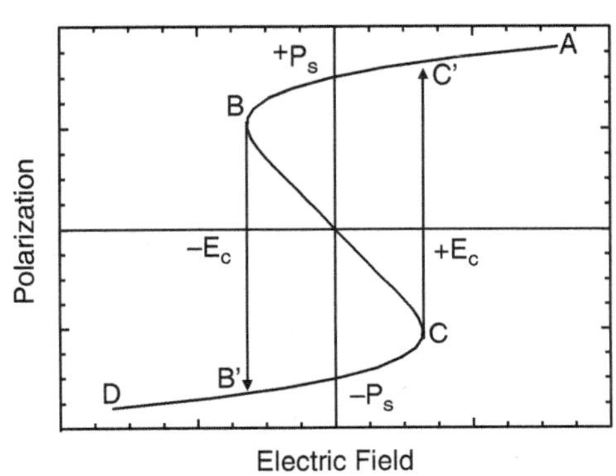

Fig. 1.6 Typical hysteresis loop as obtained from (1.15) for $T \ll T_C$ for a second-order ferroelectric system. All main features of the hysteresis loop are the same for a first-order system, except the metastable segment B–C. Definitions of spontaneous polarization and intrinsic coercive field are the same for both types of systems

component of the spontaneous polarization along the measurement axis is obtained. In the case of a polycrystalline ceramic, the remnant polarization is the apparent spontaneous polarization, but not necessarily the single crystal value. The intersection of the electric field axis at $P = 0$ gives the coercive field of the material in question. The value of the coercive field determines the field required to induce the reversal of the polarization vector. The asymmetry in the $\pm E_c$ with respect to the origin is an indication of internal bias, which is typically attributed to defects and space charge and is mostly observed in polycrystalline ceramics, multilayer structure, as well as thin films. The slope of the trace at $E = E_c$ is indicative of how well a single crystal is poled. Any substantial deviation from 90° is attributed to incomplete poling. In polycrystalline ceramics, the said deviation is always substantial no matter how well a specimen is poled and is essentially related to the charge distribution in the vicinity of grain boundaries and such.

In summary, it should be pointed out that the coercive predicted using the Landau formalism is typically two to three orders of magnitude larger than what is measured experimentally. That is so because switching in the Landau formalism is modeled as a thermodynamic instability of the polarization induced by an applied field in the reverse direction, which ignores nucleation and growth of domains. Although there have been many efforts to theoretically describe the polarization reversal based on domain wall motion and nucleation and growth of domains (Ishibashi 1992; Shur and Rumyantsev 1994; Tagantsev et al. 1994; Tagantsev et al. 1995; Chai et al. 1997; Chen et al. 1997; Shur et al. 1997; Chen and Lynch 1998; Hwang et al. 1998; Damjanovic 1998), domain switching in ferroelectrics still remains a formidable problem for which no fully matured theory exists. However, there has been some progress in this line of research by the use of a modified Kolmogorov-Avrami formalism; yet much remains to be explored (Ishibashi and Takagi 1971; Ishibashi and Orihara 1986, 1992; Ishibashi 1992; Orihara and Ishibashi 1992; Orihara et al. 1994).

1.4 Concluding Remarks

In this chapter, a brief survey of the thermodynamics of ferroelectrics is provided to establish a self-consistent foundation for the chapters that follow, which will address materials, device design, and application of ferroelectrics as a piezoelectric. It has been shown that that the Landau formalism is a time-tested macroscopic thermodynamic theory, which successfully accounts for many phenomena related to ferroelectricity such as the phase transition, phase transition temperature, temperature dependence of dielectric constant, piezoelectric coefficients, among others. The practical limitations of the Landau formalism have to be identified as well: First and foremost, as all macroscopic theories, it is devoid of any atomic interpretation of the macroscopically observed phenomena. Second, it does provide very limited insight into hysteresis phenomena. Third, the properties calculated using such thermodynamic theories are the so-called intrinsic, quasi-static, and monodomain properties. Most ferroelectrics are multidomain, and the extrinsic contribution to material properties due to domains cannot be overlooked. Fourth, frequency response of any given property cannot be analyzed. Nonetheless, the Landau formalism still remains as a powerful tool in the arsenal of ferroelectricians.

Acknowledgments The authors wish to express their gratitude to the Glenn Howatt Foundation at Rutgers University for their unfaltering financial support.

References

Chai, F. K., Brews, J. R., Schrimpf, R. D., "Domain Switching and Spatial Dependence of Permitivity in Ferroelectric Thin Films," J. Appl. Physics, 82, 2505 (1997).

Chen, W. and Lynch, C. S., "A Micro-Electro-Mechanical Model for Polarization Switching of Ferroelectric Materials," Acta Materialia 46, 5303–5311 (1998).

Chen, X., Fang, D. N., and Hwang, K. C., "Micromechanics Simulation of Ferroelectric Polarization Switching," Acta Materialia 45, 3181–3189 (1997).

Damjanovic, D., "Ferroelectric, Dielectric and Piezoelectric Properties of Ferroelectric Thin Films and Ceramics," Reports on Progress in Physics 61, 1267–1324 (1998).

de Fontaine, D., "Configurational Thermodynamics of Solid Solutions." In: Solid State Physics. H. Ehrenreich, F. Seitz, and D. Turnbull (eds.). New York, Academic (1979).

Devonshire, A. F., "Theory of Barium Titanate – Part I," Philosophical Magazine 40, 1040–1063 (1949).

Devonshire, A. F., "Theory of Barium Titanate – Part II," Philosophical Magazine 42, 1065–1079 (1951).

Fatuzzo, E. and Merz, W. J., Ferroelectricity, New York, Wiley (1967).

Ginzburg, V. L., "On the Dielectric Properties of Ferroelectric (Segnetoelectric Crystals and Barium Titanate)," Journal of Experimental and Theoretical Physics 15, 1945 (1945).

Ginzburg, V. L., "Fluctuation Phenomena near a Second Order Phase Transition," Journal of Experimental and Theoretical Physics 32, 1442 (1957).

Haun, M. J., Furman, E., Jang, S. J., McKinstry, H. A., and Cross, L. E., "Thermodynamic Theory of $PbTiO_3$," Journal of Applied Physics 62, 3331 (1987).

Haun, M. J., Furman, E., Jang, S. J., and Cross, L. E., "Thermodynamic Theory of the Lead Zirconate–Titanate Solid Solution System," Ferroelectrics 99, 13–86 (1989a).

Haun, M. J., Furman, E., Jang, S. J., and Cross, L. E., "Thermodynamic Theory of the Lead Zirconate-Titanate Solid Solution System, Part II: Tricritical Behavior," Ferroelectrics 99, 27–44 (1989b).

Haun, M. J., Furman, E., Jang, S. J., and Cross, L. E., "Thermodynamic Theory of the Lead Zirconate-Titanate Solid Solution System, Part III: Curie Constant and Sixth-Order Polarization Interaction Dielectric Stiffness Coeffcients," Ferroelectrics 99, 45–54 (1989c).

Haun, M. J., Furman, E., Jang, S. J., and Cross, L. E., "Thermodynamic Theory of the Lead Zirconate-Titanate Solid Solution System, Part IV: Tilting of the Oxygen Octahedra," Ferroelectrics 99, 55–62 (1989d).

Haun, M. J., Furman, E., Jang, S. J., and Cross, L. E., "Thermodynamic Theory of the Lead Zirconate-Titanate Solid Solution System, Part V: Theoritical Calculations," Ferroelectrics 99, 63–86 (1989e).

Hwang, S. C., Huber, J. E., McMeeking, R. M., and Fleck, N. A., "The Simulation of Switching in Polycrystalline Ferroelectric Ceramics," Journal of Applied Physics 84, 1530–1540 (1998).

Ishibashi, Y., "Theory of Polarization Reversals in Ferroelectric Based on Landau-Type Free Energy," Japanese Journal of Applied Physics 31, 2822–2824 (1992).

Ishibashi, Y. and Orihara, H., "The Characteristics of a Poly-Nuclear Growth Model," Journal of the Physical Society of Japan 55, 2315–2319 (1986).

Ishibashi, Y. and Orihara, H., "Size Effect in Ferroelectric Switching," Journal of the Physical Society of Japan 61, 4650–4656 (1992).
Ishibashi, Y. and Takagi, Y., "Note on Ferroelectric Domain Switching," Journal of the Physical Society of Japan 31, 506–510 (1971).
Jona, F. and Shirane, G., Ferroelectric Crystals, New York: Dover Publications (1962).
Landau, L. D. and Lifshitz, E. M., Statistical Physics, Oxford: Pergamon Press (1980).
Levanyuk, A. P., "Contribution to the Theory of Light Scattering near the 2nd Order Phase Transition Points," Journal of Experimental and Theoretical Physics 9, 571 (1959).
Lines, M. E. and Glass, A. M., Principles and Application of Ferroelectrics and Related Materials, Oxford: Clarendon Press (1977).
Orihara, H. and Ishibashi, Y., "A Statistical Theory of Nucleation and Growth in Finite Systems," Journal of the Physical Society of Japan 61, 1919–1925 (1992).
Orihara, H., Hashimoto, S., and Ishibashi, Y., "A Theory of D–E Hysteresis Loop Based on the Avrami Model," Journal of the Physical Society of Japan 63, 1031–1035 (1994).
Pertsev, N. A., Tagantsev, A. K., and Setter, N., "Phase Transitions and Strain-induced Ferroelectricity in SrTiO3 Epitaxial Thin Films," Physical Review B 61, R825–R829 (2000).
Salje, E. K. H., Phase Transitions in Ferroelastic and Co-Elastic Crystals, Cambridge: Cambridge Univeristy Press (1990).
Shur, V. Y. and Rumyantsev, E. L., "Kinetics of Ferroelectric Domain Structure During Switching: Theory and Experiment," Ferroelectrics 151, 171–180 (1994).
Shur, V. Y., Ponomarev, N. Y., Tonkacheva, N. A., Makarov, S. D., Nikolaeva, E. V., Shishkin, E. I., Suslov, L. A., Salashchenko, N. N., and Klyuenkov, E. B., "Fatigue in Epitaxial Lead Zirconate Titanate Films," Physics of the Solid State 39, 609–610 (1997).
Slonczewski, J. C. and Thomas, H., "Interaction of Elastic Strain with the Structural Transition of Strontium Titanate," Physical Review B 1, 3599–3608 (1970).
Strukov, B. A. and Levanyuk, A. P., Ferroelectric Phenomena in Crystals, Berlin: Springer (1998).
Tagantsev, A. K., Pawlaczyk, C., Brooks, K., and Landivar, M., "Depletion and Depolarizing Effects in Ferroelectric Thin Films and Their Manifestations in Switching and Fatigue," Integrated Ferroelectrics 6, 309–320 (1995).
Wadhawan, V. K., Introduction to Ferroic Materials, Amsterdam, Gordon and Breach (2000).
Ziman, J. M., Models of Disorder, Cambridge: Cambridge University Press (1979).

Chapter 2
Piezoelectricity and Crystal Symmetry

A.L. Kholkin,* N.A. Pertsev, and A.V. Goltsev

In this chapter, the symmetry aspects of the piezoelectric effect in various materials (single crystals, ceramics, and thin films) are briefly overviewed. First, the third-rank tensor of piezoelectric coefficients defined in the crystallographic reference frame is discussed. On this basis, the orientation dependence of the longitudinal piezoelectric response in ferroelectric single crystals is described. This dependence is especially important for relaxor single crystals, where a giant piezoelectric effect is observed. Then, the effective piezoelectric constants of polydomain crystals, ceramics, and thin films and their dependence on crystal symmetry are discussed. The domain-wall contribution to the piezoelectric properties of ferroelectric ceramics and thin films is also described. Finally, the crystallographic principles of piezomagnetic, magnetoelectric, and multiferroic materials are presented.

2.1 Historical Overview

Piezoelectricity (or, following direct translation from Greek word *piezein*, "pressure electricity") was discovered by Jacques Curie and Pierre Curie as early as in 1880 (Curie and Curie 1880). By analogy with temperature-induced charges in pyroelectric crystals, they observed electrification under mechanical pressure of certain crystals, including tourmaline, quartz, topaz, cane sugar, and Rochelle salt. This effect was distinguished from other similar phenomena such as "contact electricity" (friction-generated static charge). Even at this stage, it was clearly understood that symmetry plays a decisive role in the piezoelectric effect, as it was observed only for

A.L. Kholkin
Center for Research in Ceramics and Composite Materials (CICECO) & Department of Ceramics and Glass Engineering, University of Aveiro, 3810-193 Aveiro, Portugal
and
A. F. Ioffe Physico-Technical Institute of the Russian Academy of Sciences, 194021 St. Petersburg, Russia

A. Safari, E.K. Akdoğan (eds.) *Piezoelectric and Acoustic Materials for Transducer Applications.*

certain crystal cuts and mostly in pyroelectric materials in the direction normal to polar axis. However, Curie brothers could not predict a converse piezoelectric effect, i.e., deformation or stress under applied electric field. This important property was then mathematically deduced from the fundamental thermodynamic principles by Lippmann (1881). The existence of the converse effect was immediately confirmed by Curie brothers in the following publication (Curie and Curie 1881). Since then, the term *piezoelectricity* is commonly used for more than a century to describe the ability of materials to develop electric displacement **D** that is directly proportional to an applied mechanical stress σ (Fig. 2.1a). Following this definition, the electric charge appeared on the electrodes reverses its sign if the stress is changed from tensile to compressive. As follows from thermodynamics, all piezoelectric materials are also subject to a converse piezoelectric effect (Fig. 2.1b), i.e., they deform under applied electric field. Again, the sign of the strain **S** (elongation or contraction) changes to the opposite one if the direction of electric field **E** is reversed. Shear piezoelectric effect (Fig. 2.1c) is also possible, as it linearly couples shear mechanical stress or strain with the electric charge.

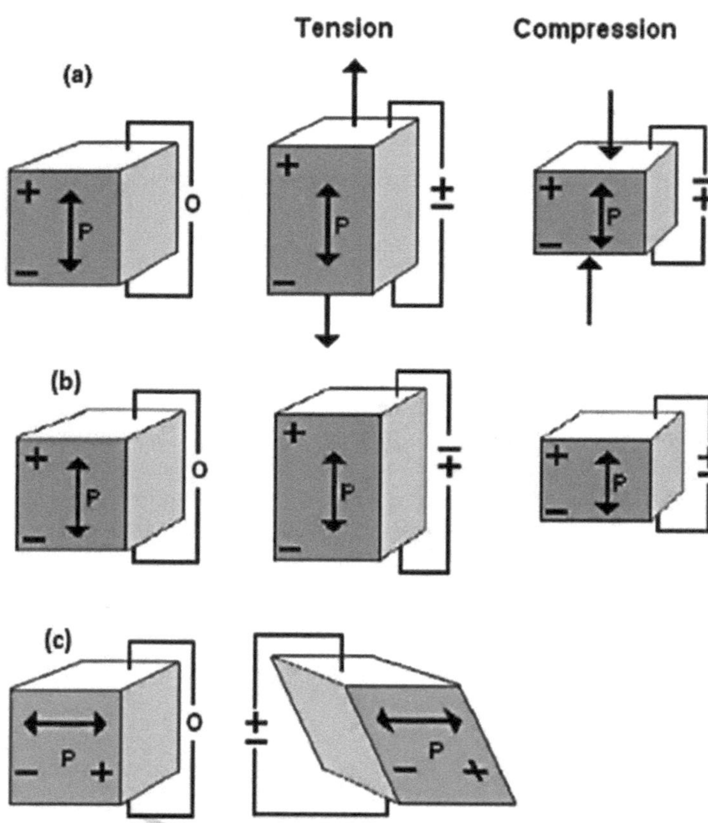

Fig. 2.1 Schematic representation of the longitudinal direct (**a**), converse (**b**), and shear (**c**) piezoelectric effects

Just after the discovery of piezoelectricity, much more work has been done to define crystallographic principles of the effect. In 1910, Voigt published the first textbook on physical crystallography (Voigt 1910), in which the correct description of the piezoelectric effect in different crystallographic classes was given. However, at that time the phenomenon of piezoelectricity was obscure because of a complicated description in crystals with low symmetry and no visible applications. Only after the Second World War the piezoelectric effect evolved from just a laboratory curiosity to a multimillion dollar industry with applications ranging from underwater sonars and medical imaging systems to car accelerometers. This was mainly due to the invention of piezoelectric ceramics (see, e.g., the textbook by Jaffe, Cook, and Jaffe (Jaffe et al. 1971)), in which the averaging of piezoelectric responses of individual crystallites (grains) in $PbZrO_3$–$PbTiO_3$ (PZT) solid solutions resulted in the high-symmetry ($\infty\infty m$) macroscopic state with only few independent piezoelectric coefficients of sufficiently high values. This research has resulted in the astonishing performance of piezoelectric materials with industrial applications in many areas. Currently, there is an immense interest in the materials with giant piezoelectric effect, so-called relaxor single crystals (Park and Shrout 1997), where the high piezoelectric activity is partly due to their symmetry.

2.2 Fundamentals of the Piezoelectric Effect in Single Crystals and Ceramics

Since the piezoelectric coupling is described by a linear relationship between the first-rank tensor or vector (**D** or **E**) and the second-rank tensor (σ or **S**), the corresponding coupling coefficients d_{kij} (also called charge piezoelectric coefficients) form a third-rank tensor. Hence, the piezoelectric equations may be written in the following form ($i,\ j,\ k = 1,\ 2,\ 3$):

$$S_{ij} = d_{kij} E_k, \tag{2.1}$$

$$D_k = d_{kij} \sigma_{ij}, \tag{2.2}$$

where the Einstein's summation rule for repeated indices is implied. Both direct and converse piezoelectric effects are frequently expressed using the reduced matrix notation d_{km}, where k denotes the component of electric displacement **D** or field **E** in the Cartesian reference frame ($x_1,\ x_2,\ x_3$), and the index $m = 1,\ldots,6$ is used to define the mechanical stress or strain. In this case, $m = 1$, 2, and 3 correspond to the normal stresses along the x_1, x_2, and x_3 axes, respectively, whereas $m = 4$, 5, and 6 denote the shear stresses S_{23}, S_{13}, and S_{12}.

As mentioned earlier, it was understood from the very beginning that the crystallographic symmetry of materials plays a decisive role in the piezoelectric phenomena. According to the definition of the piezoelectric effect, all components of the piezoelectric tensor should vanish in crystals possessing the center of symmetry.

In the remaining 21 noncentrosymmetric crystallographic classes, the piezoelectricity may exist, except for the cubic class 432, where the piezoelectric charges developed along the $\langle 111 \rangle$ polar axes cancel each other. However, the absence of piezoelectricity in this particular case does not play any significant role, because there are only few crystals that belong to this class. In this context, it should be mentioned that statistically about 30% of all materials (from about several millions known by now) are noncentrosymmetric. However, the piezoelectric properties are revealed in only few thousands of them, with about several hundreds having piezoelectric activity valuable for the applications. Therefore, it can be concluded that the absence of the center of symmetry represents the necessary but not sufficient requirement for a material to exhibit any sizeable piezoelectric effect. Though the symmetry does not determine the values of piezoelectric coefficients directly, the symmetry considerations, as it will be shown in the next sections, are indispensable for the design and fabrication of piezoelectric and acoustic devices. Table 2.1 lists the point groups that permit piezoelectricity for all crystallographic systems.

The number N of independent components of a third-rank tensor, in principle, may be as large as $3^3 = 27$. The piezoelectric tensor, however, can have maximum of 18 independent components because $d_{kij} = d_{kji}$ owing to the symmetry of the stress and strain tensors ($\sigma_{ij} = \sigma_{ji}$; $S_{ij} = S_{ji}$). The case of $N = 18$ corresponds to triclinic crystals of class 1. In crystals with higher symmetry, the number N reduces further. This feature follows from the Neumann's principle: The symmetry elements of any physical property of a crystal must include the symmetry elements of the point group of this crystal. As a result, in tetragonal crystals of the 4*mm* symmetry, for example, there are only three independent components, and the piezoelectric effect is described by the matrix shown in Fig. 2.2.

Among the 20 piezoelectric crystal classes, there are ten *pyroelectric* point groups that possess a unique polar axis. Pyroelectric crystals contain a built-in polarization, which manifests itself in temperature-induced changes of the total

Table 2.1 Centrosymmetric and noncentrosymmetric point groups in crystals with different symmetries

Crystal system	Symmetry elements	Centro-symmetric	Noncentro-symmetric
Triclinic	Center	$\underline{1}$	1
Monoclinic	Center, axis, plane	$2/m$	$2, m$
Orthorhombic	Center, axis, plane	mmm	$222, mm2$
Tetragonal	Center, axis, plane	$4/m, 4/mmm$	$4, \underline{4}, 422, 4mm, \underline{4}2m$
Trigonal	Center, axis, plane	$\underline{3}, \underline{3}m$	$3, 32, 3m$
Hexagonal	Center, axis, plane	$6/m, 6/mmm$	$6, \underline{6}, 622, 6mm, \underline{6}m2$
Cubic	Center, axis, plane	$m3, m3m$	$23, \underline{4}3m, 432$

Fig. 2.2 Matrix of the piezoelectric coefficients for crystals of the tetragonal symmetry (point group 4*mm*)

$$\begin{pmatrix} 0 & 0 & 0 & 0 & d_{15} & 0 \\ 0 & 0 & 0 & d_{15} & 0 & 0 \\ d_{31} & d_{31} & d_{33} & 0 & 0 & 0 \end{pmatrix}$$

dipole moment of the unit cell (in the absence of applied fields). If such *spontaneous polarization* can be reversed by an external (sufficiently high) electric field, the crystal is called *ferroelectric*.

Above a certain temperature T_c (often termed the Curie point), the spontaneous polarization vanishes, and the ferroelectric crystal transforms into the paraelectric state. Many ferroelectrics lose their piezoelectric properties above T_c completely, because their paraelectric phase has centrosymmetric crystallographic structure. The advantage of ferroelectrics, well understood already in the beginning of the last century (while studying Rochelle salt), is that they have much higher piezoelectric activity, especially in the vicinity of the Curie point, where the piezoelectric coefficients increase dramatically. In ferroelectrics with a centrosymmetric paraelectric phase, the piezoelectric coefficients are proportional to the product of polarization and dielectric permittivity and, therefore, should be high in materials having large polarizability and spontaneous polarization. For the tetragonal $4mm$ crystals, the longitudinal piezoelectric coefficient d_{33} can be expressed as

$$d_{33} = 2Q_{11}\varepsilon_0\varepsilon_{33}P_3, \tag{2.3}$$

where ε_{33} and P_3 are the relative permittivity and polarization along the polar x_3 axis, $\varepsilon_0 = 8.854 \times 10^{-12}\,\mathrm{F/m}$ is the permittivity of vacuum, and Q_{11} is the electrostrictive constant of the paraelectric phase, which couples longitudinal strain S_3 and polarization via the equation

$$S_3 = Q_{11}P_3^2. \tag{2.4}$$

The electrostriction coefficient involved in (2.3) practically does not depend on temperature and typically varies between 0.05 and $0.1\,\mathrm{m^4/C^2}$ for different materials.

The piezoelectric effect was discussed above for the case of single-domain ferroelectric crystals in which the spontaneous polarization is constant everywhere. Another technologically important class of materials is represented by piezoelectric ceramics, which consist of randomly oriented crystallites (grains), separated by grain boundaries. Ceramics are much less expensive in processing than single crystals and frequently offer comparable piezoelectric and electrostrictive properties (Jaffe et al. 1971). It should be emphasized that, in nonferroelectric ceramics, the piezoelectric responses of individual crystallites are canceled out after averaging over the entire sample. Hence, on the macroscopic level, the polycrystal has a center of symmetry and negligible piezoelectric properties.

In contrast, ferroelectric ceramics can be made piezoelectrically active by poling. This feature is due to the presence of so-called *ferroelectric domains* (regions with different orientations of the spontaneous polarization) in as-sintered ferroelectric ceramics. Domains appear in ceramics and single crystals when the material is cooled down through the Curie point in order to minimize the electric and elastic energy of the system.

The boundaries between ferroelectric domains (often called domain walls) can move under the action of applied electric field so that the spontaneous polarization may be reoriented in the crystallographic direction closest to the field direction. As a result of such poling process, an initially macroscopically centrosymmetric ceramic sample loses the inversion center and becomes piezoelectric. Since it acquires the $\infty\infty m$ symmetry, the poled ceramic has only three independent piezoelectric coefficients d_{33}, d_{31}, and d_{15}, which relate longitudinal, transverse, and shear deformations to the electric field applied along and perpendicular to the poling direction.

The measured piezoelectric response usually contains not only the intrinsic contribution (determined by the lattice properties and described by (2.3)), but also an extrinsic contribution caused by movements of non-180° domain walls. This domain-wall contribution, which will be discussed in detail in Sect. 2.5, depends on the symmetry of ferroelectric material too.

It should be noted that much efforts were made to create materials with enhanced piezoelectric properties. Equation (2.3) suggests that this goal can be achieved by using crystals approaching phase-transition conditions, where the increase of dielectric permittivity should result in high piezoelectric coefficients. The observation of maximum piezoelectric response in PZT solid solutions with compositions close to the morphotropic phase boundary (MPB) gave strong support to this idea long time ago.

Recently, it has been shown that the piezoelectric properties can also be improved by cutting the crystal at some angle to the polarization direction (Park and Shrout 1997). In particular, the solid solutions of $Pb(Zn_{1/3}Nb_{2/3})O_3$ (relaxor ferroelectric) with $PbTiO_3$ (normal ferroelectric) exhibit the highest piezoelectric coefficients along the pseudocubic $\langle 001 \rangle$ direction, while the spontaneous polarization is parallel to the $\langle 111 \rangle$ axis in the rhombohedral PZN–PT crystals studied.

The difference between the piezoelectric responses measured in the [001] and [111] directions is very large in these crystals (see Table 2.2), which may be explained phenomenologically by the orientation dependence of the piezoelectric effect (see Sect. 2.3). On the microscopic scale, the origin of ultrahigh piezoelectricity in PZN–PT solid solutions was attributed to the existence of a monoclinic phase near the MPB separating stability ranges of the rhombohedral and tetragonal states (Noheda et al. 2001). This intermediate monoclinic phase has the b_m axis oriented along the pseudocubic [010] direction and facilitates the polarization rotation between the $\langle 111 \rangle$ and $\langle 001 \rangle$ directions under the influence of electric field applied along the [001] crystallographic axis.

Table 2.2 Physical properties of major piezoelectric materials together with their symmetries

Parameter	Quartz	$BaTiO_3$	$PbTiO_3:Sm$	PZT 5H	LF4T	PZN–8%PT [001]	PZN–8%PT [111]
Symmetry	32	$4mm$	$4mm$	$3m/4mm$	$mm2/4mm$	$3m/4mm$	$3m/4mm$
d_{33} (pC/N)	2.3	190	65	593	410	2500	84
d_{31} (pC/N)	0.09	0.38	0	−274	−154	−1400	−20
$\varepsilon_{33}^T/\varepsilon_0$	5	1700	175	3400	2300	7000	1000
T_c (°C)		120	355	193	253	160	160

2.3 Orientation Dependence of Piezoelectric Response in Single Crystals

The piezoelectric coefficients d^*_{ijk} in an arbitrary coordinate system $(\tilde{x}_1, \tilde{x}_2, \tilde{x}_3)$ can be calculated from the coefficients d_{lmn} defined in the crystallographic reference frame (x_1, x_2, x_3) using the general relation (Cady 1964, Nye 1957)

$$d^*_{ijk} = A_{il} A_{jm} A_{kn} d_{lmn}, \tag{2.5}$$

where A_{il} are the elements of the transformation matrix $\mathbf{A}$ relating two sets of the coordinate axes. This formula makes it possible to predict the orientation dependence of the piezoelectric properties for any crystal with known piezoelectric coefficients d_{lmn}. In particular, the longitudinal piezoelectric coefficient d^*_{33} measured in the direction making an angle θ with the polar axis of the tetragonal crystal with $4mm$ symmetry can be evaluated as (Nye 1957)

$$d^*_{33} = \cos\theta[\sin^2\theta(d_{15} + d_{31}) + \cos^2\theta d_{33}]. \tag{2.6}$$

It can be seen that, when the angle θ increases, the relative contribution of the longitudinal coefficient d_{33} decreases, whereas the contribution of the shear coefficient d_{15} increases. If the sum $d_{15} + d_{31}$ is much larger than d_{33}, the maximum value of $d^*_{33}(\theta)$ corresponds to a direction different from the polar one (Davis et al. 2007). This feature is characteristic of $BaTiO_3$ at room temperature $[(d_{15} + d_{31})/d_{33} \approx 6]$, where it exhibits maximum longitudinal piezoelectric response close to the [111] direction (Damjanovic et al. 2002). In contrast, when the ratio $(d_{15} + d_{31})/d_{33}$ is smaller than some critical value, the maximum response d^*_{33} will be observed along the polar crystallographic axis ($\theta = 0$) (Davis et al. 2007). The aforementioned two different types of the orientation dependence are illustrated in Fig. 2.3, where $d^*_{33}(\theta)$ is plotted for $BaTiO_3$ and $PbTiO_3$ $[(d_{15} + d_{31})/d_{33} \approx 0.4]$ at room temperature.

The orientation dependence of d^*_{33} becomes more complex for piezoelectric crystals of other symmetries. In the case of an orthorhombic $mm2$ crystal, the coefficient d^*_{33} is a function of two Euler angles, θ and ϕ. Variation of d^*_{33} with these angles is described by the relation (Nye 1957)

$$d^*_{33} = \cos\theta[\sin^2\theta\sin^2\phi(d_{15} + d_{31}) + \sin^2\theta\cos^2\phi(d_{24} + d_{32}) + \cos^2\theta d_{33}]. \tag{2.7}$$

Again, the maximum piezoelectric response will be observed away from the polar direction if one of the shear coefficients is large compared with d_{33}. This happens, for instance, in $KNbO_3$ and PZN–9%PT crystals (Davis et al. 2007).

For the rhombohedral $3m$ crystal, the longitudinal piezoelectric coefficient measured in an arbitrary direction can be evaluated as

$$d^*_{33} = \cos\theta\sin^2\theta\sin^2\phi(d_{31} + d_{15}) + \sin^3\theta\cos\phi(3\sin^2\phi - \cos^2\phi)d_{22} + \cos^3\theta d_{33}. \tag{2.8}$$

It can be seen that at $\phi = 90°$ the contribution of the coefficient d_{22} becomes zero, and the orientation dependence of d^*_{33} reduces to that discussed above for

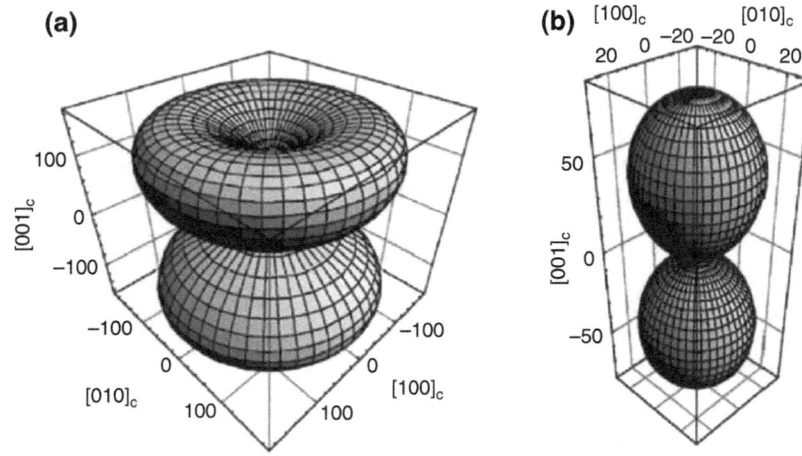

Fig. 2.3 Longitudinal piezoelectric coefficient d_{33}^* (pm/V) as a function of the measuring direction in $BaTiO_3$ (**a**) and $PbTiO_3$ (**b**) at room temperature. The spontaneous polarization is oriented along the (vertical) [001] crystallographic axis. (Reprinted with permission from Davis et al. 2007. Copyright 2007, American Institute of Physics.)

the tetragonal 4*mm* crystals. The calculations performed for PMN–33%PT crystals $((d_{15}+d_{31})/d_{33} \approx 21)$ indeed showed that the maximum value of d_{33}^* corresponds to a direction rotated by 63° away from the polarization direction (Zhang et al. 2003). The orientation dependence of this type is also expected for several other rhombohedral 3*m* crystals, including PZN–7%PT and rhombohedral PZT compositions (Davis et al. 2007).

Remarkably, the giant piezoelectric response of PZN–PT and PMN–PT relaxor-ferroelectric single crystals is observed along nonpolar directions (Park and Shrout 1997). In rhombohedral PZN–8%PT, for example, extremely large $d_{33}^* \sim 2500\,\mathrm{pC/N}$ corresponds to the $\langle 001 \rangle$ crystallographic direction, whereas the piezoelectric coefficient measured along the polar $\langle 111 \rangle$ axis is only about 80 pC/N (Park and Shrout 1997). Accordingly, the giant piezoelectric properties of PZN–PT and PMN–PT may be attributed to the field-induced polarization rotations (Fu and Cohen 2000). Such rotations are expected to be especially easy in the presence of monoclinic phases existing at the MPB regions of PZN–PT, PMN–PT, and PZT (Cox et al. 2001; Ye et al. 2001; Noheda et al. 1999). Indeed, the polarization vector in the M_A (M_B) and M_C monoclinic states can rotate freely in the $\{101\}$ and $\{010\}$ mirror planes (Vanderbilt and Cohen 2001). It is worth noting that the dominant role of polarization rotations in the high piezoelectric response of PZT ceramics with compositions close to MPB was demonstrated by in situ X-ray diffraction measurements (Guo et al. 2000).

It should be emphasized that relative magnitude of the shear piezoelectric coefficient d_{15} is directly related to the dielectric anisotropy of a ferroelectric crystal (Davis et al. 2007). The ratio d_{15}/d_{33} can be written as

$$\frac{d_{15}}{d_{33}} = 2\frac{Q_{1313}}{Q_{3333}}\frac{\eta_{11}}{\eta_{33}}, \tag{2.9}$$

where Q_{1313} and Q_{3333} represent the electrostrictive coefficients defined in the reference frame of the ferroelectric phase, and η_{11} and η_{33} are the transverse and longitudinal dielectric susceptibilities of the single-domain ferroelectric crystal. It can be seen that high dielectric anisotropy of the form $\eta_{11} \gg \eta_{33}$ will normally result in a shear piezoelectric coefficient large compared with d_{33}. The effect of dielectric anisotropy, however, is less pronounced in the tetragonal crystals than in the orthorhombic and rhombohedral ones since the ratio $2Q_{1313}/Q_{3333}$ is only about 0.3–0.4 in common $4mm$ ferroelectrics, whereas it is about 3 or even larger in the $mm2$ and $3m$ ones (Davis et al. 2007).

The dielectric anisotropy may become very large near structural phase transitions between two different ferroelectric phases (Budimir et al. 2003). This enhancement leads to strong increase of the shear piezoelectric response in the vicinity of ferroelectric–ferroelectric transformations induced by temperature or composition variations. Hence, near the morphotropic phase boundaries the longitudinal piezoelectric response d_{33}^* in nonpolar directions should increase strongly as well. The fundamental reason for such piezoelectric anomalies lies in the flattening of the free energy profile in close vicinity to a structural phase transition (Fu and Cohen 2000; Budimir et al. 2006).

2.4 Effective Piezoelectric Constants of Polydomain Crystals, Ceramics, and Thin Films

Ferroelectric crystals usually consist of many domains differing by the spatial orientation of the spontaneous polarization $\mathbf{P}_s$. Inside dissimilar domains, therefore, the local piezoelectric responses to a given external field are generally different. As a result, the piezoelectric properties of polydomain crystals are determined on the macroscopic scale by the average response of the domain ensemble. These "aggregate" material properties, in general, depend on the relative volume fractions of various ferroelectric domains formed in a crystal and on the orientation of domain walls between them.

The set of domain variants permissible in a certain crystal is defined by the crystal symmetry (Aizu 1969). The energetically most favorable orientations of ferroelastic domain walls are determined by the compatibility of spontaneous lattice strains at the wall (Fousek and Janovec 1969; Sapriel 1975). In the presence of ferroelectric polarization $\mathbf{P}$, additional electrical condition must be satisfied, ensuring the absence of polarization charges $\rho = -\mathrm{div}\,\mathbf{P}$ on the wall (Fousek and Janovec 1969). In the case of a tetragonal crystal such as $PbTiO_3$ or $BaTiO_3$ at room temperature, purely ferroelectric 180° walls and ferroelectric–ferroelastic 90° walls are only possible, the latter being oriented, e.g., parallel to the $\{101\}$ crystallographic planes of the prototypic cubic lattice. The orthorhombic phase polarized along the $\langle 110\rangle$

crystallographic direction may contain 180° walls, 90° walls parallel to the (100) crystallographic plane, 120° walls having the {110} orientation, and 60° domain walls with an irrational normal. In the rhombohedral crystal such as $Pb(Zr_{1-x}Ti_x)O_3$ with $x < 0.48$ polarized along the $\langle 111 \rangle$ direction, only the 180° walls and the 71° or 109° walls are possible.

The effective elastic, dielectric, and piezoelectric constants of a polydomain crystal may be calculated from the material constants of a single-domain single crystal for the given geometry of the domain structure. For tetragonal $BaTiO_3$ and $PbTiO_3$ crystals with the laminar 90° domain structure, the effective material constants were determined by Turik (1970) with the aid of an averaging procedure taking into account the mechanical and electrical conditions fulfilled on the domain walls. Remarkably, it was found that the matrix of these constants corresponds to the orthorhombic crystal symmetry rather than the tetragonal one. Similar calculations of the effective piezoelectric coefficients d^*_{3j} performed later for polydomain PMN–33%PT crystals showed that these coefficients strongly depend on the domain concentrations (Topolov 2004). The piezoelectric properties of "domain-engineered" perovskite single crystals were studied in detail by Davis et al. (2005). The tetragonal (point group $4mm$), orthorhombic ($mm2$), and rhombohedral ($3m$) crystal phases poled along various crystallographic directions were considered, and the effective piezoelectric coefficient d^*_{31} has been evaluated for polydomain $BaTiO_3$, $PbTiO_3$, $KNbO_3$, and PMN–33%PT crystals.

In the case of ferroelectric ceramics, the aggregate material properties are affected by the elastic, electric, and piezoelectric interactions between individual crystallites (Aleshin and Pikalev 1990; Nan and Clarke 1996; Pertsev et al. 1998a). Indeed, the piezoelectric crystallite is not free to deform under the action of an external electric field because of the elastic clamping imposed by surrounding grains. Besides, the field-induced polarization charges appear at the grain boundaries since the dielectric permittivity usually differs in the adjacent crystallites owing to different lattice orientations. As a result, internal electric fields and mechanical stresses arise (or change) during the piezoelectric measurements, thus making local responses of crystallites different from those of a stress-free electroded single crystal.

The effective piezoelectric constants of ferroelectric ceramics can be calculated theoretically using the so-called effective medium approach or self-consistent scheme (Nan and Clarke 1996; Pertsev et al. 1998a). This approach is based on the introduction of a model material system consisting of a representative crystallite (inclusion) surrounded by a dissimilar homogeneous piezoelectric medium (matrix). In the linear theory, the crystallite (usually assumed to be spherical) is characterized by the small-signal elastic, dielectric, and piezoelectric properties, independent of the internal stresses and electric fields. These properties are determined by the material constants (measured or calculated) of a single-domain or polydomain single crystal. In turn, the matrix parameters represent the unknown macroscopic elastic, piezoelectric, and dielectric constants of a polycrystalline sample, which should be calculated self-consistently by the method of successive approximations.

The numerical calculations performed for $BaTiO_3$ and $PbTiO_3$ ceramics indicated that the aggregate piezoelectric properties of ferroelectric ceramics may experience nonmonotonic variations with the increase of the remanent polarization P_r (Pertsev et al. 1998a). It was found that the piezoelectric constants d^*_{33} and d^*_{31} of $BaTiO_3$ ceramics reach their extreme values at $P_r/P_s \approx 0.87$ and 0.7, respectively. In the case of $PbTiO_3$ ceramics, these piezoelectric coefficients vary monotonically with the remanent polarization, but the constant d^*_{15} displays an absolute maximum at $P_r/P_s \approx 0.95$, where it slightly exceeds the single-crystal value of this constant. Thus, the influence of poling on the piezoelectric properties of ferroelectric ceramics may be different for dissimilar materials.

The elastic clamping of crystallites in a ferroelectric polycrystal not only changes the small-signal aggregate properties, but also modifies lattice strains and the spontaneous polarization. This mechanical effect is expected to be most pronounced in nanocrystalline ceramics, where the twinning of crystallites (formation of ferroelastic domains), which strongly reduces internal stresses, is energetically unfavorable because of a very small grain size (Pertsev and Salje 2000). The equilibrium polarization states of $BaTiO_3$ and $Pb(Zr_{1-x}Ti_x)O_3$ ceramics with single-domain grains were predicted recently using the nonlinear thermodynamic theory combined with the method of effective medium (Zembilgotov et al. 2005). The calculations showed that, owing to the elastic clamping of crystallites, the phase states of nanocrystalline ceramics may differ drastically from those of single crystals and coarse-grained materials. Remarkably, the theory predicted the coexistence of rhombohedral and tetragonal crystallites in nanocrystalline $Pb(Zr_{1-x}Ti_x)O_3$ ceramics in a wide range of compositions and temperatures. For $BaTiO_3$ ceramics, a mixture of rhombohedral and orthorhombic crystallites was found to be the energetically most favorable state at room temperature. Thus, the elastic clamping can change the symmetry of ferroelectric state in a small grain. This result indicates that nanocrystalline ferroelectric ceramics may have specific piezoelectric properties.

The current trend toward the miniaturization of electromechanical systems requires the use of piezoelectric materials in the thin-film form (Muralt 2000). The piezoelectric response of a thin film grown on a dissimilar thick substrate may be very different from that of a bulk material. The most evident reason for such a difference is the two-dimensional (2D) clamping of the film by the rigid substrate (Lefki and Dormans 1994). Since the in-plane dimensions of the film are fixed while the film thickness is free to change under the action of applied electric field E_3, the effective longitudinal piezoelectric coefficient d^f_{33} of a thin film can be evaluated as (Lefki and Dormans 1994)

$$d^f_{33} = d_{33} - \frac{2s^E_{13}}{s^E_{11} - s^E_{12}} d_{31}, \tag{2.10}$$

where d_{in} are the piezoelectric constants of a bulk material, and s^E_{mn} are the bulk elastic stiffnesses at constant electric field. The effective transverse piezoelectric coefficient e^f_{31}, which defines the stresses σ_1 and σ_2 induced in the film plane by the electric field E_3, is defined by a similar relation

$$e_{31}^{\mathrm{f}} = e_{31} - \frac{c_{13}^{\mathrm{E}}}{c_{33}^{\mathrm{E}}} e_{33}, \tag{2.11}$$

where c_{mn}^{E} are the bulk elastic compliances (Muralt 2000). It can be seen that e_{31}^{f} is always larger than the bulk coefficient e_{31}, whereas d_{33}^{f} is smaller than d_{33}. Indeed, it is implied here that the film piezoelectric and elastic constants do not differ significantly from those of the bulk material, which is a reasonable assumption for ordinary piezoelectrics, but not necessarily for ferroelectric materials (see below).

In addition to the 2D clamping, the film is usually strained by the substrate to a certain extent (e.g., because of the difference in the thermal expansion coefficients). Since in ferroelectrics the polarization is coupled to lattice strains via the electrostriction, the substrate-induced strains may change the piezoelectric properties of ferroelectric films markedly. In the case of polycrystalline thin films of multiaxial ferroelectrics, the mechanical substrate effect usually induces preferential orientations for the polar axes of crystallites, which results in the formation of a crystal texture below the phase transition temperature T_{c} (Pertsev et al. 1998a). When the substrate induces compressive in-plane stresses σ_1 and σ_2 above T_{c}, polycrystalline films of perovskite ferroelectrics such as $BaTiO_3$, $PbTiO_3$, and $Pb(Zr_{1-x}Ti_x)O_3$ acquire the c-texture with the polar axes oriented as close as possible to the substrate normal. In the films grown on "tensile" substrates $(\sigma_1, \sigma_2 > 0)$, the a-texture is formed, where directions of the polar axes have minimum possible deviations from the film plane. Evidently, the c- and a-textured films will behave quite differently during the poling process and so exhibit different piezoelectric properties.

Owing to the substrate-induced strains, the orientation of the spontaneous polarization in a ferroelectric thin film may also deviate from the polar crystallographic axis in the bulk material (e.g., from the $\langle 001 \rangle$ direction in the tetragonal crystal, the $\langle 101 \rangle$ direction in the orthorhombic crystal, and the $\langle 111 \rangle$ direction in the rhombohedral one). Such polarization rotations are expected to be especially large in the single-crystalline films free of ferroelastic domains (twins). Here the mechanical film–substrate interaction results in the formation of new phases not existing in bulk crystals (Pertsev et al. 1998b, 2000, 2003; Tagantsev et al. 2002). This phenomenon originates in the strain-induced lowering of the symmetry of the paraelectric phase.

In the case of (001)-oriented films of perovskite ferroelectrics grown on (001)-oriented cubic substrates, the strain effect lowers the symmetry of the paraelectric state from cubic to tetragonal (Pertsev et al. 1998b). As a result, five low-temperature phases become theoretically possible in the film instead of three in the bulk crystal. Among these five phases, the most important ones are the tetragonal c phase with the polarization $\mathbf{P}_{\mathrm{s}}$ orthogonal to the film surfaces, the orthorhombic aa phase with the in-plane polarization, and the monoclinic r phase with the vector $\mathbf{P}_{\mathrm{s}}$ inclined to the film surfaces (see Fig. 2.4). The aforementioned three ferroelectric phases were shown to be stable in single-domain $BaTiO_3$, $PbTiO_3$, and $Pb(Zr_{1-x}Ti_x)O_3$ $(x \geq 0.4)$ films under certain strain–temperature conditions (Pertsev et al. 1998b, 2003; Diéguez et al. 2004). The stability ranges of various possible polarization states in ferroelectric films can be conveniently described with the aid of phase diagrams, where the temperature T and the misfit strain S_{m} between the substrate

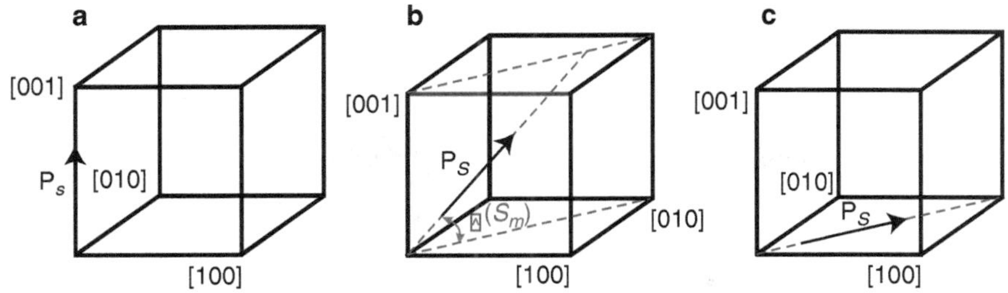

Fig. 2.4 Orientation of the spontaneous polarization $\mathbf{P}_s$ in stable ferroelectric phases forming in (001)-oriented single-domain thin films of perovskite ferroelectrics grown on (001)-oriented cubic substrates: tetragonal *c* phase (**a**), monoclinic *r* phase (**b**), and orthorhombic *aa* phase (**c**). Polarization orientations are shown relative to the prototypic cubic cell

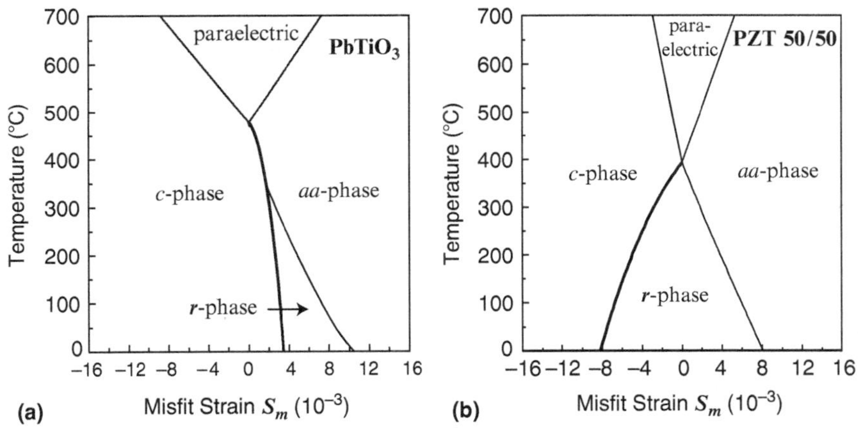

Fig. 2.5 Misfit strain–temperature phase diagrams of (001)-oriented single-domain $PbTiO_3$ (**a**) and $Pb(Zr_{0.5}Ti_{0.5})O_3$ (**b**) films epitaxially grown on (001)-oriented cubic substrates The first- and second-order phase transitions are shown by thick and thin lines, respectively. (Reprinted with permission from Pertsev et al. 1998b and 2003. Copyright 1998, 2003, American Physical Society.)

and the film prototypic cubic state are used as two independent variables (Pertsev et al. 1998b). Representative misfit strain–temperature phase diagrams of perovskite ferroelectric films are shown in Fig. 2.5. It can be seen that the diagrams of single-domain $PbTiO_3$ and $Pb(Zr_{0.5}Ti_{0.5})O_3$ films contain large stability ranges of the orthorhombic and monoclinic phases, which do not exist in the corresponding bulk materials.

The small-signal piezoelectric properties of epitaxial ferroelectric films evidently depend on the film polarization state. Hence the piezoelectric coefficients are functions of the misfit strain S_m in the film–substrate system, which can be determined using the film (S_m, T) phase diagram. Figure 2.6a shows a representative strain dependence of the longitudinal piezoelectric coefficient d_{33}, which was calculated for single-domain $Pb(Zr_{0.4}Ti_{0.6})O_3$ films using the nonlinear thermodynamic theory (Pertsev et al. 2003). Remarkably, d_{33} strongly increases near critical misfit strains

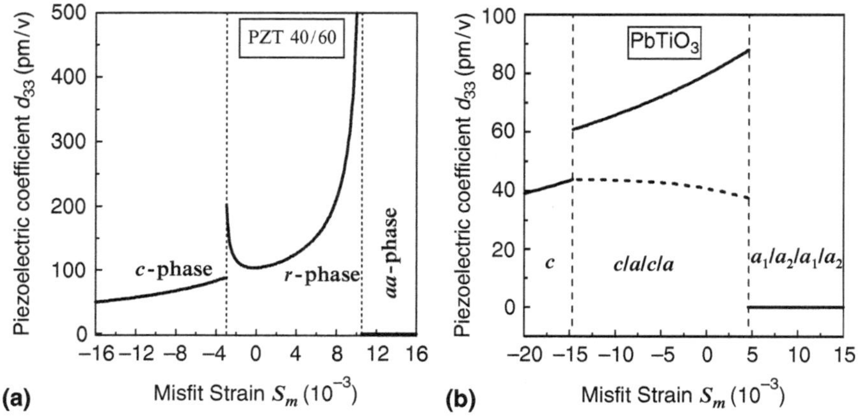

Fig. 2.6 Longitudinal piezoelectric coefficient d_{33} of single-domain $Pb(Zr_{0.4}Ti_{0.6})O_3$ (**a**) and polydomain $PbTiO_3$ (**b**) thin films as a function of the misfit strain in the epitaxial system. The dashed line in (**b**) shows the variation of d_{33} in $PbTiO_3$ film with pinned domain walls. (Reprinted with permission from Pertsev et al. 2003 and Koukhar et al. 2001. Copyright 2001, 2003, American Physical Society.)

at which the monoclinic *r* phase transforms either into the orthorhombic *aa* phase or into the tetragonal *c* phase. On the other hand, ferroelectric films grown on "strongly compressive" or "strongly tensile" substrates exhibit only small or zero piezoelectric response d_{33}.

The substrate-induced lattice strains may partially relax via the formation of elastic domains (twins) in the film (Roitburd 1976). Such relaxation modifies the (S_m, T) phase diagram and hence may change the film piezoelectric response (Koukhar et al. 2001). In $PbTiO_3$ films, the twinning removes the *r* and *aa* phases from the equilibrium diagram (see Fig. 2.7a), because the pseudo-tetragonal *c/a/c/a* and $a_1/a_2/a_1/a_2$ polydomain states appear to be energetically more favorable. As a result, the misfit-strain dependence of d_{33} weakens in a polydomain film, as shown in Fig. 2.7b. In contrast, the monoclinic phase survives in the case of $Pb(Zr_{0.5}Ti_{0.5})O_3$ films (Kukhar et al. 2006), where the monoclinic $r_1/r_2/r_1/r_2$ and $ca_1/ca_2/ca_1/ca_2$ polydomain states are stable at small misfit strains in a wide range of temperatures (see Fig. 2.7b).

2.5 Domain-Wall Contribution to Piezoelectric Coefficients of Ferroelectric Ceramics and Thin Films

The measuring electric field **E** generally exerts a driving force on the domain walls in a ferroelectric material. The field-induced displacements of these walls from their initial positions may change the average lattice strains, thus giving an extrinsic contribution to the piezoelectric response. The domain-wall contribution can be

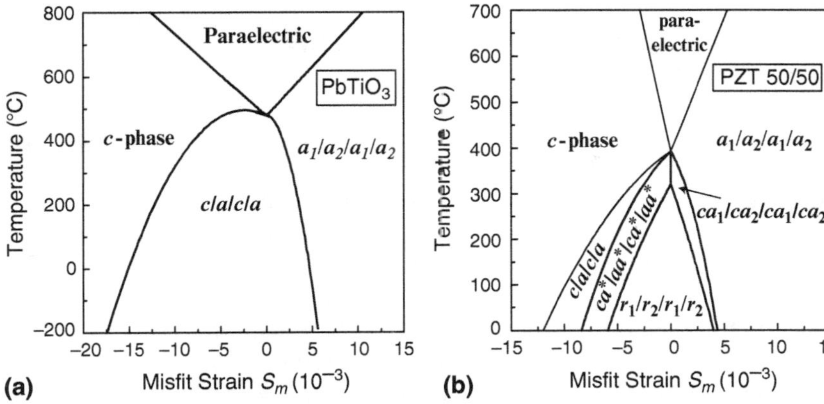

Fig. 2.7 Misfit strain–temperature phase diagrams of (001)-oriented polydomain $PbTiO_3$ (**a**) and $Pb(Zr_{0.5}Ti_{0.5})O_3$ (**b**) films epitaxially grown on (001)-oriented cubic substrates. The first- and second-order phase transitions are shown by thick and thin lines, respectively. (Reprinted with permission from Koukhar et al. 2001 and Kukhar et al. 2006. Copyright 2001, 2006, American Physical Society.)

evaluated theoretically by calculating the average wall displacement δl as a function of the field intensity E_i and determining the strain change δS_n in the volume swept by the moving walls. We shall consider below the domain-wall contribution Δd_{in} to the small-signal piezoelectric constants d_{in} only, assuming that the measuring field is weak $(\mathbf{E} \to 0)$.

Since the magnitude δl of domain-wall displacements is directly proportional to the field intensity (see below), a nonnegligible contribution Δd_{in} appears only when the strain change δS_n is independent of the applied field. This requirement demonstrates that displacements of purely ferroelectric domain walls (e.g., 180° walls in tetragonal crystals) do not affect the small-signal piezoelectric response (since $\delta S_3 \sim E_3$). In contrast, the ferroelectric–ferroelastic 90° walls create significant contribution to the piezoelectric coefficients of tetragonal ferroelectric ceramics and thin films (Bondarenko et al. 1990; Arlt and Pertsev 1991; Pertsev and Emelyanov 1997; Koukhar et al. 2001). This contribution results from the collective antiparallel motion of 90° walls forming periodic laminar patterns in these material systems (Arlt 1990; DeVeirman et al. 1993; Ramesh et al. 1993; Kwak et al. 1994). Remarkably, the shift of a 90° wall produces a local strain change proportional to the square of spontaneous polarization P_s.

The magnitude of cooperative displacements of the 90° walls is mainly restricted by the restoring forces associated with the changes of long-range elastic and electric internal fields caused by these displacements. When δl is much smaller than the domain width w, the restoring force acting per unit wall area can be written as $f_{res} = -k\ \delta l$, where k is the force constant. For tetragonal ferroelectric ceramics, the domain-wall contributions to the low-frequency piezoelectric constants can be evaluated as (Arlt and Pertsev 1991)

$$\Delta d_{33} = \frac{(Q_{11} - Q_{12})P_s^3}{k_{cer}w} J_{d_{33}}, \quad \Delta d_{31} = -\frac{1}{2}\Delta d_{33}, \quad \Delta d_{15} = \frac{(Q_{11} - Q_{12})P_s^3}{k_{cer}w} J_{d_{15}}, \tag{2.12}$$

where Q_{mn} are the electrostrictive constants of the paraelectric phase, and J_{din} are the parameters depending on distribution of the grain polarizations among various spatial orientations in a ferroelectric ceramic. The force constant k_{cer} involved in the above relations can be calculated in an analytical form (Arlt and Pertsev 1991). Since k_{cer} is inversely proportional to the domain width w, the contributions Δd_{in} are predicted to be independent of the domain-wall density in the first approximation. The numerical calculations show that Δd_{33} and Δd_{15} are considerable in poled $BaTiO_3$ and PZT ceramics, amounting to about 100 pC/N in $BaTiO_3$ and about 50 pC/N in PZT at the remanent polarization $P_r = 0.35 P_s$ (Arlt and Pertsev 1991).

In the case of epitaxial ferroelectric thin films, significant domain-wall contribution to the coefficient d_{33} appears when the pseudo-tetragonal *c/a/c/a* domain structure forms in the film (Pertsev and Emelyanov 1997; Koukhar et al. 2001). Since the electric field E_3 induced between the top and bottom electrodes covering the film surfaces interacts with the out-of-plane polarization P_3, the *c* domains change their size periodically during the piezoelectric measurements at the expense of the *a* domains having the in-plane spontaneous polarization. The force constant k_{film} determining the mechanical restoring force acting on displaced 90° c/a walls can be calculated analytically in the linear elastic approximation, which make it possible to evaluate the domain-wall contribution Δd_{33} (Pertsev and Emelyanov 1997). For relatively thick films (thickness > 100 nm), where the widths of *c* and *a* domains become much smaller than the film thickness ("dense" domain structure), Δd_{33} is given by a simple relation

$$\Delta d_{33} = \frac{(s_{11} + 2s_{12})(s_{11} - s_{12})}{s_{11}(Q_{11} - Q_{12})P_s}, \tag{2.13}$$

where s_{nm} are the elastic compliances at constant polarization. This relation shows that Δd_{33} is expected to be significant in poled thin films of conventional perovskite ferroelectrics ($\Delta d_{33} \approx 100$ pm/V in $BaTiO_3$ films and about 50 pm/V in $PbTiO_3$ films). The accurate numerical calculations performed with the aid of the nonlinear thermodynamic theory (Koukhar et al. 2001) demonstrated that actually the domain-wall contribution depends on the misfit strain in the film–substrate system (see Fig. 2.7b). Moreover, they confirmed that Δd_{33} is comparable with the intrinsic contribution to the film piezoelectric response.

Since the calculations discussed above ignore the interactions of domain walls with crystal defects and the lattice periodic potential (Peierls potential relief), they give only the upper bound for the domain-wall contribution to the piezoelectric response. Although the Peierls barriers for ferroelastic domain walls should be relatively small (because the thickness of such walls is usually much larger than the lattice period; see Chrosch and Salje 1999), at low temperatures they could reduce Δd_{33} considerably (Kukhar et al. 2006). Nevertheless, the domain-wall contribution to the piezoelectric coefficients of multiaxial polydomain ferroelectrics in general cannot be ignored. It should be noted that the field-induced displacements of the c/a domain walls were confirmed experimentally for $Pb(Zr_{0.32}Ti_{0.68})O_3$ thin films grown on MgO (Lee et al. 2001).

2.6 Piezoelectric Effect and Magnetism

Recent years witnessed the advent of novel materials that combine piezoelectric (linear in external stimulus) response with other attractive functionalities such as magnetic activity. Remarkably, the coupling between electric and magnetic variables is permitted only in crystals of certain symmetries. Such coupling is not only of high fundamental interest, but also of practical importance since it could be used to create new types of nonvolatile memories, where the data are written electrically and read magnetically (Eerenstein et al. 2006). It should be emphasized that piezoelectricity may also lead to the magnetoelectric coupling if the material simultaneously exhibits *piezomagnetism*. Indeed, the application of an electric field to such material induces strain via the converse piezoelectric effect, which, in turn, changes the magnetization because of the direct piezomagnetic effect. Similarly, the magnetic field creates lattice strains via the converse piezomagnetic effect so that the polarization also changes because of the direct piezoelectric effect.

Piezomagnetism is a linear effect relating strain and magnetic field or stress and magnetization. A strain S_{ik} produces a magnetic field $H_i = -\lambda_{ijk}S_{jk}$, where λ_{ijk} is the third-rank tensor. In turn, a magnetic field may produce strain (converse piezomagnetic effect). Similar to all magnetic tensors (see, e.g., Nye 1957), λ_{ijk} vanishes in all crystals symmetric with respect to the time inversion. Also, it is symmetric with respect to the last two indices, analogously to the piezoelectric tensor. The piezomagnetic effect is absent in the magnetic classes $m3m$, $\underline{4}3m$, and 432. Despite the piezomagnetic effect is typically small, it is used, for example, to register earthquakes. Strong coupling between piezoelectric and magnetic phenomena is expected in the materials called multiferroics that recently attracted a lot of interest in both magnetic and ferroelectric communities (Eerenstein et al. 2006; Cheong and Mostovoy 2007; Ramesh and Spaldin 2007).

Multiferroic crystals are distinguished by the simultaneous presence of two different order parameters such as polarization and magnetization (see, for example, a recent review (Fiebig 2005)). The crystal symmetry imposes a strict limitation on this class of materials (Aizu 1970, Schmid 1994). There are four major crystallographic types of multiferroics:

1. *Compounds with perovskite structure*. They have the chemical formula ABO_3 or $A_2B'B''O_6$. For example, the well-known compound $BiFeO_3$ is ferroelectric, ferroelastic, and weakly ferromagnetic at room temperature. It is rhombohedrally distorted with the crystallographic point symmetry $3m$.
2. *Compounds with hexagonal structure*. These compounds also have the chemical formula ABO_3 or $A_2B'B''O_6$. The hexagonal structure is due to relatively small

cationic radii. Ferroelectric–antiferromagnetic manganites $RMnO_3$, with $R = Sc$, Y, In, Ho, Er, Yb, or Lu, form a large group of hexagonal multiferroics. They have the crystallographic point symmetry $6mm$.

3. *Boracites*. They have the general formula $M_3B_7O_{13}X$. These crystals are ferroelectric–ferroelastic antiferromagnets (in some cases with a small ferromagnetic moment). Here M = Cr, Mn, Fe, Co, Cu, or Ni and X = Cl or I. Boracites are cubic with the point symmetry $\underline{4}3m$ at high temperatures and the $3m$ point group in the ferroelectric phase.
4. *Compounds with* $BaMF_4$ *structure*. M = Mg, Mn, Fe, Co, Ni, or Zn. These compounds are orthorhombic with the $2mm$ point symmetry at high temperatures. At low temperatures, they are ferroelastic ferroelectrics with a small ferromagnetic moment.

In addition, there is a large number of multiferroics with other structures (Eerenstein et al. 2006; Cheong and Mostovoy 2007). Basic symmetries of ferroelectrics, ferromagnets, ferroelastics, and multiferroics are compared in Table 2.3.

Appearance of spontaneous polarization and spontaneous magnetization breaks the spatial-inversion and time-reversal symmetries in multiferroics at low temperatures. This results in a linear magnetoelectric effect that involves both magnetic and electric fields: A magnetic field **H** induces an electric polarization **P** with components $P_i = \alpha_{ij}H_j$, and, in turn, an electric field **E** induces a magnetization **M** with components $M_i = \alpha_{ji}E_j$. Table 2.3 shows that the tensor α_{ij} can be nonzero only in materials that are noncentrosymmetric and time-asymmetric. There is a strict upper bound for the coefficients α_{ij} : $\alpha_{ij}^2 < \eta_{ii}\mu_{jj}$ (Brown et al. 1968). The linear magnetoelectric effect is small in most of the materials because either the dielectric (η_{ii}) or magnetic (μ_{jj}) susceptibility has a small value.

In the perovskite multiferroics, the crystal symmetry permits the linear magnetoelectric effect $(\alpha_{ij} \neq 0)$ and the existence of a spontaneous polarization **P** and a spontaneous magnetization **M**. For example, let us consider $BiFeO_3$. The free energy of this compound must be invariant with respect to the spatial inversion and rotations by the angle of 120° around the z axis. Therefore, the free energy F can be expressed through the corresponding invariants. The invariant $E_z(M_yL_x - M_xL_y)$, where **L** is the vector of the antiferromagnetic order, permits a spontaneous polarization along the z axis:

$$\mathbf{P} = -\frac{\partial F}{\partial \mathbf{E}} \propto (0, 0, M_yL_x - M_xL_y). \tag{2.14}$$

Table 2.3 Symmetry requirements for ferroics

Ferroic	Spatial-inversion symmetry?	Time-reversal symmetry?
Ferroelastic	Yes	Yes
Ferroelectric	No	Yes
Ferromagnetic	Yes	No
Multiferroic	No	No

Reprinted with permission from Eerenstein et al. 2006. Copyright 2006, Nature Publishing Group

The invariant $P_z(H_yL_x - H_xL_y)$ permits a spontaneous magnetization

$$\mathbf{M} = -\frac{\partial F}{\partial \mathbf{H}} \propto (P_zL_y, -P_zL_x, 0). \tag{2.15}$$

Using these invariants, one can also show that the magnetoelectric effect is nonzero ($\alpha_{ij} \neq 0$). However, the disadvantage of these compounds is that they may have an incommensurate magnetic structure. In $BiFeO_3$, the spins of Fe ions form a long-wavelength ($\lambda = 62\,\mathrm{nm}$) spiral structure. Averaging over the whole sample leads to zero magnetoelectric effect. This effect may be restored by applying strong magnetic field ($H \sim 20\,\mathrm{T}$) or by chemical substitution.

In contrast to perovskite multiferroics, the crystal symmetry of hexagonal manganites $R\mathrm{MnO}_3$ forbids the linear magnetoelectric effect. In these compounds, the interaction between electric and magnetic moments may arise via the piezomagnetic effect, which couples magnetization with the strain induced by electric polarization (Fiebig et al. 2002). Introduction of the strain coupling represents an attractive method for engineering an enhanced magnetoelectric effect.

In boracite $Ni_3B_7O_{13}I$, weak magnetic and electric orders emerge simultaneously below 60 K. The magnetoelectric effect is permitted. Applying a magnetic field, one can induce reversal of the magnetization, which, in turn, flips the polarization. In $BaMnF_4$, a spontaneous magnetoelectric effect is permitted owing to the presence of a weak permanent magnetization. Unfortunately, high magnitude of the electric coercive field renders the manipulation of the polarization difficult because this requires the application of a very strong magnetic field.

Recent findings showed that magnetoelectric, piezomagnetic, magnetoelastic, and other cross-coupling effects relating electrical, mechanical, and magnetic phenomena are typically weak in single-phase materials. In composites, however, they may be much stronger (Dong et al. 2002; Zheng et al. 2004; Eerenstein et al. 2006), especially near the phase-transition temperatures. The symmetry considerations can be very useful for finding a good combination of piezoelectric–ferroelectric and magnetostrictive–ferromagnetic materials. Another interesting opportunity is offered by the large coupling effects found within antiferromagnetic–ferroelectric domain walls (Fiebig et al. 2002).

2.7 Conclusion

Based on the above discussion, we can conclude that the piezoelectric effect in crystalline materials to a large extent is defined by the crystal symmetry. The symmetry considerations are indispensable for understanding the piezoelectric properties of single crystals, ceramics, and thin films. By varying the orientation of the applied field with respect to the crystallographic axes, it is possible to enhance the piezoelectric response of some crystals drastically. This orientation dependence of the piezoelectric properties is especially strong in crystals of the relaxor ferroelectrics. The effective piezoelectric coefficients of ferroelectric ceramics and thin films strongly depend on the spatial orientations of polar axes in crystallites and domains as well. Finally, it should be emphasized that crystallographic symmetry imposes strict limitations on the piezoelectric crystals, which could be simultaneously magnetically active.

Acknowledgments This work was partly supported by the EU-funded project "Multiceral" (NMP3-CT-2006-032616). N.A.P. acknowledges the Erasmus Mundus programme of the European Commission for financial support.

References

Aizu K (1969) Possible species of "ferroelastic" crystals and of simultaneously ferroelectric and ferroelastic crystals. J Phys Soc Jpn 27:387

Aizu K (1970) Possible species of ferromagnetic, ferroelectric, and ferroelastic crystals. Phys Rev B 2:754–772

Aleshin VI, Pikalev EM (1990) Effect of internal mechanic tensions on properties of ferroceramics. Zh Tekh Fiz 60:129–134

Arlt G (1990) Twinning in ferroelectric and ferroelastic ceramics – stress relief. J Mater Sci 25:2655–2666

Arlt G, Pertsev NA (1991) Force constant and effective mass of 90° domain walls in ferroelectric ceramics. J Appl Phys 70:2283–2289

Bondarenko EI et al. (1990) The effect of 90° domain wall displacements on piezoelectric and dielectric constants of perovskite ferroelectric ceramics. Ferroelectrics 110:53–56

Brown WF et al. (1968) Upper bound on the magnetoelectric susceptibility. Phys Rev 168:574–577

Budimir M et al. (2003) Piezoelectric anisotropy–phase transition relations in perovskite single crystals. J Appl Phys 94:6753–6761

Budimir M et al. (2006) Piezoelectric response and free-energy instability in the perovskite crystals $BaTiO_3$, $PbTiO_3$, and $Pb(Zr,Ti)O_3$. Phys Rev B 73:174106

Cady WG (1964) Piezoelectricity. Dover, New York

Cheong SW, Mostovoy M (2007) Multiferroics: A magnetic twist for ferroelectricity. Nat Mater 6:13–20

Chrosch J, Salje EKH (1999) Temperature dependence of the domain wall width in $LaAlO_3$. J Appl Phys 85:722–727

Cox DE et al. (2001) Universal phase diagram for high-piezoelectric perovskite systems. Appl Phys Lett 79:400–402

Curie P, Curie J (1880) Développement, par pression, de l'électricité polaire dans les cristaux hémièdres à faces inclinées. Comptes Rendus (France) 91:294–295

Curie P, Curie J (1881) Contractions et dilatations produites par des tensions électriques dans les cristaux hémièdres à faces inclinées. Comptes Rendus (France) 93:1137–1140

Damjanovic D et al. (2002) Crystal orientation dependence of the piezoelectric d_{33} coefficient in tetragonal $BaTiO_3$ as a function of temperature. Appl Phys Lett 80:652–654

Davis M et al. (2005) Domain engineering of the transverse piezoelectric coefficient in perovskite ferroelectrics. J Appl Phys 98:014102

Davis M et al. (2007) Rotator and extender ferroelectrics: Importance of the shear coefficient to the piezoelectric properties of domain-engineered crystals and ceramics. J Appl Phys 101:054112

DeVeirman AEM et al. (1993) TEM and XRD characterization of epitaxially grown $PbTiO_3$ prepared by pulsed-laser deposition. Philips J Res 47:185–201

Diéguez O et al. (2004) Ab initio study of the phase diagram of epitaxial $BaTiO_3$. Phys Rev B 69:212101

Dong SX et al. (2002) Circumferentially magnetized and circumferentially polarized magnetostrictive/piezoelectric laminated rings. J Appl Phys 96:3382–3387

Eerenstein W et al. (2006) Multiferroic and magnetoelectric materials. Nature 442:759–765

Fiebig M (2005) Revival of the magnetoelectric effect. J Phys D: Appl Phys 38:R123–152

Fiebig M et al. (2002) Observation of coupled magnetic and electric domains. Nature 419:818–820

Fousek J, Janovec V (1969) The orientation of domain walls in twinned ferroelectric crystals. J Appl Phys 40:135–142

Fu H, Cohen RE (2000) Polarization rotation mechanism for ultrahigh electromechanical response in single-crystal piezoelectrics. Nature 403:281–283

Guo R et al. (2000) Origin of the high piezoelectric response in $PbZr_{1-x}Ti_xO_3$. Phys Rev Lett 84:5423–5426

Jaffe B, Cook WR, Jaffe H (1971) Piezoelectric Ceramics. Academic Press, New York

Koukhar VG et al. (2001) Thermodynamic theory of epitaxial ferroelectric thin films with dense domain structures. Phys Rev B 64:214103

Kukhar VG et al. (2006) Polarization states of polydomain epitaxial $Pb(Zr_{1-x}Ti_x)O_3$ thin films and their dielectric properties. Phys Rev B 73:214103

Kwak BS et al. (1994) Domain formation and strain relaxation in epitaxial ferroelectric heterostructures. Phys Rev B 49:14865–14879

Lee KS et al. (2001) In situ observation of ferroelectric 90°-domain switching in epitaxial $Pb(Zr, Ti)O_3$ thin films by synchrotron X-ray diffraction. Appl Phys Lett 79:2444–2446

Lefki K, Dormans G J M (1994) Measurement of piezoelectric coefficients of ferroelectric thin films. J Appl Phys 76:1764–1767

Lippmann G (1881) Principe de la conservation de l'elecricité. Ann de Chemie e de Physique (5 serie) 24:145

Muralt P (2000) Ferroelectric thin films for micro-sensors and actuators: A review. J Micromech Microeng 10:136–146

Nan CW, Clarke DR (1996) Piezoelectric moduli of piezoelectric ceramics. J Am Ceram Soc 79:2563–2566

Noheda B et al. (1999) A monoclinic ferroelectric phase in the $Pb(Zr_{1-x}Ti_x)O_3$ solid solution. Appl Phys Lett 74:2059–2061

Noheda B et al. (2001) Polarization rotation via a monoclinic phase in the piezoelectric $92\%PbZn_{1/3}Nb_{2/3}O_3$–$8\%PbTiO_3$. Phys Rev Lett 86:003891

Nye JF (1957) Physical Properties of Crystals. Clarendon Press, Oxford

Park S-E, Shrout TR (1997) Ultrahigh strain and piezoelectric behavior in relaxor based ferroelectric single crystals. J Appl Phys 82:1804–1811

Pertsev NA, Emelyanov AYu (1997) Domain-wall contribution to the piezoelectric response of epitaxial ferroelectric thin films. Appl Phys Lett 71:3646–3648

Pertsev NA, Salje EKH (2000) Thermodynamics of pseudoproper and improper ferroelastic inclusions and polycrystals: Effect of elastic clamping on phase transitions. Phys Rev B 61: 902–908

Pertsev NA et al. (1998a) Aggregate linear properties of ferroelectric ceramics and polycrystalline thin films: Calculation by the method of effective piezoelectric medium. J Appl Phys 84:1524–1529

Pertsev NA et al. (1998b) Effect of mechanical boundary conditions on phase diagrams of epitaxial ferroelectric thin films. Phys Rev Lett 80:1988–1991

Pertsev NA et al. (2000) Phase transitions and strain-induced ferroelectricity in $SrTiO_3$ epitaxial thin films. Phys Rev B 61:R825–R829

Pertsev NA et al. (2003) Phase diagrams and physical properties of single-domain epitaxial $Pb(Zr_{1-x}Ti_x)O_3$ thin films. Phys Rev B 67:054107

Ramesh R, Spaldin N (2007) Mutliferroics: Progress and prospects in thin films. Nat Mater 6:21–28

Ramesh R et al. (1993) Effect of crystallographic orientation on ferroelectric properties of $PbZr_{0.2}Ti_{0.8}O_3$ thin films. Appl Phys Lett 63:731–733

Roitburd AL (1976) Equilibrium structure of epitaxial layers. Phys Status Solidi A 37:329–339

Sapriel J (1975) Domain-wall orientations in ferroelastics. Phys Rev B 12:5128–5140

Schmid H (1994) Multiferroic magnetoelectrics. Ferroelectrics 162:317–338

Tagantsev AK et al. (2002) Strain-induced diffuse dielectric anomaly and critical point in perovskite ferroelectric thin films. Phys Rev B 65:012104

Topolov VYu (2004) The remarkable orientation and concentration dependences of the electromechanical properties of $0.67Pb(Mg_{1/3}Nb_{2/3})O_3$–$0.33PbTiO_3$ single crystals. J Phys Condens Matter 16:2115–2128

Turik AV (1970) Elastic, piezoelectric, and dielectric properties of single crystals of $BaTiO_3$ with a laminar domain structure. Sov Phys Solid State 12:688

Vanderbilt D, Cohen MH (2001) Monoclinic and triclinic phases in higher-order Devonshire theory. Phys Rev B 63:094108

Voigt W (1910) Lerbuch der Kristallphysik. Teubner, Leipzig–Berlin

Ye ZG et al. (2001) Monoclinic phase in the relaxor-based piezoelectric/ferroelectric $Pb(Mg_{1/3}Nb_{2/3})O_3$–$PbTiO_3$ system. Phys Rev B 64:184114

Zembilgotov AG et al. (2005) Phase states of nanocrystalline ferroelectric ceramics and their dielectric properties. J Appl Phys 97:114315

Zhang R et al. (2003) Orientation dependence of piezoelectric properties of single domain $0.67Pb(Mn_{1/3}Nb_{2/3})O_3$–$0.33PbTiO_3$ crystals. Appl Phys Lett 82:3737–3739

Zheng H et al. (2004) Multiferroic $BaTiO_3$–$CoFe_2O_4$ nanostructures. Science 303:661–663

Chapter 3
Crystal Chemistry of Piezoelectric Materials

Susan Trolier-McKinstry

3.1 Symmetry Hierarchy

Piezoelectricity linearly relates an induced polarization to an applied stress, as shown in (3.1),

$$P_i = d_{ijk}\sigma_{jk}, \tag{3.1}$$

where σ_{jk} is the applied stress, P_i is the induced polarization, and d_{ijk} is the piezoelectric charge coefficient. Einstein notation is used, where repeated indices are summed. Because piezoelectricity is a third-rank tensor property, a good starting point to understanding the crystal chemistry of piezoelectric materials is to consider the impact of symmetry on such a property.

Neumann's law states that the geometrical representation of any physical property contains the symmetry of the point group of the material. As shown in Fig. 3.1, of the 32 crystallographic point groups, only 21 are noncentrosymmetric. Odd-rank tensor properties are symmetry forbidden in centrosymmetric structures, making piezoelectricity a null property for such materials. In the same way, in point group 432, the combination of symmetry elements eliminates piezoelectricity. The remaining 20 point groups are potentially piezoelectric. Of these 20 point groups, ten are polar, that is, they have a vector direction in the material that is not symmetry-related to other directions. Such materials can have a spontaneous polarization, which is typically a function of temperature. Thus, these materials are pyroelectric. Ferroelectric materials are a subset of pyroelectric materials in which the spontaneous polarization can be reoriented between crystallographically-defined directions by a realizable electric field. Thus, all ferroelectric materials are both piezoelectric and pyroelectric.

While symmetry considerations can describe whether a material is potentially piezoelectric, they provide no information on the magnitude of the piezoelectric

S. Trolier-McKinstry
Materials Science and Engineering Department & Materials Research Institute, The Pennsylvania State University, University Park, PA 16802, USA

A. Safari, E.K. Akdoğan (eds.) *Piezoelectric and Acoustic Materials for Transducer Applications.*

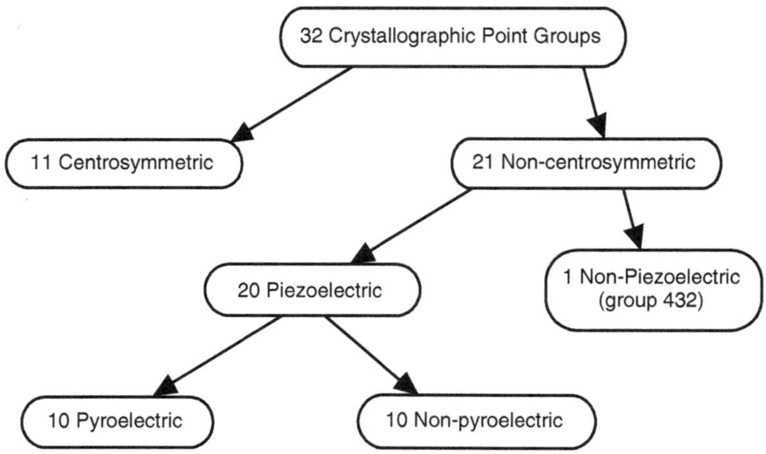

Fig. 3.1 Symmetry hierarchy for piezoelectricity

response. Insight into the latter is provided via structure–property relationships that entail a marriage of crystallographic information with functional property data. The remainder of this chapter attempts to describe some of the relevant structure–property relations for important piezoelectric materials.

3.2 Piezoelectricity in Nonferroelectric Materials

There are a number of commercially important piezoelectrics, including wurtzite and quartz, which are not ferroelectric. In such materials, the polarization direction is set by the crystal orientation, and cannot be reversed after processing. Thus, they are employed either as single crystals or oriented polycrystalline samples (including films).

It is particularly easy to visualize the origin of the piezoelectric response in wurtzite-structured compounds such as ZnO and AlN – two piezoelectrics of increasing importance in thin-film bulk acoustic resonators. The crystal structure of ZnO (zincite) is shown in Fig. 3.2. It can be seen there that the fundamental building block of the crystal is a puckered hexagonal ring in which both Zn and O have tetrahedral coordination. In zincite, the puckered rings are superimposed on each other, stacked parallel to the *c* axis. The result is a structure in which all of the ZnO_4 tetrahedra have the same orientation; in the figure, all of the Zn have one oxygen neighbor directly above it along the *c* axis, and three neighbors below it. When ZnO is compressed along the *c* axis, the material deforms by changing the O–Zn–O bond angle, since this angle is more readily compressed than the ZnO bond (Newnham 1975). As the tetrahedron deforms, the center of positive and negative charge of the unit are displaced, producing the polarization that is the hallmark of piezoelectricity.

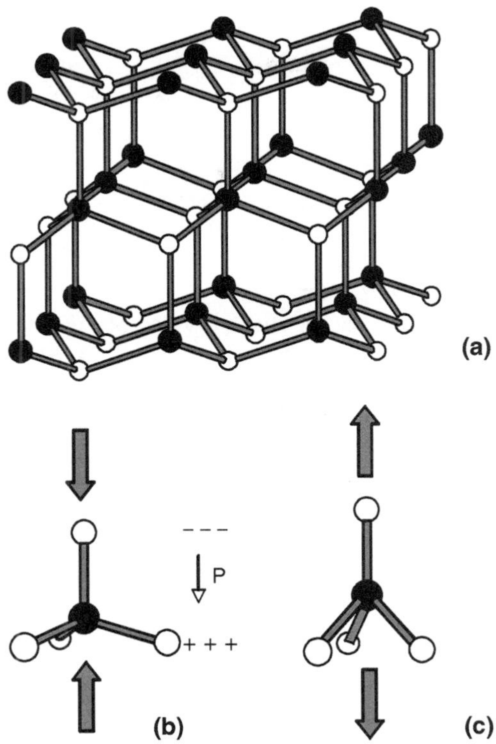

Fig. 3.2 (**a**) Zincite crystal structure showing the tetrahedral coordination of Zn and O. The *c* axis is vertical in the figure. (**b**) On application of a compressive stress parallel to the *c* direction, the legs of the tetrahedra spread, because the ZnO bond length is less compressible than O–Zn–O bond angle. Thus, the Zn moves closer to the three basal O; the base of the pyramid then becomes positively charged. (**c**) A tensile stress applied parallel to *c* elongates the ZnO_4 tetrahedra, inducing the opposite polarization

If, instead, a tensile stress is applied parallel to the *c* axis, the tetrahedra elongate, producing a surface charge of opposite polarity.

This simple example also points out that the piezoelectric effect is anisotropic. That is, the material will not develop identical polarizations for stresses applied parallel to the *c* and parallel to the *a* axes. Since the stress-induced deformation in the tetrahedra will differ, so will the induced polarization and the piezoelectric coefficient. It is also essential to note that if the orientation of the ZnO_4 tetrahedra was reversed (i.e., to make a pyramid balanced on its point, rather than its base), the sign of the piezoelectric response would be reversed. Thus, the piezoelectric coefficients of ZnO or AlN films are maximized when there is a single orientation. In a randomly oriented polycrystalline sample, the polarizations induced in various grains would tend to cancel each other out, so that there would be no net piezoelectric effect in the ensemble, even though each grain is piezoelectric.

Because the piezoelectric coefficients of ZnO and AlN are set by the crystal orientation during growth, they cannot be reoriented without bond breaking. Since the bonds in these compounds are quite strong, the piezoelectric response in these materials is thus stable against excursions in stress, electric field, or time.

3.3 Crystal Structures of Important Ferroelectric Piezoelectrics

Although the piezoelectric coefficients of nonferroelectric materials show excellent stability, they also tend to be quite low. Thus, in many applications, it is not possible to generate sufficient displacement in an actuator or sensitivity in a sensor. Substantially larger piezoelectric coefficients are demonstrated in many ferroelectric materials. This section describes the crystal structures of many of the important ferroelectric piezoelectrics. Emphasis is placed on the possible orientations for the spontaneous polarization.

From a crystal-chemistry perspective, ferroelectricity is often associated with materials with d^0 transition metal cations in octahedral coordination, and with cations with lone electron pairs, such as Pb^{2+} or Bi^{3+}. The d^0 ions tend toward more covalent character in their bonding, which can lead to asymmetric coordination geometries that favor the development of a spontaneous polarization. Of course, not all titanates demonstrate ferroelectricity (e.g., rutile TiO_2), and so it is also important to consider the details of the crystal structure and the symmetry in searching for new ferroelectric compounds. Lone electron pairs can also produce structural distortions that favor ferroelectricity.

3.3.1 *Perovskites*

Most of the commercially important ferroelectric materials possess some distortion of the perovskite crystal structure. Perovskite electroceramics are widely used in bulk capacitors, microwave resonators, and piezoelectric transducers, and are finding increased usage as thin films. The prototype perovskite ABO_3 crystal structure is shown in Fig. 3.3. The cubic parent (prototype) phase is adopted by many compounds at elevated temperatures. In cubic perovskites, the larger cation

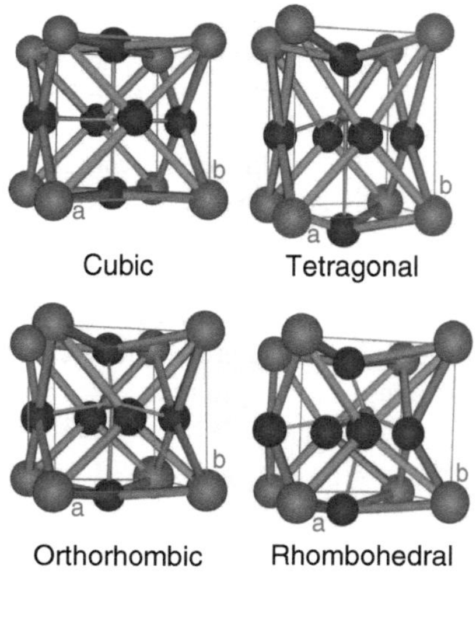

Fig. 3.3 Cubic perovskite structure and several of the common ferroelectric distortions of the perovskite unit cell (shown for $BaTiO_3$ with exaggerated ionic displacements)

(e.g., Pb^{2+}, Ba^{2+}, Sr^{2+}, Bi^{3+}, or Na^{+}) is 12-fold coordinated with respect to oxygen. The smaller cation (e.g., Ti^{4+}, Zr^{4+}, Nb^{5+}, Zn^{2+}, Mg^{2+}) is octahedrally coordinated by oxygen. Oxygen adopts a distorted six coordination, with two near B neighbors, and four more distant A neighbors. The point group symmetry is $m3m$.

The perovskite crystal structure is enormously tolerant to variations in composition and distortions (elastic strain). Thus, as temperature is decreased, many compounds with the perovskite structure undergo one of two general families of phase transitions. The most common phase transitions in perovskite-structured compounds occur because of cooperative motions of oxygen octahedra, resulting in coupled tilts or rotations of the oxygen octahedral framework. These are collectively known as tilt transitions, and they have a profound impact on the temperature coefficient of resonance frequency in microwave dielectrics and on piezoelectric constants. The second general family of phase transitions is displacive transitions, in which one or more cations move off high-symmetry positions in the lattice, typically producing ferroelectric or antiferroelectric characteristics to the material. Often, tilt transitions compete with other instabilities such as those inducing ferroelectricity (Zhong and Vanderbilt 1995; Seshadri and Hill 2001; Ghosez et al. 1995). These two families of transitions will be discussed next.

Many perovskite-structured compounds (including the mineral perovskite, $CaTiO_3$, itself) undergo phase transitions associated with tilting of the octahedral units (with or without distortion of the octahedra). Fundamentally, these transitions often occur when the A-site ion is too small for 12 coordination; the tilt transitions lead to some of the oxygen neighbors moving to larger distances, effectively reducing the cation coordination.

An example of a tilted perovskite is given in Fig. 3.4. Many possible tilt transitions are known, depending on how the coupled rotations or twists of one octahedra occur with respect to its neighbors (Glazer 1972). When such a tilt transition occurs

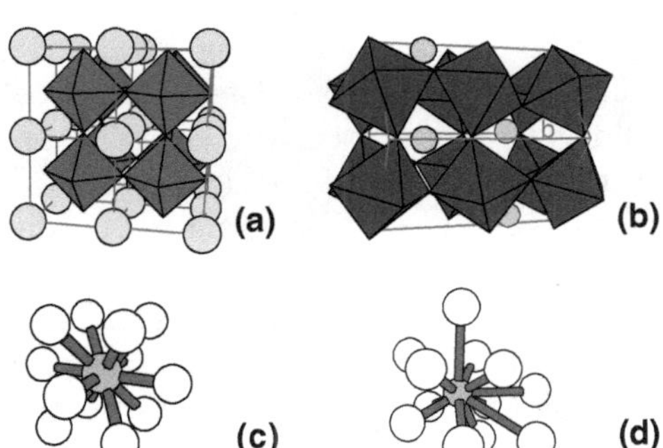

Fig. 3.4 Perovskite crystal structure with the A-site ions shown in *light grey circles*, and the B-site octahedra in *darker grey*. The oxygen atoms (not shown) are at the corners of the octahedra. (**a**) Untilted perovskite. (**b**) A perovskite with tilted octahedra. (**c**) Undistorted 12 coordination of A-site ion in cubic perovskite. (**d**) The distorted coordination of Gd in $GdScO_3$, showing that some of the 12 oxygen neighbors are at a greater distance, so that the coordination is effectively reduced by the tilt transition

without the simultaneous development of a polarization, the material is referred to as ferroelastic, rather than ferroelectric, since the domain states differ only in the spontaneous distortion.

Alternatively, the perovskite prototype phase can distort via a phase transition that leads to the development of a spontaneous polarization and ferroelectricity. An illustration of several of the common perovskite distortions is given in Fig. 3.3 (where the ion displacements have been exaggerated to facilitate visualization). In each case it is clear that the center of positive and negative charge within the unit cell are no longer coincident, which is the origin of the spontaneous polarization (P_S). The number of possible orientations for the spontaneous polarization for the ferroelectric depends on the symmetry change between the paraelectric and ferroelectric phases. For a tetragonal perovskite such as $BaTiO_3$ at room temperature, the polarization develops as the Ti^{4+} ion moves along [001] of the cubic lattice, closer to one of the adjacent oxygen ions. Since the Ti is octahedrally coordinated, there are six possible oxygen atoms it could have displaced toward, and six possible orientations for the spontaneous polarization. At the phase transition, the material elongates parallel to the spontaneous polarization and contracts laterally. In orthorhombic perovskites, the Ti^{4+} displaces toward two of its oxygen neighbors along the $\langle 101 \rangle$ family of directions in the prototype cell, yielding 12 possible directions for P_S. In rhombohedral perovskites, the Ti^{4+} moves close to three of the adjacent oxygens by displacing along one of the body diagonals of the prototype cube, toward one of the corners of the unit cell. This produces eight possible directions for P_S, oriented along $\langle 111 \rangle$. There are several possible monoclinic phases in the perovskite family, as discussed by Vanderbilt and Cohen (2001). The monoclinic A phase, denoted M_A, occurs when the polarization develops at an angle intermediate between the tetragonal and rhombohedral distortions. M_B places the polarization between orthorhombic and rhombohedral distortions, and M_C between tetragonal and orthorhombic. For each of the monoclinic phases, 24 polarization directions are possible.

In ferroelectric materials, it is possible to reorient the direction of the spontaneous polarization between the allowed states with an applied electric field. It is important to realize that the structural distortions associated with ferroelectricity tend to be rather modest, with atoms moving a fraction of an Ångstrom from the prototype phase to the ferroelectric phase. While all the domain states are energetically degenerate when the sample is under zero-field, strain-free conditions, the free energy of the system is lowered when the polarization is parallel (or at least as nearly so as possible) to the applied electric field. As a rule of thumb, for atom displacements under an Ångstrom, large applied electric fields can permanently reorient the polarization between the crystallographically-allowed states. When the field is removed, there is no restoring force for the polarization to return to its original orientation unless there are stress or strain fields acting. As a result, even randomly oriented polycrystals can be electrically poled to produce net piezoelectric and pyroelectric coefficients. The larger the number of possible orientations for the spontaneous polarization, the more likely there is an allowed direction that is nearly parallel to the applied electric field. As a result, poling tends to be more complete in systems with more polarization directions.

Consider also the link between the ferroelectric distortions of the crystal structure and the piezoelectric response. In all of the ferroelectric distortions of $BaTiO_3$, the crystal elongates parallel to the spontaneous polarization, and contracts perpendicular to it. This is an important point, as it underlies the link between strain and polarization in the lattice. In much the same way, when an electric field is applied that increases the polarization, the material will elongate further in that direction, and contract laterally. Similarly, when applied stresses deform the material, the polarization will change. This is the origin of the intrinsic converse and direct piezoelectric coefficients, respectively.

The $PbZr_{1-x}Ti_xO_3$ (PZT) solid solution is a perovskite system showing both ferroelectric phases and octahedral tilt transitions (see Fig. 3.5) (Alberta et al. 2001). It is also the most commercially important piezoelectric ceramic. For Ti concentrations above about 48 mol%, the material adopts a tetragonally distorted ferroelectric phase. At the morphotropic phase boundary (MPB, the nearly temperature-independent boundary between the two ferroelectric distortions), there is a transition to the rhombohedral distortion, with a sliver of monoclinic phase that appears at low temperatures. As described earlier, the rhombohedral ferroelectric phases entail displacement of the Zr and Ti ions along [111], which induces the spontaneous polarization. Two rhombohedral distortions are known, which differ as to whether the oxygen octahedra are tilted; in the high temperature, rhombohedral-I phase, they are not. At low temperatures, compositions in a broad section of the phase diagram undergo a tilt transition of the $a^-a^-a^-$ type associated with rotation of the octahedra about [111] (using Glazer notation (Glazer 1972)). This results in cell doubling for each of the lattice parameters (Corker et al. 1998). A second example of a perovskite ferroelectric in which tilt transitions are important is $Na_{1/2}Bi_{1/2}TiO_3$ (Jones and Thomas 2002).

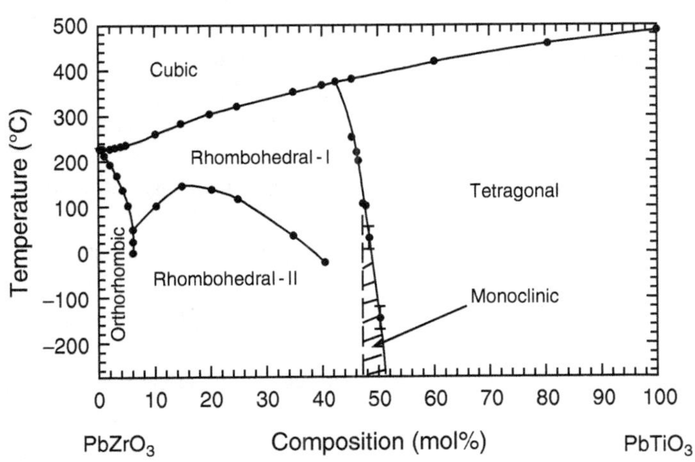

Fig. 3.5 PZT phase diagram (modified from Alberta et al. 2001). Note that the $PbZrO_3$-rich side of the diagram has subsequently been redrawn by Woodward, Knudsen and Reaney, 2005

3.3.2 Perovskite Tolerance Factor

As can be seen in Fig. 3.4, for a cubic perovskite, the lattice parameter a should be $2(R_B + R_O)$, where R_B is the radius of the B-site ion, and R_O is the radius of the oxygen ion. The length of the face diagonal is $2(R_A + R_O)$, where R_A is the radius of the A ion. Simple geometry then points to the observation that the face diagonal should be $\sqrt{2}a$. Goldschmidt (1926) defined a tolerance factor as:

$$t = \frac{R_A + R_O}{\sqrt{2}(R_B + R_O)}. \tag{3.2}$$

It is generally found that the perovskite structure is stable for tolerance factors between ~0.80 and 1.06 (Muller and Roy 1974; Shrout and Halliyal 1987). It is observed that tolerance factors above 1 generally favor tetragonally distorted ferroelectrics, while tolerance factors below 1 favor rhombohedral distortions (Eitel et al. 2001). Thus, it is not surprising that morphotropic phase boundaries are often found near perovskite tolerance factors of 1 (0.97–1.02). This can be used as a guide to developing new families of perovskite-based piezoelectrics.

3.3.3 Domains and Domain Walls

Because there are multiple possible directions for the spontaneous polarization, it is not surprising that in different volumes of the material, the polarization nucleates in different orientations. One can then define a domain as a volume of material in which the direction of the spontaneous polarization is uniform (or at least nearly so). The number of allowed domain states is given by the number of possible polarization directions.

Domain walls demark the boundaries between domains. In many ferroelectrics, the domain walls lie along low index crystallographic planes, which are mechanically compatible across the wall (Fousek and Janovek 1969). This is a consequence of the need to match the spontaneous deformation that occurs as the ferroelectric polarization is generated in order to avoid developing strains that would fracture the material at a domain wall.

As described earlier, the perovskite unit cell typically elongates parallel to P_S, and contracts laterally. For a tetragonal perovskite, the zero-strain condition across the domain wall occurs either when the polarization vectors make an angle of 180° with respect to each other (in which case the domain wall is parallel to the polarization vector), or when the polarization vectors differ by ~90° (where the domain wall lies along one of the {110} planes of the pseudocubic cell). Domain walls are named for the angle made by the polarization across the wall. Thus, in a tetragonally-distorted perovskite, both 90° domain walls and 180° domain walls exist. A schematic illustration of a 90° domain wall in a perovskite is shown in Fig. 3.6. By using comparable logic, the allowed domain states and domain-wall orientations can be determined for other cases; for example, in rhombohedral perovskites, 180°, 71°, and 109° walls occur.

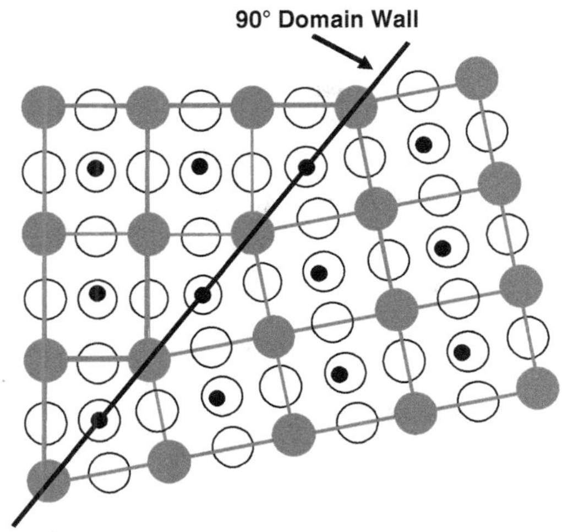

Fig. 3.6 (100) projection of a 90° domain wall in a tetragonally distorted perovskite showing atom positions. Note that domain walls are thicker than indicated in the diagram. The A ion is shown in grey, oxygen in white, and the B cation in black

Domain walls can be moved by electric fields, stresses, or by combinations of the two. This has significant consequences in ferroelectric piezoelectrics, since domain-wall motion contributes to the electric-field-induced strain. In soft PZT ceramics, roughly 50% of the net piezoelectric coefficient arises from the motion of non-180° domain walls. In particular, domains in which the polarization directions differ by some angle other than 180° have different spontaneous strains. Motion of a domain wall between these domains changes the relative volume fraction of the material with different strain states, and so contributes to the observed shape change (Zhang et al. 1983). It has recently been shown that under certain circumstances, 180° domain-wall motion can also contribute to the piezoelectric coefficients and their electric-field dependence (Trolier-McKinstry et al. 2006; Bassiri Gharb et al. 2006).

Domain-wall motion is very useful in increasing the magnitude of the achievable piezoelectric response. It does, however, introduce a variety of complications, including time dependence of the piezoelectric constants via aging, hysteresis in the strain response, and a drive field dependence of the material constants (Damjanovic 1997; Zhang et al. 1994, 1988; Li et al. 1991; Hall 1999; Hall and Stevenson 1999). Consequently, it is typical for high-power applications, in which an actuator will be driven under high electric or elastic fields, to employ hard piezoelectrics, where domain-wall motion is limited. In contrast, where large piezoelectric coefficients are required and in sensors, soft piezoelectric ceramics, with their attendant nonlinearities, are employed.

Interestingly, it is possible in some piezoelectric single crystals to engineer stable domain states, which also have extremely large piezoelectric response (Park and Shrout 1997). In perovskite single crystals, stable domain states can be achieved in one of two ways. First, a single domain state can be achieved by an electric field parallel to one of the spontaneous polarization directions. Although this enables full poling of the material, it is not uncommon for the material to be comparatively stiff for electric fields applied in this same direction. The result is often large, but unremarkable piezoelectric constants. Alternatively, stable domains in which two or more domain states are equally favorable can also be developed. For example, in a rhombohedrally-distorted perovskite poled along [001] crystallographic direction, there are four equivalent polarization directions: [111], $[1\bar{1}1]$, $[\bar{1}11]$, and $[\bar{1}\bar{1}1]$. Such a crystal will be polydomain, but there is no reason for an electric field applied parallel to the [001] to move any of the domain walls. This is sometimes referred to as an "engineered domain state," in which the macroscopic symmetry imparted by the domain structure may differ from the local symmetry. Comparable situations exist for a number of different distortions of the perovskite structure (Wada et al. 1999; Park et al. 1999). Such a domain structure can result in very large piezoelectric coefficients if the energy surface constraining the direction of the spontaneous polarization is shallow, so that the polarization vectors can be rotated by an electric field (Du et al. 1998).

Large strains and piezoelectric coefficients accompany the polarization rotation, without engendering substantial hysteresis in the response. This behavior has been linked to the presence of multiple phase transitions that "soften" the lattice along particular directions at some temperatures, yielding enhancements in the shear piezoelectric response (Damjanovic et al. 2002). The large d_{15} piezoelectric response increases the effective longitudinal and transverse piezoelectric coefficients in rotated cuts. In general, the optimized properties are observed in [001] oriented materials in rhombohedrally distorted perovskites (Du et al. 1998, Bell 2001). Large responses have also been shown in [101] oriented rhombohedral perovskite crystals (Bell 2001), or in orthorhombic perovskites (Lu et al. 2001).

It is generally observed that the piezoelectric properties go through a maximum near the morphotropic phase boundary, where the polarizability of the structure is high (see Fig. 3.7). Comparable increases in piezoelectric coefficient are available at other polymorphic (ferroelectric–ferroelectric) phase transitions. However, in the case where the phase boundary is strongly dependent on temperature, the piezoelectric properties also show large temperature dependences, and may be hysteretic as the phase transformation takes place and the domain structures are disturbed. Moreover, the upper use temperature for the material is generally governed by the *lowest* phase transition temperature. Curved phase boundaries lead to ferroelectric–ferroelectric phase transitions at temperatures that may be well below the Curie temperature. Consequently, in designing piezoelectric actuator materials, nearly temperature-independent boundaries are greatly preferred. Ferroelectric–ferroelectric phase transitions can also be induced by applied electric fields or mechanical stresses. Figure 3.8 shows examples of field-induced ferroelectric–ferroelectric phase transition in rhombohedral [001] oriented $Pb(Zn_{1/3}Nb_{2/3})O_3$-$PbTiO_3$ crystals. The linear portions of the strain-field curve define the converse piezoelectric coefficient, while the steps in the response denote the field-forced transition from a rhombohedral to a tetragonal distortion. These discontinuities are present in both the dielectric and electromechanical response. Not surprisingly, the closer the composition is to the MPB at room temperature, the smaller the electric field is required to induce the tetragonal mate-

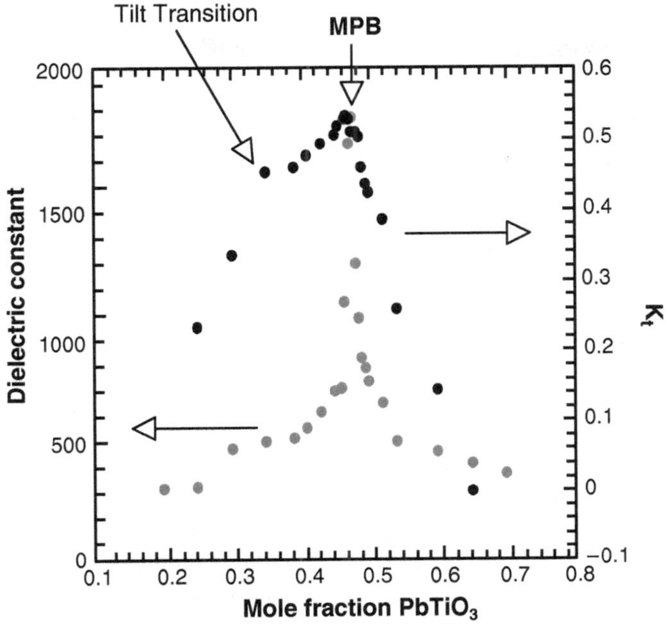

Fig. 3.7 Variations in properties across the PZT phase diagram (data from Yamamoto 1996)

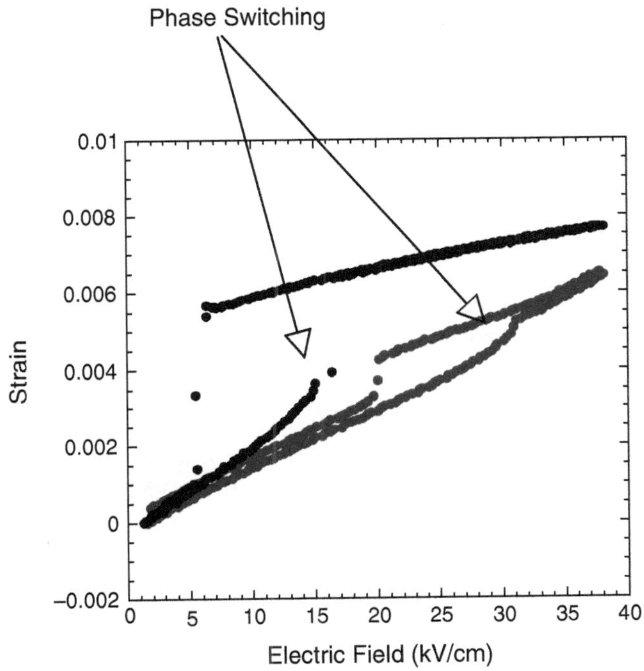

Fig. 3.8 Evidence for field-induced phase transformations in [001] $PbZn_{1/3}Nb_{2/3}O_3$–$PbTiO_3$ crystals where the *black symbols* are for $0.92PbZn_{1/3}Nb_{2/3}O_3$–$0.08PbTiO_3$ and the *grey symbols* for $0.955PbZn_{1/3}Nb_{2/3}O_3$–$0.045PbTiO_3$. The latter is further from the morphotropic phase boundary and so requires a higher field for phase switching

rial. In some cases, once the tetragonal phase is introduced, it is partially retained when the driving electric field is reduced (Durbin et al. 2000). This is problematic, as the appearance of retained tetragonal material is coupled with both a decrease in the small signal piezoelectric response as well as an increase in hysteresis in these crystals.

Tilt transitions have considerable importance in governing the properties of the materials. For example, it has recently been demonstrated that the effectiveness of morphotropic phase boundaries in increasing the piezoelectric properties of perovskites is strongly affected by the tilt transitions (Zheng et al. 2002). The increase in piezoelectric and dielectric properties of PZT ceramics at the MPB is well known (see for example, Fig. 3.7) (Yamamoto 1996). However, it was shown that when the tilt transition is pushed over to the MPB, the piezoelectric activity drops nearly a factor of 2 (Zheng et al. 2002). Similarly, when the MPB was bounded by an untilted rhombohedral phase in several of the $BiMO_3$–$PbTiO_3$ solid solutions, the piezoelectric coefficients were substantially higher than when the tilted phase was present (Randall et al. 2004). In the case of perovskite ferroelectrics, it is likely that both the intrinsic and the extrinsic (domain-wall or phase boundary motion related) contributions to the properties will be affected by tilt transitions. For example, in the case of the piezoelectric response, the intrinsic piezoelectric coefficient, d, is given by:

$$d \approx 2\varepsilon_0 \varepsilon Q P_r, \tag{3.3}$$

where ε_0 and ε are the permittivities of free space and the material, respectively, P_r is the remanent polarization, and Q is the electrostrictive constant. In many perovskites going through the R3m–R3c phase transformation, P_r increases modestly, ε is approximately constant, but Q decreases.

As a result, the intrinsic piezoelectric response drops. In addition, it is possible that the 109° ferroelectric domain walls can be pinned by antiphase boundaries in the tilted phase. This decreases the extrinsic piezoelectric constant (Eitel and Randall 2007).

3.3.4 Bismuth-Layer Structure Ferroelectrics

The family of compounds with the general chemical formula given by $Bi_2M_{n-1}R_nO_{3n+3}$, where $n = 1, 2, 3, 4, 5$ is known as the bismuth layer structure ferroelectrics, or Aurivillius phases (Aurivillius 1950). Here the R cation is typically a ferroelectrically active transition metal (such as Ti^{4+}, Nb^{5+}, Ta^{5+}) octahedrally coordinated with oxygen, while M is a larger ion with distorted 12 coordination. Structurally, the crystals can be regarded as an interleaving of layers of the perovskite structure with Bi_2O_2 layers, as shown in Fig. 3.9. The parameter n measures the number of octahedra in a perovskite layer. It is observed that when n is an even number, the spontaneous polarization develops along the a axis, and can be switched between [100], $[\bar{1}00]$, [010], and $[0\bar{1}0]$. No component of the polarization

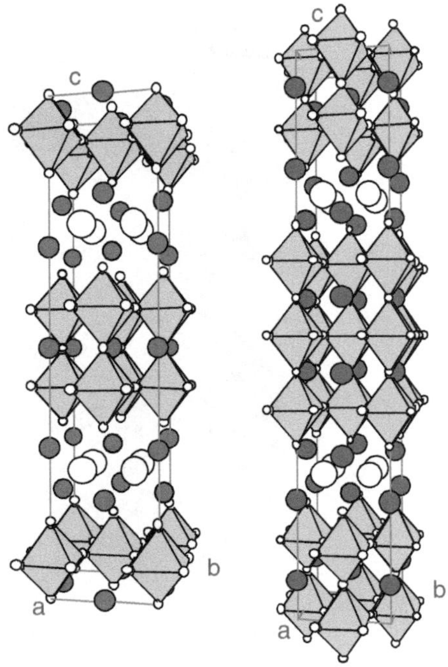

Fig. 3.9 Two examples of bismuth layer structures. Bismuth ions are shown as *grey circles*, oxygen atom positions are marked by *white circles*, and the transition metal-centered oxygen octahedra are shown. The octahedra belong to the perovskite layer structure. It is clear that the perovskite slabs are broken up by Bi_2O_2 layers

develops perpendicular to the perovskite layers, as polarization is symmetry forbidden in this case (Newnham et al. 1971). In contrast, when n is an odd number, finite value of switchable polarization can develop parallel to [001], although the principal component of the polarization still lies in-plane.

The strong anisotropy of the polarization leads to significant anisotropy in the piezoelectric coefficient in single crystals (Sa Neto and Cross 1982). In ceramics, the confinement of the polarization to the $a-b$ plane (or nearly so) makes polarization switching difficult. This occurs because misoriented grains can block the switching of other grains, decreasing the remanent polarization and piezoelectric coefficient (Takenaka and Sakata 1980). Significant improvements in the piezoelectric coefficients can be attained if the ceramic is textured (e.g., by hot-forging, templated grain growth, hot-pressing, centrifugal sintering, etc.) (Takenaka and Sakata 1980; Messing et al. 2004; Horn et al. 1999; Xiang et al. 2006). Among the members of this family investigated for piezoelectric applications are $Bi_4Ti_3O_{12}$ and $SrBi_2Nb_2O_9$. It should be noted that some of the members of this family also undergo octahedral tilt transitions (Reaney et al. 1995).

3.3.5 $LiNbO_3$

$LiNbO_3$ and $LiTaO_3$ are more widely used for their electro-optic properties, rather than their piezoelectric response. Exceptions to this occur in high-frequency

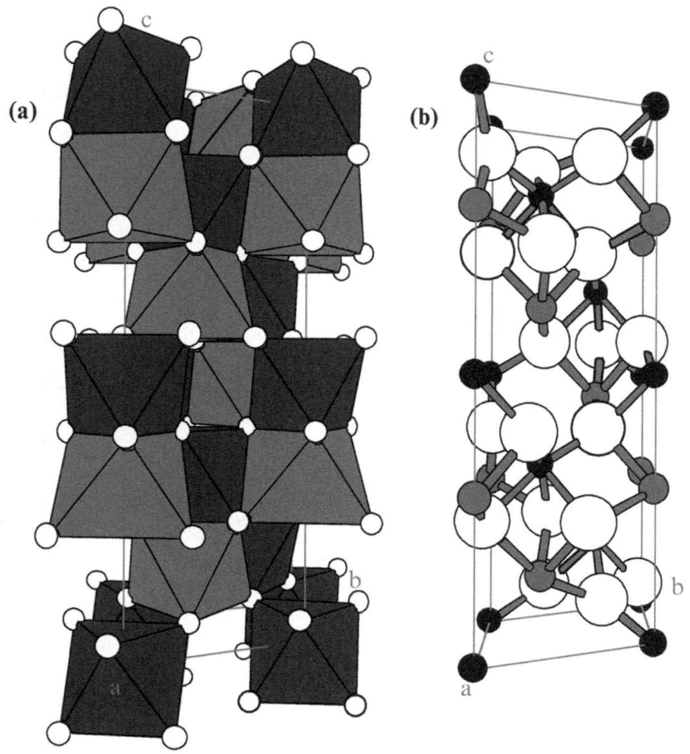

Fig. 3.10 Crystal structure of $LiNbO_3$ showing the nearly close-packed oxygen sublattice and the alternation of Li, Nb, and vacancies in the octahedral interstices. Nb is *dark*, Li is *grey*, and the O is *white*. (**a**) Polyhedral model; (**b**) bonds connecting nearest neighbors are shown

actuators or high-temperature sensors. The high Curie temperatures of these compounds, ~1200 °C and 620 °C, respectively (Lines and Glass 1977), reduce the likelihood of depoling at elevated temperatures. $LiNbO_3$ and $LiTaO_3$ are isostructural. The unit cell is a trigonal one, which can be thought of either as a derivative of the corundum structure, or of the perovskite structure (see Fig. 3.10). Counting the nearest neighbors, the coordination of the cations are both six; both adopt distorted octahedral coordinations with respect to oxygen. Along the trigonal axis, the cations stack as Nb, Li, vacancy, Nb, Li, vacancy. Off-centering of the cations from their high-symmetry positions leads to the ferroelectric polarization along the *c* axis. The polarization can be reversed if the Li^+ ions are forced through a small interstice between three oxygens, so that the stacking becomes Nb, vacancy, Li, Nb, vacancy, Li. As a result, the domain structures here are far simpler than they are in the perovskite family. Only 180° domain walls exist.

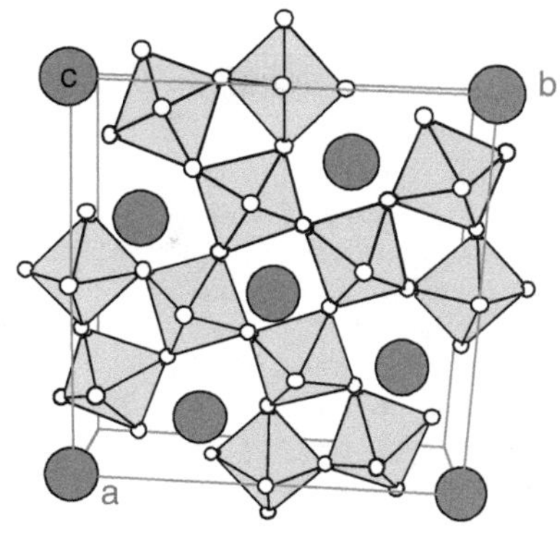

Fig. 3.11 Tungsten bronze ferroelectrics. The octahedra around the smaller cations are shown. Large interstices are filled by the largest cations in the structure (e.g., Ba^{2+} or Sr^{2+})

3.3.6 Tungsten Bronzes

Less commercially important, at present, is the family of tungsten bronzes, typified by materials such as $Sr_{1-x}Ba_xNb_2O_6$. In these materials, as in the perovskites, the bismuth layer structure compounds, and $LiNbO_3$, the key structural feature is the oxygen octahedra. The octahedra are corner-shared here, although the arrangement is less symmetric than that shown in the perovskites. The large ions occupy the interstices in the structure. In the case of the tetragonal tungsten bronzes, the polarization develops perpendicular to the page for the orientation shown in Fig. 3.11. Again, the limited number of polarization directions accessible leads to inefficient poling in polycrystalline ceramics. That is, if the orientation distribution for the grains is measured, the switchable polarization could be estimated by assuming that the appropriate vector component of the polarization that is aligned most closely to the field will switch. Experimentally, the remnant polarization is well below this calculated number until a significant degree of texture is achieved in the ceramic (Duran et al. 2003). This implies that grains with their ferroelectric axis at a large angle to the applied electric field must block the switching of adjacent, well-oriented grains. Once percolation of oriented grains is achieved, the switchable polarization rises toward the calculated value.

3.4 SbSI and Piezoelectric Composites

In the majority of the crystal structures described heretofore, the bonding is three dimensional. It is not surprising that piezoelectric deformations along one axis couple to motion along other axes. In many cases, this leads to a low hydrostatic d constant; i.e., $d_h = d_{31} + d_{32} + d_{33}$. Thus, in many perovskites, the ratio of d_{31}/d_{33} is approximately -0.4 to -0.45. The d_{33} and d_{31} are opposite in sign since a polarization is enhanced if a tensile stress is applied parallel to the polarization axis, but is reduced if a tensile stress is applied perpendicular to the same axis. The net result is that d_h tends to be modest in many perovskite ferroelectrics.

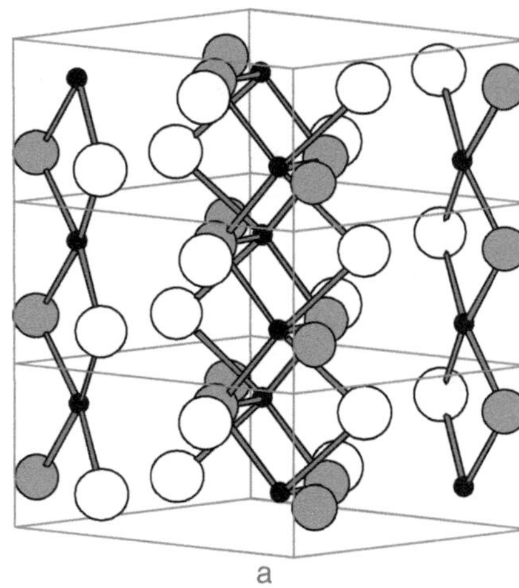

Fig. 3.12 Crystal structure of SbSI showing the chain-like bonding. The Sb atoms are shown with small *black circles*; I is shown in *white* and S in *grey*. While each chain is fairly well-bonded along the vertical axis in the drawing, the bonding between chains is quite weak. Thus, SbSI can be regarded as a prototype of a 1–3 piezoelectric–polymer composite in which the lateral mechanical coupling is deliberately weakened

To increase the d_h value, one approach is to decrease the coupling between displacements in these orthogonal directions. This can be achieved by reducing the three-dimensional nature of the structure. An excellent example of this for a single-phase material is given by SbSI. As shown in Fig. 3.12, the crystal structure consists of parallel chains that are weakly coupled to each other. As a result, strain that develops parallel to the chain direction is weakly transmitted between chains. The result is that the ratio of d_{31}/d_{33} is approximately -0.125. Unfortunately, the electrical conductivity of SbSI is too high for many applications (Lines and Glass 1977). The general principle illustrated here, however, was one of the factors that lead to development of piezoelectric composites (Newnham 2003). In fact, the 1–3 piezoelectric/polymer composite is a macroscopic analog of SbSI. The polymer separating the piezoelectric pillars serves to decouple the piezoelectric laterally, increasing the d_h value.

References

Alberta EF, Guo R, Bhalla AS (2001) Structure–property diagrams of ferroic solid solutions. Part I: Perovskite relaxor ferroelectrics with morphotropic phase boundaries. Ferro. Rev. 4: 1–327.

Aurivillius B (1950) Mixed oxides with layer lattices: III. Structure of $BaBi_4Ti_4O_{15}$. Arkiv Kemi 2 (37): 519–527.

Bassiri Gharb N, Trolier-McKinstry S, Damjanovic D (2006) Piezoelectric nonlinearity in ferroelectric thin films. J. Appl. Phys. 100, 044107.

Bell AJ (2001) Phenomenologically derived electric field-temperature phase diagrams and piezoelectric coefficients for single crystal barium titanate under fields along different axes. J. Appl. Phys. 89: 3907–3914.

Corker DL, Glazer AM, Whatmore RW, Stallard A, Fauth F (1998) A neutron diffraction investigation into the rhombohedral phases of the perovskite series $PbZr_{1-x}Ti_xO_3$. J. Phys.: Condens. Matter 10: 6251–6269.

Damjanovic D (1997) Stress and frequency dependence of the direct piezoelectric effect in ferroelectric ceramics. J. Appl. Phys. 82: 1788–1797.

Damjanovic D, Brem F, Setter N (2002) Crystal orientation dependence of the piezoelectric d_{33} coefficient in tetragonal $BaTiO_3$ as a function of temperature. Appl. Phys. Lett. 80: 652–654.

Du XH, Zheng JH, Belegundu U, Uchino K (1998) Crystal orientation dependence of piezoelectric properties of lead zirconate titanate near the morphotropic phase boundary. Appl. Phys. Lett. 72: 2421–2423.

Duran C, Trolier-McKinstry S, Messing GL (2003) Dielectric and piezoelectric properties of textured $Sr_{0.53}Ba_{0.47}Nb_2O_6$ ceramics prepared by templated grain growth. J. Mat. Res. 18: 228–238.

Durbin MK, Hicks JC, Park SE, Shrout TR (2000) X-ray diffraction and phenomenological studies of the engineered monoclinic crystal domains in single crystal relaxor ferroelectrics. J. Appl. Phys. 87: 8159–8164.

Eitel R, Randall CA (2007) Octahedral tilt-suppression of ferroelectric domain wall dynamics and the associated piezoelectric activity in $Pb(Zr,Ti)O_3$. Phys. Rev. 75.

Eitel RE, Randall CA, Shrout TR, Rehrig PW, Hackenberger W, Park SE (2001) New high temperature morphotropic phase boundary piezoelectrics based on $Bi(Me)O_3$-$PbTiO_3$ ceramics. Jpn. J. Appl. Phys., 40(Pt 1): 5999–6002.

Fousek J, Janovek V (1969) The orientation of domain walls in twinned ferroelectric crystals. J. Appl. Phys. 40: 135–142.

Ghosez P, Cockayne E, Waghmare UV, Rabe KM (1995) Lattice dynamics of $BaTiO_3$, $PbTiO_3$, and $PbZrO_3$: A comparative first principles study. Phys. Rev. B 60: 836–843.

Glazer AM (1972) The classification of tilted octahedra in perovskites. Acta Crystallogr., Sect. B: Struct. Crystallogr. Cryst. Chem. B28: 3384–3392.

Goldschmidt VM (1926) The laws of crystal chemistry. Naturwissenschaften 14: 477–485.

Hall DA (1999) Rayleigh behaviour and the threshold field in ferroelectric ceramics. Ferroelectrics 223: 319–328.

Hall DA, Stevenson PJ (1999) High field dielectric behaviour of ferroelectric ceramics. Ferroelectrics 228: 139–158.

Horn JA, Zhang XC, Selvaraj U, Messing GL, and Trolier-McKinstry S (1999) Templated grain growth of textured bismuth niobate. J. Am. Ceram. Soc. 82: 921–926.

Jones GO, Thomas PA (2002) Investigation of the structure and phase transitions in the novel A-site substituted distorted perovskite compound $Na_{0.5}Bi_{0.5}TiO_3$. Acta Crystallogr., Sect. B: Struct. Crystallogr. Cryst. Chem. 58: 168–178.

Li SP, Cao WW, Cross LE (1991) The extrinsic nature of nonlinear behavior observed in lead zirconate titanate ferroelectric ceramic. J. Appl. Phys. 69: 7219–7224.

Lines ME, Glass AM (1977) Principles and Applications of Ferroelectric and Related Materials. Clarendon Press, Oxford.

Lu Y, Jeong DY, Cheng ZY, Zhang QM, Luo HS, Yin ZW, Viehland D (2001) Phase transitional behavior and piezoelectric properties of the orthorhombic phase of $Pb(Mg_{1/3}Nb_{2/3})O_3$–$PbTiO_3$ single crystals. Appl. Phys. Lett. 78: 3109–3111.

Messing GL, Trolier-McKinstry S, Sabolsky EM, Duran C, Kwon S, Brahmaroutu B, Park P, Yilmaz H, Rehrig PW, Eitel KB, Suvaci E, Seabaugh M (2004) Templated grain growth of textured piezoelectric ceramics. Crit. Rev. Sol. State Mat. Sci. 29: 45–96.

Muller O, Roy R (1974) The Major Ternary Structural Families. Springer-Verlag, New York.

Newnham RE (1975) Structure–Property Relations. Springer-Verlag, New York.

Newnham RE, Private communication (2003).

Newnham RE, Wolfe RW, Dorrian JF (1971) Structural basis of ferroelectricity in bismuth titanate family. Mat. Res. Bull. 6: 1029.

Park SE, Shrout TR (1997) Ultrahigh strain and piezoelectric behavior in relaxor based ferroelectric single crystals. J. Appl. Phys. 82: 1804–1811.

Park SE, Wada S, Cross LE, Shrout TR (1999) Crystallographically engineered $BaTiO_3$ single crystals for high-performance piezoelectrics. J. Appl. Phys. 86: 2746–2750.

Randall CA, Eitel R, Jones B, Shrout TR, Woodward DI, Reaney IM (2004) Investigation of a high T_c piezoelectric system: $(1-x)Bi(Mg_{1/2}Ti_{1/2})O_3$–$(x)PbTiO_3$. J. Appl. Phys. 95: 3633–3639.

Reaney IM, Roulin M, Shulman HS, Setter N (1995) In situ observations of octahedral tilt transitions in strontium bismuth titanate layered perovskites. Ferroelectrics 165: 295–305.

Sa Neto A, Cross LE (1982) Electro-mechanical behavior of single domain single crystals of bismuth titanate ($Bi_4Ti_3O_{12}$). J. Mat. Sci. 17: 1409–1412.

Seshadri R, Hill NA (2001) Visualization of the role of Bi 6s "lone pair" electrons in the off-center distortions in ferromagnetic $BiMnO_3$, Chem. Mat. 13: 2892–2899.

Shrout TR, Halliyal A (1987) Preparation of lead-based ferroelectric relaxors for capacitors. Am. Ceram. Soc. Bull. 66: 714–711.

Takenaka T, Sakata K (1980) Grain-orientation and electrical-properties of hot-forged $Bi_4Ti_3O_{12}$ ceramics. Jpn. J. Appl. Phys. 19: 31–39.

Trolier-McKinstry S, Bassiri Gharb N, Damjanovic D (2006) Piezoelectric nonlinearity due to motion of 180° domain walls in ferroelectric materials at subcoercive fields: A dynamic poling model. Appl. Phys. Lett. 88: 202901.

Vanderbilt D, Cohen MH (2001) Monoclinic and triclinic phases in higher-order Devonshire theory. Phys. Rev. B 63: 094108.

Wada S, Suzuki S, Noma T, Suzuki T, Osada M, Kakihana M, Park SE, Cross LE, Shrout TR (1999) Enhanced piezoelectric property of $BaTiO_3$ single crystals with engineered domain configurations. Jpn. J. Appl. Phys. Pt. 1 38: 5505–5511.

Woodward DI, Kundsen J, Reaney IM (2005) Review of crystal and domain structures in the $PbZr_xTi_{1-x}O_3$ solid solution. Phys. Rev. B 72: 104–110.

Xiang PH, Kinemuchi Y, Watari K (2006) Effective grain alignment in bismuth titanate ceramic by centrifugal force. J. Mat. Res. 21: 1830–1835.

Yamamoto T (1996) Ferroelectric properties of $PbZrO_3$–$PbTiO_3$ system. Jpn. J. Appl. Phys. 35: 5104–5108.

Zhang QM, Pan WY, Jang SJ, Cross LE (1988) Domain-wall excitations and their contributions to the weak-signal response of doped lead zirconate titanate ceramics. J. Appl. Phys. 64: 6445–6451.

Zhang QM, Wang H, Kim N, Cross LE (1994) Direct evaluation of domain-wall and intrinsic contributions to the dielectric and piezoelectric response and their temperature dependence on lead zirconate–titanate ceramics. J. Appl. Phys. 75: 454–459.

Zhang XL, Chen ZX, Cross LE, Schulze WA (1983) Dielectric and piezoelectric properties of modified lead titanate zirconate ceramics from 4.2K to 300K. J. Mater. Sci. 18: 968–972.

Zheng H, Reaney IM, Lee WE, Jones N, Thomas H (2002) Effects of octahedral tilting on the piezoelectric properties of strontium/barium/niobium-doped soft lead zirconate titanate ceramics. J. Am. Ceram. Soc. 85: 2337–2344.

Zhong W, Vanderbilt D (1995) Competing structural instabilities in cubic perovskites. Phys. Rev. Lett. 74: 2587–2590.

Part II
Piezoelectric and Acoustic Materials for Transducer Technology

Chapter 4
Lead-Based Piezoelectric Materials

Dragan Damjanovic

This chapter discusses properties of lead-based piezoelectric materials, the most versatile and the most widely used piezoelectrics. Majority of these materials were discovered in 1950s and 1960s, and their properties and applications are described in classical textbooks, e.g. (Jaffe et al. 1971; Lines and Glass 1979). After giving essential background, this chapter will focus on recent developments. Lead titanate is discussed first, followed by modified lead titanate compositions. Lead zirconate titanate is then discussed in some details, focusing on mechanisms of hardening and softening and properties at morphotropic phase boundary. The subsequent sections discuss field-induced piezoelectric effect in relaxors, relaxor-ferroelectric ceramics, and crystals. Other lead-based materials and environmental issues are briefly discussed in the closing sections of the chapter.

4.1 Lead Titanate ($PbTiO_3$)

Pure lead titanate, $PbTiO_3$, is not commercially used as a piezoelectric material, but can be either modified or form solid solutions to obtain materials with excellent piezoelectric properties. Most of piezoelectrics presently exploited commercially are solid solutions based on lead titanate. In this section, the structure and basic properties of $PbTiO_3$ are presented.

$PbTiO_3$ belongs to the perovskite family, ABO_3. At room temperature, $PbTiO_3$ is ferroelectric with tetragonal $C^1_{4v} - P4mm$ space group. Lead titanate undergoes a first-order phase transition at 493 °C into cubic $O^1_4 - Pm3m$ paraelectric form. This phase transition was considered in the past to be a prototypic displacive phase transition (Lines and Glass 1979) but recent results have suggested that it exhibits mixed displacive and order-disorder character (Nelmes et al. 1990).

D. Damjanovic
Ceramics Laboratory, Swiss Federal Institute of Technology–EPFL, Lausanne, Switzerland

A. Safari, E.K. Akdoğan (eds.) *Piezoelectric and Acoustic Materials for Transducer Applications.*

Although $PbTiO_3$ is isomorphous at room temperature with another widely used ferroelectric perosvkite, barium titanate ($BaTiO_3$), there are many important differences between the two materials. The tetragonal lattice distortion of lead titanate is much larger than in $BaTiO_3$. The lattice constants of $PbTiO_3$ at room temperature are $a = 3.902$ Å and $c = 4.156$ Å (Nelmes and Kuhs 1985). This gives $c/a = 1.063$ as opposed to only 1.01 in $BaTiO_3$ (Megaw 1957). Another difference is that $BaTiO_3$ goes through a sequence of ferroelectric phases (cubic-tetragonal-orthorhombic-rhombohedral) on cooling from the paraelectric cubic phase, while only cubic and tetragonal ferroelectric phase seem to be present in lead titanate (Mestric et al. 2005). Recent first principles calculations suggest that increasing hydrostatic pressure may induce in $PbTiO_3$ phase transitions from tetragonal to monoclinic to rhombohedral to cubic phase (Wu and Cohen, 2005).

In the tetragonal phase of $PbTiO_3$, the coordinates of the ions may be described in the following way, taking the Pb ion at the origin of the coordinate system [Pb (0,0,0), Ti(1/2, 1/2, $1/2 + dz$(Ti)), O1(1/2,1/2, dz(O1)), O2 ($1/2, 0, 1/2 + dz$(O2)) and ($0, 1/2, 1/2 + dz$(O2)]. The dz(X) denote shifts of ion X from the ideal cubic position, in fractional units. In contract to $BaTiO_3$ (Harada et al. 1970) where Ti and O ions move in the opposite direction, in $PbTiO_3$, Ti and both types of O ions move in the same direction. Refinement of the structure using neutron powder diffraction (Glazer and Mabud 1978) and neutron single-crystal diffraction (Nelmes and Kuhs 1985) have shown that a significant distortion of the oxygen octahedron takes place, with $\delta z(O2) > \delta z(O1)$, Fig. 4.1.

The covalent character of Pb and its role in ferroelectric and piezoelectric properties of Pb-based compounds has been discussed often in the literature (Cohen 1992). The recent progress in ab-initio calculations has given first clear indication

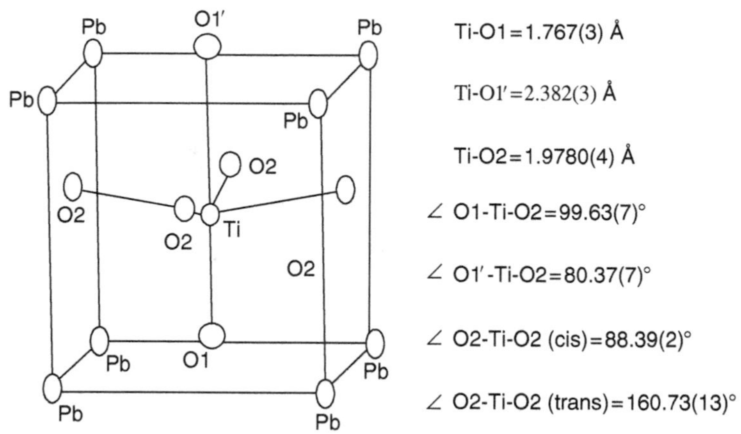

Fig. 4.1 The perovskite structure of lead titanate with distorted oxygen octahedron. The lattice parameters, bond lengths, and angles are determined at room temperature using neutron diffraction of $PbTiO_3$ powder prepared by the author. The neutron diffraction experiment was performed by F. Rotella at Argonne National Laboratory (1986). Anisotropy of thermal oscillations is represented by ellipsoids at each ionic site

of the role of Pb in the ferroelectricity of lead-based materials and why $PbTiO_3$ and lead-free perovskites ($BaTiO_3$, $KNbO_3$) have such markedly different properties. Cohen has shown that in both $BaTiO_3$ and $PbTiO_3$ hybridization of Ti–O bond is essential for ferroelectricity by weakening short-range repulsions. If this hybridization is inhibited, the ferroelectric instability disappears and the material becomes cubic. The calculations show that 6s states of lead and 2p states of oxygen are strongly hybridized in $PbTiO_3$, while 5p states of Ba do not hybridize with the valence band. The large strain in $PbTiO_3$ is then related to the Pb–O bonding interaction and smaller radius of Pb compared with Ba. Hybridization of Pb–O also indirectly affects Ti–O interactions. The calculations of Cohen are confirmed experimentally by several authors including work of Tanaka et al. (2006), who demonstrated evidence of dipolar polarization of Pb in $PbTiO_3$, using maximum entropy method analysis of X-ray diffraction data.

Because of the difficulties in growing good single crystals of $PbTiO_3$ of a size suitable for measurements, there have been relatively few attempts to systematically study dielectric, elastic, and piezoelectric properties of this crystal. The experimental studies are further complicated by the fact that crystals exhibit large conductivity, possibly related to high concentration of lead vacancies (Lines and Glass 1979). The most complete data can be found in the works by Gavrilyachenko and Fesenko (1971) and more recently by Li et al. (1993). For example, $d_{33} = 117\ (84)\ \mathrm{pC\,N^{-1}}$, $d_{31} = -25\ (-27)\ \mathrm{pC\,N^{-1}}$, $d_{15} = 65\ (60)\ \mathrm{pC\,N^{-1}}$, $s^{\sigma}_{33} = 32\ (21)\ \mathrm{m^2N^{-1}}$, $\kappa^{\sigma}_{11} = 210\ (130)$, $\kappa^{\sigma}_{33} = 126\ (80)$, where $\kappa = \varepsilon/\varepsilon_0$ is the relative permittivity and data are those of Gavrilyachenko and Fesenko with Li et al.'s data given in brackets. The dielectric and piezoelectric properties of single domain crystals of $PbTiO_3$ were predicted as a function of temperature by Haun et al. (1987) using Landau-Ginzburg-Devonshire phenomenological thermodynamic theory. Interestingly, polycrystalline thin films of lead titanate have been prepared without too many difficulties and d_{33} of about $50\,\mathrm{pm\,V^{-1}}$ has been reported (Kighelman et al. 2002). Kushida and Takeuchi (1987) have reported large thickness coupling coefficient ($k_t = 0.8$) in c-oriented $PbTiO_3$ films, but that result has not been confirmed independently.

The most striking difference between lead titanate and most other perovskite ferroelectrics is a low anisotropy of the permittivity and relatively low shear piezoelectric coefficients in $PbTiO_3$. The ratio between the permittivity perpendicular and parallel to the polar axis, $\kappa_{\perp}/\kappa_{\parallel} = \kappa_{11}/\kappa_{33}$, is less than 2 in lead titanate at room temperature, while in $BaTiO_3$, depending on the source, $\kappa_{11}/\kappa_{33} > 20$–30. The dielectric anisotropy ratio is relatively large in most other ferroelectrics (Davis et al. 2007). The large κ_{11} (and, consequently, a large shear piezoelectric coefficient, d_{15}) signifies propensity for a large polarization rotation by external electric field applied along off-polar directions or by shear stress applied out of plane containing the polar axis. This is of a considerable interest for applications, because in materials with a large d_{15} the longitudinal piezoelectric coefficient d_{33} can become unusually large along nonpolar directions (Park and Shrout 1997; Davis et al. 2007) (Sect. 4.4). The low dielectric anisotropy (or, equivalently, a low d_{15}/d_{33} ratio) in $PbTiO_3$ has nevertheless been of practical interest as it is related to a low transverse piezoelectric response (low d_{31}) in polycrystalline samples and domain-engineered single crystals (Davis et al. 2007; Davis et al. 2005). This is discussed in more details in the following section.

4.2 Modified Lead Titanate Ceramics

A large spontaneous strain and thermal expansion anisotropy of the tetragonal phase of lead titanate cause the mechanical fracture of lead titanate ceramics as they are cooled from the sintering temperature through the cubic to tetragonal phase transition at 490 °C. In addition, often observed increase in the conductivity in lead-based ceramics is probably related to the loss of lead during sintering, which, together with high coercive fields, makes poling of these ceramics very difficult and results in their poor dielectric and piezoelectric properties. Thus, no extensive data on properties of pure lead titanate ceramics are available.

A large number of different additives have been tried to obtain lead titanate ceramics with useful electromechanical properties. A high transition temperature into the paraelectric state and a low dielectric constant make lead titanate a particularly interesting material for the high temperature applications and for high frequency filters and transducers.

The main role of modifiers and dopants in $PbTiO_3$ is to improve mechanical strength by reducing tetragonality and increase electrical resistivity. The reduction of tetragonality leads to a lower Curie temperature and lower spontaneous strain and consequently less risk to cracking on cooling from the sintering temperature. An unexpected bonus and common characteristic of many modified compositions are the small transverse and planar coupling coefficients, k_{31} and k_p (in general significantly less than 0.1), and a relatively large thickness coupling coefficient (0.4–0.5) as shown in Fig. 4.2. The high electromechanical anisotropy (a high k_t/k_p or high d_{33}/d_{31} ratio) is useful in array transducers for reducing cross talk between elements, for reducing interference of higher harmonics of transverse mode with thickness mode vibrations, and in all cases where a good directivity of the acoustic signal is needed (Takeuchi et al. 1985). With a relatively small relative permittivity (200–300), transducers fabricated with modified lead titanates where of particular interest for high frequency applications. The samples with somewhat higher relative permittivity (500) and/or d_{33} ($>90\,pC\,N^{-1}$) but preserving the high electromechanical anisotropy were also developed (Troilo et al. 1994). The most widely used and studied compositions are those where modifiers are rare earths $(Pb_{1-3/2x}Re_x)TiO_3$ (Takeuchi et al. 1985) and alkaline earths $(Pb_{1-x}Ca_x)TiO_3$ (Yamashita et al. 1981), developed, respectively, at Hitachi and Toshiba.

The origin of the low transverse piezoelectric properties, first discussed by Turik et al. (1975), became a topic of considerable interest during 1980s and 1990s (Turik and Topolov 1997). The effect in ceramics is related to averaging of the properties of individual grains and domains and the low anisotropy of the permittivity; it was thus believed that the effect is possible only in polycrystalline materials. Some

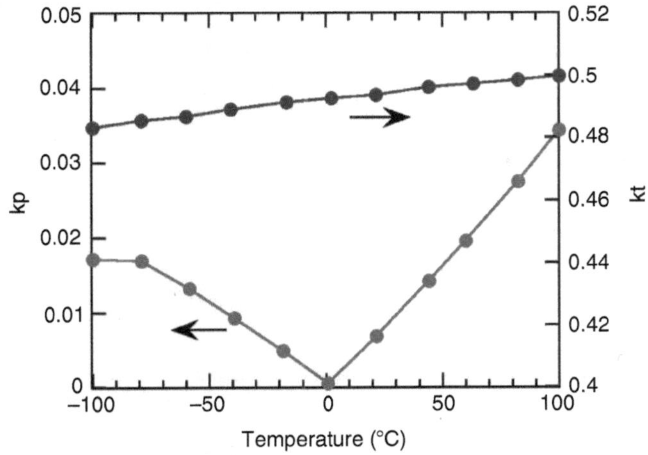

Fig. 4.2 Planar and thickness coupling coefficients in Ca-doped $PbTiO_3$ ceramics as a function of temperature. For details see (Damjanovic et al. 1987)

ceramic samples exhibited a vanishing transverse piezoelectric effect that was explained by a change of the sign of d_{31} coefficient with temperature and the poling degree (Damjanovic et al. 1987). Only recently it became clear that the zero and positive d_{31} can be observed along certain crystal directions even in single crystals, providing that d_{15} is small enough (Davis et al. 2005).

4.3 Lead Zirconate Titanate

Lead zirconate, $PbZrO_3$, is antiferroelectric (therefore not piezoelectric) (Lines and Glass 1979), and it has been seen in Sect. 4.1 that pure $PbTiO_3$ itself is of a limited interest as a piezoelectric material. It is thus surprising that the solid solution of these two compositions, $Pb(Zr_{1-x}Ti_x)O_3$ or PZT, exhibits outstanding piezoelectric properties. Its widespread use in industry, science, medicine, transportation, communications and information technologies has also made PZT one of the most studied ferroelectrics. Despite this, many aspects of the high piezoelectric response in PZT still remain obscure. Perhaps nothing attests better to the complex nature of PZT than the fact that 50 years after its discovery (Shirane et al. 1952) and numerous in-depth studies (Jaffe et al. 1971) modifications of the composition-temperature phase diagram of this ternary system are still being proposed (Noheda et al. 1999), raising new questions about origins of the large piezoelectric activity in this and related ferroelectric materials.

Lead zirconate and lead titanate are soluble in all proportions. The first phase diagram of $Pb(Zr_{1-x}Ti_x)O_3$ was derived by Sawaguchi (Sawaguchi 1953) showing the most prominent feature of the system, the compositionally driven structural transformation from tetragonal ($P4mm$) to rhombohedral phase ($R3m$) at $x \approx 0.47$ at

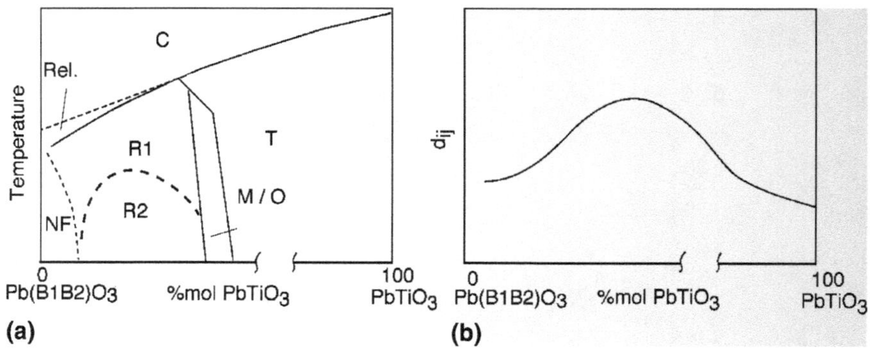

Fig. 4.3 (**a**) Simplified composition-temperature phase diagram of PZT or $PbTiO_3$-relaxor ferroelectrics, and (**b**) schematic illustration of variation of piezoelectric properties with composition. For PZT, B1 = B2 = Zr. T is tetragonal phase, M/O is a monoclinic or orthorhombic phase, R1 and R2 are rhombohedral phases differing in rotation of oxygen octahedra, NF is either relaxor or antiferroelectric phase, and Rel. is a relaxor phase. In PZT, relaxor phase is absent and NF phase is antiferroelectric. For details see (Woodward et al. 2005, Jaffe et al. 1971, Eitel and Randall 2007, Noheda et al. 1999)

room temperature. This work was soon followed by the discovery of a large piezoelectric response in compositions close to this phase boundary (Jaffe et al. 1954). The composition-temperature phase diagram of Sawaguchi was refined by Jaffe and coworkers (Jaffe et al. 1971), and most recently by several groups.

A simplified phase diagram and evolution of the properties across it are schematically shown in Fig. 4.3. The most significant modification of the Jaffe et al.'s phase diagram is addition by Noheda et al. (Noheda et al. 1999) of the monoclinic phase separating the tetragonal and rhombohedral regions. The nature of this phase is still a subject of many studies. The Zr-rich rhombohedral region is divided into two subregions with $R3m$ symmetry at high and $R3c$ symmetry at low temperatures. These two phases differ essentially by appearance of rotation of oxygen octahedra about the polar axis in the low temperature phase (Eitel and Randall 2007; Michel et al. 1969; Woodward et al. 2005). Antiferroelectric phases appearing for $x < 0.1$ are of no interest for the present discussion.

The most striking feature of the phase diagram is nearly temperature-independent boundary between tetragonal and rhombohedral (monoclinic) phases, called "morphotropic phase boundary" or MPB. The term was used by Jaffe et al. to designate "an abrupt structural change within a solid solution with variation in composition" (p. 135 of Ref. (Jaffe et al. 1971)), following the nomenclature introduced by Goldschmidt et al. (1926).

It should be said here that morphotropic phase boundary in PZT is not perfectly vertical. Thus, rhombohedral PZT samples with compositions in vicinity of the MPB will undergo phase change from $R3c$ to $R3m$ to $P4mm$ symmetries on heating from room temperature. The monoclinic compositions will transform to $P4mm$ phase on heating, and only tetragonal compositions can be heated all the way to cubic phase without phase transformation, Fig. 4.3. Crossing the phase transitions leads to appreciable temperature dependence of the properties in PZT, as shown in early works of Berlincourt and colleagues (Berlincourt et al. 1964). However, this is less of a problem in PZT than in materials in which polymorphic character of the phase transition is strong (e.g., lead free Li and Ta modified $(K, Na)NbO_3$, (Zhang and Shrout 2007). In PZT, the thermodynamic variable (composition) responsible for the high properties does not change as temperature varies. In compositions with polymorphic phase transitions, the temperature is the thermodynamic variable that is both responsible for the high properties and which may change during exploitation.

PZT is available in form of ceramics (although commercial samples are always doped) and polycrystalline, textured and epitaxial thin films, while there are only few reports on successful growth of bulk single crystals (Fesenko et al. 1986). Growth of high quality single crystals of PZT with compositions at or close to the morphotropic phase boundary remains one of the biggest challenges for the ferroelectrics community.

4.3.1 *Morphotropic Phase Boundary and Enhanced Piezoelectric Response*

The origin of the enhanced dielectric, elastic, and piezoelectric properties of PZT at the morpohotropic phase boundary has been investigated in detail by many authors and is still a topic of intense interest (Isupov 2002). Many mechanisms have been invoked to explain the enhanced piezoelectric properties of PZT at MPB. There is no consensus in the literature on this problem and this section will merely outline the main points of different approaches without attempting to be exhaustive:

(a) The most common explanation for enhanced properties in PZT ceramics at MPB is related to increased ease of polarization reorientation during poling (Jaffe et al. 1971; Isupov 2002). The tetragonal phase has six possible orientations of polarization (along pseudocubic <100> directions), while in rhombohedral phase, polarization can orient along eight <111> directions. At MPB, the tetragonal and rhombohedral phases coexist (Isupov 2001), their free energies are nearly degenerate and switching between the phases can be accomplished by external fields (e.g., by poling field) so that polarization within each grain effectively has 14 directions along which it can orient. The better poling results in larger piezoelectric properties. Indeed, it has been shown experimentally that remnant polarization in PZT has a maximum near MPB (Jaffe et al. 1971). If the MPB region contains a monoclinic phase, as recent results indicate (Noheda et al. 1999), the number of available orientations for polarization would be further increased because in a monoclinic phase polarization can adopt 24 orientations.

(b) Displacement of domain walls and phase boundaries, and their contribution to the electro-mechanical properties, is facilitated at the MPB region (Isupov 2002; Cross 1993). This argument is closely related to the first one, but relates to operation under dynamic fields. With multiple domain states available in the

MPB region, the energy barrier between different states should be relatively small, facilitating displacement of domain walls even by relatively weak driving fields. In a similar fashion, the boundary between crystalline phases with degenerate energies should be easily displaced. It is known that displacement of domain walls can greatly enhance the piezoelectric and dielectric properties of ferroelectric ceramics (Bondarenko et al. 1991). Although experimental results clearly show that domain wall contributions in PZT indeed increase as the MPB is approached (Damjanovic and Demartin 1997), there is no conclusive experimental evidence that displacement of phase boundaries participate in the enhanced response.

(c) Intrinsic properties of PZT crystals at MPB region are high. Because of unavailability of PZT single crystals, the information on the intrinsic (lattice) properties of PZT is almost exclusively available through theoretical calculations, based on phenomenological Landau-Ginzburg-Devonshire (LGD) approach (Haun et al. 1989), and more recently ab-initio calculations (Bellaiche et al. 2000; SaghiSzabo and Cohen 1997). Using LGD thermodynamic approach, Carl and Härdtl (1971) were able to show qualitatively that the dielectric permittivity perpendicular to polarization direction (transverse permittivity, $\varepsilon_{\perp}$ or ε_{11} in the crystallographic coordinate system) diverges as the MPB is approached from both tetragonal and rhombohedral side, while the permittivity along the polar axis ($\varepsilon_{\|}$ or ε_{33}) exhibits comparatively weaker anomaly, Fig. 4.4. Why transverse permittivity should diverge as tetragonal-rhombohedral phase transition is approached can be understood intuitively in the following way. In the tetragonal phase, polarization is oriented along the [001] pseudocubic direction, while in the rhombohedral phase it appears along [111] pseudocubic axis. In the phase transition region, the crystal dielectrically softens along directions dictated by

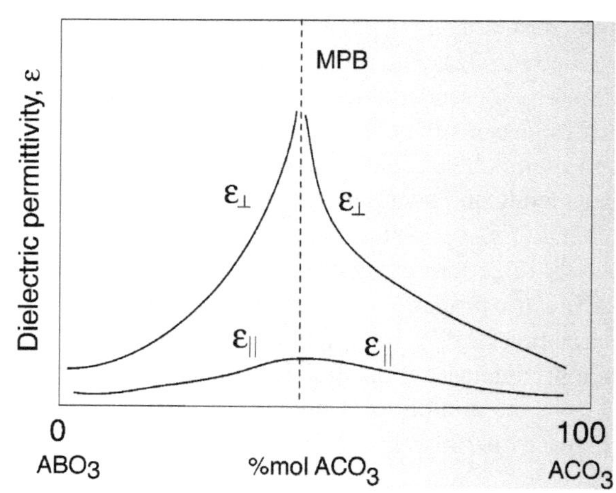

Fig. 4.4 Behavior of permittivity perpendicular and parallel to polar axis as a function of concentration in systems exhibiting a morphotropic phase boundary. For detail see (Carl and Härdtl, 1971)

the symmetry of the incipient crystal structure (Budimir et al. 2003). Thus, on the tetragonal side of the MPB, the crystal dielectrically softens in such a way to anticipate new direction of the polarization along [111] pseudocubic axis in the rhombohedral phase. This means that permittivity ε_{11} will increase much more than ε_{33}. Similar effect will take place on the rhombohedral side.

The result of Carl and Härdtl was confirmed in a detailed quantitative study of Haun et al. (1989) and later extended by Ishibashi and Iwata (1998) to derive general conditions for appearance of an MPB. The maximum in piezoelectric properties is thus a direct consequence of the anomaly in the permittivity; this can bee seen through phenomenological relations relating permittivity and piezoelectric coefficients, $d \propto \varepsilon QP$, where Q is the electrostrictive constant. Since divergence at MPB is stronger for the transverse than for longitudinal component of the permittivity, the shear piezoelectric coefficient ($d_{\text{shear}} \propto \varepsilon_{\perp} QP$) will in all phases be enhanced more than the longitudinal coefficient ($d_{\text{long}} \propto \varepsilon_{\parallel} QP$) (Haun et al. 1989). This means that propensity of crystals for polarization rotation is enhanced at MPB. The anomalous behavior of monodomain single crystals at MPB is reflected in ceramics through averaging of the properties over all grains and domain orientations. The ceramic samples thus exhibit a maximum of the properties at MPB, which is essentially intrinsic, single crystal property. Improved poling of ceramics at MPB and facilitated domain wall displacement in ceramics add to this intrinsic mechanism. However, as shown by Jones et al. (2007) and Hall et al. (2005), internal stresses in ceramics, interactions among grains and domains, and local texture affect the overall intrinsic as well as extrinsic response of ceramics and should not be neglected.

(d) Role of the monoclinic phase and polarization rotation. After the discovery of the monoclinic phase in PZT by Noheda et al. (1999), many attempts were made to explain MPB behavior in PZT and other ferroelectrics by enhanced piezoelectric response within this phase. It was indicated in the previous sections that propensity of PZT crystals for polarization rotation is augmented as MPB is approached. The field induced polarization rotation takes places through monoclinic mirror planes (Davis et al. 2006; Fu and Cohen 2000), and it is believed that if this phase is actually present the polarization rotation is facilitated, giving high dielectric and piezoelectric response. However, it should be remembered that the propensity of a crystal for polarization rotation is determined not by its crystal structure but by the value of the transverse permittivity, (see point (c) above). Thus, LGD calculations of Iwata and Ishibashi (2005), which include monoclinic phase, suggest that properties enhancement is not the largest within the monoclinic phase, but at the boundary between monoclinic and tetragonal or monoclinic and rhombohedral phases. This also agrees with results of phenomenological simulations by Bell (2006). The enhancement of the properties associated with presence of a monoclinic phase is thus likely related to pretransitional phenomena. One such example of pretransitional phenomena is Curie-Weiss behavior of permittivity in the cubic phase of ferroelectric material. In the case of MPB, if polarization changes direction at a phase transition point between two phases (e.g., tetragonal-rhombohedral), the transverse permittivity will increase in both phases around the transition point and enhance propensity of material for polarization rotation. In PZT (and other MPB materials), a monoclinic phase may enhance these pretransitional effects by providing a structural bridge between tetragonal and rhombohedral phases and an extended rotation path. However, the mechanisms of enhancement of the properties via divergence of the transverse permittivity are not qualitatively different, whether monoclinic phase is present or not (Budimir et al. 2006).

As already mentioned, in ceramic samples, presence of a monoclinic phase has additional beneficial effect: it increases the number of available domain state and effectively leads to a better poling of samples. Presently, there is no consensus in the literature on the existence of a monoclinic phase at MPB in PZT. The disagreement could be a mere question of the quality of samples examined by different authors but also a consequence of the scale on which the material is examined. Recent studies by Schonau et al. (2007) suggest that local tetragonal and rhombohedral nanostructure of PZT in MPB region may appear to X-ray diffraction as monoclinic. It is suggested that the high properties in the MPB region are then at least in part related to the high mobility of these nanodomains. The facilitated rotation of polarization (as evidenced by high transverse permittivity) is accomplished through easy reorientation of the nanodomains. Finally, it should be pointed out that the local atomic structure of PZT is investigated extensively and those studies are likely to shed more light on mechanisms of high piezoelectricity at MPB in PZT (Grinberg et al. 2002).

4.3.2 Hard and Soft PZT

PZT ceramics are rarely used as chemically "pure." Commercially available samples are prepared with dopants and additives to adjust properties for specific applications. Many examples of how this is accomplished and what is the role of different additives can be found in classical textbooks, such as (Jaffe et al. 1971). In this section, the properties of so called hard and soft PZT will be briefly reviewed, and some recent insight into mechanisms of softening and hardening will be discussed.

Chemically "pure" PZT ceramics (or better said ceramics that are not intentionally doped) exhibit overall good piezoelectric properties (Jaffe et al. 1954). However, it was discovered early that addition of substitutions that produce Pb vacancies lead to materials with improved resistivity, small ageing, higher piezoelectric coefficients, higher losses, lower mechanical quality factor, low coercive field, and small ageing; these materials are known as "soft" (Jaffe et al. 1971). The result is qualitatively the same whether Pb vacancies are created by introducing donor (higher valence) cations on A or B site of the perovskite cell. Typical cations used are Nb^{+5} on B site or La^{+3} on A site. Ionic substitutions that produce oxygen vacancies lead to materials with a lower permittivity, low dielectric loss, lower resistivity, high mechanical quality factor, and more difficult poling and depoling (Jaffe et al. 1971). They are know as "hard" materials, and typical acceptor dopants are Fe^{+3} on B site

Table 4.1 Typical properties of generic soft and hard PZT ceramics

	d_{33}	k_t	Q_m	ε_{33}	Tan δ
Soft	450	0.47	<100	1,800	0.02
Hard	290	0.47	1,000	1,300	0.005

Besides acceptor and donor dopants other modifications are used in commercial samples to tune the properties

or K^{+1} on A-site. Anionic doping has been tried with F^- (Guiffard et al. 2005). Properties of typical soft and hard PZT materials are shown in Table 4.1.

The mechanisms of hardening are relatively well understood, while most interpretations of softening use "hand waving" arguments (Cross, 1993). It is interesting that for this family of materials that dominates piezoelectric applications for over 50 years, the basic physical processes and even crystal structure are still not well understood. However, as all soft compositions on one and all hard materials on the other hand behave qualitatively alike, regardless of whether, for instance, the valence of the B-site acceptor is +3 or +2, it appears that both hardening and softening are mean effects, not varying much qualitatively within either group. Thus, it may be possible to propose a general picture of hard and soft materials without knowing exact microscopic details of the underlying mechanisms.

The most widely accepted model of hardening is that proposed by Carl and Haerdtl (1978), later developed by Arlt and coworkers (Robels and Arlt, 1993). In this model, the acceptor dopants, say Fe^{+3} on (Ti, Zr) site, creates oxygen vacancies for charge neutrality. For simplicity, we shall consider Ti ions only, and write the resulting defects as Fe'_{Ti} and $V_O^{\bullet\bullet}$. We shall return to the question of charge neutrality later in this section. The resulting electric dipoles are randomly distributed at sintering temperature, but when sample is cooled below Curie temperature (around 300–400 °C for a typical PZT), the dipoles interact with ferroelectric polarization within grains and domains and reorient to align with polarization and minimize their potential energy. This alignment is realized by displacement of oxygen vacancy through continuous network of oxygen octahedra. The rotation can happen spontaneously or by action of electric field.

Reorientation of dipoles by electric field has been well documented by EPR (electron paramagnetic resonance) experiments performed at Sandia National Laboratory (Warren et al. 1996). Spontaneous reorientation of dipoles happens each time the equilibrium is affected (e.g., after cooling sample through the T_C, application of external stress or field) and is usually called aging. Macroscopic evidence of aging is characteristic (pinched) shape of the ferroelectric-polarization-electric field loop, Fig. 4.5. In samples in which dipoles are not aligned with polarization within domains, the loop is unpinched (relaxed). The reason for the pinching is presence of effective restoring force that pushes domain walls back to equilibrium position when electric field is removed. This point was examined in detail by Ren and coworkers using single crystals of Mn-doped $BaTiO_3$ (Zhang and Ren, 2006). Their work is also of importance as it established that the pinching and aging effects are present

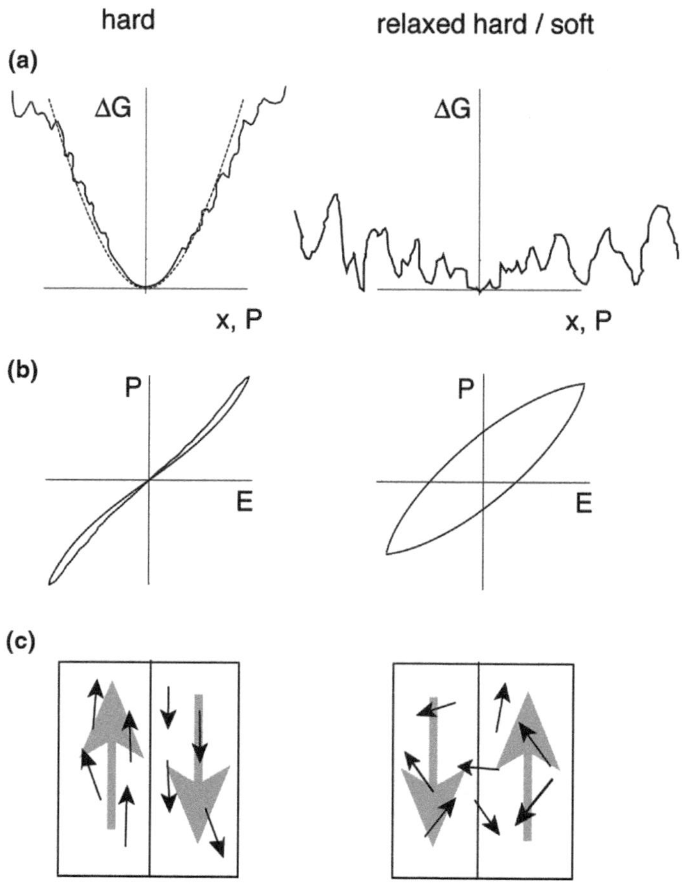

Fig. 4.5 (**a**) Energy potential of a domain wall, (**b**) typical subswitching field hysteresis loops, and (**c**) orientation of defect dipoles within domain for hard (*left*) and soft (*right*) PZT. For details see (Damjanovic 2005)

in single crystals. In Carl and Härdtl work, three mechanisms were suggested for hardening: (a) volume effect discussed earlier, in which dipoles are reorientable and align with polarization within domains; (b) grain boundary effect, where charges related to dopants are concentrated at grain boundaries creating internal fields, which pin polarization; and (c) domain wall effect, in which charges within domain walls inhibit domain wall movement. Most authors (Lambeck and Jonker 1986; Zhang and Ren 2006; Robels and Arlt 1993) but not all (Lupascu et al. 2006) seem to agree that the volume effect is the main mechanism of hardening.

It is interesting examining in some detail ionic species that create acceptor-oxygen vacancy dipole. Examination of Fe-doped $PbTiO_3$ and PZT with EPR (Eichel, 2007) suggests that (1) Fe is trivalent and V_O is doubly ionized; (2) that orientation of dipole is colinear with polarization in $PbTiO_3$ but that, as Zr content increases, the probability of the dipole being formed with oxygen vacancy in the equatorial plane of the oxygen octahedra increases, and finally, (3) that oxygen vacancy is situated in the first sphere of coordination of Fe_{Ti}. These results thus indicate that V_O-Fe_{Ti} dipole is itself charged, requiring additional charges somewhere nearby to maintain the charge neutrality. Considering that the same hardening effect can be obtained in a variety of ferroelectrics ($BaTiO_3$, $KNbO_3$, PZT) with acceptors of different valence state again suggests that these atomistic details are perhaps not important for the hardening process. That is, the hardening effects will occur when there are orientable dipoles in the material, regardless of atomistic details.

Assuming that small amounts of acceptor dopants (typically on the order of 1%) do not appreciably affect ferroelectric intrinsic properties, then the main effect of hardening is through blocking movement of the domain walls. This assumption seems to be experimentally confirmed by Zhang et al. (1983). They showed that PZT with different concentration of dopants all have the same properties at cryogenic temperatures, where domain wall displacement is frozen, while properties differ vastly at room temperature, where domain wall contribution is active.

In soft (donor doped) materials, the defect dipole is created between the donor atom and lead vacancies, e.g. Nb_{Ti}-V_{Pb}. Again, charge compensation is not directly provided so that other charge species or cluster of dipoles may be present to ensure the charge neutrality. Only at very high concentrations of dopant, the charge neutrality is assured by electrons (Jaffe et al. 1971). This is possible because of the high volatility of Pb. In lead-free compositions, donor doping usually leads to strong semiconducting character of the material (Lines and Glass 1979). As opposed to acceptor-oxygen dipole pair, donor-lead vacancies dipoles are less mobile; this could be due to a low mobility of lead vacancies, covalent character of Pb and/or relatively large distances between Pb sites in the perosvkite structure. In any case, no ordering of dipoles occurs in soft materials, which is macroscopically manifested by the absence of hysteresis loop pinching, Fig. 4.5. Special kind of soft ceramics are La-doped PZT compositions ($(Pb,La)(Zr,Ti)O_3$), with interesting piezoelectric and electro-optical properties (Haertling 1994).

Interestingly, undoped PZT exhibit weak hard characteristics, probably due to natural presence of acceptor dopants (e.g., Fe) in oxides used for preparation of ceramics. One can speculate that donor doping, at least at low concentrations, compensates acceptor effects; thus, one may speculate that slightly doped soft ceramics simply exhibit characteristics of what would be seen in pure PZT. It appears reasonable to assume that at high concentrations of donor dopant, the random orientation of dipoles creates conditions for facilitated displacement of domain walls.

Schematically, potential energy of a domain wall in a hard and a soft material may be presented as shown in Fig. 4.5. A domain wall in a hard material moves in a V-potential, originally proposed by Robels and Arlt (1993), while energy of a domain wall in a soft material is best described by a random potential (Damjanovic 1997). This qualitative picture may account for most experimental macroscopic phenomena, including field relaxation of pinched hystereses loops and ageing (evolution from disordered to ordered state in hard ceramics with time). For more details, the reader can consult (Damjanovic 2005).

4.4 Piezoelectricity in Relaxors and Relaxor-Ferroelectric Solid Solutions

The term relaxor-ferroelectrics is used for solid solutions between relaxors (or "relaxor ferroelectrics") such as $Pb(B'_{1/3}B''_{2/3})O_3$ or $Pb(C'_{1/2}C''_{1/2})O_3$ and ferroelectrics such as $PbTiO_3$. The Bs and Cs are metals of appropriate charge and size (Cross 1987). Relaxor themselves are of interest because of their large, but strongly frequency and temperature-dependent dielectric permittivity, Fig. 4.6. For the most part, relaxors exhibit proper ferroelectric characteristics (hysteresis and domains) only at low temperatures. The physical origin of relaxor behavior is still not clear, and its discussion is out of the scope of this chapter.

A strong piezoelectric effect may be induced in nonferroelectric phase of relaxors by external bias field. The electric bias field induces polar character in any centrosymmetric material. The piezoelectric-like effect is phenomenologically obtained simply from the field-biased electrostrictive effect: starting with $x = QP_{\text{ind}}^2$ and $d = \partial x/\partial E$ it follows that, for example, $d_{33} = 2Q_{11}\varepsilon_{33}P_{\text{ind}} = 2Q_{11}\varepsilon_{33}^2 E_{\text{dc}}$. Huge, electric field tunable piezoelectric response can thus be obtained in relaxor materials (Fig. 4.6). Exploitation of the induced piezoelectric effect has been reported for ceramic-polymer composite array transducers with field tunable properties (Takeuchi et al. 1990). Besides tunability, an advantage of the field induced piezoelectric effect in relaxors is absence of the charge-strain hysteresis and creep.

The origin of the large induced piezoelectric effect has been interpreted in terms of the large permittivity in relaxor materials (Takeuchi et al. 1990). The maximum in $d(E)$ is then explained by the competing effects of the field dependence of polarization (which increases with increasing field) and permittivity (which first increases and then decreases with increasing bias field). Recent studies, however, suggest that the peak in the piezoelectric properties may be related to a field-induced critical point (Kutnjak et al. 2007).

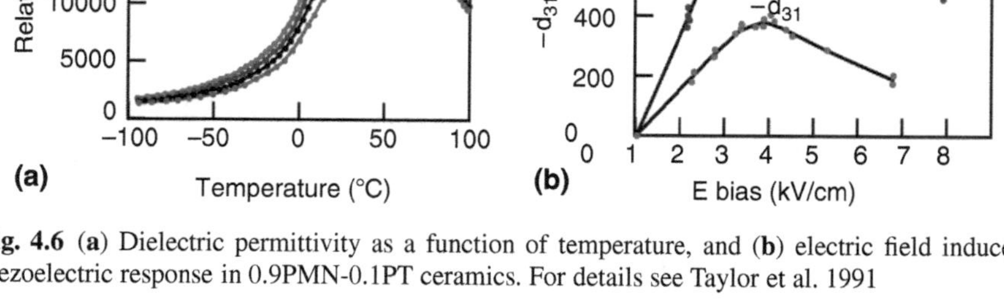

Fig. 4.6 (**a**) Dielectric permittivity as a function of temperature, and (**b**) electric field induced piezoelectric response in 0.9PMN-0.1PT ceramics. For details see Taylor et al. 1991

4.4.1 Relaxor-Ferroelectric Ceramics and Single Crystals

Properties of relaxor-ferroelectric solid solutions are of a more interest than those of relaxor end-member alone. Among many compositions that were investigated, including complex solid solutions of several relaxors and ferroelectrics (Yamashita and Hosono 2005), the most widely studied ones are $Pb(Zn_{1/3}Nb_{2/3})O_3$–$PbTiO_3$ (PZN–PT) (Kuwata et al. 1981), $Pb(Mg_{1/3},Nb_{1/3})O_3$–$PbTiO_3$ (PMN–PT) (Choi et al. 1989), $Pb(Ni_{1/3}Nb_{1/3})O_3$–$Pb(Zr,Ti)O_3$ (PNN–PZT) (Chung et al. 1989). Note that in general relaxors of type $Pb(B'_{1/2}B''_{1/2})O_3$-$PbTiO_3$ (e.g., $Pb(Sc_{1/2}Nb_{1/2})O_3$-$PbTiO_3$) were not found experimentally to exhibit exceptional piezoelectric properties. In PZN-PT, PMN-PT, and PNN-PZT ceramics, the permittivity, electromechanical coupling, and piezoelectric coefficients are exceptionally high (d_{33} up to 800 pC N^{-1}) in compositions close to the MPB. Like in PZT, the MPB region is situated between rhombohedral and tetragonal phases, and depending on crystal state, this region can exhibit a monoclinic or an orthorhombic symmetry (Lu et al. 2001). MPB in PZN-PT is situated at approximately 9–10% PT and at about 35%PT in PMN-PT.

The MPB in these materials is curved near Curie temperature and phase transition temperature from the low (e.g. rhombohedral) to the high temperature ferroelectric phase (usually tetragonal) is below 400 K (Noheda 2002). The Curie temperature in these materials is low (below 400–450 K), and much of the enhancement of the piezoelectric and dielectric response in ceramics is due to the low T_C. It has been a challenge to produce ceramics with a high piezoelectric response, a high permittivity and $T_C > 200°C$. Such properties would be especially important for medical transducers, which require sterilization at about 140 °C and can lead to partial depoling if T_C is too low. The new trend to use disposable medical transducers may relax requirement for high T_C because sterilization is no longer necessary and materials with a lower T_C and thus better performance may be used.

In contrast to PZT where only ceramic samples are available, both PZN-PT and PMN-PT can be grown by variety of methods as large single crystals (Trolier-McKinstry et al. 2004). Discovery of the large piezoelectric response in PZN-PT single crystals (Kuwata et al. 1982) was not exploited until 1997 when unexpectedly large piezoelectric properties of single crystals were "rediscovered" (Park and Shrout, 1997) causing a small revolution in the field of ferroelectrics and unprecedented activities in ab-initio calculations, crystal growth, and structural studies of ferroelectrics.

Clearly the properties of PMN-PT and PZN-PT ceramics are a combination of the averaged intrinsic properties of single crystal grains and domain structure, domain walls displacement and degree of poling of ceramics with compositions that are nearly always close to the MPB. Thus availability of single crystals in these complex systems gave an opportunity, which did not exist earlier with PZT, to study in details intrinsic properties of MPB systems.

The most extraordinary property of these complex crystals is the large shear piezoelectric coefficient ($d_{15} > 4{,}000$ pC N^{-1}) and large permittivity perpendicular to polarization ($\varepsilon_{11} > 3{,}000$ at constant stress) (Zhang et al. 2003b) in samples

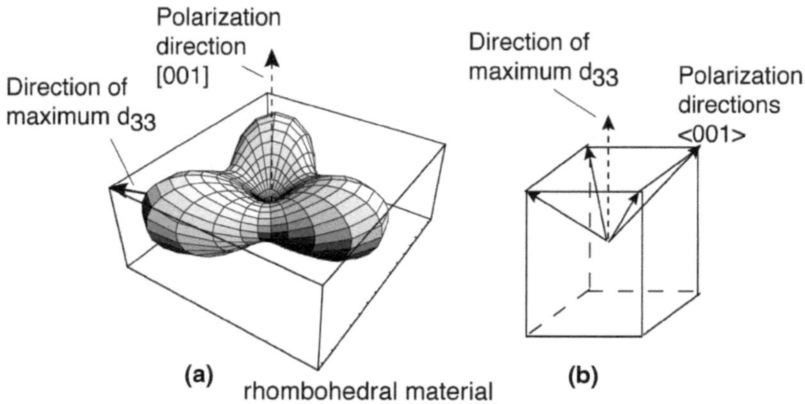

$$d=\begin{bmatrix} 0 & 0 & 0 & 0 & 4100 & -2680 \\ -1340 & 1340 & 0 & 4100 & 0 & 0 \\ -90 & -90 & 190 & 0 & 0 & 0 \end{bmatrix} \quad \text{single domain state rhombohedral symmetry}$$

$$d=\begin{bmatrix} 0 & 0 & 0 & 0 & 146 & 0 \\ 0 & 0 & 0 & 146 & 0 & 0 \\ -1338 & -1338 & 2820 & 0 & 0 & 0 \end{bmatrix} \quad \text{engineered domain state tetragonal symmetry}$$

Fig. 4.7 (**a**) Crystal orientation with maximal piezoelectric effect in rhombohedral crystals, and (**b**) matrices of piezoelectric coefficients for PMN-33PT crystals in single domain and domain engineered state. For details see (Zhang et al. 2003b, Zhang et al. 2001, Zhang et al. 2003a)

with compositions on the rhombohedral (or relaxor) side of the MPB. Both large ε_{11} and d_{15} indicate large propensity of these compositions for polarization rotation (see section on PZT). This leads to a large d_{33} ($>2{,}000\,\mathrm{pC\,N^{-1}}$) (Zhang et al. 2001) along an off-polar directions, which lies approximately but not exactly along the <001> pseudocubic direction of the rhombohedral cell (Damjanovic et al. 2003). Crystals are usually poled along <001> pseudocubic direction, which in rhombohedral crystals ideally produces domain configuration with four equivalent domain states with polarization oriented along <111> pseudocubic axes (Fig. 4.7) and average macroscopic tetragonal symmetry. Additional benefit of this domain structure (so called, engineered domain state) is that domain walls do not contribute to the piezoelectric effect (Park and Shrout 1997). This is of a significant interest for actuator applications: this single crystal not only give a large strain (few times higher than in PZT ceramics or ceramics of the same composition) but the strain is nearly anhysteretic.

The application of large fields in off-polar directions leads to polarization rotation within monoclinic mirror plane and to switching of the crystal into monoclinic/orthorhombic and eventually tetragonal phase (Davis et al. 2006). The large strains developed during phase switching are hysteretic and not usable for a large spectrum of devices. Another interesting property of relaxor-ferroelectric single crystals is large coupling coefficients along nonpolar axis. Giant coupling coefficients ($k > 90\%$) have been reported for transverse (k_{31}) (Ogawa et al. 2002) and longitudinal mode (k_{33}) (Park and Shrout 1997), depending on crystal orientation and poling direction.

The mechanisms of the large piezoelectric response in relaxor-ferroelectric single crystals and ceramics have been subject of many studies in the last ten years. The proposed origins include polarization rotation (Fu and Cohen 2000), adaptive domain states (Jin et al. 2003), and critical phenomena (Kutnjak et al. 2007). In the latter case, the proposed model explains enhancement of the response induced by the electric field and not the large response itself, which is essentially a weak (zero) field property. The same can be said about polarization rotation mechanism: the polarization need not be rotated by a large angle for a material to have a large d_{33} along nonpolar directions; what is essential is that the material has propensity for a large polarization rotation, which is measured by a large d_{15} and large ε_{11} (expressed in crystallographic coordinate system). These issues are discussed in detail in (Davis et al. 2007) and the section on PZT of this chapter.

Evidence of relaxor like behavior often reported above T_C in rhombohedral compositions is sometimes linked to the high properties. However, this relaxor character appears to be related to crystals quality and some authors report very high properties in crystals without any "background" relaxor character (Sehirlioglu et al. 2006).

4.5 Other Lead-Based Materials

Other lead-based materials that have some practical relevance or exhibit promising properties include lead metaniobate, $Pb_2Nb_2O_6$, in the first and $BiMeO_3$-$PbTiO_3$ solid solutions in the second category.

Lead metaniobate, $Pb_2Nb_2O_6$, has potassium tungsten bronze structure (Jaffe et al. 1971), and is of interest because of its high T_C, low mechanical quality factor Q_m, and low transverse coupling factor k_p. Ba-modified material exhibits an MPB between an orthorhombic and tetragonal phase at about 30% Ba, but the T_C is much lower for these compositions and advantages of low transverse coupling and low Q_m are lost. However, compositions on the tetragonal side (e.g., with 70% Ba) exhibit only 180° domain walls and anhysteretic piezoelectric response.

$BiMeO_3$-$PbTiO_3$ solid solution has properties comparable to that of PZT, and a higher T_C and is of interest as high performance high temperature material. The most studied composition is the one where Me = Sc (Eitel et al. 2002). The additional advantage of this family is a low Pb content.

4.6 Environmental Issues

Lead is a toxic element and recently there was much concern about its use in piezoelectric materials. The chapter of this book by M. Kosec discusses lead free piezoelectric materials and more discussion can be found elsewhere (Zhang and

Shrout 2007). Although toxicity of metallic lead and lead oxide powder is notorious, no comprehensive studies are presently available on toxicity of lead-based piezoelectric ceramics. One study indicates relatively easy transfer of lead from modified lead titanate ceramics into water (Kosec et al. 1998).

References

Bell AJ (2006) Factors influencing the piezoelectric behaviour of PZT and other "morphotropic phase boundary" ferroelectrics. J Mat Sci 41: 13–25.

Bellaiche L et al. (2000) Finite-temperature properties of $Pb(Zr_{1-x}Ti_x)O_3$ alloys from first principles. Phys Rev Lett 84: 5427–30.

Berlincourt D et al. (1964) Stability of phases in modified lead zirconate with variation in pressure, electric field, temperature and composition. J Phys Chem Solids 25: 659–74.

Bondarenko EI et al. (1991) The role of 90° domain wall displacements in forming physical properties of perovskite ferroelectric ceramics. Ferroelectr Lett 13: 13.

Budimir M et al. (2003) Piezoelectric anisotropy-phase transition relations in perovskite single crystals. J Appl Phys 94: 6753–61.

Budimir M et al. (2006) Piezoelectric response and free-energy instability in the perovskite crystals BaTiO3, PbTiO3, and $Pb(Zr,Ti)O_3$. Phys Rev B 73: 174106

Carl K, Haerdtl KH (1978) Electrical after-effects in $Pb(Ti,Zr)O_3$ ceramics. Ferroelectrics 17: 473–86.

Carl K, Härdtl KH (1971) On the origin of the maximum in the electromechanical activity in $Pb(Zr_xTi_{1-x})O_3$ ceramics near the morphotropic phase boundary. Phys Stat Sol (a) 8: 87.

Choi SW et al. (1989) Morphotropic phase boundary in $Pb(Mg_{1/3}Nb_{2/3})O_3$-$PbTiO_3$ system. Mat Lett 8: 253–55.

Chung ST et al. (1989) Piezoelectric and dielectric properties of $Pb(Ni,Nb)O_3$–$Pb(Zn,Nb)O_3$–$PbZrO_3$–$PbTiO_3$ system ceramics. Ferroelectrics 94: 243–7.

Cohen RE (1992) Origin of ferroelectricity in perovskite oxides. Nature 358: 136–8.

Cross LE (1987) Relaxor ferroelectrics. Ferroelectrics 76: 241–67.

Cross LE (1993) Ferroelectric ceramics: Tailoring properties for specific applications. In Setter N, Colla EL (Eds.) Ferroelectric Ceramics. Basel, Birkhäuser p. 1.

Damjanovic D (1997) Stress and frequency dependence of the direct piezoelectric effect in ferroelectric ceramics. J Appl Phys 82: 1788–97.

Damjanovic D (2005) Hysteresis in piezoelectric and ferroelectric materials. In Bertotti G, Mayergoyz I (Eds.) Science of Hysteresis. Amsterdam, Elsevier p. 337.

Damjanovic D, Demartin M (1997) Contribution of the irreversible displacement of domain walls to the piezoelectric effect in barium titanate and lead zirconate titanate ceramics. J Phys: Condens Matter 9: 4943–53.

Damjanovic D et al. (1987) Anisotropy in piezoelectric properties of modified lead titanate ceramics. Am Ceram Soc Bull 66: 699–703.

Damjanovic D et al. (2003) Monodomain versus polydomain piezoelectric response of 0.67 $Pb(Mg_{1/3}Nb_{2/3})O_3$-0.33$PbTiO_3$ single crystals along nonpolar directions. Appl Phys Lett 83: 527–9.

Davis M et al. (2005) Domain engineering of the transverse piezoelectric coefficient in perovskite ferroelectrics. J Appl Phys 98: 014102.

Davis M et al. (2006) Electric field-, temperature-, and stress-induced phase transitions in relaxor ferroelectric single crystals. Phys Rev B 73: 014115.

Davis M et al. (2007) Rotator and extender ferroelectrics: Importance of the shear coefficient to the piezoelectric properties of domain-engineered crystals and ceramics. J Appl Phys 101: 054112.

Eichel RA (2007) Defect structure of oxide ferroelectrics – valence state, site of incorporation, mechanisms of charge compensation and internal bias fields. J Electroceramics 19: 9–21.
Eitel R, Randall CA (2007) Octahedral tilt-suppression of ferroelectric domain wall dynamics and the associated piezoelectric activity in $Pb(Zr,Ti)O_3$. Phys Rev B 75: 094106.
Eitel RE et al. (2002) Preparation and characterization of high temperature perovskite ferroelectrics in the solid-solution $(1-x)BiScO_3$-$xPbTiO_3$. Jpn J Appl Phys Part 1 41: 2099–104.
Fesenko EG et al. (1986) Phase (x,T) diagram of $PbZr_{1-x}Ti_xO_3$ crystals. Sov Phys Solid State 28: 181.
Fu H, Cohen RE (2000) Polarization rotation mechanism for ultrahigh electromechanical response in single-crystal piezoelectrics. Nature 403: 281–3.
Gavrilyachenko VG, Fesenko EG (1971) Piezoelectric effect in lead titanate single crystals. Sov Phys Crystallogr 16 p. 549.
Glazer AM, Mabud SA (1978) Powder profile refinement of lead zirconate titanate at several temperatures, Part II: Pure $PbTiO_3$. Acta Cryst B 34: 1065.
Goldschmidt VM et al. (1926) Geochemische Verteilunggestze der Elemente VII Die Gesetze der Krystallochemie. Srkrifter Utgitt av der Norske Videnskaps-Akademi i Oslo, I Matem-Naturvid Klasse p. 2.
Grinberg I et al. (2002) Relationship between local structure and phase transitions of a disordered solid solution. Nature 419: 909–11.
Guiffard B et al. (2005) Influence of donor co-doping by niobium or fluorine on the conductivity of Mn doped and Mg doped PZT ceramics. J Eur Ceram Soc 25: 2487–90.
Haertling GH (1994) Chemically reduced PLZT ceramics for ultra-high displacement actuators. Ferroelectrics 154: 101–6.
Hall DA et al. (2005) Micromechanics of residual stress and texture development due to poling in polycrystalline ferroelectric ceramics. J Mech Phys Solids 53: 249.
Harada J et al. (1970) X-ray and neutron diffraction study of tetragonal barium titanate. Acta Cryst A 26: 336.
Haun MJ et al. (1987) Thermodynamic theory of $PbTiO_3$. J. Appl. Phys. 62: 3331–3338.
Haun MJ et al. (1989) Thermodynamic theory of the lead zirconate-titanate solid solution system, Part V: Theoretical calculations. Ferroelectrics 99: 63–86.
Ishibashi Y, Iwata M (1998) Morphotropic phase boundary in solid solution systems of perovskite-type oxide ferroelectrics. Jpn J Appl Phys 37: L985–L987.
Isupov VA (2001) Phase coexistence in lead zirconate titanate solid solutions. Phys Solid State 43: 2262–6.
Isupov VA (2002) Phases in the PZT ceramics. Ferroelectrics 266: 91–102.
Iwata M, Ishibashi Y (2005) Phenomenological theory of morphotropic phase boundary with monoclinic phase in solid-solution systems of perovskite-type oxide ferroelectrics. Jpn J Appl Phys 44: 3095–8.
Jaffe B et al. (1954) Piezoelectric properties of lead zirconate-lead titanate solid-solution ceramics. J Appl Phys 25: 809–10.
Jaffe B et al. (1971) Piezoelectric Ceramics. New York, Academic.
Jin YM et al. (2003) Conformal miniaturization of domains with low domain-wall energy: Monoclinic ferroelectric states near the morphotropic phase boundaries. Physl Rev Lett 91: 197601.
Jones JL et al. (2007) Time-resolved and orientation-dependent electric-field-induced strains in lead zirconate titanate ceramics. Appl Phys Lett 90:172909.
Kighelman Z et al. (2002) Properties of ferroelectric $PbTiO_3$ thin films. J Appl Phys 91: 1495–501.
Kosec M et al. (1998) Effect of a chemically aggressive environment on the electromechanical behaviour of modified lead titanate ceramics. J Kor Phys Soc 32: S1163–S1166.
Kushida K, Takeuchi H (1987) Piezoelectricity in c-axis oriented $PbTiO_3$ thin films. Appl Phys Lett 50: 1800–1.
Kutnjak Z et al. (2007) Electric field induced critical points and polarization rotations in relaxor ferroelectrics. Phys Rev B 76: 104102.
Kuwata J et al. (1981) Phase transitions in the $Pb(Zn_{1/3}Nb_{2/3})O_3$–$PbTiO_3$ system. Ferroelectrics 37: 579–82.

Kuwata J et al. (1982) Dielectric and piezoelectric properties of $0.91Pb(Zn_{1/3}Nb_{2/3})\ O_3$–0.09 $PbTiO_3$ single crystals. Jpn J Appl Phys 21: 1298–302.

Lambeck PV, Jonker GH (1986) The nature of domain stabilization in ferroelectric perovskites. J Phys Chem Solid 47: 453–61.

Li Z et al. (1993) The elastic, piezoelectric and dielectric constants of tetragonal $PbTiO_3$ single crystals. Ferroelectrics 141: 313–25.

Lines ME, Glass AM (1979) Principles and Applications of Ferroelectrics and Related Materials. Oxford, Clarendon.

Lu Y et al. (2001) Phase transitional behavior and piezoelectric properties of the orthorhombic phase of $Pb(Mg_{1/3}Nb_{2/3})O_3$–$PbTiO_3$ single crystals. Appl Phys Lett 78: 3109–11.

Lupascu DC et al. (2006) Aging in ferroelecrtrics. J Am Ceram Soc 89: 224–9.

Megaw HD (1957) Ferroelectricity in Crystals. London, Methuen.

Mestric H et al. (2005) Iron-oxygen vacancy defect centers in $PbTiO_3$: Newman superposition model analysis and density functional calculations. Phys Rev B 71: 134109.

Michel C et al. (1969) Atomic structures of two rhombohedral ferroelectric phases in the $Pb(Zr, Ti)O_3$ solid solution series. Solid State Comm 7: 865–8.

Nelmes RJ, Kuhs WF (1985) The crystal structure of tetragonal $PbTiO_3$ at room temperature and at 700 K. Solid State Comm 54: 721.

Nelmes RJ et al. (1990) Order-disorder behaviour in the transition of $PbTiO_3$. Ferroelectrics 108: 165–70.

Noheda B (2002) Structure and high-piezoelectricity in lead oxide solid solutions. Curr Opin Solid State Mat Sci 6: 27–34.

Noheda B et al. (1999) A monoclinic ferroelectric phase in the $Pb(Zr_{1-x}Ti_x)O_3$ solid solution. Appl Phys Lett 74: 2059–61.

Ogawa T et al. (2002) Giant electromechanical coupling factor of k_{31} mode and piezoelectric d_{31} constant in $Pb[(Zn_{1/3}Nb_{2/3})(0.91)Ti_{0.09}]O_3$ piezoelectric single crystal. Jpn J Appl Phys Part 2: Letters 41: L55–L57.

Park SE, Shrout TR (1997) Ultrahigh strain and piezoelectric behavior in relaxor based ferroelectric single crystals. J Appl Phys 82: 1804–11.

Robels U, Arlt G (1993) Domain wall clamping in ferroelectrics by orientation of defects. J Appl Phys 73: 3454–60.

SaghiSzabo G, Cohen RE (1997) Long-range order effects in $Pb(Zr_{1/2}Ti_{1/2})O_3$. Ferroelectrics 194: 287–98.

Sawaguchi E (1953) Ferroelectricity versus antiferroelectricity in the solid solutions of $PbZrO_3$ and $PbTiO_3$. J Phys Soc Japan 8: 615–29.

Schonau KA et al. (2007) Nanodomain structure of $Pb(Zr_{1-x}Ti_x)O_3$ at its morphotropic phase boundary: Investigations from local to average structure. Phys Rev B 75: S184117–S184200.

Sehirlioglu A et al. (2006) Effect of poling on dielectric anomalies at phase transitions for lead magnesium niobate-lead titanate crystals in the morphotropic phase boundary region. J Appl Phys 99: 064101.

Shirane G et al. (1952) Phase transitions in solid solutions of $PbZrO_3$ and $PbTiO_3$, Part II: X-ray study. J Phys Soc Jpn 7: 12.

Takeuchi H et al. (1985) Highly anisotropic piezoelectric ceramics and their application in ultrasonic probes. IEEE Ultrasonics Symposium. San Francisco.

Takeuchi H et al. (1990) Relaxor ferroelectric transducers. IEEE Ultrasonics Symposium. Honolulu.

Tanaka H et al. (2006) Electrostatic potential of ferroelectric $PbTiO_3$: Visualized electron polarization of Pb ion. Physical Review B 74: 172105.

Taylor DJ et al. (1991) Large hydrostatic piezoelectric coefficient in lead magnesium niobate: Lead titanate ceramics. J Mater Sci Lett 10: 668.

Troilo LM et al. (1994) Modified lead titanate ceramics with relatively large dielectric constant for hydrophone applications. J Am Ceram Soc 77: 857.

Trolier-McKinstry S et al. (Eds.) (2004) Ferroelectric single crystals and their application.

Turik AV et al. (1975) Anisotropy of the dielectric and piezoelectric properties of lead titanate. Sov Phys Crystallogr 19: 677–78.

Turik AV, Topolov VY (1997) Ferroelectric ceramics with a large piezoelectric anisotropy. J Phys D: Appl Phys 30: 1541–9.

Warren WL et al. (1996) Oxygen vacancy motion in perovskite oxides. J Am Ceram Soc 79: 536–8.

Woodward DI et al. (2005) Review of crystal and domain structures in the $PbZr_xTi_{1-x}O_3$ solid solution. Phys Rev B 72: 104110.

Wu ZG, Cohen RE (2005) Pressure-induced anomalous phase transitions and colossal enhancement of piezoelectricity in $PbTiO_3$. Phys Rev Lett 95: 037601.

Yamashita Y, Hosono Y (2005) Material design of high-dielectric-constant and large-electromechanical-coupling-factor relaxor-based piezoelectric ceramics. Jpn J Appl Phys Part 1: Regular Pap Brief CommRevPap 44: 7046–9.

Yamashita Y et al. (1981) $(Pb,Ca)((Co_{1/2}W_{1/2})Ti)O_3$ piezoelectric ceramics and their applications. Jpn J Appl Phys 20: 183.

Zhang LX, Ren X (2006) Electro-shape-memory effect in Mn-doped $BaTiO_3$ single crystals and in situ observation of the reversible domain switching. Mater Sci Eng A: Struct Mater 438: 354–9.

Zhang R et al. (2001) Elastic, piezoelectric, and dielectric properties of multidomain $0.67Pb(Mg_{1/3}Nb_{2/3})O_3$–$PbTiO_3$ single crystals. J Appl Phys 90: 3471–75.

Zhang R et al. (2003a) Orientation dependence of piezoelectric properties of single domain $0.67(Mg_{1/3}Nb_{2/3})O_3$–$0.33PbTiO_3$ crystals. Appl Phys Lett 82: 3737–9.

Zhang R et al. (2003b) Single-domain properties of $0.67Pb(Mg_{1/3}Nb_{2/3})O_3$–$0.33PbTiO_3$ single crystals under electric field bias. Appl Phys Lett 82: 787–9.

Zhang S, Shrout TR (2007) Lead-free piezoelectric ceramics: Alternatives for PZT? J Electroceramics 19: 111–24.

Zhang XL et al. (1983) Dielectric and piezoelectric properties of modified lead titanate zirconate ceramics from 4.2 to 300 K. J Mater Sci 18: 968–72.

Chapter 5
KNN-Based Piezoelectric Ceramics

Marija Kosec,* Barbara Malič, Andreja Benčan, and Tadej Rojac

5.1 The Crystal Structure of $K_{0.5}Na_{0.5}NbO_3$ and Phase Relations in the $KNbO_3$–$NaNbO_3$ System

Alkaline niobates and, more particularly, the sodium potassium niobate solid solution became the topic of much research at the end of the 1990s, because of increased environmental awareness. Prior to this, a lot of work on these materials was carried out in the 1950s and 1960s. The compositions with the highest electromechanical coupling coefficients are those close to the morphotropic phase boundary (MPB) at 52.5% Na, and the most studied composition has been $K_{0.5}Na_{0.5}NbO_3$, subsequently referred to as KNN (Shirane et al. 1954; Egerton and Dillon 1959; Jaeger and Egerton 1962; Jaffe et al. 1971).

$KNbO_3$ and $NaNbO_3$ form a solid solution across the whole compositional range. The pseudobinary phase diagram, which is a compilation of early work on this system, published in Jaffe's Piezoelectric ceramics, is shown in Fig. 5.1. $KNbO_3$ is cubic above 435 °C, tetragonal between 435 and 225 °C, orthorhombic at room temperature and rhombohedral below −12 °C. The sequence of phase transitions upon cooling from the Curie point is similar to that observed in $BaTiO_3$. $NaNbO_3$ is cubic above 630 °C; however, upon cooling it undergoes a sequence of phase transitions and is antiferroelectric at room temperature (Jaffe et al. 1971).

All the compositions within the $KNbO_3$–$NaNbO_3$ system have an orthorhombic perovskite structure at room temperature (Shirane et al. 1954; Jaffe et al. 1971; Tennery and Hang 1968). Thermal and XRD studies revealed boundaries between the different orthorhombic phases at 52.5%, 67.5%, 82.5% and 98% $NaNbO_3$ (Tennery and Hang 1968; Megaw and Wells 1958; Ahtee and Glazer 1974; Ahtee and Hewat 1978). It should be noted that although nominally orthorhombic, the solid solutions exhibit a monoclinic distortion of a few tenths of a degree (Tennery and Hang 1968).

M. Kosec
Jožef Stefan Institute, Jamova 39, 1000 Ljubljana, Slovenia

A. Safari, E.K. Akdoğan (eds.) *Piezoelectric and Acoustic Materials for Transducer Applications.*

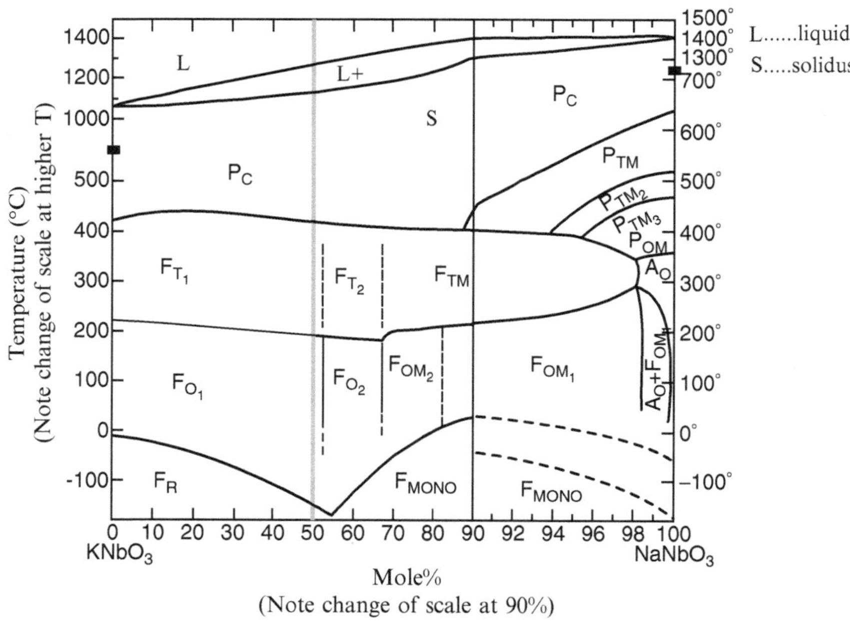

Fig. 5.1 Binary phase diagram $KNbO_3$–$NaNbO_3$ (F_R–ferroelectric rhombohedral with high potassium, F_{MONO}–ferroelectric monoclinic, F_{O1}–ferroelectric orthorhombic, F_{O2}–ferroelectric orthorhombic, F_{OM1}–ferroelectric orthorhombic field enforced in pure $NaNbO_3$, F_{OM2}–ferroelectric orthorhombic, F_{T1}–ferroelectric tetragonal with high potassium, F_{T2}–ferroelectric tetragonal, F_{TM}–ferroelectric tetragonal obtained with additives, A_O–antiferroelectric orthorhombic, P_C–paraelectric cubic, P_{OM}–paraelectric orthorhombic, P_{TM}–paraelectric tetragonal, P_{TM2}–paraelectric pseudotetragonal, P_{TM3}–paraelectric pseudotetragonal) (Jaffe et al. 1971)

The $K_{0.5}Na_{0.5}NbO_3$ composition is, according to the phase diagram, isostructural with $KNbO_3$ (Bmm2, $a = 0.5697$ nm, $b = 0.3971$ nm, $c = 0.5721$ nm), (JCPDS; Katz and Megaw 1967). There is no JCPDS data for the specific composition $K_{0.5}Na_{0.5}NbO_3$.

The Curie point of $K_{0.5}Na_{0.5}NbO_3$ is at about 400 °C, upon cooling the tetragonal phase transforms at about 200 °C to orthorhombic and at about −160 °C to a rhombohedral phase (Fig. 5.1). The melting points of $KNbO_3$ and $NaNbO_3$ are at 1,040°C and 1,420°C, respectively. For $K_{0.5}Na_{0.5}NbO_3$, the solidus and liquidus lines are at about 1,140°C and 1,280°C. To the best of the authors' knowledge the phase relations in the ternary diagram Na_2O–K_2O–Nb_2O_5 have not been reported.

The binary phase diagram $K_2O(K_2CO_3)$–Nb_2O_5 contains, in addition to the perovskite $KNbO_3$ phase with an incongruent melting point at 1,039°C, the phases K_3NbO_7, $K_4Nb_6O_{17}$, KNb_3O_8, and $K_6Nb_{44}O_{113}$ (Reisman and Holtzberg 1955). Studies of the niobium-rich region ($KNbO_3$–Nb_2O_5) also revealed $K_{5.75}Nb_{10.85}O_{30}$, $K_3Nb_7O_{19}$, $K_2Nb_8O_{21}$, KNb_7O_{18}, and $KNb_{13}O_{33}$ (Roth 1980; Irle et al. 1991). The stoichiometric $KNbO_3$ is stable; however, even a slight substoichiometry in potassium will cause separation of the phases, some of them strongly hygroscopic (Reisman et al. 1959). $K_4Nb_6O_{17}$ easily forms stable hydrates, with water molecules intercalated in its layered structure (Bizeto and Constantino 2004). The binary phase diagram $Na_2O(Na_2CO_3)$–Nb_2O_5 from 1959 covers the complete compositional range and contains five sodium niobate phases: Na_3NbO_4, $NaNbO_3$, $Na_2Nb_8O_{21}$, $NaNb_7O_{18}$, and $NaNb_{10}O_{25}$ (Shafer and Roy 1959). The phase diagrams of the Nb_2O_5-rich part of this system, i.e., $NaNbO_3$–Nb_2O_5, contain the following sodium niobate phases: $Na_2Nb_4O_{11}$, $Na_{1\pm x}Nb_{3\pm x}O_{8\pm 2x}$, $Na_2Nb_{12}O_{31}$, $NaNb_7O_{18}$, and $NaNb_{13}O_{33}$ as well as $NaNbO_3$ (Irle et al. 1991; Appendino 1973; Roth et al. 1974).

There are some discrepancies in the compositions, the solid solubility regions and/or the temperature-stability ranges of the niobium-rich phases present in the individual K_2O–Nb_2O_5 and Na_2O–Nb_2O_5 phase diagrams, which may stem from the inexact stoichiometry of the starting mixtures due to the high hygroscopicity of the alkaline carbonates, especially potassium carbonate, and also from the volatility of both carbonates at the typical temperatures of the solid-state reactions, between 700 and 950 °C. It is reasonable to expect that polyniobate phases also form in the ternary Na_2O–K_2O–Nb_2O_5 system.

5.2 Powder Synthesis

5.2.1 KNbO₃

From the first studies on phase equilibria in the K_2CO_3–Nb_2O_5 system, it was realized that the formation of potassium niobate phases of other than K/Nb 1:1 ratio, especially $2K_2O \cdot 3Nb_2O_5$, coupled with the significant volatility of K_2O above 800 °C, would represent one of the greatest problems in the synthesis of pure $KNbO_3$ (Jaffe et al. 1971; Reisman and Holtzberg 1955; Reisman et al. 1959; Reisman et al. 1956; Fluckiger and Arend 1977). To overcome this, a leaching procedure was used (Reisman et al. 1956). The method consists of selectively dissolving the nonperovskite phases in a 2% K_2CO_3 solution after calcination. Alternatively, a stoichiometric composition can be hot-pressed (Haertling 1967).

Kodaira et al. prepared $KNbO_3$ by solid-state synthesis using $KHCO_3$ and Nb_2O_5 as the reagents (Kodaira et al. 1982). The synthesis steps, i.e., from mixing the starting powders to sintering the pellets, were conducted under flowing air dried by P_2O_5. The relative densities obtained ranged between 90 and 95%, depending on the sintering temperature (740–990 °C). During sintering at 990 °C, however, a weight loss was observed, which was attributed to the volatilization of K_2O.

Since the K_2CO_3–Nb_2O_5 system can easily form undesired phases if the stoichiometric ratio K/Nb 1:1 is not satisfied, Birol et al. stressed the importance of controlling the humidity conditions, especially before the calcination, due to the hygroscopic nature of K_2CO_3 (Birol et al. 2005). Moreover, to obtain dense and stable samples, sintering in a dry oxygen atmosphere was proposed to avoid the K_2O

volatilization above 800 °C, which is enhanced in the presence of humidity. Because of the great difficulty of obtaining single-phase $KNbO_3$, alternative methods were explored. To achieve a high degree of homogeneity, Nazeri-Eshghi et al. used the sol–gel processing of a solution of niobium and potassium ethoxides in methanol to prepare polycrystalline $KNbO_3$ (Nazeri-Eshghi et al. 1990). The gel powders completely crystallized after 2 h at 600 °C. Densities of up to 99% were achieved by hot-pressing at 950–1,000°C and pressures less than half those required in conventional methods (2,500–3,000 p.s.i.). Similarly, Amini and Sacks studied the preparation of $KNbO_3$ from metal alkoxides (Amini and Sacks 1991). A bimetallic ethoxide was first prepared by reacting potassium and niobium ethoxides in ethanol and subsequently hydrolyzed using several water concentrations (0.75–6.0 mol of water/(mol of ethoxide)). A similar procedure was adopted for bimetallic propoxide using 1-propanol. Rapid precipitation of potassium-deficient particles occurred when higher concentrations of water were used. However, single-phase $KNbO_3$ was obtained by reducing the water concentration to 1 mol/(mol of alkoxide).

As alternatives to the classical solid-state reaction route, hydrothermal and glycothermal processes were found to be successful for the synthesis of pure $KNbO_3$ at low temperatures (Lu et al. 1998, 2002). The hydrothermal synthesis was carried out by dispersing different amounts of Nb_2O_5 in an aqueous solution of KOH (Lu et al. 1998). The mixed slurries were heated in an autoclave at temperatures of 200–250 °C for 2–11 h. The authors found that the concentration of the KOH solution is the critical factor controlling the synthesis of $KNbO_3$. A complete crystallization of $KNbO_3$ was obtained after 2 h at 200 °C using an 8 M KOH solution. Increasing the concentration of KOH tends to increase the solubility of Nb_2O_5 in water under supercritical conditions and thereby increases the formation of $KNbO_3$. The glycothermal process, using isopropanol as a reaction medium, allows a further reduction of the concentration of KOH (Lu et al., 2002). Using a 0.5 M KOH solution and a molar ratio of K/Nb greater than 2:1, well crystallized single-phase $KNbO_3$ was obtained at 250 °C.

5.2.2 $NaNbO_3$

Dating back to the work of Dungan et al. (Dungan and Golding 1964) in 1964, when the first studies on $NaNbO_3$ took place, the perovskite phase was prepared by a conventional solid-state reaction route. The powder mixture of Na_2CO_3 and Nb_2O_5 was milled in methanol, calcined at 950 °C for 16 h, and then sintered at up to 1,400°C for 1 h. Using this classical method, however, it is rather difficult to achieve a homogeneous mixture of the components. Moreover, high sintering temperatures, which are usually required because of the low surface area of the raw powders, can also enhance the volatilization of the alkali species, leading to stoichiometry variations in the sintered material. Consequently, a lot of efforts have been made in the time since to prepare fine and single-phase $NaNbO_3$ powder with high sinterability. This includes solution synthesis (Lanfredi et al. 2000; Camargo et al.

2002; Calderon-Moreno and Camargo 2003; Nobre et al. 1996), microemulsion-mediated synthesis (Shiratori et al. 2005a), and mechanochemical synthesis (Castro et al. 2004; Hungria et al. 2005; Rojac et al. 2005a, 2006, 2007).

Lanfredi et al. reported on the synthesis of $NaNbO_3$ using the so-called amorphous-complex method (Lanfredi et al. 2000). This method uses a precursor powder, obtained by evaporation of a starting solution containing sodium nitrate ($NaNO_3$), niobium ammonium oxalate ($NH_4H_2(NbO(C_2O_4)_3) \cdot 3H_2O$), oxalic acid and ammonium hydroxide. The last of these was used to adjust the pH of the solution to prevent the precipitation of sodium oxalate and niobium pentoxide. The crystallization of orthorhombic $NaNbO_3$ from the precursor powder occurred at 400 °C, which is much lower than normally observed using the classical solid-state reaction route (700 °C) (Lanfredi et al. 2000). For calcination temperatures ranging between 450 and 900 °C, the mean crystallite size was found to be between 32 and 85 nm. High sinterability was achieved within a narrow temperature range with a maximum relative density of 98% after sintering at 1,250°C for 2 h. Similar results were obtained by Camargo et al. using Na_2CO_3, niobium acid and DL-malic acid instead of oxalic acid (Camargo et al. 2002; Calderon-Moreno and Camargo 2003).

Nobre et al. prepared $NaNbO_3$ by a polymeric precursor method (the Pechini method) using sodium nitrate and niobium ammonium oxalate as precursor salts (Nobre et al. 1996). Niobium and sodium citrates were formed by dissolution of the precursor salts in a water solution of citric acid. Ethylene glycol was added to promote the polymerization with a polyestherification reaction at 250 °C. After annealing the amorphous polymeric precursor, $NaNbO_3$ with high crystallinity was obtained at temperatures of 400–500 °C, which is comparable with the amorphous-complex method (Lanfredi et al. 2000) and microemulsion-mediated synthesis (Shiratori et al. 2005a). After calcination at 600 °C for 5 h, they obtained nanocrystalline single-phase $NaNbO_3$ with a mean particle size of 46 nm. The relative density after sintering at 1,190°C for 3 h was 98%.

Castro et al. reported on the synthesis of $NaNbO_3$ by mechanochemical synthesis using a starting powder mixture of Na_2CO_3 and Nb_2O_5 (Castro et al. 2004). After 30 days of activation in a vibration mill, a powder mixture with increased reactivity was achieved. Single-phase $NaNbO_3$ was obtained by heating the activated mixture at a lower temperature, i.e., 600 °C, than usually required for conventional solid-state synthesis. To decrease the mechanochemical activation time, more reactive sodium reagents were used, such as NaOH and Na_2O (Hungria et al. 2005). Because of the high reactivity and, therefore, the difficulty in handling the Na_2O an off-stoichiometry resulted, and a secondary phase was formed upon heating the activated precursor. To further improve the synthesis of $NaNbO_3$, a combined wet-chemistry and mechanochemical activation technique was proposed (Hungria et al. 2005). In this case, single-phase $NaNbO_3$ was prepared by heating the 2-days-activated mixture at 700 °C. In contrast, Rojac et al. prepared nanocrystalline $NaNbO_3$ with a crystallite size in the range of a few nanometers to 25 nm directly by mechanochemical synthesis, starting from a Na_2CO_3–Nb_2O_5 mixture and using a planetary mill (Rojac et al. 2005a). It was found that a carbonate complex is formed during the mechanochemical reaction between Na_2CO_3 and Nb_2O_5, which represents an intermediate stage in the synthesis of $NaNbO_3$ (Rojac et al. 2007). To estimate the milling time necessary to obtain $NaNbO_3$ under specific milling conditions, a new approach was made using a milling map (Rojac et al. 2006). The mechanochemical synthesis of $NaNbO_3$ requires a minimum cumulative kinetic energy, which is constant irrespective of the ball-impact energy and ball-impact frequency applied during the milling operation.

5.2.3 (K, Na)NbO_3

Early investigations on the $(K,Na)NbO_3$ (KNN) solid solution were focused on improving the KNN's piezoelectric properties by enhancing its sinterability. To synthesize the solid solution, a classical solid-state reaction route was applied, starting from a powder mixture of K_2CO_3, Na_2CO_3, and Nb_2O_5, with typical calcination temperatures between 900 and 950 °C (Egerton and Dillon 1959; Tennery and Hang 1968; Haertling 1967; Kosec and Kolar 1975). Since one of the major problems in the synthesis of the KNN solid solution involves the volatilization of alkaline species at high temperatures, Egerton and Dillon used an 0.1 mol% excess of alkaline carbonates in the initial mixture (Egerton and Dillon 1959). To remove possible unreacted alkali, leaching in a 2% K_2CO_3 solution was performed after the calcination in accordance with the work of Reisman et al. (Reisman et al. 1956).

The solid-state synthesis of KNN involves several problems, one of them being the appropriate choice of sufficiently stable reagents, in this case, the alkali elements. Alkaline carbonates are typically used in the starting mixture, but K_2CO_3, in particular, is quite moisture-sensitive (Ringaard and Wurlitzer 2005). Alternatively, Matsubara et al. synthesized a solid solution using alkaline hydrogencarbonates (Matsubara et al. 2004). Furthermore, sodium potassium tartrate tetrahydrate (Rochelle salt) was found to be a good alternative, especially because of its inherent K/Na 1:1 ratio (Ringaard and Wurlitzer 2005; Attia et al. 2005). Pithan et al. reported on the synthesis of KNN by microemulsion-mediated synthesis (Pithan et al. 2005). Initially, an amorphous powder with an average particle size of about 30 nm was obtained by hydrolytic decomposition of the appropriate mixed alkaline metal–niobium ethoxide solutions in water/oil microemulsions (Shiratori et al. 2004). By annealing the amorphous powder at different temperatures, the particle size effect on the crystal structure symmetry of $(K_{0.5}Na_{0.5})NbO_3$ was studied (Shiratori 2005b). The thermodynamically stable monoclinic structure at room temperature was refined for the coarse particles (>200 nm), while for finer particles (<200 nm) a new metastable triclinic polymorph was found. Nobre and Lanfredi (Nobre and Lanfredi 2003) synthesized the $(K_{0.2}Na_{0.8})NbO_3$ solid solution using the amorphous-complex method (or the chemical evaporation method), using potassium nitrate, sodium nitrate, niobium ammonium oxalate, oxalic acid and ammonium hydroxide, similar to the case of $NaNbO_3$ synthesis (Lanfredi et al. 2000). A single phase was obtained with a crystallization temperature of 440 °C. Mechanochemical synthesis was also used to synthesize the $(K_{0.5}Na_{0.5})NbO_3$ solid solution (Rojac et al. 2005b). Amorphisation of the initial alkaline carbonates occurred after 10 h of milling, whereas the formation of the solid solution was not observed.

Bomlai et al. studied the effect of the calcination conditions and an excess of alkali carbonates on the phase formation and particle morphology of $(K_{0.5}Na_{0.5})NbO_3$ powders (Bomlai et al. 2007). The samples were prepared by the conventional solid-state reaction method with calcination temperatures ranging from 600 to 900 °C and an excess of Na_2CO_3 and K_2CO_3 of 1, 3, and 5 mol% (composition of the excess $Na/K = 1:1$). In comparison with the stoichiometric powder, the calcination temperature, needed to obtain a well-crystallized KNN, was lowered, i.e., from 800 to 700 °C, by adding 5 mol% excess of alkali carbonates. Moreover, unreacted Nb_2O_5 disappeared from the mixture after calcination at 700 °C for 2 h, which is 200 °C lower than observed for the nonexcess powder. Coarser particles, with a cuboid shape, were observed after calcination (1.0–2.5 µm; 900 °C, 2 h; 5 mol% excess of alkali carbonates) compared with the stoichiometric samples (0.15–0.3 µm; 900 °C, 2 h), suggesting a liquid-phase formation at temperatures higher than 800 °C. The authors made reference to the phase diagram of the Na_2CO_3–K_2CO_3 system, which shows a minimum melting temperature of 710 °C for the alkali carbonate mixture at a composition of 55 mol% Na_2CO_3. Because of the liquid-phase formation, an increased rate of mass transport is expected, which explains the formation of a highly homogeneous solid solution at lower temperatures compared with the KNN solid solution synthesized from the stoichiometric mixture. However, the lowest density was obtained in the case of 5 mol% excess of alkali carbonates, due to the particles' growth and inter-particle necking, which took place during the calcination step.

In the latest paper, Hagh et al. report on the property-processing relationship in the Li, Ta, Sb-modified KNN solid solution (Hagh et al. 2007). The synthesis was carried out using two different solid-state synthesis methods, denoted as the "mixed-oxide" and "perovskite" route. In the former case, the solid solution was synthesized directly from a mixture of carbonates and oxides. In the latter case, the binary compounds ($KNbO_3$, $NaNbO_3$, $KTaO_3$, $NaSbO_3$, and $LiSbO_3$) were first synthesized and then mixed together to form the final composition of $(K_{0.44}Na_{0.52}Li_{0.04})(Nb_{0.84}Ta_{0.10}Sb_{0.06})O_3$. The effect of humidity was studied by comparing the results using as-received and preheated raw materials (24 h at 220 °C). For further comparison, the powder precursors were also prepared in a laboratory and inert atmosphere (glove box, Ar). In the sintering step, the ceramics were fired in covered (Pt foil) and uncovered conditions (direct exposure to sintering atmosphere) under the oxygen atmosphere. This made possible to analyze the influence of the oxygen atmosphere on the final electromechanical properties. By preheating of the as-received raw materials and using an inert-atmosphere preparation of powder precursors, a substantial improvement in the electromechanical properties was achieved for both synthesis methods. This shows the importance of the careful control of humidity conditions during the preparation of the KNN-based solid solution, which includes moisture-sensitive raw materials. Since both preparation techniques gave similar electromechanical properties the authors concluded that the "mixed-oxide" technique is more suitable as it eliminates the extra processing steps taken in the "perovskite" technique. Moreover, high-oxygen flow rate along with the direct exposure were found to be crucial to obtain higher piezoelectric properties.

5.3 Sintering and Microstructure of $(K_{0.5}Na_{0.5})NbO_3$ Ceramics

One of the major problems with $K_{0.5}Na_{0.5}NbO_3$ is the sintering: according to early reports the stoichiometric material is extremely difficult to consolidate. Egerton and Dillon prepared $K_{0.5}Na_{0.5}NbO_3$ ceramics with 94% relative density by atmosphere sintering (the temperature and time of the annealing are not given) and measured the dielectric permittivity and the radial coupling coefficient to be 290 and 0.34–0.39, respectively (Table 5.1). The authors stressed the need for careful

Table 5.1 Density (in g/cm^3 and as percentage of theoretical density TD = 4.51 g cm^{-3}), dielectric permittivity (ε) and dielectric loss (tg δ), piezoelectric d_{33} coefficient and electromechanical planar coefficient (k_p) of $(K_{0.5}Na_{0.5})NbO_3$ ceramics prepared by atmosphere sintering, hot pressing or spark-plasma sintering

Temperature/ time	ρ (g cm^{-3})	% density	ε	tan δ	d_{33} [pC/N]	k_p	References[a]
1,115°C/6 h (AS)	4.25–4.30	94.24–95.34	290	0.04	80	0.34–0.39	[1–3]
1,115°C/2 h (AS)	4.26	94.46	450	0.06	80	–	[4]
	–	94–95	400	0.025	70–90	0.39	[5]
1,114°C/2 h (AS)	4.30	95.34	~500	<0.05	110	0.39	[6]
1,120°C/2 h (AS)	4.40	97.56	~500	–	120	0.40	[7]
1,100°C/4 h[b] (AS)		98.4	605	0.04	102	0.38	[8]
1,100°C/20 min (HP)	4.46	98.89	420	–	160	0.45	[2]
850–1,150°C/1 h (HP)	4.50	99.78	470	0.035	–	0.45	[9]
1080°C/30 min (HIP)	4.50	99.78	400	–	–	0.45	[3]
1080°C/45 min (HP)	4.48	99.33	700	0.06	–	0.47	[10]
1,040–1,100°C/ 3 min. (SPS)+ 950°C/5h (postanneal)	–	98	550	–	–	0.38	[11]
920°C/5 min. (SPS)+ 900°C/4 h (postanneal)	4,48		≈700	≈0.03	≈145	≈0.35	[4]

AS, atmosphere sintering; *HP*, hot press; *HIP*, hot isostatic press; *SPS*, sparl plasma sintering
[a][1] Egerton and Dillon (1969), [2] Jaeger and Egerton (1962), [3] Egerton and Bieling (1968), [4] Zhang et al. (2006a), [5] Ringgard and Wurlitzer (2005), [6] Birol et al. (2005), [7] Du et al. (2006), [8] Zuo et al. (2006), [9] Haertling (1967), [10] Kosec and Kolar (1975), [11] Wang et al. (2004)
[b]The powder synthesis included 48 h of attritor milling

control of the powder synthesis, as explained in the previous part, and the firing conditions (Egerton and Dillon 1959). The densification curve of stoichiometric $K_{0.5}Na_{0.5}NbO_3$ demonstrates an extremely narrow sintering interval close to the solidus temperature 1140 °C (Kosec and Kolar 1975). Dense ceramics can therefore be obtained by sintering at 1,100–1,120 °C (Table 5.1). By applying 48 h of high-energy attrition milling, the density of $K_{0.5}Na_{0.5}NbO_3$ sintered at 1,100 °C was as high as 98.4% (Zuo et al. 2006).

The sintering of $K_{0.5}Na_{0.5}NbO_3$ could be promoted by introducing A-site vacancies into the perovskite lattice. KNN ceramics sintered at 1,125 °C for 24 h reached 97.1% relative density by introducing Mg ions (1%) as the A-site dopant, when compared with 94.5% for the stoichiometric composition. Similarly, an improvement in density was achieved by introducing excess Nb (Kosec and Kolar 1975). Hot pressing or hot isostatic pressing at temperature of 1,080–1,100 °C has been very successful in achieving relative densities above 99% as early as the 1960s (Jaeger and Egerton 1962; Haertling 1967; Kosec and Kolar 1975; Egerton and Bieling 1968). Spark-plasma sintering (SPS) has also been used to obtain almost theoretically dense $K_{0.5}Na_{0.5}NbO_3$ ceramics (Wang et al. 2004; Zhang et al. 2006a).

The dielectric permittivity, the piezo electromechanical planar coupling coefficient, and d_{33} of atmosphere-sintered $K_{0.5}Na_{0.5}NbO_3$ ceramics are in the range 200–700, 0.34–0.40, and 80–120 pC N^{-1}, respectively (Zhang et al. 2006a; Du et al. 2006). The hot-pressed samples with almost theoretical densities had higher k_p factors, 0.45–0.47, and the highest reported d_{33} value of 160 pC N^{-1} for $K_{0.5}Na_{0.5}NbO_3$ (Jaeger and Egerton 1962; Kosec and Kolar 1975). Spark-plasma sintering also yielded a high value of d_{33}, i.e., 145 pC N^{-1}, and k_p, comparable to that of air-sintered ceramics (Wang et al. 2004; Zhang et al. 2006a), Table 5.1 summarizes some of the reported densities, dielectric and piezoelectric properties of $K_{0.5}Na_{0.5}NbO_3$ ceramics, prepared by air-sintering, hot pressing, and spark-plasma sintering.

To the best of the authors' knowledge, there have not been many studies of the microstructure of $K_{0.5}Na_{0.5}NbO_3$ ceramics. The typical features of the microstructure are cube-shaped grains that are seen on the fracture surfaces. (Fig. 5.2a) (Zuo et al. 2006; Malič et al. 2005; Jenko et al. 2005; Zhen and Li 2006). A study of the microstructure of $K_{0.5}Na_{0.5}NbO_3$ ceramics sintered at 1,100 °C with 95% theoretical density by Jenko et al. revealed a bimodal distribution of grain sizes with a fraction of large grains of about 20 μm across within the fine-grained matrix, i.e., the average grain size was 1.71 μm, in agreement with the observation of the fracture surface (Fig. 5.2 a, c, d). Trapped angular pores were observed within the large grains, parallel to the grain faces (Fig. 5.2b). The crystal planes of the grain to which the pore was attached are of the {100} family, with respect to the simple perovskite cell (Jenko et al. 2005). No data on the surface energy of $K_{0.5}Na_{0.5}NbO_3$ were found, so it was assumed that the {100} surfaces had a lower surface energy than the other crystal faces, similar to that reported for $SrTiO_3$ (Sano et al. 2003). The SEM-EDS analysis revealed that the microstructure consisted of the matrix phase in which the $(Na+K)/Nb$ ratio corresponded to the nominal composition and of the sodium-deficient secondary-phase inclusions with a Na/K ratio of 0.30 and a $(Na+K)/Nb$ ratio of 0.6 (Jenko et al. 2005).

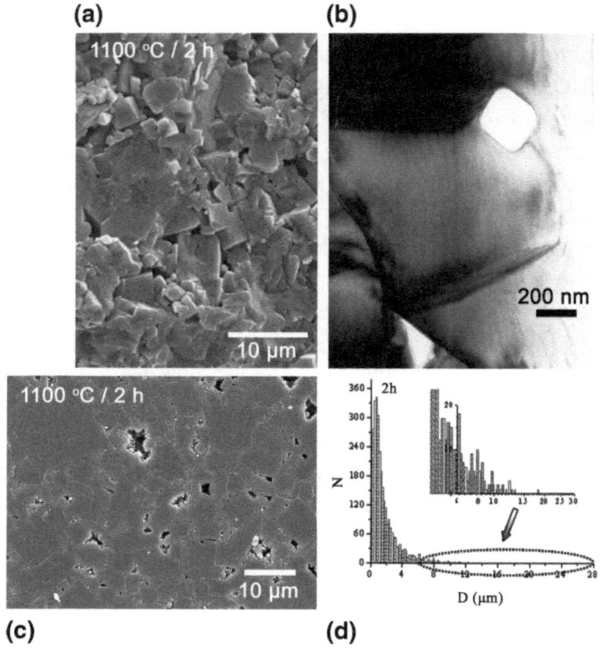

Fig. 5.2 Microstructure of KNN ceramics sintered at 1,100 °C for 2 h: (**a**) SEM micrographs of the fracture surface, (**b**) TEM image, (**c**) thermally etched surface and (**d**) grain size distribution (Jenko et al. 2005)

Zhen and Li observed that the abnormal grain growth in Li and Li, Ta-doped $K_{0.5}Na_{0.5}NbO_3$ ceramics occurred when the temperature exceeded the optimal temperature, i.e., when the density started to decrease. The deficiency of sodium was determined by EDXS and it increased with increasing sintering temperature. The authors proposed that the exaggerated grain growth was due to the formation of a potassium-enriched liquid phase. Exaggerated grain growth was hindered by using alkaline–niobate-based packing powder (Zhen and Li 2006). Du et al. studied the effect of heating rate on the densification and microstructure development in $K_{0.5}Na_{0.5}NbO_3$. The fracture surfaces of the ceramics sintered at 1,100 °C with a heating rate of $3\,K\,min^{-1}$ revealed a bimodal grain-size distribution, and a more homogeneous microstructure at a heating rate of $5\,K\,min^{-1}$ (Du et al. 2006).

5.4 Chemical Modifications of $K_{0.5}Na_{0.5}NbO_3$

A possible method to improve the sinterability of KNN ceramics at atmospheric pressure is the addition of sintering aids, and in fact the addition of CuO (Takao et al. 2006; Matsubara et al. 2005a), $K_4CuNb_8O_{23}$ (KCN) (Matsubara et al. 2004, 2005a), or $K_{5.4}Cu_{1.3}Ta_{10}O_{29}$ (KCT) (Matsubara et al. 2005b,c,d) in small amounts was shown to be effective in improving the densification of KNN-based ceramics.

Matsubara et al. found that the sinterability and piezoelectric properties of KNN depended upon the amount of CuO addition as well as on the A/B ratio (Matsubara et al. 2005a). The CuO and the excess of the B-site ions in the ABO_3 lattice were effective in forming the $K_4CuNb_8O_{23}$ phase, which was found to be the key to a higher density for the KNN ceramics. The highest density of $4.40\,g\,cm^{-3}$ after sintering at 1,200 °C for 1 h was obtained for the A-site-poor KNN with an A/B ratio <1 and the addition of 0.4 mol% CuO. The authors also reported that the Ta-modified (5 mol%) KNN ceramics with 0.4 mol% of CuO sintered at ∼1,090 °C showed the maximum value of k_p, i.e., 0.41. The CuO was also an effective sintering aid for textured KNN ceramics. After sintering at 1,075 °C for 1 h in O_2 and the addition of 1 mol% CuO to crystalline-oriented KNN, the relative density of the KNN ceramic was 95%, with a k_p of 0.54, d_{33} equal to $123\,pC\,N^{-1}$ and tan δ (at 1 kHz) equal to 0.020 (Takao et al. 2006).

Consequently, $K_4CuNb_8O_{23}$ (KCN) was also used as a sintering aid. The addition of 0.5 mol% of KCN to the KNN ceramic resulted in a density of $4.40\,g\,cm^{-3}$ after sintering at 1,090 °C (Matsubara et al. 2004) or at 1,100 °C (Matsubara et al. 2005a) for 1 h. These KNN ceramic samples with 0.5 mol% of KCN showed a bimodal grain size distribution. The enhanced densification was attributed to the low melting point of KCN; i.e. 1,050 °C. KCN also provided excess B-site ions, which additionally promoted the sintering. Adding more than 0.5 mol% of KCN caused the density of the KNN to decrease (Matsubara et al. 2004, 2005a). The $K_{5.4}Cu_{1.3}Ta_{10}O_{29}$ (KCT) also improved the sinterability of the KNN (Matsubara et al. 2005b), Ta-(Matsubara et al. 2005c) and Li-modified (Matsubara et al. 2005d) KNN ceramics. With the addition of 0.38 mol% KCT, after sintering at 1,120 °C, the density of the KNN ceramics was $4.6\,g\,cm^{-3}$, which is even higher than the theoretical density of pure KNN ($4.51\,g\,cm^{-3}$). As the melting point of KCT is well above the sintering temperature of KNN, at about 1,250 °C, the formation of the liquid phase was attributed to the reaction between the KNN and the KCT. The authors reported that the KCT was the source of the liquid phase, the source of excess B-site ions, and it also inhibited the abnormal grain growth of the matrix phase (Matsubara et al. 2005b). In the case of 0.38 mol% KCT added to Ta-modified KNN, the ceramic samples were sintered, depending on Ta content, from 1,050 °C to 1,270 °C. All the specimens exhibited a high density of 94–99%. In this system, a content of 20–30 mol% of Ta was found to promote a high k_p value of 0.44 (Matsubara et al. 2005c). When 0.38 mol% KCT was added to Li-modified (2 mol%) KNN, after sintering at 1,120 °C for 1 h in O_2 the ceramic was dense (99% TD) with a k_p of 0.43.

As already mentioned, the sinterability of KNN can be improved by the creation of vacancies on the A sites. A wide range of ions were introduced on the A site in the KNN perovskite structure, e.g., Mg^{2+} (Kosec and Kolar 1975; Malič et al. 2005a), Ca^{2+} (Malič et al. 2005a), Sr^{2+} (Malič et al. 2005a,b; Maeder et al. 2004), Ba^{2+} (Malič et al. 2005a; Maeder et al. 2004; Ahn and Schulze 1987; Tashiro and Ishii 2006), Eu^{3+} (Murty and Bhanumathi 1989), and La^{3+} (Murty et al. 1988). The improvement in the sinterability of KNN was also observed when B-site oxide additives were add to KNN, e.g., Zn^{2+} (Zuo et al. 2006; Park et al. 2004) as well as Sn^{4+}, Cd^{2+}, Sc^{2+}, W^{6+}, Y^{3+}, and Ce^{4+} (Zuo et al. 2006).

Kosec and Kolar prepared KNN ceramics sintered at 1,125 °C for 24 h with 97% TD by introducing Mg ions (1%); this should be compared with the 94.5% TD for the stoichiometric composition. In Mg-doped samples, the authors observed a slight retardation of grain growth, some lowering of ε, a large increase in the resistivity and an increase in k_p to 0.45, which are comparable to the values reported for hot-pressed KNN ceramics (Table 5.1) (Kosec and Kolar 1975). Malič et al. reported a 91.2% TD for KNN ceramics with the addition of 0.5 mol% of Mg, after sintering at 1,150 °C for 2 h (Malič et al. 2005a). On the basis of the observations with an SEM, the Mg-doped samples were single phase. Mg doping decreases the O–T transition by 9 °C (for pure KNN it is 190 °C) but it has no influence on the T–C transition (for pure KNN it is 400 °C).

Ca doping of 0.5 mol% resulted in an almost unchanged density for KNN ceramics, while the addition of Sr^{2+} promoted the densification of KNN. After sintering KNN with 0.5 mol% Sr at 1,115 °C for 2 h, the obtained density was 4.33 g cm^{-3}. Both ions, Ca^{2+} and Sr^{2+}, decrease the phase-transition temperatures of KNN, i.e., 15–17 °C for T_{O-T} and 7–8 °C for T_{T-C}. The dielectric constant and losses measured at 10 kHz and the piezo d_{33} constant of Ca- and Sr-doped KNN sintered at 1,150 °C for 2 h are 495, 0.12, 95 and 500, 0.04, 95, respectively (Malič et al. 2005a). With an increasing content of strontium, the orthorhombic distortion of KNN is gradually reduced, and a niobium-rich secondary phase was observed in KNN ceramics with 10 mol% Sr added. The fracture surface of KNN–0.5 mol% Sr sintered at 1,115 °C revealed a few 10-μm-large cube-shaped uniform grains, while in the ceramics with 2, 5, and 10 mol% Sr, the grains were micrometer-sized, without the distinct cube-shaped morphology (Malič et al. 2005b).

According to Ahn and Schulze, additions of up to 4 mol% Ba decrease the grain size and increase the density of KNN; however, a Ba-rich secondary phase was observed after adding more than 1.5 mol% of Ba (Ahn and Schulze 1987). The authors reported that KNN ceramics with 2 mol% Ba sintered at 1,100 °C for 7 h had a relative density of 95%TD, a dielectric constant of 760, a d_{33} of 115 pC N^{-1}, and losses of 0.024. According to Malič et al., the secondary phase is formed with 0.5 mol% of Ba^{2+} in the KNN after sintering at 1,115 °C for 2 h, and the density was 4.24 g cm^{-3} (94.1% TD) (Malič et al. 2005a). In addition, the phase transitions were markedly broader, compared with the undoped KNN. Tashiro and Ishii reported that the KNN ceramics with 2 mol% Ba sintered between 1,110 and 1,170 °C for 2 h consisted of small and large grains (>30 μm), had a density of 4.48 g cm^{-3}, an orthorhombic crystal structure and a k_p of 0.431, while adding 3 mol% of Ba to KNN resulted in a fine grain size of less then 1 μm, a density of 4.57 g cm^{-3}, a pseudocubic crystal structure and a lower k_p of 0.141 (Tashiro and Ishii 2006). The addition of 0.5 mol% of MnO to 2 mol% Ba KNN ceramics suppressed the grain growth, resulting in the formation of dense ceramics with uniform grain size from 3 to 7 μm. After sintering at 1,160 °C for 2 h, the resulting density was 4.59 g cm^{-3}, the k_p was 0.443, the dielectric constant was 471 and the d_{33} was 124 pC N^{-1} (Tashiro and Ishii 2006). In the latest paper written by Hagh et al. the properties on $K_{0.5}Na_{0.5}NbO_3$–$LiTaO_3$–$LiSbO_3$ (KNN–LT–LS) ceramics modified with up to 2 mol% Ba^{2+} were studied [Hagh et al. in press]. Authors verified the A site substitution of donor-doped barium using electron–hole compensation model, which fitted well with the experimental results. Similar to barium modified KNN ceramics, the authors observed a decrease in the grain size with increasing Ba^{2+} content from 3 µm to 0.5 µm. The addition of 1 mol% of Ba^{2+} to KNN–LT–LS improved the d_{33} and the k_{33} up to 36% ($d_{33} = 210\,pC\,N^{-1}$) and 58% ($k_{33} = 0.491$), respectively.

Murty et al. reported that the addition of 1 mol% Eu^{3+} (Murty and Bhanumathi 1989) or La^{3+} (Murty et al. 1988) decreased the average grain size of the KNN matrix and reduced the cubic transition temperature to 355 °C or 342 °C, respectively. Eu doping also reduced the O–T phase transition by ~20 °C, but the addition of La^{3+} to the KNN did not affect the O–T phase transition.

Zuo et al. studied the effect of adding a wide range of oxides (Zn^{2+}, Sn^{2+}, Cd^{2+}, Sc^{2+}, W^{6+}, Y^{3+}, and Ce^{4+}) on the sinterability and functional response of KNN (Zuo et al. 2006). The authors assumed that these are B-site oxide additives, causing oxygen vacancies in the lattice of the KNN. Before sintering the calcined powders were attrition milled for 6 h. After sintering at 1,100 °C for 4 h, the densification of KNN was improved by adding 1 mol% of ZnO (97%TD), CdO (95% TD), Sc_2O_3 (98% TD), and SnO_2 (94% TD), while the addition of CeO_2, Y_2O_3, and WO_3 inhibited the sintering. The oxides had a different effect on the microstructure of KNN. The addition of ZnO and CdO increased the grain size, while the addition of Y_2O_3 and CeO_2 decreased the grain size. The abnormal grain growth observed in CdO-doped KNN was most probably due to the low melting point of CdO (900 °C). The maximum dielectric permittivity measured at 1 kHz, the k_p and d_{33} values (652, 44% and 117 pC N^{-1}, respectively) were obtained for KNN ceramics with 1 mol% ZnO.

5.5 Ternary and Quaternary Systems

A number of papers on alkaline-earth (AE) titanate-modified KNN ceramics, where AE is Mg^{2+} (Chang et al. 2007), Ca^{2+} (Chang et al. 2007), Sr^{2+} (Chang et al. 2007; Shimojo and Wang 2005; Wang et al. 2005; Bobnar et al. 2004; Kosec et al. 2004; Guo et al. 2004a), or Ba^{2+} (Chang et al. 2007; Shimojo and Wang 2005; Guo et al. 2004b; Lee et al. 2006; Park et al. 2006; Ahn et al. 2005) can be found in the open literature. On the basis of similar ionic radii, the authors assume that AE ions occupy the A sites of the KNN perovskite lattice, while Ti ions occupy the B sites. KNN ceramics were also modified with $PbTiO_3$ (Wang et al. 2002; Tashiro and Nagata 2004). According to Chang et al., when a small amount (0.5 mol%) of $MgTiO_3$ is added to KNN, a secondary phase of alkaline polyniobates is formed after sintering at 1,115 °C for 2 h, resulting in a low density and poor dielectric and piezoelectric properties (Chang et al. 2007). Adding $MgTiO_3$ to KNN decreased $T_{O\text{–}T}$ (~172 °C) and increased T_c (~406 °C). Adding 0.5 mol% of $CaTiO_3$ to KNN after sintering at 1,115 °C for 2 h promoted the densification to 97.8% TD and enhanced the electrical properties (Table 5.2). Both transition temperatures of the KNN–$CaTiO_3$ ceramics were decreased ($T_c = \sim 397\,°C$, $T_{O\text{–}T} = \sim 182\,°C$) (Chang et al. 2007). When 0.5 mol% $SrTiO_3$ is added to KNN and sintered at 1,115 °C for 2 h, the density

Table 5.2 Dielectric and piezoelectric properties of some AE titanate-modified KNN ceramics

Dopant	$(1-x)$ KNN–x $AETiO_3$ $x=$	Temp. and time	Density ($g\,cm^{-3}$)	ε	$\tan\delta$ (10 kHz)	d_{33} [$pC\,N^{-1}$]	k_p	References[a]
$MgTiO_3$	0.005	1,115°C/2h	4.1	480 (10 kHz)	0.037	51	0.15	[1]
$CaTiO_3$	0.005	1,115°C/2h	4.41	411 (10 kHz)	0.024	116	0.39	[1]
$SrTiO_3$	0.005	1,115°C/2h	4.38	448 (10 kHz)	0.024	110	0.35	[1]
$SrTiO_3$	0.05	–	–	–	–	195	0.37	[2]
$BaTiO_3$	0.005	1,115°C/2h	4.32	393 (10 kHz)	0.024	85	0.26	[1]
$BaTiO_3$	0.020	1,100–1,250°C	4.44	1,003	0.038	104	0.29	[3]
$BaTiO_3$	0.050	1,060°C/2h	–	1,058	–	225	0.36	[4]

Samples were prepared by spark-plasma sintering method
[a][1] Chang et al. (2007), [2] Wang et al. (2005), [3] Guo et al. (2004b), [4] Park et al. (2006)

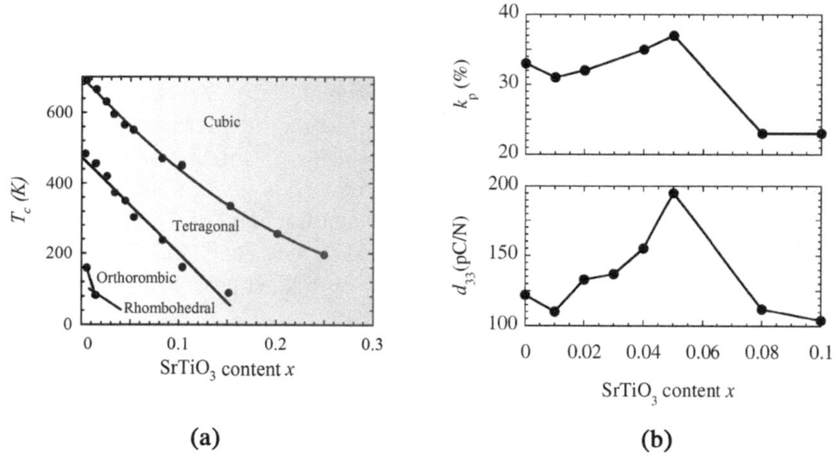

Fig. 5.3 (**a**) Phase diagram for $(1-x)$KNN–$x$$SrTiO_3$ solid solution and (**b**) variations of k_p and d_{33} with x (Wang et al. 2005)

is increased to 97.1% of the theoretical density. In addition, both transition temperatures were decreased ($T_C = \sim179$°C, $T_{O\text{–}T} = \sim390$°C) (Chang et al. 2007). With increasing $SrTiO_3$ content, the grain size of the KNN matrix decreases (Shimojo and Wang 2005). A typical relaxor-like dielectric response was detected in KNN–$SrTiO_3$ ceramics (Wang et al. 2005; Bobnar et al. 2004; Kosec et al. 2004; Guo et al. 2004a). Kosec et al. reported a solid solution with a pseudocubic structure for the $(1-x)$KNN–$x$$SrTiO_3$ system, where x was between 0.15 and 0.25 (Kosec et al. 2004). The high density (96.8% TD) and the uniform distribution of fine grains and pores were confirmed by the translucency of these ceramics. Wang et al. published the phase diagram, established from the temperature dependence of the dielectric constant for the $(1-x)$KNN–$x$$SrTiO_3$ solid solution (Wang et al. 2005). The samples were prepared using spark-plasma sintering and the maximum k_p of 0.37, d_{33} of 195 $pC\,N^{-1}$ were found at $x = 0.05$ (Fig. 5.3).

When 0.5 mol% of $BaTiO_3$ is added to KNN, and after sintering at 1,115 °C for 2 h, the density is reduced compared with pure KNN. Phase transitions were observed for T_c at 400 °C and $T_{O\text{–}T}$ at 185 °C (Chang et al. 2007). In 5 mol%, $BaTiO_3$-modified KNN ceramics, sintered at 1,040–1,075 °C, abnormal grain growth and a liquid phase were observed. Above 1,075 °C, the grain size decreased due to the formation of a large amount of liquid phase. The dielectric constant and d_{33} values increased for the specimens sintered between 1,050 and 1,075 °C, the highest d_{33} obtained was 225 pC N^{-1} and the k_p was 0.36 (Park et al. 2006). Between 6 mol% and 20 mol% of $BaTiO_3$ in KNN, the relaxor-like behavior was observed. The MPB between the orthorhombic and tetragonal phase transition appeared at 6 mol% of $BaTiO_3$ in KNN (Guo et al. 2004b). In Table 5.2 the dielectric, piezoelectric, and ferroelectric properties of some AE titanate-modified KNN ceramics are shown.

A report that textured Li,Ta, and Sb-modified KNN ceramics exhibit piezoelectric properties (d_{33}~416 pC N^{-1}) comparable to PZT (Saito et al. 2004) stimulated a number of systematic investigations of the KNN system, where $LiSbO_3$ (Zhang et al. 2006b; Yang et al. 2006; Zang et al. 2006), $LiNbO_3$ (Du et al. 2007; Guo et al. 2004c; Kakimoto et al. 2005; Tang et al. 2006), and $LiTaO_3$ (Guo et al. 2006) were added to form new solid solutions. Randomly oriented ceramics of KNN–LT–LS in the vicinity of the morphotropic phase boundary typically exhibit a d_{33} of ~300 pC N^{-1}, while textured ceramics of the same show a substantial increase to ~416 pC N^{-1}, surpassing PZT thereby. The mode and manner by which texturing has been accomplished has not been disclosed by Saito et al., although one expects that the use of a suitable transient phase sintering has been used (Saito et al. 2004). In Fig. 5.4a d_{31} as a function of Li and Ta contents for $(K_{0.5}Na_{0.5})_{1-x}Li_xNb_{1-y}Ta_yO_3$ ceramics is shown. In Fig. 5.4b, the d_{33} values for lead-free (LF) and PZT ceramics as a function of the Curie temperature are shown (Saito et al. 2004).

When 2 mol% of $LiSbO_3$ is added to KNN, characteristic cubic-shaped grains ranging from 0.3 to 1 μm were observed. The particle size of the KNN matrix

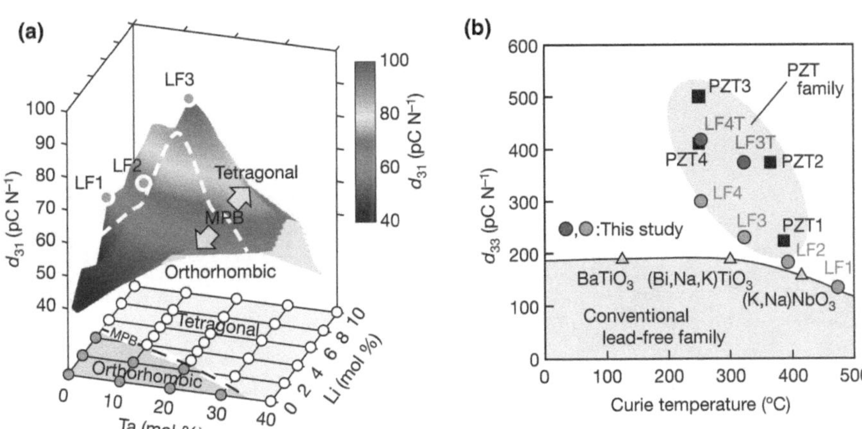

Fig. 5.4 Piezoelectric performances for lead-free (LF) piezoelectric ceramics (Saito et al. 2004)

increased and the porosity decreased with increasing $LiSbO_3$ content. After sintering at 1,100 °C for 2 h, the density of KNN–6 mol% $LiSbO_3$ was 4.44 g cm^{-3}. Above 6 mol% of $LiSbO_3$, second phases were formed and the porosity began to increase. In the $(1-x)$KNN–$x$$LiSbO_3$ system the MPB was found in the compositional range of 0.06–x–0.1 (Yang et al. 2006).

Zhang et al. observed that the enhancement of the piezoelectric properties in the KNN–$LiSbO_3$ system occurs at $T_{O\text{–}T}$ and is strongly temperature-dependent (Zhang et al. 2006b). The addition of 5.2 mol% $LiSbO_3$ to KNN shifted the T_c to 386 °C and the $T_{(O\text{–}T)}$ to ∼35 °C. With the increasing $LiNbO_3$ content added to KNN, the pore size significantly decreased and the average grain size increased. After sintering at 1,080 °C for 2 h, the density of KNN–6 mol% $LiNbO_3$ was 4.41 g cm^{-3}, and the average grain size was about 8 μm (Du et al. 2007). In $LiNbO_3$-modified KNN, the two phase transitions were observed: the T_c was shifted to a higher temperature (above 450 °C for $LiNbO_3$ between 4 and 6 mol%) and the $T_{O\text{–}T}$ to a lower temperature (Guo et al. 2004c). Guo et al. found a morphotropic phase boundary (MPB) between the orthorhombic and tetragonal phases in the compositional range 0.05–x–0.07 for $1-x(Na_{0.5}K_{0.5})$–$x$$LiNbO_3$ ceramics (Guo et al. 2004c; Kakimoto et al. 2005). Similarly, when $LiTaO_3$ was added to KNN ceramics two phase transitions were observed. With increasing content of $LiTaO_3$, the T_c was shifted to a higher temperature while the $T_{T\text{–}O}$ was shifted to a lower temperature. A MPB between the orthorhombic and tetragonal phases was found in the compositional range 0.05–x–0.06 for $1-x(Na_{0.5}K_{0.5})$–$x$$LiTaO_3$ (Guo et al. 2005). In Table 5.3, some dielectric and piezoelectric properties of modified KNN ceramics are shown.

Table 5.3 Dielectric and piezoelectric properties of modified KNN ceramics

Composition	x	Temp. and time	Density (g cm^{-3})	ε	tan δ (10 kHz) [/]	d_{33} (pC N^{-1})	k_p/k_t	References[a]
$(1-x)$ KNN–$x$$LiSbO_3$	0.052	1,100°C 2 h	4.35	1380 (1 kHz)	0.02	265	0.5/0.46	[1]
$(1-x)$ KNN–$x$$LiSbO_3$	0.060	1,100°C 2 h	4.44	708 (10 kHz)	0.037	171	0.38	[2]
$(1-x)$ KNN–$x$$LiNbO_3$	0.060	1,080°C 2 h	4.31	530 (10 kHz)	0.034	215	0.41	[3]
$(1-x)$ KNN–$x$$LiTaO_3$	0.050	1,110°C	4.44	570 (10 kHz)		200	0.36	[4]
(K, Na, Li_x) $NbTa_yO_3$	$x=3$, $y=20$	1,065–1, 135 °C/ 10–60 min	4.8	929 (10 kHz)	0.024	190	0.46/0.46	[5]

[a][1] Zhang et al. (2006b), [2] Yang et al. (2006), [3] Kakimoto et al. (2005), [4] Guo et al. (2005), [5] Hollenstein et al. (2005)

5.6 Single Crystals

Since texture engineering can lead to an improvement in piezoelectric properties (Saito et al. 2004), single crystals of KNN are also expected to have superior properties than their polycrystalline equivalents. Domain-engineered single crystals of $KNbO_3$ have been found to possess good piezoelectric properties (Wada et al. 2004). A few crystal-growth studies have been carried out on the mixed $(K_xNa_{1-x})NbO_3$ system. Shirane et al. grew single crystals containing up to 10 mol% $KNbO_3$ using NaF as a flux (Shirane et al. 1954); Ahtee and Glazer grew single crystals containing up to 25 mol% $KNbO_3$ using NaF as a flux (Ahtee and Glazer 1976); and Raevskiĭ et al. grew single crystals containing up to 40 mol% $KNbO_3$ using $NaBO_2$ as a flux (Raevskiĭ et al. 2003). In these studies, compositional variations within individual crystals and the difficulty in growing crystals with high $KNbO_3$ contents were encountered. Recently, single crystals of pure KNN and doped KNN $[(Na_{0.5}K_{0.5})_{0.97}Li_{0.03}(Nb_{0.8}Ta_{0.2})O_3]$ have been grown using the solid-state crystal-growth (SSCG) method (Fig. 5.5) (Fisher et al. 2007a,b). This method allows the chemical composition of the single crystal to be carefully controlled.

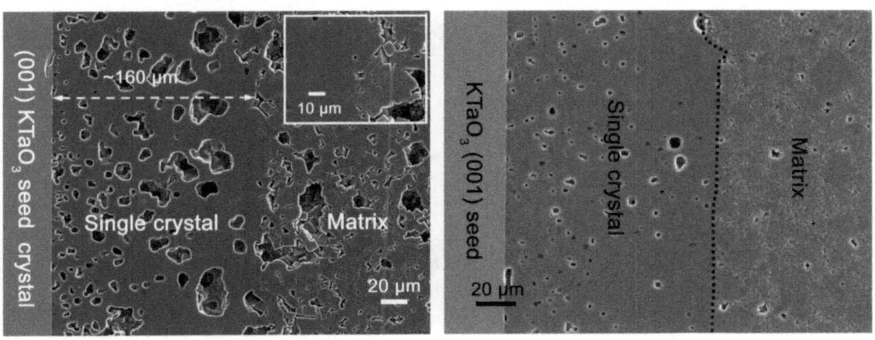

(a) (b)

Fig. 5.5 SEM micrographs of (**a**) KNN (the inset shows the boundary between the matrix and the single crystal) and (**b**) a KNLNT $[(Na_{0.5}K_{0.5})_{0.97}Li_{0.03}](Nb_{0.8}Ta_{0.2})O_3$ single-crystal layer grown on a (001) $KTaO_3$ seed crystal. The sample shown in (**a**) was annealed in air at 1,100 °C for 10 h. The sample shown in (**b**) was hot-pressed at 975 °C and 50 MPa for 2 h, followed by annealing at 1,135 °C for 50 h in a tube furnace (Fisher et al. 2007a,b)

5.7 Thick Films

To the best of the authors' knowledge, there have been only occasional reports on lead-free thick-film structures. Oh et al. used commercial LF4 = $(K_{0.44}Na_{0.52}Li_{0.04})(Nb_{0.86}Ta_{0.10}Sb_{0.04})O_3$ powder to prepare dense, about 5-μm-thick, films on glass, stainless steel and $Pt/Ti/ZrO_2$ substrates using an aerosol deposition method (ADM) (Oh et al. 2006). The films from the as-received powders exhibited high dielectric losses at low frequencies. To decrease them, the powder was dried prior to deposition to remove the moisture. The dielectric permittivity of the as-deposited films on zirconia substrates was increased by a postannealing process at 1,000 °C from 81 to 439 at 1 MHz, which was attributed to crystallite growth and stress relaxation.

Ryu et al. also used aerosol deposition to prepare dense 7-μm-thick $K_{0.5}Na_{0.5}NbO_3$ films on platinized silicon substrates (Ryu et al. 2007). The powder was prepared using a double calcination at 850 °C from the alkaline carbonates and niobium oxides. The dielectric permittivity of the films, postannealed at 700 °C was 545 at 1 kHz, which was higher than the value of about 340, obtained for the 91.3% dense ceramic, prepared from the same powder and sintered at 1,070 °C. The enhancement of ε was attributed to the higher density, the better crystallinity and to the slight (h00) texture of the perovskite phase.

References

Ahn CW, Song HC, Nahm S, Park SH, Uchino K (2005) Effect of MnO_2 on the piezoelectric properties of (1-x) $(Na_{0.5}K_{0.5})NbO_3$-$xBaTiO_3$ ceramics. Jpn J Appl Phy 44:1361–1364.

Ahn ZS, Schulze WA (1987) Conventionally sintered $(Na_{0.5}K_{0.5})NbO_3$ with barium additions. J Am Ceram Soc 70(1):C18–C21.

Ahtee M, Glazer AM (1974) Phase transitions in sodium niobate-potassium niobate solid solutions. Ferroelectrics 7(1–4):93–95.

Ahtee M, Glazer AM (1976) Lattice parameters and tilted octahedra in sodium-potassium niobate solid solutions. Acta Cryst A32:434–446.

Ahtee M, Hewat AW (1978) Structural phase transitions in sodium-potassium niobate solid solutions by neutron powder diffraction. Acta Cryst A34:309–317.

Amini MM, Sacks MD (1991) Synthesis of potassium niobate from metal alkox ides. J Am Ceram Soc 74(1):53–59.

Appendino P (1973) Contribution of study with system $Na_2O - Nb_2O_5$. Ann Chim (Rome) 63: 547–556.

Attia J, Bellaiche L, Gemeiner P, Dkhil B, Malič B (2005) Study of potassium-sodium-niobate alloys: A combined experimental and theoretical approach. J Phys IV Fr 128:55–60.

Birol H, Damjanović D, Setter N (2005) Preparation and characterization of $KNbO_3$ ceramics. J Am Ceram Soc 88(7):1754–1759.

Bizeto MA, Constantino VRL (2004) Structural aspects and thermal behavior of the proton-exchanged layered niobate $K_4Nb_6O_{17}$. Mater Res Bull 39(11):1729–1736.

Bobnar V, Bernard J, Kosec M (2004) Relaxor-like dielectric properties and history-dependent effects in the lead-free $K_{0.5}Na_{0.5}NbO_3$-$SrTiO_3$. Appl Phy Lett 85(6):994–996.

Bomlai P, Wichianrat P, Muensit S, Milne SJ (2007) Effect of calcination and excess alkali carbonate on the phase formation and particle morphology of $Na_{0.5}K_{0.5}NbO_3$ powders. J Am Ceram Soc 90(5):1650–1655.

Calderon-Moreno JM, Camargo ER (2003) Electron microscopy studies on the formation and evolution of sodium niobate nanoparticles from a polymeric precursor. Catal Today 78:539–542.

Camargo ER, Popa M, Kakihana M (2002) Sodium niobate $(NaNbO_3)$ powders synthesized by a wet-chemistry method using a water-soluble malic acid complex. Chem Mater 14:2365–2368.

Castro A, Jimenez B, Hungria T, Moure A, Pardo L (2004) Sodium niobate ceramics prepared by mechanical activation assisted methods. J Eur Ceram Soc 24(6):941–945.

Chang Y, Yang Z, Chao X, Zhang R, Li X (2007) Dielectric and piezoelectric properties of alkaline earth titanate doped $(K_{0.5}Na_{0.5})NbO_3$ ceramics. Mater Lett 61:785–789.

Du H, Li Z, Tang F, Qu S, Pei Z, Zhiou W (2006) Preparation and piezoelectric properties of $(K_{0.5}Na_{0.5})NbO_3$ lead-free piezoelectric ceramics with pressure-less sintering. Mater Sci Eng B131:83–87.

Du H, Tang F, Liu D, Zhu D, Zhou W, Qu S (2007) The microstructure and ferroelectric properties of $(K_{0.5}Na_{0.5})NbO_3$-$LiNbO_3$ lead free piezoelectric ceramics. Mater Sci Eng B136:165–169.

Dungan RH, Golding RD (1964) Metastable ferroelectric sodium niobate. J Am Ceram Soc 47(2): 73–76.

Egerton L, Bieling CA (1968) Isostatically hot-pressed sodium-potassium niobate transducer material for ultrasonic devices. Am Ceram Soc Bull 47(12):1151–1156.

Egerton L, Dillon DM (1959) Piezoelectric and dielectric properties of ceramics in the system potassium-sodium niobate. J Am Ceram Soc 42(9):438–442.

Fisher JG, Benčan A, Bernard J, Holc J, Kosec M, Vernay S, Rytz D (2007a) Growth of $(Na,K,Li)(Nb,Ta)O_3$ single crystals by solid state crystal growth. J Eur Ceram Soc 27(13–15): 4103–4106.

Fisher JG, Benčan A, Holc J, Kosec M, Vernay S, Rytz D (2007b) Growth of potassium sodium niobate single crystals by solid state crystal growth. J Cryst Growth 303:487–492.

Fluckiger U, Arend H (1977) Synthesis of $KNbO_3$ powder. J Am Ceram Soc Bull 56(6):575–577.

Guo Y, Kakimoto K, Ohsato H (2004a) Dielectric and piezoelectric properties of lead free $(Na_{0.5}K_{0.5})NbO_3$-$SrTiO_3$ ceramics. Solid State Commun 129:279–284.

Guo Y, Kakimoto K, Ohsato H (2004b) Structure and electrical properties of lead-free $(K_{0.5}Na_{0.5})NbO_3$-$BaTiO_3$ ceramics. Jpn J Appl Phys 43(9B):6662–6666.

Guo Y, Kakimoto K, Ohsato H (2004c) Phase transitional behaviour and piezoelectric properties of $(Na_{0.5}K_{0.5})NbO_3$-$LiNbO_3$ ceramics. Appl Phys Lett 85(18):4121–4123.

Guo Y, Kakimoto K, Ohsato H (2005) $(K_{0.5}Na_{0.5})NbO_3$-$LiTaO_3$ lead-free piezoelectric ceramics. Mater Lett 59:241–244.

Hagh NM, Jadidian B, Safari A (2007) Property-processing relationship in lead-free (K,Na,Li) NbO_3-solid solution system. J Electroceram 18:339–346.

Hagh NM, Jadidian B, Ashbahian E, Safari A (2008) Lead-free piezoelectric ceramics transducer in the donor-doped $K_{1/2}Na_{1/2}NbO_3$ solid solution system.IEEE Trans Ultrason Ferroelectr Freq Control 55(1):214–224.

Haertling GH (1967) Properties of hot-pressed ferroelectric alkali niobate ceramics. J Am Ceram Soc 50:329–330.

Hollenstein E, Davis M, Damjanovic D, Setter N (2005) Piezoelectric properties of Li and Ta modified $(K_{0.5}Na_{0.5})NbO_3$ ceramics. Appl Phys Lett 87:182905/1–182905/3.

Hungria T, Pardo L, Moure A, Castro A (2005) Effect of mechanochemical activation on the synthesis of $NaNbO_3$ and processing of environmentally friendly piezoceramics. J All Comp 395(1–2):166–173.

Irle E, Blachnik R, Gather B (1991) The phase diagrams of Na_2O and K_2O with Nb_2O_5 and the ternary system Nb_2O_5-Na_2O-Y_2O_3. Thermochim Acta 179:157–169.

Jaeger RE, Egerton L (1962) Hot pressing of potassium-sodium niobates. J Am Ceram Soc 45(5): 209–213.

Jaffe B, Cook WR, Jaffe H (1971) Piezoelectric ceramics. Academic Press, London. JCPDS 71–2171.

Jenko D, Benčan A, Malič B, Holc, J, Kosec M (2005) Electron microscopy studies of potassium sodium niobate ceramics. Microsc Microanal 11:572–580.

Kakimoto K, Akao K, Guo Y, Ohasto H (2005) Raman scattering study of piezoelectric $(Na_{0.5}K_{0.5})NbO_3$-$LiNbO_3$ ceramics. Jpn Soc Appl Phy 44(9B):7064–7067.

Katz L, Megaw HD (1967) The structure of potassium niobate at room temperature: The solution of a pseudosymmetric structure by Fourier methods. Acta Crystallogr 22:639–648.

Kodaira K, Shioya J, Shimada S, Matsushita T (1982) Sintering and dielectric properties of $KNbO_3$. J Mater Sci Lett 1:277–278.

Kosec M, Bobnar V, Hrovat M, Bernard J, Malič B, Holc J (2004) New lead free relaxors based on the $K_{0.5}Na_{0.5}NbO_3$-$SrTiO_3$ solid solution. J Mater Res 19(6):1849–1853.

Kosec M, Kolar D (1975) On activated sintering and electrical properties of $NaKNbO_3$. Mater Res Bull 10:335–340.

Lanfredi S, Dessemond L, Martins Rodrigues AC (2000) Dense ceramics of $NaNbO_3$ produced from powders prepared by a new chemical route. J Eur Ceram Soc 20:983–990.

Lee HJ, Bae MS, Cho YK (2006) Transmission electron microscopy of $0.95(Na_{0.5}K_{0.5})NbO_3$–$0.05BaTiO_3$ ceramics. J Am Ceram Soc, 89(11):3529–3532.

Lu CH, Lo SY, Lin HC (1998) Hydrothermal synthesis of nonlinear optical potassium niobate ceramic powder. Mater Lett 34:172–176.

Lu CH, Lo SY, Wang YL (2002) Glycothermal preparation of potassium niobate ceramic particles under supercritical conditions. Mater Lett 55:121–125.

Maeder MD, Damjanovic D, Setter N (2004) Lead free piezoelectric materials. J Electroceram 13:385–392.

Malič B, Bernard J, Holc J, Jenko D, Kosec M (2005a) Alkaline-earth doping in $(K,Na)NbO_3$ based piezoceramics. J Eur Ceram Soc 25:2707–2711.

Malič B, Bernard J, Holc J, Kosec M (2005b) Strontium doped $K_{0.5}Na_{0.5}NbO_3$ based piezoceramics. Ferroelectrics 314:149–156.

Matsubara M, Yamaguchi T, Sakamoto W, Kikuta K, Yogo T, Hirano S (2005a) Processing and piezoelectric properties of lead free $(K,Na)(Nb,Ta)O_3$. J Am Ceram Soc 88(5):1190–1196.

Matsubara M, Yamaguchi T, Kikuta K, Hirano S (2005b) Sintering and piezoelectric properties of potassium sodium niobate ceramics with newly developed sintering aid. Jpn J Appl Phy 44(1):258–263.

Matsubara M, Kikuta K, Hirano S (2005c) Piezoelectric properties of $(K_{0.5}Na_{0.5})(Nb_{1-x}Ta_x)O_3$-$K_{5.4}CuTa_{10}O_{29}$ ceramics. J Appl Phys 97:114105/1–114105/7.

Matsubara M, Yamaguchi T, Kikuta K, Hirano S (2005d) Effect of Li substitution on the piezoelectric properties of potassium sodium niobate ceramics. Jpn J Appl Phys 44(8):6136–6142.

Matsubara M, Yamaguchi T, Kikuta K, Hirano SI (2004) Sinterability and piezoelectric properties of $(K,Na)NbO_3$ ceramics with novel sintering aid. Jpn J Appl Phys 43(10):7159–7196.

Megaw HD, Wells M (1958) The Space Group of $NaNbO_3$ and $(Na_{0.995}K_{0.005})NbO_3$, Acta Cryst 11:858–862.

Murty SN, Bhanumathi A (1989) Phase transitions in europium doped $(Na,K)NbO_3$ ceramics. Ferroelectrics 94:285–290.

Murty SN, Umakantham K, Bhanumathi A (1988) Ferroelectric behaviour of lanthanum doped $(Na,K)NbO_3$ ceramics. Ferroelectrics 82:141–147.

Nazeri-Eshghi A, Kuang AX, Mackenzie JD (1990) Preparation and properties of $KNbO_3$ via the sol-gel method. J Mater Sci 25:3333–3337.

Nobre MAL, Lanfredi S (2003) Dielectric loss and phase transition of sodium potassium niobate ceramic investigated by impedance spectroscopy. Catal Today 78:529–538.

Nobre MAL, Longo E, Leite ER, Varela JA (1996) Synthesis and sintering of ultrafine $NaNbO_3$ powder by use of polymeric precursors. Mater Lett 28:215–220.

Oh SW, Akedo J, Park JH, Kawakami Y (2006) Fabrication and evaluation of lead-free piezoelectric ceramic LF4 thick film deposited by aerosol deposition method. Jpn J Appl Phys 45(9B):7465–7470.

Park SH, Ahn CW, Nahm S, Song JS (2004) Microstructure and piezoelectric properties of ZnO added $(Na_{0.5}K_{0.5})NbO_3$. Jpn J Appl Phys 43(8B):L1072–L1074.

Park HY, Ahn CW, Song HC, Lee JH, Nahm S (2006) Microstructure and piezoelectric properties of $0.95(Na_{0.5}K_{0.5})NbO_3$–$0.05BaTiO_3$ ceramics. Appl Phys Lett 89:062906/1–062906/3.

Pithan C, Shiratori Y, Dornseiffer J, Haegel FH, Magrez A, Wasser R (2005) Microemulsion mediated synthesis of nanocrystalline $(K_xNa_{1-x})NbO_3$ powders. J Cryst Growth 280:191–200.

Raevskiĭ IP, Reznichenko LA, Ivliev MP, Smotrakov VG, Eremkin VV, Malitskaya MA, Shilkina LA, Shevtsova SI, Borodin AV (2003) Growth and study of single crystals of $(Na,K)NbO_3$ solid solutions. Crystall Rep 48(3):486–490.

Reisman A, Holtzberg F (1955) Phase equilibria in the system K_2CO_3-Nb_2O_5 by the method of differential thermal analysis. J Am Ceram Soc 77:2115–2118

Reisman A, Holtzberg F, Triebwasser S, Berkenblit M (1956) Preparation of pure potassium metaniobate. J Am Ceram Soc 78:719–720.

Reisman A, Holtzberg F, Triebwasser S, Berkenblit M (1959) Metastability in niobate system. J Am Ceram Soc 81:1292–1295.

Ringgaard E, Wurlitzer T (2005) Lead-free piezoceramics based on alkali niobates. J Eur Ceram Soc 25:2701–2706.

Rojac T, Kosec M, Malič B, Holc J (2005a) Mechanochemical synthesis of $NaNbO_3$. Mater Res Bull 40:341–345.

Rojac T, Kosec M, Malič B, Holc J (2005b) Mechanochemical synthesis of $NaNbO_3$, $KNbO_3$ and $K_{0.5}Na_{0.5}NbO_3$. Sci Sintering 37:61–67.

Rojac T, Kosec M, Malič B, Holc J (2006) The application of a milling map in the mechanochemical synthesis of ceramic oxides. J Eur Ceram Soc 26:3711–3716.

Rojac T, Kosec M, Šegedin P, Malič B, Holc J (2007) The formation of a carbonate complex during the mechanochemical treatment of a Na_2CO_3-Nb_2O_5 mixture. Solid State Ionics 177: 2987–2995.

Roth RS (1980) Thermal stability of long range order in oxides. Prog Solid St Chem 13(2): 159–192.

Roth RS, Parker HS, Brower WS, Minor DB (1974) Alkali oxide-tantalum oxide and alkali oxide-niobium oxide ionic conductors. NASA Contract. Rep., Report No. NASA-CR-134599, Contract No. C-29933C. NASA, Cleveland, pp 1–59.

Ryu J, Choi JJ, Hahn BD, Park DS, Yoon WH, Kim HK (2007) Fabrication and ferroelectric properties of highly dense lead-free piezoelectric $K_{0.5}Na_{0.5}NbO_3$ thick films by aerosol deposition. Appl Phys Lett 90:1–3.

Saito Y, Takao H, Tani T, Nonoyama T, Takatori K, Homma T, Nagaya T, Nakamura M (2004) Lead free Piezoceramics. Nature 432:84–87.

Sano T, Saylor DM, Rohrer GS (2003) Surface energy anisotropy of $SrTiO_3$ at 1400 °C in air. J Am Ceram Soc 86(11):1933–1939.

Shafer MW, Roy R (1959) Phase equilibria in the system Na_2O-Nb_2O_5. J Am Ceram Soc 42(10): 482–486.

Shimojo Y, Wang R (2005) Dielectric and piezoelectric properties of $MeTiO_3$ (Me = Ba and Sr) modified $(K,Na)NbO_3$. J Kor Phy Soc 46(1):48–51.

Shirane G, Newnham R, Pepinsky R (1954) Dielectric properties and phase transition of $NaNbO_3$ and $(Na,K)NbO_3$. Phys Rev 96(3):581–588.

Shiratori Y, Magrez A, Pithan C (2004) Phase transformation of $KNaNb_2O_6$ induced by size effect. Chem Phys Lett 391:288–292.

Shiratori Y, Magrez A, Dornseiffer J, Haegel FH, Pithan C, Waser R (2005a) Polymorphism in micro-, submicro- and nanocrystalline $NaNbO_3$. J Phys Chem B 109:20122–20130.

Shiratori Y, Magrez A, Pithan C (2005b) Particle size effect on the crystal structure symmetry of $K_{0.5}Na_{0.5}NbO_3$. J Eur Ceram Soc 25:2075–2079.

Takao H, Saito Y, Aoki Y, Horibuchi K (2006) Microstructural evolution of crystalline-orientated $(K_{0.5}Na_{0.5})NbO_3$ piezoelectric ceramics with a sintering aid of CuO. J Am Ceram Soc 89(6):1951–1956.

Tang F, Du H, Li Z, Zhou W, Qu S, Pei Z (2006) Preparation and properties of $(K_{0.5}Na_{0.5})NbO_3$-$LiNbO_3$ ceramics. Trans Nonferrous Met Soc China 16:s466–s469.

Tashiro S, Ishii K (2006) Grain size and piezoelectric properties of $(Ba,K,Na)NbO_3$ lead-free ceramics. J Ceram Soc Jpn 114(5):386–391.

Tashiro S, Nagata K (2004) Infuence of mixing condition and nonstoichiometry on piezoelectric properties of $(K,Na,Pb)NbO_3$ ceramics. Jpn J Appl Phys 43(9B):6711–6715.

Tennery VJ, Hang KW (1968) Thermal and X-ray diffraction studies of the $NaNbO_3$-$KNbO_3$ system. J Appl Phys 39(10):4749–4753.

Wada S, Muraoka K, Kakemoto H, Tsurumi T, Kumagai H (2004) Enhanced piezoelectric properties of potassium niobate single crystals by domain engineering. Jpn J Appl Phys 43(9B):6692–6700.

Wang R, Xie R, Sekiya T, Shimojo Y, Akimune Y, Hirosaki N, Itoh M (2002) Piezoelectric properties of spark plasma sintered $(Na_{0.5}K_{0.5})NbO_3$-$PbTiO_3$ ceramics. Jpn J Appl Phys 41: 7119–7122.

Wang R, Xue RJ, Hanada K, Matsusaki K, Bando H, Itoh M (2005) Phase diagram and enhanced piezo-electricity in the strontium titanate doped potassium-sodium niobate solid solution. Phys Stat Sol 202(6):R57–R59.

Wang RP, Xie RJ, Sekiya T, Shimoyo Y (2004) Fabrication and characterization of potassium-sodium niobate piezoelectric ceramics by spark-plasma sintering method. Mater Res Bull 39:1709–1715.

Yang Z, Chang Y, Liu B, Wei L (2006) Effect of composition on phase structure, microstructure and electrical properties of $(K_{0.5}Na_{0.5})NbO_3$-$LiSbO_3$ ceramics. Mater Sci Eng A432:292–298.

Zang GZ, Wang JF, Chen HC, Su WB, Wang CM (2006) Perovskite $(Na_{0.5}K_{0.5})_{1-x}(LiSb)_x Nb_{1-x}O_3$ lead free piezoceramics. Appl Phys Lett 88:212908/1–212908/3.

Zhang BP, Li JF, Wang K, Zhang H (2006a) Compositional dependence of piezoelectric properties in $Na_xK_{1-x}NbO_3$ lead-free ceramics prepared by spark plasma sintering. J Am Ceram Soc 89(2):706–709.

Zhang S, Xia R, Shrout TR, Zang G, Wang J (2006b) Piezoelectric properties in perovskite $0.948(K_{0.5}Na_{0.5})NbO_3$–$0.052LiSbO_3$ lead-free ceramics. J Appl Phys100:104108/1–104108/6.

Zhen Y, Li JF (2006) Normal sintering of $(K,Na)NbO_3$-based ceramics: Influence of sintering temperature on densification, microstructure, and electrical properties. J Am Ceram Soc 89(12):3669–3675.

Zuo R, Roedel J, Chen R, Li L (2006) Sintering and electrical properties of lead-free $Na_{0.5}K_{0.5}NbO_3$ piezoelectric ceramics. J Am Ceram Soc 89:2010–2015.

Chapter 6
Bismuth-based Piezoelectric Ceramics

Tadashi Takenaka

6.1 Introduction

Important ferroelectric or antiferroelectric oxide ceramics for dielectric, ferroelectric, piezoelectric, electrostrictive, and/or pyroelectric applications are restricted to perovskite-type, tungsten bronze-type, and bismuth layer-structured compounds. A recent trend in the study on piezoelectric and/or pyroelectric ceramic compounds is the use of lead-free materials. The other trend is the use of grain orientation techniques as the ceramic fabrication methods.

Recently, bismuth layer-structured ferroelectrics (BLSF), which form one of BO_6 octahedral ferroelectric groups, have been extensively studied in the form of thin films because they seem to be an excellent candidate for nonvolatile FeRAM applications. For example, $SrBi_2Ta_2O_9$ shows fatigue-free properties, which are very desirable in such applications.

The BLSF family constitutes an important ferroelectric system for dielectric, ferroelectric, piezoelectric, and pyroelectric applications because ceramics such as $Bi_4Ti_3O_{12}$ (BIT), $PbBi_2Nb_2O_9$ (PBN), and $Na_{0.5}Bi_{4.5}Ti_4O_{15}$ (NBT) exhibit (Subbarao 1962; Ikegami and Ueda 1974; Takenaka and Sakata 1982) low dielectric constant ε_s, high Curie temperature T_c, and large anisotropy in electromechanical coupling factors k_t/k_p or k_{33}/k_{31} compared with the PZT family. Furthermore, these characteristics can be easily enhanced by texturing techniques such as the hot-forging (HF) method (Takenaka et al. 1976, 1977; Takenaka and Sakata 1980). Therefore, the grain-oriented BLSF ceramics seem to be a superior candidate for the lead-free or low-lead-content piezoelectric materials with high T_c and anisotropic characteristics (Takenaka and Sakata 1984, 1989), or for pyroelectric sensor materials with a large figure of merit (Takenaka and Sakata 1983, 1991). Moreover BLSF

T. Takenaka
Department of Electrical Engineering (EE), Faculty of Science and Technology, Tokyo University of Science, Yamazaki 2641, Noda, Chiba-ken 278-8510, Japan

A. Safari, E.K. Akdoğan (eds.) *Piezoelectric and Acoustic Materials for Transducer Applications.*

ceramics can be used as stable and high-Q piezoelectric materials (Nagata et al. 2003) to be operated at high temperatures and at high frequencies.

Piezoelectric properties of conventionally sintered BLSF ceramics, however, are not as large as that of single crystals (Cummins and Cross 1967, 1968). One reason for the low piezoelectric activity in such ceramics may be due to two-dimensional restriction on the permissible rotations of the spontaneous polarization by electric poling, in comparison with the three-dimensionally permitted ones in perovskite-type ferroelectrics. Thus, ferroelectric ceramics whose grain crystallites have such lower crystal symmetry as bismuth layer-type compounds may have an essentially weak point in that a satisfactorily large polarization cannot be obtained by electrical poling.

To solve this problem, new fabrication processes of piezoelectric ceramics are desirable through which the texture in the ceramic can be well controlled to give preferred grain orientations. There are usually two fabrication methods for orientating the grains in oxide ceramics: (1) Hot working technique, which makes use of the overturn of dislocations in the grains and of the slip in the grain boundaries at high temperatures, and (2) topotaxial reaction, which is based on the shape of the powder of the raw material.

The grain orientation technique, which is one of the several ceramic preparation methods with a controlled grain structure during the ceramic sintering process, gives the ceramics an anisotropic pseudo-single crystal behavior. Of many hot-working methods such as hot-forging, hot-rolling, hot-extrusion, and super plastic deformation, etc., the hot-forging (hereafter HF) method (Takenaka et al. 1976, 1977; Takenaka and Sakata 1980) seems to be the most advantageous grain orientation technique for lower symmetry materials with a large anisotropic crystal structure such as the bismuth layer-structured type and the tungsten bronze type.

The HF samples have a higher plastic deformation than those produced by hot-pressing. A slip between Bi_2O_2 layers and perovskite-like layers in the c-plane of BLSF ceramics occurs during the hot-forging process, and finally the c-axis is almost aligned parallel to the forging axis with a well-oriented grain structure. The HF-BLSF ceramics, whose spontaneous polarization lies in the grain-oriented c-plane normal to the forging axis, display the piezoelectric and pyroelectric properties that are enhanced by the grain orientation effects. The studies of the grain orientations in piezoelectric and pyroelectric ceramics have been systematically carried out for more than 30 years. The author of this chapter has been able to obtain ceramics with a high degree of preferred grain orientation by using HF in many BLSF, which have shown greatly enhanced dielectric, ferroelectric, piezoelectric, and pyroelectric properties.

Among a great number of the BLSF compounds, the following solid solution systems were selected in this chapter on the basis of the results of the electrical properties: BIT-based systems including $Bi_4Ti_{3-x}M_xO_{12}$ (M = Nb [BITN-x], V [BITV-x]) and $(Bi_{4-y}Nd_y)_{1-(x/12)}(Ti_{3-x}V_x)O_{12}$ [BNTV-y], $(Na_{1/2}Bi_{1/2})_{1-x}Ca_xBi_4Ti_4O_{15}$ [NCBT-100x], $SrBi_2Nb_{2-x}V_xO_9$ [SBNV], and $(Sr_{1-x}Ca_x)_2Bi_4Ti_5O_{18}$ [SCBT], where the abbreviation of the respective systems is given in [ ].

This chapter describes the effects of grain orientation on electrical properties such as dielectric, ferroelectric, and piezoelectric properties of some grain-oriented BLSF ceramics by the hot-forged (HF) method, comparing them with ordinarily fired (OF) nonoriented ones (Takenaka 2002, 2006).

6.2 Bismuth Layer-Structured Ferroelectrics

Fifty years ago, Aurivillius (Aurivillius 1949, 1950) discovered a family of bismuth compounds with a layer structure, and to date more than 50 compounds belonging to this group have been widely investigated by many workers (Smolenskii et al. 1959; Subbarao 1961, 1962a; Fang et al. 1962; Aurivillius and Fang 1962; Newnham et al. 1971; Armstrong and Newnham 1972). Especially the dielectric properties have been examined in detail by Subbarao (Subbarao 1962b), and ferroelectric and piezoelectric studies by some Japanese scientists (Ikegami and Ueda 1974; Takenaka and Sakata 1982, 1984, 1989).

The bismuth layer-structured oxides (BLSO) have a generalized chemical formula given by $A_{m-1}Bi_2B_mO_{3m+2}$, where A is mono-, di-, trivalent ions, or a mixture of them and B is tetra-, penta-, hexavalent ions, or a mixture of them, and the integer m takes any of the values from 1 to 5. A large number of them are found to be ferroelectric. Examples of their crystal structures are shown in Fig. 6.1, where A shows $(A_{m-1}B_mO_{3m+1})$, which consists of $(m-1)$ layers of perovskite-like units (B) and

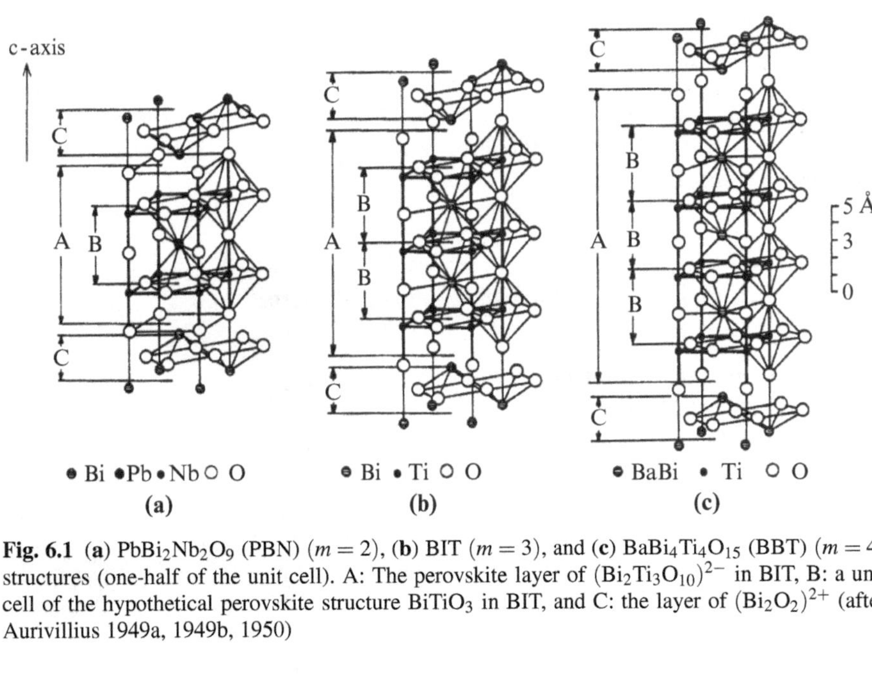

Fig. 6.1 (**a**) $PbBi_2Nb_2O_9$ (PBN) ($m = 2$), (**b**) BIT ($m = 3$), and (**c**) $BaBi_4Ti_4O_{15}$ (BBT) ($m = 4$) structures (one-half of the unit cell). A: The perovskite layer of $(Bi_2Ti_3O_{10})^{2-}$ in BIT, B: a unit cell of the hypothetical perovskite structure $BiTiO_3$ in BIT, and C: the layer of $(Bi_2O_2)^{2+}$ (after Aurivillius 1949a, 1949b, 1950)

is sandwiched between (Bi_2O_2) layers (C). Bismuth titanate, BIT (Fig. 6.1b), is a prototype of many bismuth layer structure oxides with the simplest chemical composition ($m = 3$) and has a high Curie point of 680 °C (Cummins and Cross 1967, 1968).

BLSO have BO_6 octahedra-like perovskite, pyrochlore, and tungsten-bronze structures and many of them are ferroelectric. An electrically neutral condition gives the following formula $(m-1)N(\mathrm{A}) + m\,N(\mathrm{B}) = 6m$, where $N(\mathrm{A})$ and $N(\mathrm{B})$ indicate the mean valencies of the A and the B ions, respectively. However, BLSO have a great variety of available combinations of A and B ions, respectively. For example, the basic BLSF are BIT ($m = 3$), PBN ($m = 2$), Bi_3TiNbO_9 (BTN) ($m = 2$), NBT ($m = 4$), $PbBi_4Ti_4O_{15}$ (PBT) ($m = 4$), and $Pb_2Bi_4Ti_5O_{18}$ (P2BT) ($m = 5$). Many derivative solid solutions were also studied, mainly in grain-oriented piezoelectric ceramics.

6.3 Grain Orientation Method and Electrical Measurements

6.3.1 Ceramic Sample Preparations

Ordinarily fired (OF) and randomly oriented ceramic samples were prepared by the conventional ceramic processing. Reagent grade oxide or carbonate powders with 99 + % purity were used as starting materials. These materials were mixed by ball milling and were calcined at 600–1,000 °C for 1–2 h. After calcining, the ground and ball-milled powders were pressed into a disc of 20 mm diameter and 10 mm thickness by uniaxial pressing, followed by cold isostatic pressing (CIP) at 150 MPa. Pressed discs were finally sintered at 900–1,250 °C for 1–2 h in air. The weight loss during the sintering process was assessed by a TG-DTA spectrometer. Samples were polished and thermally etched for microstructures analysis by scanning electron microscopy.

6.3.2 Grain Orientation by Hot-Forging Method

Grain-oriented samples were prepared by the hot-forging (HF) method as described in detail elsewhere (Takenaka et al. 1976, 1977; Takenaka and Sakata 1980). Cylinder specimens 15 mm in diameter and 30–40 mm in height were prepressed in a steel die at a pressure of $800\,\mathrm{kg/cm^2}$, and were used in the HF process. The prepressed cylinders (green body) were sandwiched between two platinum plates to prevent the sample from reacting with the alumina plungers. A uniaxial compressive load of $200\,\mathrm{kg/cm^2}$ acting over an area of $177\,\mathrm{mm^2}$ was applied along the thickness of the sample (the forging axis) using oil pressure.

After ordinary firing at 1,000–1,200 °C for 2 h, the sample was gradually pressed by two alumina plungers, since abrupt pressing will produce cracks in the sample. Then a constant pressure and a maximum temperature were maintained. Four conditions are most critical: (1) the maximum temperature, (2) the maximum total pressure, (3) the soaking time when maximum pressure is imposed, and (4) annealing time.

A degree of deformation in hot-forged (HF) ceramics is expressed in terms of an area ratio γ, and a thickness contraction rate λ, i.e., $\gamma = S_f/S_i$, and $\lambda = (h_i - h_f)/h_i$, where S_i and S_f are cross-section areas, and h_i, h_f thickness, of pre-sintered and hot-forged samples, respectively. The value of h_i is estimated by a relation of $h_i = (1-\alpha)h_{gb}$, where h_{gb} is the thickness of the green body and α is the linear shrinkage of the ordinarily fired pellet, which was sintered at the same temperature.

The plastic deformations, namely γ and λ, depend heavily on the forging temperature and the composition of the sample. For example, HF BLSF ceramics showed an area ratio, λ, of about 2–10, and a thickness contraction rate, λ, of about 50–90%, respectively. A measured density of HF samples was higher than 95% of the theoretical X-ray density, and it increased to 99% if further hot-pressing was applied after the hot-forging.

The grain orientation factor, (F), can be estimated by comparing the X-ray diffraction patterns with those of nonoriented (OF) ceramics as follows (the Lotgering method) (Lotgering 1959). First, a ratio p of the sum of the intensities $I(00l)$ of the $(00l)$ reflections to the sum of the intensities $I(hkl)$ including $(00l)$ is calculated over the range of 2θ. Namely, it is

$$p = \frac{\Sigma I(00l)}{\Sigma I(hkl)}. \tag{6.1}$$

As the degree of orientation is increased, the value of p is increased from p_0 (the value for a nonoriented case, $p_0 = 0.1$–0.3, depends on samples) to 1 (a value for a completely oriented case). Hence, if F is defined as

$$F = \frac{p - p_0}{1 - p_0}, \tag{6.2}$$

F varies from 0 (nonoriented) to 1 (completely oriented), and F may be used as a measure of the grain orientation.

6.3.3 Electrical Measurements

The sintered ceramics were cut into discs and polished. Electrodes were applied using fired-on silver–palladium (Ag-Pd) paste for studying the temperature dependence of dielectric properties. The anisotropies in the electrical properties of HF samples were measured by applying an electric field in perpendicular [$\perp$] and parallel [//] directions to the forging axis, that is, the electrodes of the HF samples (Takenaka and Sakata 1980, 1984) were adhered to the faces to which were applied electric fields perpendicular and parallel to the forging axis, respectively. The temperature dependence of dielectric constant, (ε_s), and dielectric loss tangent, ($\tan\delta$), were measured at 1 or 10 MHz by an automatic dielectric measurement system with a multifrequency LCR meter in the temperature range from RT to 800 °C.

Temperature dependence of resistivity, ρ, was measured by using a high-resistance meter. Resistivity measurements were made on unpoled specimens. In general, conductivity of most dielectric materials increases with increasing temperature. It is essential for the high-temperature poling process that the conductivity of the material should somehow be suppressed; otherwise, it becomes very difficult to apply a larger electric field at high temperatures.

Also, piezoceramics having high Curie temperatures or large coercive fields such as BLSF or $PbTiO_3$ must have high resistivity at high temperatures to avoid thermal breakdown during the poling process. The remnant polarization, (P_r), and coercive field, (E_c), of the ferroelectric properties were determined from the *D–E* hysteresis loop, obtained using a standard Sawyer-Tower circuit at 50 Hz or a ferroelectric tester with a high-voltage amplifier.

Specimens for piezoelectric measurements were poled in a stirred silicone oil bath and at $E_p = 3$–12 kV/mm, $T_p = 100$–250 °C, and $t_p = 5$–30 min. The piezoelectric properties of the poled ceramics were measured by a resonance–antiresonance method on the basis of IEEE standards, using a precision impedance analyzer. A longitudinal vibration of the (33) mode was measured using a rectangular specimen of $4 \times 2 \times 2\,\mathrm{mm}^3$. The electromechanical coupling factor, (k_{33}), was calculated using series and parallel resonance frequencies, f_s and f_p. The θ_{max} is the maximum phase of the inductance (L) between f_r and f_a. The $Q_{e\text{-}max}$ is defined by the tangent of the maximum phase, θ_{max}.

In the case of the HF sample, the poling direction is perpendicular to the forging axis, as shown in Fig. 6.2, with the (33), (15), and (24) mode specimens in the HF sample. The temperature dependences of piezoelectric properties were measured in the temperature range from −25 °C to 125 °C by using a temperature controller. The temperature coefficient of the resonance or antiresonance frequencies, *TCf*, is defined as follows:

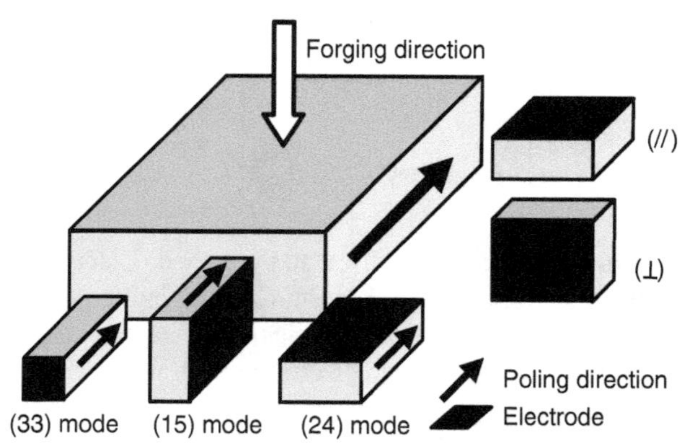

Fig. 6.2 HF sample of (33), (15), and (24) modes, and (//) or (⊥) direction

$$TCf = \frac{f_{[125^\circ\mathrm{C}]} - f_{[-25^\circ\mathrm{C}]}}{f_{[20^\circ\mathrm{C}]} \times \Delta T}, \tag{6.3}$$

where, $f_{[-25^\circ\mathrm{C}]}$ and $f_{[125^\circ\mathrm{C}]}$ are the resonance frequencies at $-25\,^\circ$C and $125\,^\circ$C, respectively, and ΔT is the measured temperature range ($150\,^\circ$C in this study).

6.4 Electric Properties of Bismuth-based Ceramics

6.4.1 BIT-based Ceramics

6.4.1.1 $Bi_4Ti_{3-x}M_xO_{12}$ (M = Nb [BITN-x] and V [BITV-x]) ($0 \le x \le 0.2$, $x = 0$: BIT)

Bismuth titanate, BIT, is a typical well-known BLSF (Subbarao 1962a, 1962b; Armstrong and Newnham 1972; Cross and Pohanka 1971; Smolenskii et al. 1959, 1961), and the BIT single crystal has a good piezoelectricity. However, it is difficult to measure piezoelectric properties on the BIT single crystal because the shape of the grown BIT single crystals is always in the form of very thin platelets. On the other hand, fully reliable piezoelectric properties of BIT ceramics have not been reported because of some problems such as the low resistivity and the large coercive field (Ikegami and Ueda 1974; Takenaka and Sakata 1982, 1980; Cummins and Cross 1968; Takeuchi et al. 1999; Nagata et al. 2000; Dorrian et al. 1971; Villegas et al. 1999). To solve these problems, Nb^{5+} and V^{5+} ions were doped into BIT ceramic to obtain higher resistivities (Shulman et al. 1996; Noguchi and Miyayama 2001). Recently, $Bi_4Ti_{3-x}Nb_xO_{12}$ [BITN-x] and $Bi_4Ti_{3-x}V_xO_{12}$ [BITV-x] ceramics (Nagata et al. 2004) are studied for their dielectric, ferroelectric, and piezoelectric properties. Furthermore, the grain orientation effects of BITN and BITV ceramics on their piezoelectric properties are discussed using the textured ceramics prepared by HF.

X-ray diffraction patterns for BITN and BITV ceramics (OF) show single phase of bismuth layer-structured compounds with $m = 3$. No peaks of Nb_2O_5 and V_2O_5 were observed for $x \le 0.12$. The density of both BITN and BITV ceramics was higher than 95% of the X-ray density. Figure 6.3a shows the Curie temperature, T_c, as a function of Nb and V concentration. The T_c of the BIT ($x = 0$) ceramic is $683\,^\circ$C, and gradually decreases lower with increasing Nb and V concentration. Therefore, it is thought that the Nb and V ions occupy the B-site of the pseudoperovskite cell in the bismuth layer-structure. However, the T_c of BITV has the tendency to saturate with increasing V concentration, compared to the BITN ceramics. Therefore, it is thought that V ions are difficult to substitute for Ti ions. Excess V ions seem to exist on grain boundaries or the triple point when the V concentration increases. This consideration arises from comparison of ionic radii of Ti, Nb, and V ions. The ionic radii of Ti^{4+} (IV coordination), Nb^{5+} (IV coordination), and V^{5+} (IV coordination) are 0.605, 0.64, and 0.54 Å, respectively (Shannon 1976). It is thought that the ionic radius of the V ion is too small to substitute on the B-site of

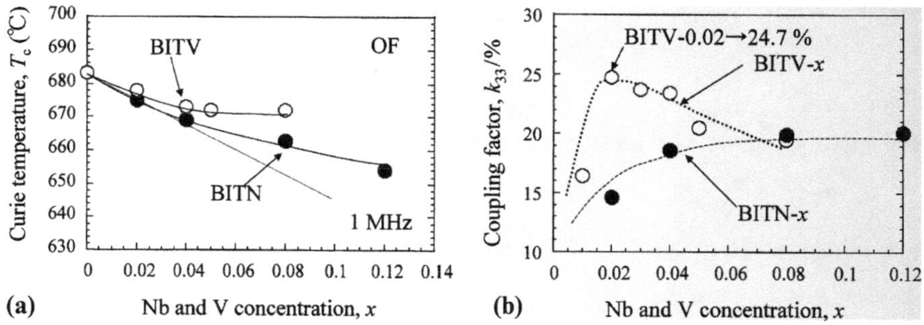

Fig. 6.3 (**a**) Curie temperature, T_c, of BITN and BITV ceramics as a function of Nb and V concentrations, (**b**) Electromechanical coupling factor, k_{33}, of BITN (OF) and BITV (OF) as a function of the dopant concentration (x)

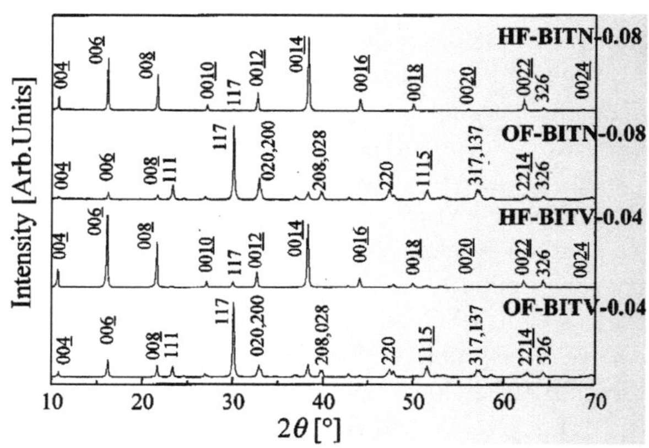

Fig. 6.4 X-ray diffraction patterns of BITN-0.08 and BITV-0.04 ceramics for the perpendicular plane (polished) of the HF and the OF

the pseudoperovskite cell in the bismuth-layered structure. The resistivity, ρ, of BIT $(x = 0)$ is about 10^{10}–$10^{11}\Omega\cdot$cm, and, those of BITN and BITV ceramics are about 10^{13}–$10^{14}\Omega\cdot$cm. It is clear that ρ is enhanced by some donor dopings. The optimum charge neutrality was observed for each composition of BITN-0.08 and BITV-0.01, respectively.

Figure 6.3b shows the compositional dependences of the electromechanical coupling factor, (k_{33}), on the dopant Nb and V concentration (x). The k_{33} of BITN reaches the saturated value (0.20) for the composition of $x = 0.08$, while the k_{33} of BITV-0.02 is 0.25, which is a high value for BLSF ceramics with random orientations. Figure 6.4 compares X-ray diffraction patterns of BITN-0.08 and BITV-0.04 ceramics for the perpendicular plane (polished) of the HF ceramics and the OF ones. It is very clear that grains in the HF ceramics were oriented along the c-axis because the intensities of the $(00l)$ planes of the HF samples are very high. The grain orientation factor (F) of BITN-0.08 and BITV-0.04 was 0.91 and 0.75, respectively.

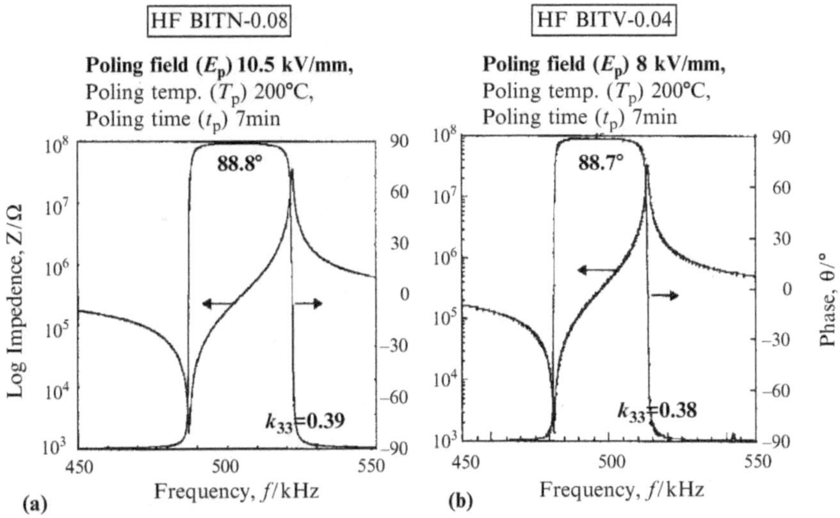

Fig. 6.5 Frequency dependence of the impedance, Z (magnitude $|Z|$, and phase θ), of (**a**) HF BITN-0.08 and (**b**) HF BITV-0.04

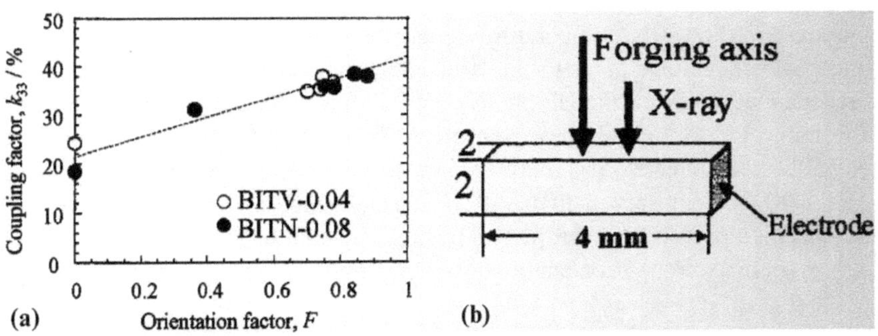

Fig. 6.6 (**a**) Coupling factor, k_{33}, of HF BITN-0.08 and BITV-0.04 as a function of the orientation factor, F, (**b**) shape for measuring of the orientation factor, F, and the coupling factor, k_{33}

Figure 6.5 shows the frequency dependence of the impedance, Z (magnitude $|Z|$, and phase θ), of HF BITN-0.08 and HF BITV-0.04 ceramics. BITN-0.08 samples were poled with 10.5 kV/mm at 200 °C for 7 min. On the other hand, BITV-0.04 samples were poled with 8 kV/mm at the same temperature, and for the same time as before. Good profiles were obtained, and their k_{33} values were enhanced to 0.39 and 0.38 for BITN-0.08 and BITV-0.04, respectively. These values are about twice as large as those of nonoriented (OF) ones, and are larger than the reported value ($k_{33} = 0.27$) (Takenaka and Sakata 1986) of HF BIT ceramics. In this investigation, a saturated k_{33} value of the BIT ceramic was not observed because of electrical breakdown during the poling.

Figure 6.6a shows the k_{33} of the HF BITN-0.08 and the BITV-0.04 as a function of the orientation factor (F). Accurate F was obtained using the X-ray diffraction

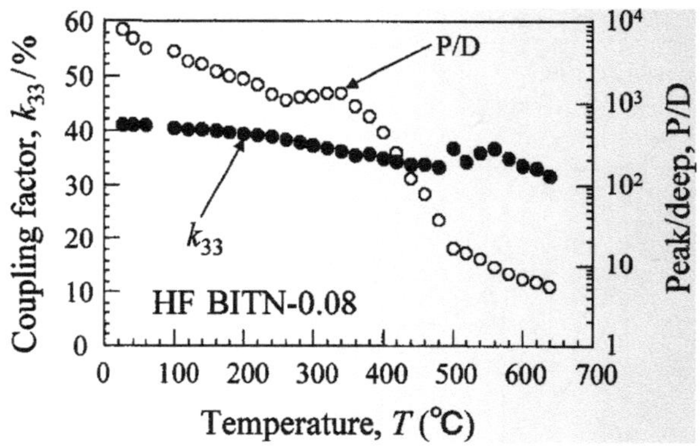

Fig. 6.7 Temperature dependence of the k_{33} and the ratio of impedance peak/deep, P/D, of the HF BITN-0.08

patterns for the k_{33} specimen, that is, XRD was performed directly on the k_{33} specimen shown in Fig. 6.6b. From this measurement, the relationship between k_{33} and F is clear: k_{33} increases linearly with the increasing orientation factor (F). From this figure, the k_{33} value for the specimen with a perfect orientation $(F = 1)$ could be extrapolated to be almost 0.42.

Figure 6.7 shows the temperature dependence of the k_{33} and the ratio (P/D) of the peak impedance at f_a to the deep at f_r, obtained from the resonance and antiresonance curves for the HF BITN-0.08 ceramic. The k_{33} was higher than 0.35 in the range RT to 650 °C. However, the P/D decreased rapidly at temperatures over 350 °C. In other words, sharp resonance and antiresonance peaks with high $P/D > 1{,}000$ were observed up to 350 °C. It is clear that the HF BITN-0.08 ceramic maintains high piezoelectric properties from RT up to 350 °C. The donor-doped BIT ceramics seem to be superior candidates for lead-free high-temperature applications.

6.4.1.2 $(Bi_{4-y}Nd_y)_{1-\delta}(Ti_{3-x}V_x)O_{12}$ (BNTV-y)

The piezoelectric properties of Nd and V cosubstituted bismuth titanate ceramics, $(Bi_{4-y}, Nd_y)_{1-(x/12)}(Ti_{3-x}, V_x)O_{12}$ [BNTV-y ($x = 0.01$, $y = 0.00$–1.00)], $Bi_{4-y}Nd_yTi_3O_{12}$ [BIT-Nd ($y = 0.75$)] and $Bi_{4-x/3}Ti_{3-x}V_xO_{12}$ [BIT-V ($x = 0.01$)], were investigated (Matsuzawa et al. 2006; Nagata et al. 2006). Furthermore, the grain orientation effects of BNTV-y ceramics on their piezoelectric properties were studied by the hot-forging (HF) method. The electromechanical coupling factor, k_{33}, of ordinarily fired (OF) BNTV-y0.75 ceramic was 0.21, which is one of the highest k_{33} values for BLSFs. On the other hand, the k_{33} of the grain oriented (HF) BNTV-y0.25 ceramic was 0.30.

X-ray diffraction patterns for (OF) BIT-Nd, BIT-V, and BNTV ceramics show single phase of bismuth layer-structured compounds with the layer number $m = 3$. In BNTV-y ($y = 0.0$–1.0) ceramics, the lattice distortion ratio of a- and b-lattice parameters, a/b, decreased to $a/b - 1$ with increasing amount of Nd. This tendency was in good agreement with the results for rare-earth-substituted bismuth titanates

Table 6.1 Sintering temperatures and resistivity of BNTV ceramics

	BIT	BIT-Nd (y0.75)	BIT-V ($x = 0.01$)	BNTV-y0.75
Sintering Temp. (°C)	1,100	1,150	1,010	1,090
Resistivity $\rho\,(\Omega \cdot \mathrm{cm})$	1.6×10^{10}	1.83×10^{13}	4.57×10^{13}	8.32×10^{13}

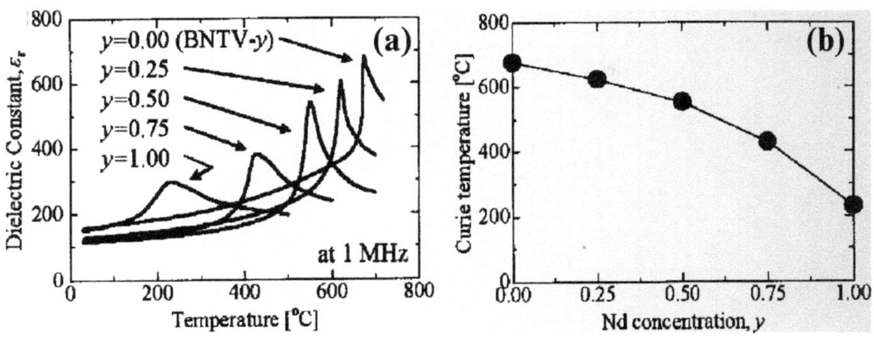

Fig. 6.8 (**a**) Temperature dependence of dielectric constant, ε_r, for BNTV-y ceramics measured at 1 MHz, (**b**) Curie temperature, T_c, of BNTV-y ceramics as a function of Nd concentration

reported by Wolfe and Newnham (1969). Prepared samples have high relative density, typically more than 95% of the theoretical density.

Table 6.1 shows sintering temperatures and resistivities of BIT, BIT-Nd, BIT-V, and BNTV-y0.75 ceramics. The resistivities, ρ, of (OF) BIT-Nd, BIT-V, and BNTV ceramics were about 10^{13}–$10^{14}\Omega \cdot \mathrm{cm}$ at room temperature. It is thought that the cause of the low resistivity of BIT ceramic is the Bi vaporization during sintering at a temperature around 1,100 °C during sintering. As V substitution has the effect of lowering the sintering temperature, Bi vaporization is expected to be suppressed (Cummins and Cross 1967). On the other hand, the sintering temperature of Nd-substituted BIT (BIT-Nd) was 1,150 °C, and yet the BIT-Nd ceramic maintained high resistivity (about $10^{13}\Omega \cdot \mathrm{cm}$). This result implied that Nd ions preferentially substitute for unstable Bi ions, which are easily vaporized from the perovskite block. The optimum charge neutrality compositions were observed for $x = 0.01$ (BIT-V) and $y = 0.75$ (BNTV-y).

Figure 6.8a shows the temperature dependence of the dielectric constant (ε_r) for BNTV-y ceramics measured at 1 MHz. Figure 6.8b shows the Curie temperature, T_c, of BNTV-y ceramics as a function of Nd concentration. T_c decreased with increasing amount of Nd, which also agreed with the result for rare-earth-substituted bismuth titanates (Wolfe and Newnham 1969). It is thought that the decrease of T_c was attributed to the decrease in the lattice distortion ratio (a/b).

Figure 6.9 shows the frequency dependence of impedance Z (magnitude, $|Z|$, and phase, θ) for (OF) BIT-Nd, BIT-V, and BNTV-y0.75 ceramics, respectively. Using this data set, it was found that the k_{33} of (OF) BNTV-y0.75 ceramics increased with increasing amount of Nd, except when $y = 1.00$ as shown in Fig. 6.10. The

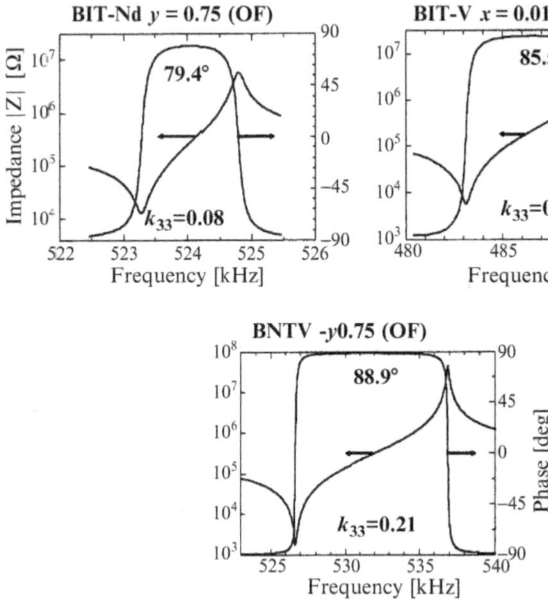

Fig. 6.9 Frequency dependence of impedance Z (magnitude $|Z|$, and phase θ) for (OF) BIT-Nd, BIT-V, and BNTV-y0.75 ceramics

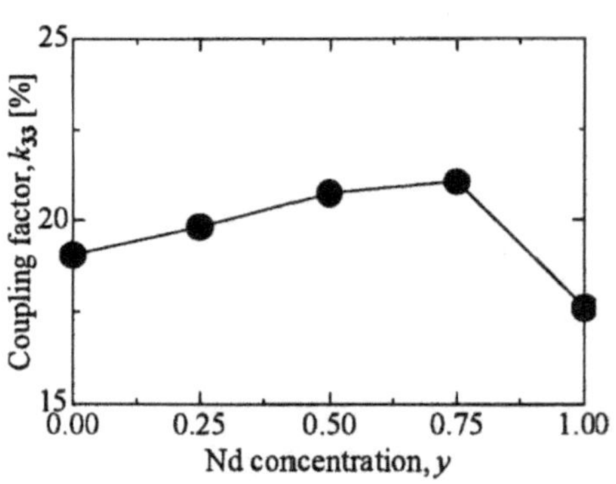

Fig. 6.10 Electromechanical coupling factor, k_{33}, of (OF) BNTV-y ceramics as a function of Nd concentration

largest k_{33} of BNTV-y ceramic was 0.21 at $y = 0.75$, which is one of the highest k_{33} values of (OF) BLSF ceramics. However, the k_{33} of the BIT-Nd ceramic was 0.08. These results suggest that domain movement in the BIT-Nd ceramic is more difficult than that in BIT-V and BNTV ceramics. It is inferred that these differences are due to defect formation occurring at the sintering temperature (~1,150°C). In other words, as Bi ions are vaporized from the BIT-Nd ceramic during the sintering process, vacancies are found and domain pinning occurred in the BIT-Nd ceramic more easily than in BIT-V and BNTV ceramics. Bi vaporization also causes oxygen vacancies, which cause low resistivity at high temperatures.

Table 6.2 Piezoelectric constants for (OF) and (HF) BNTV-y0.25 and BNTV-y0.75 ceramics

	BNTV-y0.25		BNTV-y0.25	
	OF	HF	OF	HF
F (%)	–	74.4	–	70.5
K_{33}	0.19	0.30	0.21	0.28
θ_{max}	87.3	88.2	88.9	88.4
Q_m	3,000	1,800	4,200	2,200
d_{33} [pC/N]	18.5	29.2	19.5	28.4

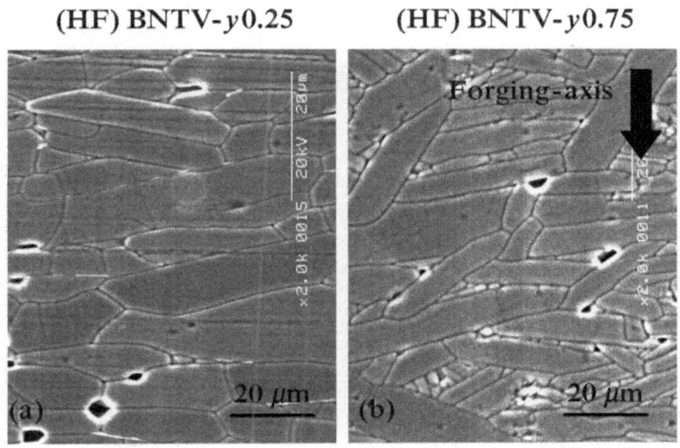

Fig. 6.11 SEM micrographs of (**a**) HF BNTV-y0.25 and (**b**) HF BNTV-y0.75 ceramics

Furthermore, the maximum phase (θ_{max}) in the frequency dependence of impedance is one of a standard value for the degree of full poling. The θ_{max} of BIT-Nd, BIT-V, and BNTV-y0.75 ceramics was 79.4°, 85.5°, and 88.9°, respectively.

The BNTV-y0.75 ceramic was easier to pole than BIT-Nd and BIT-V ceramics. Table 6.2 shows piezoelectric constants of ordinarily fired (OF) and hot-forged (HF) samples. The k_{33} of (HF) BNTV-y0.25 and BNTV-y0.75 ceramics was 0.30 and 0.28, respectively. The k_{33} of (HF) samples is larger than that of (OF) ones. However, the orientation factors are not substantially high.

Figure 6.11 shows SEM micrographs of the plan parallel (//) to the forging axis of HF-BNTV-y0.25 and BNTV-y0.75 ceramics sintered at 1,100 °C. The grain orientation factors, F, for BNTV-y0.25 and BNTV-y0.75 ceramics were 71.9 and 69.5%, respectively. The values of F were insufficient. As shown in Fig. 6.12b, the (HF) BNTV-y0.75 ceramic has numerous small grains less than 5 µm. It is considered that small grains prevent large grains from being orienting. On the other hand, the number of small grains in the (HF) BNTV-y0.25 ceramic is fewer than those of the (HF) BNTV-y0.75 ceramic (Fig. 6.12a). This is because the optimum sintering temperature for BNTV-y ceramics is higher for increasing amounts of Nd. Therefore, it is necessary to carry out sintering over 1,100 °C to obtain high F, but Bi vaporization becomes a problem.

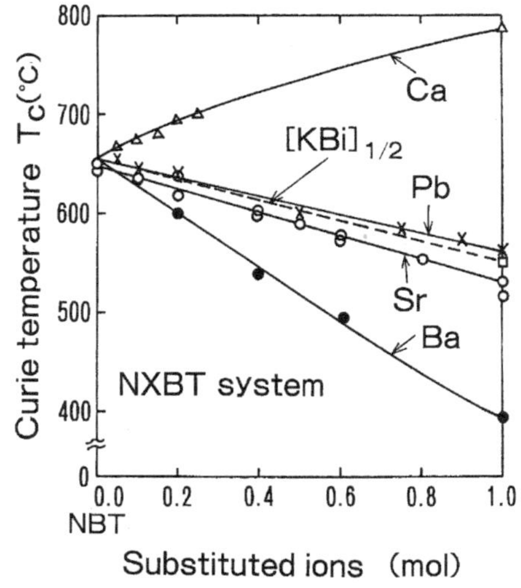

Fig. 6.12 Curie temperature, T_c, of the $(Na_{1/2}Bi_{1/2})_{1-x}M_xBi_4Ti_4O_{15}$ (M=Ba, Ca, Pb, Sr, and $K_{1/2}Bi_{1/2}$) [NXBT] system as a function of the amount (x) of the substituted ions

The piezoelectric properties of Nd and V cosubstituted bismuth titanate ceramics, $(Bi_{4-y},Nd_y)_{1-(x/12)}(Ti_{3-x},V_x)O_{12}$ [BNTV-y ($x = 0.01$, $y = 0.00$–1.00)], were investigated. The k_{33} of ordinarily fired (OF) BNTV-y0.75 ceramic was 0.21, which is one of the highest values for (OF) BLSFs. Furthermore, the grain orientation effects of BNTV-y ceramics on their piezoelectric properties were studied by hot forging (HF). The k_{33} of the (HF) BNTV-y0.25 ceramic was 0.30.

Vanadium (V) substitution has the effect of lowering the sintering temperature. This is expected to suppress the Bi vaporization, which causes low resistivity, ρ. On the other hand, the BIT-Nd ceramic maintained high ρ (about $10^{13}\Omega$ cm). This result implies that Nd ions preferentially substitute for unstable Bi ions, which are easily vaporized from the perovskite block. The k_{33} and ρ of BNTV-y0.75 ceramics are larger than those of BIT-Nd and BIT-V ceramics. Therefore, the effect of Nd and V cosubstitution is advantageous.

6.4.1.3 $(Na_{1/2}Bi_{1/2})_{1-x}Ca_xBi_4Ti_4O_{15}$ [NCBT-100x]

Among a great number of the BLSF compounds, the NBT ($m = 4$) ceramic displays excellent piezoelectric properties (Ikegami and Ueda 1974; Takenaka and Sakata 1985, 1988, 1989). On the other hand, substituted perovskite compounds by Ca for A-site ions are expected to have large anisotropic piezoelectric properties because the calcium ionic radius is the smallest in the A-site ion group among the perovskite compounds. Because of an enhanced anisotropy, the electromechanical coupling factor k_p or k_{31} in the radial or transverse mode of 24 mol% calcium-modified $PbTiO_3$ (PT) ceramics is almost negligible (Yamashita et al. 1983), in contrast with the large coupling factor k_t or k_{33} in the thickness expansion or longitudinal expansion mode, resulting in a very large anisotropy in the coupling factor, k_t/k_p or k_{33}/k_{31}. The Ca-modified NBT, namely, $(Na_{1/2}Bi_{1/2})_{1-x}Ca_xBi_4Ti_4O_{15}$

Table 6.3 Elastic, piezoelectric, and dielectric (EPD) constants of the ordinarily fired (OF) and the hot-forged (HF) $(Na_{1/2}Bi_{1/2})_{1-x}Ca_xBi_4Ti_4O_{15}$ (NCBT-100x) ceramics

Sample	NBT + Mn (0.1)		NCBT-5 + Mn(0.1)		NCBT-5 + Mn (0.2)	
	HF	OF	HF	OF	HF	OF
Grain orientation factor F_i	0.98	–	0.91	–	0.93	–
Measured density ρ_0 (g/cm^3)	7.29	7.01	6.92	6.88	7.20	6.65
Curie temperature T_c (°C)	660	658	665	680	–	677
Dielectric constant $\varepsilon_{33}{}^T/\varepsilon_0$	149	140	134	148	132	135
Coupling factor						
k_{33} (%)	32.5	14.7	35.6	16.1	40.0	18.8
k_{31}	2.8	3.3	2.75	2.85	2.34	2.39
Frequency constant						
N_{33} (Hz m)	2,170	2,000	2,260	1,950	2,290	1,950
N_{31}	2,130	1,990	2,180	1,980	2,170	1,970
Piezoelectric constant						
d_{33} (10^{-12} C/N)	33.7	15.6	34.9	18.3	38.3	20.8
d_{31}	2.80	3.49	2.61	3.15	2.17	2.58
g_{33} (10^{-3} V·m/N)	25.6	12.6	29.4	13.9	32.8	17.4
g_{31}	2.12	2.81	2.20	2.40	1.86	2.16
d_hg_h (10^{-15} m^2/N)	600	60	742	109	988	205
Elastic constant						
$s_{33}{}^E$ (10^{-12} m^2/N)	817	9.13	8.11	9.84	7.85	102
$s_{11}{}^E$	7.56	9.01	7.58	9.32	7.38	9.72

[NCBT-100x] ceramics can be expected to show the most excellent piezoelectric and/or pyroelectric properties. The HF NCBT-5 was almost smoothly deformed at temperatures of 1,100–1,150 °C, and a thickness contraction rate λ of about 90% was obtained with good reproducibility. The grain orientation factor F of the Mn-doped NCBT system is more than 90%. The F values are given in Table 6.3, along with other relevant data.

Figure 6.12 shows the Curie temperature (T_c) of $(Na_{1/2}Bi_{1/2})_{1-x}M_xBi_4Ti_4O_{15}$ (M = Ba, Ca, Pb, Sr, and $K_{1/2}Bi_{1/2}$) [NXBT] system as a function of dopant concentration (x). The T_c, except the case of M = Ca [NCBT], decreases as the concentration, x, of the substituted ions increases. On the other hand, the T_c of M = Ca [NCBT] increases as the Ca concentration increases. This suggests (Takenaka and Sakata 1988) that the T_c is related to the change in the lattice distortion b/a as mentioned previously. The T_c of $CaBi_4Ti_4O_{15}$ [CBT] ($x = 1$: Ca substituting for $Na_{1/2}Bi_{1/2}$) reaches temperatures as high as 785 °C.

The electromechanical properties of poled ferroelectric ceramics depend strongly on the poling conditions, namely, poling temperature, poling field, and poling time. The high-temperature poling seems to be very effective for activating piezo- and pyroceramics with a high Curie temperature or large coercive field such as BLSF or $PbTiO_3$-based materials. To pole at high temperatures, the BLSF ceramics must have high resistivity to avoid thermal breakdown. Empirical data show that Mn is a very effective dopant in reducing conductivity at high temperatures.

Since the grain orientation effects on the piezoelectric properties by HF give almost proportional enhancements to piezoelectric properties compared with those of

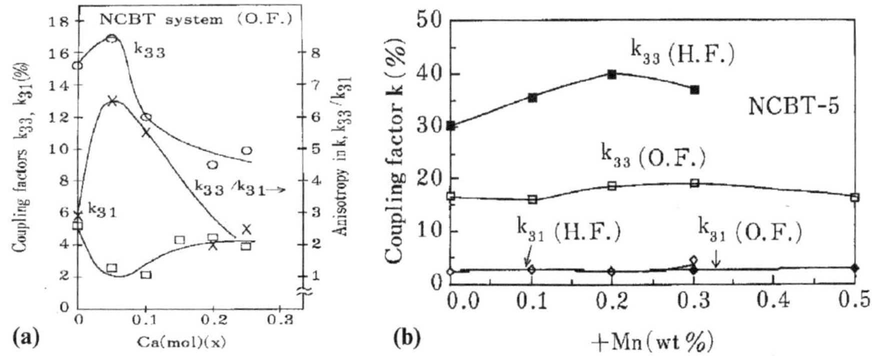

Fig. 6.13 Electromechanical coupling factors, k_{33} and k_{31}, for each vibration mode and the anisotropy k_{33}/k_{31} in the k of the (**a**) NCBT system (OF) as a function of Ca mol (x), and (**b**) NCBT-5 (HF and OF) as a function of Mn doping

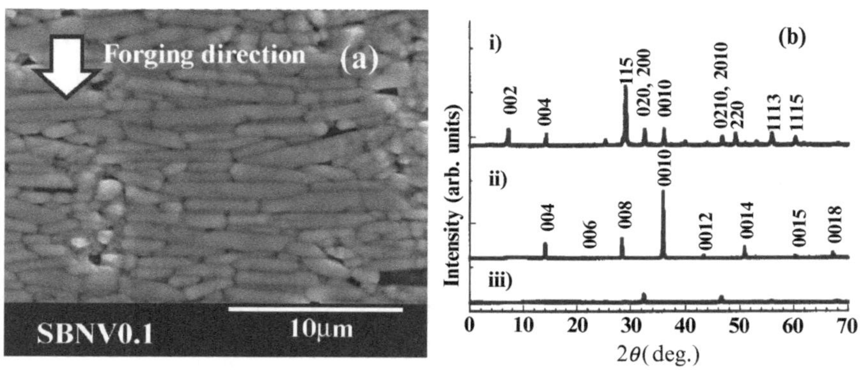

Fig. 6.14 (**a**) SEM image of grain-oriented SBNV0.1 ceramics, (**b**) XRD patterns of SBNV0.1 ceramics, for (i) ordinary fired ceramics, and for (ii) perpendicular plane, and (iii) parallel plane to forging axis of HF sample

nonoriented (OF) samples, the preparation of grain-oriented samples was limited to only composition with 5 mol% Ca concentration ($x = 0.05$). The HF samples almost smoothly deformed at temperatures of 1,100–1,200°C. The grain orientation factor (F) of the HF ceramics is higher than 90%, and each F_i (internal grain orientation factor) is given in Table 6.3.

Figure 6.13a shows electromechanical coupling factors of the longitudinal k_{33} and the transverse k_{31} modes, along with the anisotropy, k_{33}/k_{31} of the NCBT (OF) system as a function of Ca concentration. The k_{33} shows a peak at $x = 0.05$, corresponding to the maximum of ε_s (at T_c), while the k_{31} has a minimum value at this composition, resulting in a high value for the anisotropy, k_{33}/k_{31}. This composition ($x = 0.05$) seems to be the most suitable for piezoelectric properties. Figure 6.14b shows the coupling factors, k_{33} and k_{31}, of the Mn-doped NCBT-5 [+Mn (wt%)] (HF and OF) as a function of Mn doping concentration in wt%. The k_{33} of the HF sample has a maximum value of about 0.4 at 0.2 wt% of Mn. On the other hand, the k_{31} of HF or OF sample showed low values of about 0.03 or less. As a result, the anisotropy, (k_{33}/k_{31}), in the coupling factor k of the HF NCBT-5 + Mn(0.2) reaches about 0.17, which is three or more times larger than that of the nonoriented (OF) samples or the usual PZT piezoelectric ceramics.

Table 6.3 gives typical values of the elastic, piezoelectric, and dielectric (EPD) constants of the NCBT + Mn ceramics, comparing grain-oriented (HF) samples with nonoriented (OF) ones. The T_c is higher than 660 °C for NCBT-5 ceramics. The $\varepsilon_{33}{}^{T}$ is one-third of that of the conventional PZT ceramics. The values of k_{33}, d_{33}, or g_{33} of the grain-oriented (HF) ceramics are twice as large as that of nonoriented (OF) ceramics, while their values of the (31) mode show little change. The large anisotropy, k_{33}/k_{31} or d_{33}/d_{31}, was enhanced by the grain orientation. This means that the HF NCBT ceramics have a larger $d_h g_h = (d_{33} - 2d_{31})(g_{33} - 2g_{31})$ constant than that of the OF ones or the conventional PZT ceramics. Therefore, these piezoceramic materials are preferable for the hydrophone application at high temperatures, or for high-frequency resonators.

6.4.1.4 $SrBi_2Nb_{2-x}V_xO_9$ ($0 \leq x \leq 0.2, x = 0$: SBN)

Vanadium-substituted strontium bismuth niobate, ($SrBi_2Nb_{2-x}VxO_9$ (SBNVx)), ceramics (Inai et al. 2006) were synthesized by a low-temperature ceramic fabrication process. Their electric properties were characterized, and grain orientation effects on their electric properties were also investigated using a hot-forging (HF) method. Resonance characteristics of the (33) and (15) modes of grain-oriented HF-SBNV0.1 showed relatively high electromechanical coupling factors of $k_{33} = 0.272$ and $k_{15} = 0.151$ as well as piezoelectric constants of $d_{33} = 31.7$ and $d_{15} = 24.8$ pC/N. The temperature dependences of the resonance frequency (TCf) of the (33) and (15) modes of the SBNV0.1(HF) ceramics were −16.5 and −25.5 ppm/°C, respectively.

Among many BLSF materials, $SrBi_2Nb_2O_9$ (SBN)-based compositions necessitate attention as lead-free piezoelectric materials for resonator and filter applications because of their high mechanical quality factors (Q_m) and low temperature coefficients of resonance frequency (TCf) (Ando et al. 2003; Ogawa et al. 2001, 2005). For piezoelectric resonator applications, a multilayered structure and a cofiring technique are required for making an internal electrode on SBN-based ceramics (Ando et al. 2003). Also, a high sintering temperature in excess of 1,200 °C is necessary to obtain dense SBN ceramics, so that the internal electrode material is limited to Pt metal and Ag–Pd alloy. For SBN-based ceramics, the Takenaka group reported that a vanadium substitution for niobium sites in SBN ($SrBi_2Nb_{2-x}V_xO_9$) ceramics lowered the sintering temperature to 950 °C, improving ferroelectric and piezoelectric properties thereby (Aoyagi et al. 2005).

On the other hand, the direction of spontaneous polarization is restricted to two dimensions for BLSFs. Grain orientation techniques are well known to be effective for achieving high ferroelectricity and piezoelectricity in BLSFs, without degrading the anisotropy of a single crystal. There are several possible ways, such as the hot-forging (HF) method and the templated grain growth (TGG) method, to obtain grain-oriented samples. The grain orientation of BLSF can be easily performed because the anisotropy of a crystal structure is large. In this section, the grain orientation effects on electric properties in hot-forged (HF) $SrBi_2Nb_{2-x}V_xO_9$ (SBNVx) ceramics are described as compared with that in the case of ordinarily fired (OF) ones.

The observed densities of the SBNV0.1 and SBN ceramics as functions of sintering temperature were also studied. In the case of the OF samples, SBN ceramics prepared at 1,160 °C showed the highest density of 7.07 g/cm^3 and a relative density of 96.7%. On the other hand, V-substituted SBNV0.1 ceramics, which were sintered at even 950 °C, showed a density of 7.06 g/cm^3 and a higher relative density of 97.0%. V-substitution helped the sinterability of SBN ceramics by lowering the optimum sintering temperature. Similarly, in the case of HF samples, a high relative density was obtained for SBNV0.1 ceramics at the sintering temperature of 950 °C. However, in the case of SBN ceramics, it was difficult to obtain good HF samples because the sample was fragile during the HF process.

An SEM image depicting the surface parallel to the forging direction of SBNV0.1(HF) ceramics is shown in Fig. 6.14a. It is seen that almost all grains were aligned along the direction perpendicular to the forging axis. Figure 6.14b shows XRD patterns of SBNV0.1 ceramics, for (a) OF ceramics, and for (b) a perpendicular plane and (c) a parallel plane with respect to the forging axis of the HF sample. The XRD pattern shows that in the HF sample, the reflection intensities of each $(00l)$ peak in pattern (b) were stronger than those in the OF sample. On the other hand, the reflection intensities of other peaks were weak. Therefore, the HF sample was oriented perpendicular to the forging direction. The F was 0.98. High values of relative density and F were obtained in SBNV0.1(HF) ceramics. Hence, we focused our investigation to the properties of SBNV0.1 ceramics (Fig. 6.14b).

Figure 6.15a shows the D–E hysteresis loops of SBNV0.1 ceramics, for the OF sample and for the direction perpendicular to the forging axis of the HF sample at 160 °C. The remnant polarization (P_r) for the HF sample was 15.2 μC/cm^2, which is about two times larger than that for the OF sample (6.2 μC/cm^2). Figure 6.15b shows the frequency dependences of the impedance, $|Z|$, and phase, θ, of (15) vibration modes for the SBNV0.1(HF) ceramics. Fine resonant characteristics were obtained. Table 6.4 summarizes the piezoelectric properties of each vibration mode for SBN and SBNV0.1 ceramics.

The electromechanical coupling coefficient, k_{33}, and piezoelectric strain constant, d_{33}, of the longitudinal (33) vibration mode on HF SBNV0.1 ceramics were 0.272 and 31.7 pC/N, respectively. These k and d values are about two times larger than those of OF ceramics. This is due to the grain orientation. The maximum phase, θ_{max}, and electrical quality factor, $Q_{e\text{-}max}$ ($= \tan\theta_{max}$), of the HF sample were 88.8° and 52.1, respectively, which are excellent values for resonator applications.

The coupling coefficient, k_{15}, and piezoelectric strain constant, d_{15}, of the thickness shear (15) vibration mode on HF SBNV0.1 ceramics were 0.151 and 24.8 pC/N, respectively, which are about 1.5 times larger than those of the OF ceramics. θ_{max} and Q_m were 87.8° and 4,420, respectively. On the other hand, k_{24} and d_{24} were about half of the k_{15} and d_{15} of the OF ceramics. There was a large

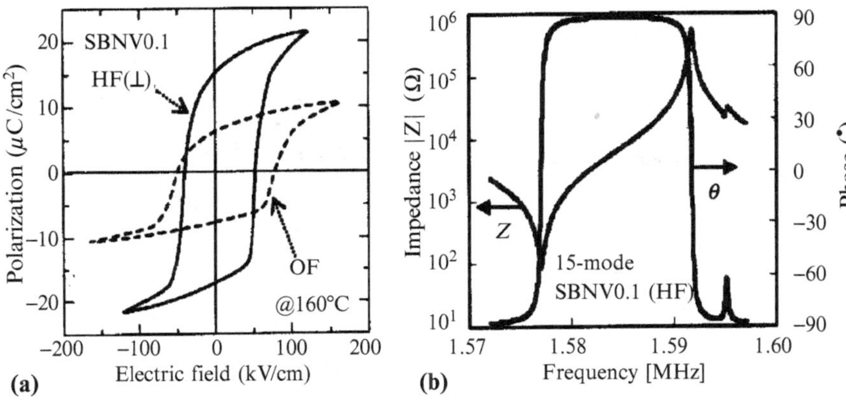

Fig. 6.15 (**a**) D–E hysterisis loops of OF and HF SBNV0.1 ceramics at 160°C, (**b**) frequency characteristics of impedance, *Z*, of SBNV0.1(HF) ceramics

Table 6.4 Piezoelectric properties of each mode for SBN and SBNV0.1 ceramics

Composition	Mode	k (%)	Q_m	Q_e	d (pC/N)	$TC\text{-}f_r$ (ppm/°C)
SBN	33(OF)	14.3	1900	6.7	14.9	−34.6
	33(OF)	16.9	5500	31.8	17.4	−44.7
	33(HF)	27.2	2010	52.1	31.7	−16.5
SBNV0.1	15(OF)	9.52	4600	11.7	15.2	−48.6
	15(HF)	15.1	4420	26.0	24.8	−25.5
	24(HF)	5.2	6450	3.3	10	−84.2

difference between k_{15} and k_{24}, which can be explained by the anisotropy of the crystal structure (Ogawa et al. 2001).

Figure 6.16 shows the temperature coefficient of the resonant frequencies, *TCf*, in the thickness shear (15) mode of the SBNV0.1(OF) ceramics, and the thickness shear (15) and (24) modes of the SBNV0.1(HF) ceramics. It was clear that there was a large difference in the TCf of the (15) and (24) modes on the HF sample. The TCf of the (15) and (24) modes on the HF sample were −25.5 and −84.2 ppm/°C, respectively. The *TCf* of the (15) mode on the OF sample was between those of the (15) and (24) modes on the HF sample. This result is in agreement with the case of $CaBi_4Ti_4O_{15}$ (CBT) and Nd-substituted $Sr_2Bi_2Nb_2O_9$ (SNBN) ceramics reported by Ogawa et al. (2001, 2005). However, Ogawa et al. reported a positive *TCf* for the SNBN ceramics textured by the TGG method.

In this study, the resonant frequency decreased with temperature for the hot-forged SBNV. It is thought that this difference is caused by V substitution. The *TCf* of the (33) mode on the HF sample was −16.5 ppm/°C, which is better than that of the OF sample. Here, in the case of the (33) mode, antiresonance frequency, f_a, can be described by the following equation that uses elastic compliance, ${s_{33}}^D$:

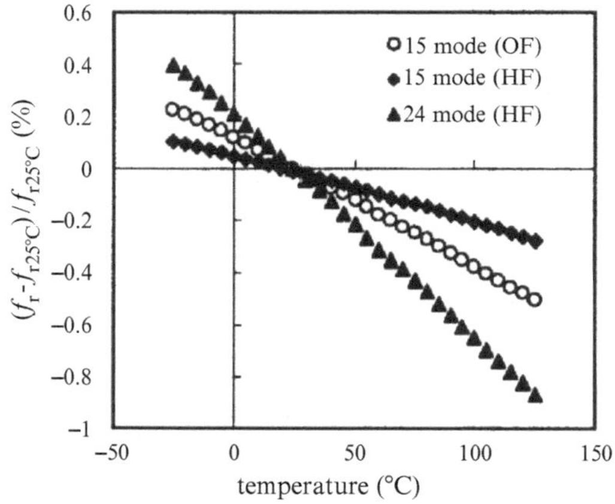

Fig. 6.16 Temperature dependence of resonant frequency, *TCf*

$$f_a = \frac{1}{2x}\sqrt{\frac{1}{\rho \times s_{33}{}^{D}}}, \tag{6.4}$$

where x and ρ are the vibration length of the (33) mode sample and observed density, respectively. The $s_{33}{}^{D}$ of the HF sample was smaller than that of the OF sample, which means that the HF sample was harder than the OF sample. The difference in $s_{33}{}^{D}$ might relate to the difference in TCf, but it is still under investigation. The substitutions of Ba and Nd for some of the Sr sites in SBN ceramics were effective for improving TCf (Ando et al. 2003, 2003). By applying this method to SBNV, it is thought that SBNV ceramics with an additional smaller TCf could be obtained. As a result, SBNV-based materials are expected to be lead-free piezoelectric resonator materials that can be prepared at low sintering temperatures.

6.4.1.5 $(Sr_{1-x}Ca_x)_2Bi_4$ Ti_5O_{18} $(0 \le x \le 1, x = 0$: S2BT, $x = 1$: C2BT)

Piezoelectric properties of $Sr_2Bi_4Ti_5O_{18}$(SBT)–$Ca_2Bi_4Ti_5O_{18}$ (CBT) solid solution, also designated as $(Sr_{1-x}Ca_x)_2Bi_4Ti_5O_{18}$ [SCBT-x] $(0 \le x \le 1)$ with $m = 5$, were investigated (Nagata et al. 2006). Effects of grain orientation in SCBT system on their piezoelectric properties were studied using the hot-forging (HF) method. In the case of ordinarily fired (OF) samples with no texturing, electromechanical coupling factor, k_{33}, and piezoelectric constant, d_{33}, of the $MnCO_3$ (0.4 wt%) doped SCBT-0 were 0.15 and 26 pC/N, respectively. On the other hand, the HF-SCBT-0 showed relatively high k_{33} and d_{33} values of 0.29 and 51 pC/N, respectively.

Temperature coefficients of the resonance frequency, *TCf*, in the (33) mode of SCBT-0 (OF) and HF-SCBT-0 were −89.2 and −63.0 ppm/°C, respectively. HF-SCBT-0.5 ceramics have excellent mechanical and electrical quality factor, $Q_m =$ 2,320, and $Q_{e\text{-max}} = 31.7$. Temperature coefficient of the resonance frequency,

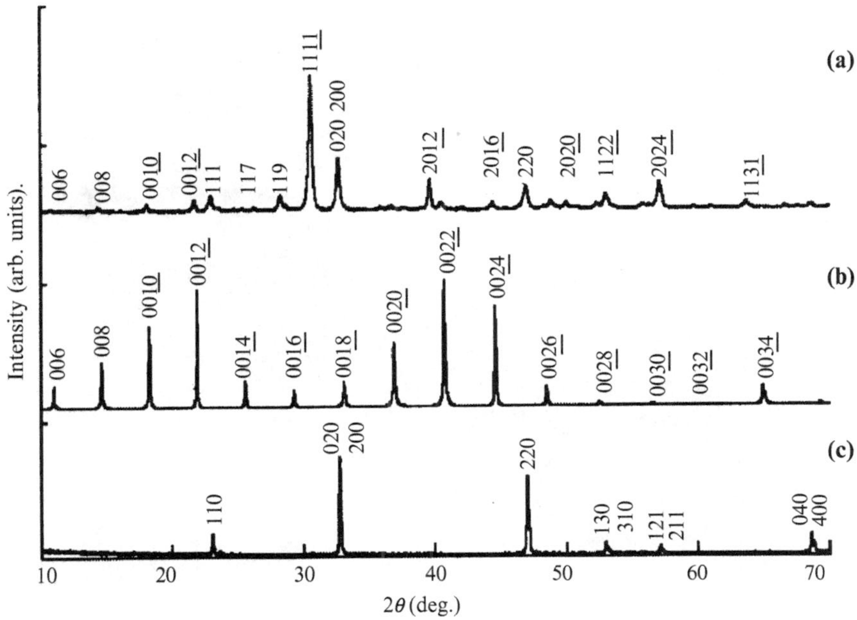

Fig. 6.17 X-ray diffraction patterns SCBT-0 (S2BT) ceramics: (**a**) nonoriented ceramics and grain-oriented ceramics, (**b**) the perpendicular [⊥], and (**c**) the parallel [//] planes to the forging axis

TCf, in the (33)-mode of HF-SCBT-0.5 was −50.0 ppm/°C. These properties are almost similar to commercialized hard PZT ceramics used in resonator applications.

X-ray diffraction patterns for SCBT-*x* ceramics show a single phase of bismuth layer-structured compounds ($m = 5$) with orthorhombic symmetry. Figure 6.17 shows the XRD patterns of the SCBT-0 (S2BT) ceramics (a) fabricated by the conventional method (OF), (b) in the perpendicular [⊥], and (c) in the parallel [//] planes to the forging axis of HF specimen. In HF specimens, the intensities of each $(00l)$ peak on the parallel to the forging axis was stronger than those in the OF specimens. On the other hand, the intensities of other reflections were weak.

Therefore, it was recognized that the c-axis of HF specimens was preferentially oriented, and the direction of c-axis was almost perpendicular to the forging axis. The degree of grain orientation (F) was determined to be ~100%.

The SEM micrographs of SCBT-0 and SCBT-0.5 are depicted in Fig. 6.18, where (a) and (c) are micrographs of the surface of nonoriented SCBT-0 and SCBT-0.5 ceramics. Figures 6.18b, d are those of the surface parallel to the stacking direction of HF-SCBT-0 and HF-SCBT-0.5 ceramics, respectively. Most grains' major faces were found to be perpendicular to the forging axis.

As shown in Fig. 6.19, the anisotropic behaviors of electrical properties in HF specimen were measured by applying an electric field perpendicular [⊥] and parallel [//], respectively, to the forging axis. The dielectric constant ε_s[⊥] and ε_s[//] are 392 at $F = \sim 0.90$ and 84 at $F = \sim 1$, respectively, at room temperature. The dielectric

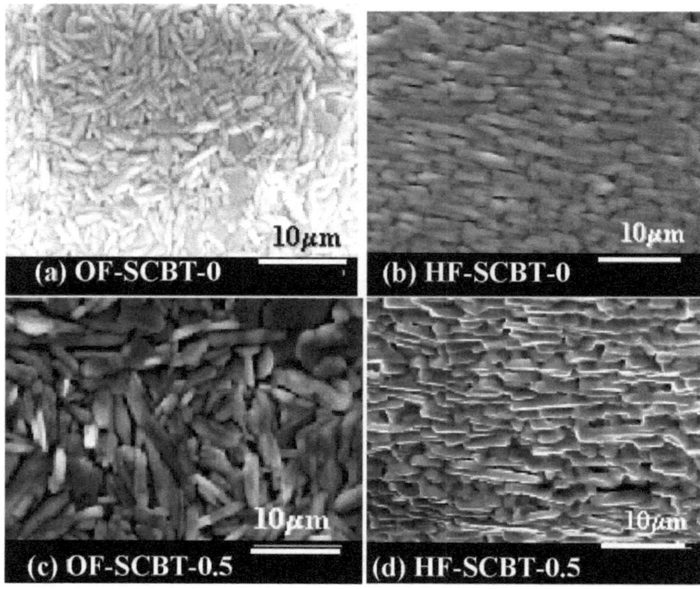

Fig. 6.18 SEM micrographs of the surface of the nonoriented (**a**) SCBT-0 and (**c**) SCBT-0.5 ceramics, and the grain-oriented (HF) (**b**) SCBT-0 and (**d**) SCBT-0.5 ceramics

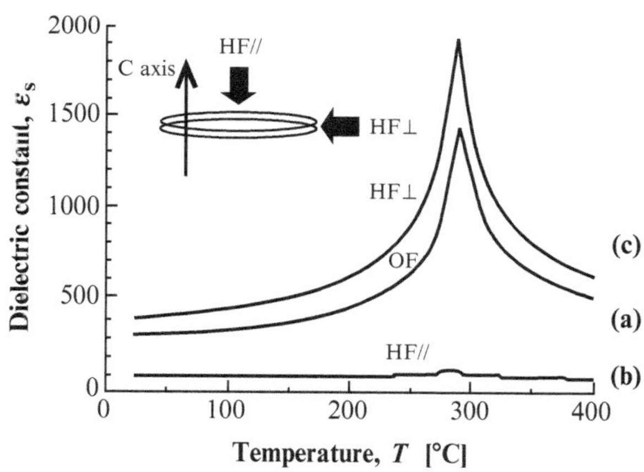

Fig. 6.19 Temperature dependence of the dielectric constant, ε_s, of SCBT-0. (**a**) nonoriented (OF) ceramics, and (**b**) the parallel [//], and (**c**) the perpendicular $[\perp]$ of grain-oriented samples to the forging axis

constant, ε_s, of nonoriented specimen is 296. Dielectric constant, ε_s, of nonoriented specimen shows an intermediate value between $\varepsilon_s[\perp]$ and $\varepsilon_s[//]$. The $\varepsilon_s[\perp]$ and $\varepsilon_s[//]$ are 1,932 and 118, respectively, at the Curie point, T_c. The ratio of $\varepsilon_s[\perp]$ to $\varepsilon_s[//]$ is about 12 at the T_c. It is considered that SCBT-x ceramics show a large anisotropy. The Curie point, T_c, of SCBT-0 was about 290°C. The Curie point, T_c, became higher with increasing the amount (x) of concentration Ca (Fig. 6.19).

Figure 6.20 shows the frequency dependence of impedance, Z (magnitude, $|Z|$, and phase, θ), of nonoriented (a) SCBT-0 and (b) SCBT-0.5 ceramics. The electromechanical coupling factors, k_{33}, of nonoriented SCBT-0 and SCBT-0.5 ceramics are 0.15 and 0.10; respectively. The maximum θ of nonoriented SCBT-0 and SCBT-0.5 are 81.2° and 82.2°; respectively as shown in Fig. 6.20. The mechanical quality factors, Q_m, of SCBT-0 and SCBT-0.5 are 1390 and 3,810, respectively (Table 6.5). The SCBT-0 samples were poled with 11.25 kV/mm, while the same for SCBT-0.5 samples was done at 10 kV/mm in the case of HF-SCBT-0 and HF-SCBT-0.5 ceramics.

The k_{33} of grain-orientated SCBT-x ceramics with $m = 5$ could be synthesized for the first time in this study. Good profiles were obtained, and k_{33} values of SCBT-0 and SCBT-0.5 ceramics were enhanced to 0.301 and 0.267, respectively. These

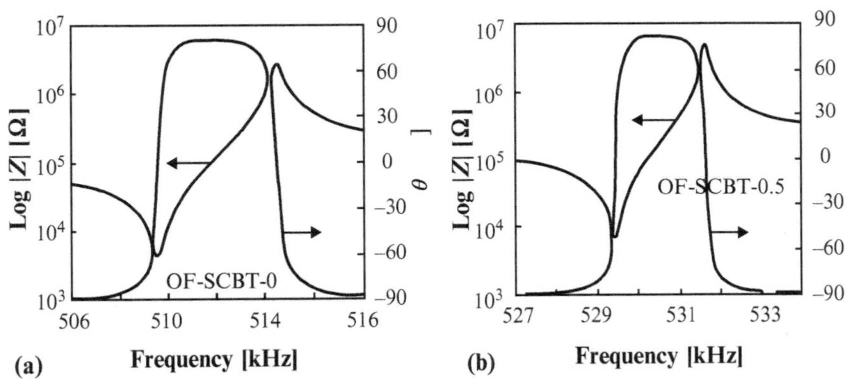

Fig. 6.20 Frequency dependence of impedance, $|Z|$, and phase, θ, of (**a**) SCBT-0 and poled with 11.25 kV/mm, (**b**) SCBT-0.5 ceramics poled with 10 kV/mm

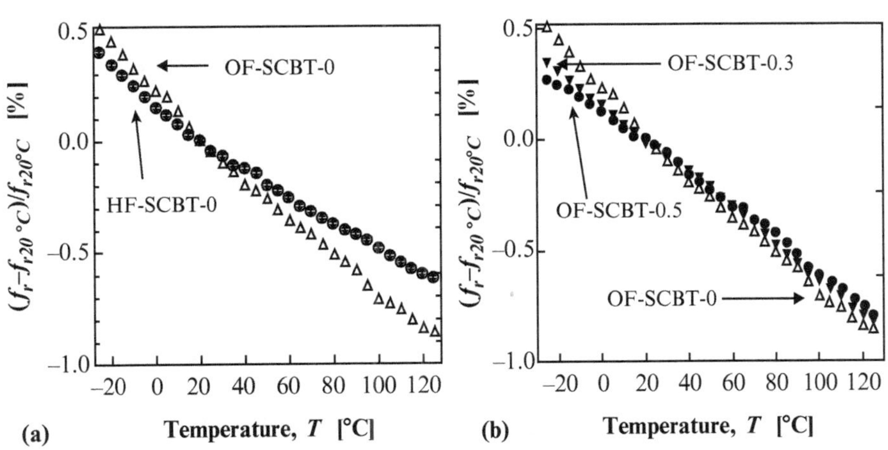

Fig. 6.21 Temperature dependence of resonant frequency of (**a**) SCBT-0 and (**b**) OF-SCBT-x ceramics

Table 6.5 Piezoelectric properties of SCBT-x ($x = 0, 0.3, 0.5$)

Samples	Electromechanical coupling factor k_{33} (%)		Piezoelectric constant d_{33} (pC/N)		Free permittivity $\varepsilon_{33}{}^{T}/\varepsilon_0$		Mechanical quality factor Q_m	
	OF	HF	OF	HF	OF	HF	OF	HF
SCBT-0	15.1	30.1	26	50	315	432	1,390	1,570
SCBT-0.3	13.3	24.5	17	39	236	316	3,730	2,170
SCBT-0.5	10.2	26.7	13	35	207	344	3,810	2,320

Table 6.6 Temperature coefficients of the resonant frequency of SCBT-x ($x = 0, 0.3, 0.5$)

Samples	Temperature coefficient TC-f_r (ppm/°C)	
	OF	HF
SCBT-0	−85.6	−63.0
SCBT-0.3	−77.6	−60.1
SCBT 0.5	−66.6	−50.0

values are about two times larger than those of nonoriented ceramics. The maximum phase, θ, of SCBT-0 and SCBT-0.5 is 88.1° and 87.9°. The mechanical quality factors, Q_m, of SCBT-0 and SCBT-0.5 are 1,570 and 2,320, respectively.

Piezoelectric constants, d_{33}, of HF specimen are about two times lager than those of nonoriented ceramics. The d_{33} decreased with increasing concentration Ca. On the other hand, mechanical quality factors, Q_m, become higher with increasing the amount (x) of concentration Ca as shown in Table 6.5.

Figure 6.21 shows the temperature dependence of the resonance frequencies (TCf) of the longitudinal 33-mode of OF and HF SCBT-0 as well as OF-SCBT-x. These temperature coefficients (TCf) were obtained from the data by regression analysis and are summarized in Table 6.6. It is noted that the TCf becomes lower as the grain orientation is enhanced, while it decreases when the Ca concentration is increased.

6.5 Conclusion and Future Trends

The results of our investigations on hot-forging (HF) of ferroelectric ceramics indicate that strongly preferred grain orientation can be obtained. The degree of grain orientation F calculated from X-ray diffraction patterns increases commensurately with the area ratio γ (or the thickness contraction rate λ). Considerable anisotropies in the dielectric constant of HF ceramics were observed, and the values of $\varepsilon_s[\perp]/\varepsilon_s[//]$ were 5–20 at the Curie temperature, T_c. Moreover, we have noted that remnant polarizations $P_r[\perp]$ and $P_r[//]$ in the HF ceramics correspond to the components *P*sa along the a-axis and *P*sc along the c-axis, respectively, of the spontaneous polarization *P*s of BIT single crystal. It was made clear that the remnant polarization $P_r[\perp]$ in the HF ceramics is much larger than that of the nonoriented (OF) ones. For example, the ratio $P_r[\perp]/P_r[//]$ and $P_r[\perp]/P_r[\mathrm{OF}]$ were about 11 and 4 in BIT, respectively, for $F = 0.95$ at room temperature. The value of the remnant polarization $P_r[\perp]$ of the grain-oriented (HF) BIT ceramic for $F = 0.95$ is estimated to be 82% of the theoretical value.

In $Bi_4Ti_{3-x}Nb_xO_{12}$ [BITN-*x*] and $Bi_4Ti_{3-x}V_xO_{12}$ [BITV-*x*], for the ordinarily fired (OF) samples, the electromechanical coupling factors, k_{33}, of BITN-0.08 and BITV-0.02 were enhanced to 0.20 and 0.25, respectively, which are very large as compared with BLSF with a random orientation. For HF samples, the k_{33} values of BITN-0.08 and BITV-0.04 ceramics were enhanced to 0.39 and 0.38, respectively. The BITN and BITV ceramics seem to be good candidates as lead-free high-temperature piezoelectric materials.

In the case of $(Bi_{4-y},Nd_y)_{1-(x/12)}(Ti_{3-x},V_x)O_{12}$ [BNTV-*y* ($x = 0.01$, $y =$ 0.00–1.00)] systems, the k_{33} of ordinarily fired (OF) BNTV-*y*0.75 ceramic was 0.21, which is one of the highest values for (OF) BLSFs. The k_{33} of the (HF) BNTV-*y*0.25 ceramic was 0.30. Vanadium substitution has the effect of lowering the sintering temperature. BIT-Nd ceramics maintained high ρ (about $10^{13}\Omega$ cm). This result implied that Nd ions suppressed Bi loss at high temperatures. The effect of Nd and V cosubstitution was found to be very advantageous.

Grain orientation of calcium-modified $(Na_{1/2}Bi_{1/2})_{1-x}Ca_xBi_4Ti_4O_{15}$ [NCBT] system enhanced their piezoelectric properties by 200–300%. From the viewpoint of piezoelectric applications, 5 mol% Ca-modified, Mn-doped, and grain-oriented NCBT (HF. NCBT-5 + Mn) ceramics are superior, with a lower free permittivity, $\varepsilon_{33}{}^{T}/\varepsilon_0$ ($\sim$130) and a higher electromechanical coupling factor, k_{33} (= 0.36–0.40), along with a higher anisotropy, $k_{33}/k_{31} = 13$–17). Therefore, the HF NCBT-5 + Mn ceramics are candidates for hydrophone or high temperature and frequency ultrasonic applications.

In the $SrBi_2Nb_{2-x}V_xO_9$ (SBNV*x*) system, dense grain-oriented ceramics, sintered at 950°C. The P_r of SBNV0.1(HF) ceramics was 15.2 μC/cm^2, which was larger than the P_r of OF samples. The *k* and *d* of HF ceramics were larger than those of OF ceramics. The *TCf* values of SBNV0.1(HF) were −25.5 and −16.5 ppm/°C in the (15) and (33) modes, which were lower than those of the OF ceramics. Hence, SBNV-based are low-temperature sintering materials good for resonator applications.

In the $(Sr_{(1-x)}Ca_x)_2Bi_4Ti_5O_{18}$ [SCBT-*x*] ($0 \le x \le 1$) system, the following results were obtained: (a) Q_m becomes higher with increasing the amount of Ca, that *TCf* becomes lower with increasing the amount of Ca, (b) TC_f of grain-oriented samples is lower than nonoriented ones, and (c) HF-SCBT-0.5 showed high Q_m and low *TCf* in the (33)-mode. From these results, the SCBT-*x* ceramics are considered to be a superior candidate as lead-free piezoelectric ceramics in applications such as ceramic filters and resonators.

Future trends in the research of BLSF are likely to focus on environmentally gentle compositions for high-temperatures sensors, high-frequency applications, and temperature stable ceramic resonators. Also templated grain growth (TGG) and seeded polycrystal conversion (SPC) methods seem to be promising technologies in addition to HF. The use of such Pb-free ceramics might also find applications in thick films in FBAW and MEMS devices.

Acknowledgements This work was partially supported by a Grant-in-Aid for Scientific Research (C) (No: 07650790, 09650737) and (B) (No: 11555168, 13555176, 15360352, and 17360327) from the Japan Society for the Promotion of Science. The author would like to express sincere thanks to the late Professor Masakazu Marutake of University of Electro-Communications for valuable discussion, particularly on the calculation of the remanent polarization. He also would like to thank Professor Emeritus Koichiro Sakata of Tokyo University of Science for encouraging this work and for his continued support over many years.

References

Ando A et al. (2003a) Piezoelectric Resonance Characteristics of $SrBi_2Nb_2O_9$-based Ceramics, Jpn. J. Appl. Phys. 42: 150–156.

Ando A et al. (2003b) Piezoelectric Properties of Ba and Ca doped $SrBi_2Nb_2O_9$ Based Ceramic Materials, Jpn. J. Appl. Phys. 42: 520–525.

Aoyagi R et al. (2005) Piezoelectric Properties of Vanadium-Substituted Strontium Bismuth, Jpn. J. Appl. Phys. 44: 7055–7058.

Armstrong RA, Newnham RE (1972) Bismuth Titanate Solid Solutions, Mater. Res. Bull. 7(10): 1025–1034.

Aurivillius B (1949a) Mixed Bismuth Oxides with Layer Lattices. I. The Structure Type of $CaNb_2Bi_2O_9$, Arkiv Kemi 1: 463–480.

Aurivillius B (1949b) Mixed Bismuth Oxides with Layer Lattices. II. Structure of $Bi_4Ti_3O_{12}$, Arkiv Kemi 1: 499–512.

Aurivillius B (1950) Mixed Bismuth Oxides with Layer Lattices. III. Structure of $BaBi_4Ti_4O_{15}$, Arkiv Kemi 2: 519–527.

Aurivillius B, Fang PH (1962) Ferroelectricity in the Compound $Ba_2Bi_4Ti_5O_{18}$, Phys. Rev. 126: 893–896.

Cross LE, Pohanka RC (1971) Ferroelectricity in Bismuth Oxides Type Layer Structure Compounds, Mater. Res. Bull. 6: 939–949.

Cummins SE, Cross LE (1967) Crystal Symmetry, Optical Properties, and Ferroelectric Polarization of $Bi_4Ti_3O_{12}$ Single Crystal, Appl. Phys. Lett. 10(1): 14–16.

Cummins SE, Cross LE (1968) Electrical and Optical Properties of Ferroelectric $Bi_4Ti_3O_{12}$ Single Crystals, Appl. Phys. 39(5): 2268–2274.

Dorrian JF et al. (1971) Crystal Structure of Bi4Ti3O12, Ferroelectrics 3: 17–27.

Fang PH et al. (1962) Ferroelectricity in the Compound $Bi_4Ti_3O_{12}$, Phys. Rev. 126(3): 892–896.

Ikegami S, Ueda I (1974) Piezoelectricity in Ceramics of Ferroelectric Bismuth Compound with Layer Structure, Jpn. J. Appl. Phys. 13(10): 1572–1579.

Inai S et al. (2006) Electrical Properties of Grain-Oriented $SrBi_2Nb_{2-x}V_xO_9$ Ceramics, Key Eng. Mater. 320: 31–34.

Lotgering FK (1959) Topotactical Reactions with Ferrimagnetic Oxides Having Hexagonal Crystal Structures – I. J. Inorg. Nucl. Chem. 9(2): 113–123.

Matsuzawa S et al. (2006) Piezoelectric Properties of Nd and V co-substituted $Bi_4Ti_3O_{12}$ Ceramics, Key Eng. Mater. 320: 39–42.

Nagata H et al. (2000) Piezoelectric Anisotropies of Bismuth Layer-Structured Ferroelectrics, Trans. Mater. Res. Jpn, 25(1): 273–276.

Nagata H et al. (2003) Piezoelectric Properties of Bismuth Layer-Structured Ferroelectric $SrBi_2Ta_2O_9$-Bi_3TiTaO_9 Ceramics, Ferroelectrics, 286: 85–92.

Nagata H et al. (2004) Piezoelectric Properties of Nb and V Substituted $Bi_4Ti_3O_{12}$ Ferroelectric Ceramics, Ceram. Trans. 150: 253–263.

Nagata H et al. (2006a) Bismuth Layer-Structured Ferroelectric $(Sr,Ca)_2Bi_4Ti_5O_{18}$ Ceramics for Lead-Free Piezoelectric Resonator Applications, Proc. 2005 IEEE Ultrason. Symp. pp. 1077–1082.

Nagata H et al. (2006b) Piezoelectric Properties of Nd and V co-substituted $Bi_4Ti_3O_{12}$ Ceramics for resonator applications, Proc. 2006 IEEE Ultrason. Symp. 355–358.

Newnham RE et al. (1971) Structural Basis of Ferroelectricity in the Bismuth Titanate Family, Mater. Res. Bull. 6: 1029–1039.

Noguchi Y, Miyayama M (2001) Large remnant polarization of vanadium-doped $Bi_4Ti_3O_{12}$. Appl. Phys. Lett. 78(13): 1903–1905.

Ogawa H et al. (2001) Temperature Dependence of Piezoelectric Properties of Grain-Oriented $CaBi_4Ti_4O_{15}$ Ceramics, Jpn. J. Appl. Phys. 40(9B): 5715–5718.

Ogawa H et al. (2005) Piezoelectric Properties of $SrBi_2Nb_2O_9$ Textured Ceramics, Jpn. J. Appl. Phys. 44: 7050–7054.

Shannon RD (1976) Revised Effective Ionic Radii and Systematic Studies of Interatomic Distances in Halides and Chalcogenides, Acta Crystallogr. A 32(5): 751–767.

Shulman HS et al. (1996) Microstructure, Electrical Conductivity, and Piezoelectric Properties of Bismuth Titanate, J. Am. Ceram. Soc. 79(12): 3124–3128.

Smolenskii GA et al. (1959) A New Group of Ferroelectrics (with Layered Structure), Sov. Phys. Solid State 1: 149–150.

Smolenskii GA et al. (1961) Ferroelectrics of the Oxygen-Octahedral Type with Layered Structure, Sov. Phys.-Solid State 3: 651–655.

Subbarao EC (1961) Ferroelectricity in $Bi_4Ti_3O_{12}$ and Its Solid Solutions, Phys. Rev. 122(3): 804–807.

Subbarao EC (1962a) Crystal Chemistry of Mixed Bismuth Oxides with Layer-Type Structure, J. Am. Ceram. Soc. 45(4): 166–169.

Subbarao EC (1962b) A Family of Ferroelectric Bismuth Compounds, J. Phys. Chem. Solids 23: 665–676.

Takenaka T et al. (1976) Ferroelectric and Dielectric Properties of Press Forged $Bi_4Ti_3O_{12}$ Ceramics, Proc. 19th Jpn. Cong. Materials Research, Tokyo, 1975 (The Society of Materials Science, Kyoto, 1976), pp. 230–233.

Takenaka T et al. (1977) Grain Orientation and Microstructure of Hot-Forged $Bi_4Ti_3O_{12}$ Ceramics, Proc. 20th Jpn. Cong. Materials Research, Kyoto, 1976 (The Society of Materials Science, Kyoto, 1977), pp. 212–214.

Takenaka T, Sakata K (1980) Grain Orientation and Electrical Properties of Hot-Forged Bi4Ti3O12 Ceramics, Jpn. J. Appl. Phys. 19(1): 31–39.

Takenaka T, Sakata K (1982) Dielectric and Piezoelectric Properties of Some Bismuth Layer-Structured Ferroelectric Ceramics, Jpn. J. IEEE (C)J65–C(7): 514–521 (in Japanese).

Takenaka T, Sakata K (1983) Pyroelectric Properties of Grain-Oriented Bismuth Layer-Structured Ferroelectric Ceramics, Jpn. J. Appl. Phys. 22 (Suppl. 22–2): 53–56.

Takenaka T, Sakata K (1984) Grain Orientation Effects on Electrical Properties of Bismuth Layer-Structured Ferroelectric $Pb_{1-x}(Na,Ce)_{x/2}Bi_4Ti_4O_{15}$ Solid Solution, J. Appl. Phys. 55(4): 1092–1099.

Takenaka T et al. (1985) Piezoelectric Properties of Bismuth Layer-Structured Ferroelectric $Na_{0.5}Bi_{4.5}Ti_4O_{15}$ Ceramic, Jpn. J. Appl. Phys. 24 (Suppl. 24–2): 730–732.

Takenaka T, Sakata K (1986) Piezoelectric Properties of Grain-Oriented Bismuth Layer-Structured Ferroelectric Ceramics, Proc. Sixth Int. Symp. Appl. Ferroelectr. IEEE, pp. 414–417.

Takenaka T, Sakata K (1988) Grain-Oriented and Mn-Doped $(NaBi)_{(1-x/2)}Ca_xBi_4Ti_4O_{15}$ Ceramics for Piezo- and Pyrosensor Materials, Sens. Mater. 1: 35–46.

Takenaka T, Sakata K (1989) Piezoelectric and Pyroelectric Properties of Calcium-Modified and Grain-Oriented $(NaBi)_{1/2}Bi_4Ti_4O_{15}$ Ceramics, Ferroelectrics 94: 175–181.

Takenaka T, Sakata K (1991) Pyroelectric Properties of Bismuth Layer-Structured Ferroelectric Ceramics, Ferroelectrics 118: 123–133.

Takenaka T (2002) Grain Orientation Effects on Electrical Properties of Bismuth Layer-Structured Ferroelectric Ceramics, J. Cer. Soc. Jpn 110: 215–224.

Takenaka T, Nagata H (2006) Grain Orientation and Electrical Properties of Some Bismuth Layer-Structured Ferroelectrics for Lead-Free Piezoelectric Applications, Ferroelectrics 336: 119–136.

Takeuchi T et al. (1999) Piezoelectric Properties of Bismuth Layer-Structured Ferroelectric Ceramics with a Preferred Orientation Processed by the Reactive Templated Grain Growth Method, Jpn. J. Appl. Phys. 38(9B): 5553–5556.

Villegas M et al. (1999) Factors Affecting the Electrical Conductivity of Donor-Doped $Bi_4Ti_3O_{12}$ Piezoelectric Ceramics, J. Am. Ceram. Soc. 82(9): 2411–2416.

Wolfe RW, Newnham RE (1969) Rare Earth Bismuth Titanates, J. Electrochem. Soc; SOLD STATE SCIENCE: 116, 832–835.

Yamashita Y et al. (1983) Effects of MnO Additive on Piezoelectric Properties in Modified $(Pb, Ca)TiO_3$ Ferroelectric Ceramics, Jpn. J. Appl. Phys. 22 (Suppl. 22-2): 40–42.

Chapter 7
Electropolymers for Mechatronics and Artificial Muscles

Zhongyang Cheng,* Qiming Zhang, Ji Su, and Mario El Tahchi

7.1 Introduction

This chapter covers the main features and material examples of the electric field-activated electropolymers used for electromechanical applications. This class of electropolymers is very attractive in performing the energy conversion between the electric and mechanical forms and hence can be utilized as both solid-state electromechanical actuators and motion sensors. These polymers are also attractive for artificial muscles and for energy-harvesting applications. As will be discussed, the electromechanical response in this class of polymers can be linear, as in typical piezoelectric polymers and electrets, or nonlinear, as the electrostrictive polymers and Maxwell stress-induced response.

Most of the piezoelectric polymers under investigation and in commercial use are based on poled ferroelectric (FE) polymers, including poly(vinylidene fluoride) (PVDF) and related copolymers. This chapter will discuss in detail the properties of these FE polymers. In comparison with the electromechanical responses in inorganic materials, the electromechanical activity in these polymers is relatively low. To significantly improve the electromechanical properties in these electropolymers, new avenues or approaches have to be explored. From the basic material consideration, these approaches include the strain change accompanied with the molecular conformation change, due to the polar vector reorientation, from the morphology change due to the ordering degree change in the interfacial layer between crystalline and amorphous regions, and from the Maxwell stress effect in soft polymer elastomers. This chapter will discuss the recent advances based on those approaches, which have produced remarkable improvements in terms of the electric field-induced strain level, elastic energy density, and electromechanical conversion efficiency in the electropolymers.

Z. Cheng
Materials Research and Education Center, 275 Wilmore Lab, Auburn University, Auburn, AL 36849-5341, USA

A. Safari, E.K. Akdoğan (eds.) *Piezoelectric and Acoustic Materials for Transducer Applications.*

7.2 General Terminology of Electromechanical Effects

The piezoelectric effect, which only exists in 20 classes of crystal point symmetry groups (Nye 1987), is a linear electromechanical effect where the mechanical strain (x) and stress (X) are coupled to the electric field (E) and displacement (or charge density D) linearly, i.e., $x = dE$ and $D = dX$, where d is the piezoelectric coefficient. The former effect is often known as the converse piezoelectric effect, while the latter as the direct piezoelectric effect. Adding the linear elastic (Hook's law) and dielectric responses to these expressions, and writing it out in the full tensor form, we get the piezoelectric constitutive equations (Nye 1987; IEEE Standard on Piezoelectricity 1988):

$$x_{ij} = d_{kij}E_k + s^{\mathrm{E}}_{ijkl}X_{kl} \tag{7.1a}$$

and

$$D_i = \varepsilon^{\mathrm{X}}_{ik}E_k + d_{ikl}X_{kl}, \tag{7.1b}$$

where s^{E}_{ijkl} is the elastic compliance, $\varepsilon^{\mathrm{T}}_{ik}$ is the dielectric permittivity, and $i,\ j,\ k,\ l =$ 1–3. In the equation, the convention in which repeated indices are summed is used. The superscripts E and X refer to the condition under which these quantities are measured. The material properties can be written in matrix form by using a compressed notation. That is, ij or kl in tensor form is replaced by p or q according to Nye (1987) and IEEE Standard on Piezoelectricity (1988): $11 \rightarrow 1,\ 22 \rightarrow 2,\ 33 \rightarrow 3,\ 23$ or $32 \rightarrow 4,\ 13$ or $31 \rightarrow 5,\ 12$ or $21 \rightarrow 6$. Therefore, (7.1a) in matrix form is:

$$x_p = d_{kp}E_k + s^{\mathrm{E}}_{pq}X_q, \quad (p,k,q) = 1-6, k = 1-3, \tag{7.2a}$$

and

$$D_i = \varepsilon^{\mathrm{X}}_{ij}E_j + d_{iq}X_q, \quad (i,j) = 1-3, q = 1-6. \tag{7.2b}$$

For a given piezoelectric material, the number of independent elements for each material property can be reduced based on the symmetry of crystal using Neumann's principle (Nye 1987). For instance, for an unstretched PVDF polymer poled along 3-direction, which has a point group ∞m, the piezoelectric coefficient, the dielectric permittivity, and elastic compliance matrices are

$$\begin{pmatrix} 0 & 0 & 0 & 0 & d_{15} & 0 \\ 0 & 0 & 0 & d_{15} & 0 & 0 \\ d_{31} & d_{31} & d_{33} & 0 & 0 & 0 \end{pmatrix}, \begin{pmatrix} \varepsilon_{11} & 0 & 0 \\ 0 & \varepsilon_{22} & 0 \\ 0 & 0 & a_{33} \end{pmatrix}, \text{ and } \begin{pmatrix} s_{11} & s_{12} & s_{13} & 0 & 0 & 0 \\ s_{12} & s_{22} & s_{23} & 0 & 0 & 0 \\ s_{13} & s_{23} & s_{33} & 0 & 0 & 0 \\ 0 & 0 & 0 & s_{44} & 0 & 0 \\ 0 & 0 & 0 & 0 & s_{55} & 0 \\ 0 & 0 & 0 & 0 & 0 & s_{66} \end{pmatrix},$$

where $s_{66} = 2(s_{11} - s_{12})$. For a PVDF stretched along 1-direction and poled along 3-direction, which has a point group $mm2$, the d_{ip}, ε_{ij}, and s_{pq} are

$$\begin{pmatrix} 0 & 0 & 0 & 0 & d_{15} & 0 \\ 0 & 0 & 0 & d_{24} & 0 & 0 \\ d_{31} & d_{32} & d_{33} & 0 & 0 & 0 \end{pmatrix}, \begin{pmatrix} \varepsilon_{11} & 0 & 0 \\ 0 & \varepsilon_{22} & 0 \\ 0 & 0 & \varepsilon_{33} \end{pmatrix}, \quad \text{and} \quad \begin{pmatrix} s_{11} & s_{12} & s_{13} & 0 & 0 & 0 \\ s_{12} & s_{22} & s_{23} & 0 & 0 & 0 \\ s_{13} & s_{23} & s_{33} & 0 & 0 & 0 \\ 0 & 0 & 0 & s_{44} & 0 & 0 \\ 0 & 0 & 0 & 0 & s_{55} & 0 \\ 0 & 0 & 0 & 0 & 0 & s_{66} \end{pmatrix}.$$

For a piezoelectric, the dielectric property depends on the mechanical condition and the elastic property depends on the electric condition. For example,

$$s^{D}_{ijkl} - s^{E}_{ijkl} = -d_{mij}d_{nkl}\beta^{X}_{mn}, \quad (i,j,k,l,m,n = 1-3) \tag{7.3}$$

and

$$\varepsilon^{x}_{ij} - \varepsilon^{X}_{ij} = -d_{ikl}d_{imn}c^{E}_{klmn}, \quad (i,j,k,l,m,n = 1-3), \tag{7.4}$$

where β_{mn} is defined as $\varepsilon_{ik}\beta_{jk} = \delta_{ij}$ and c_{klmn} is defined as $s_{ijkl}c_{klmn} = \delta_{(ij)(mn)}$.

The electrostrictive effect, which originates from the change in the dipole density in the material, is a nonlinear electromechanical effect and occurs in all substances, whether crystalline or amorphous, solid or liquid. The electrostrictive strain can be written as:

$$x_{ij} = Q_{ijkl}P_kP_l = M_{ijkl}E_kE_l, \quad (i,j,k,l = 1-3), \tag{7.5}$$

where Q_{ijkl} and M_{ijkl} are the charge- and field-related electrostrictive coefficient, respectively. It is also known that (Uchino et al. 1984):

$$Q_{ijkl} = -\frac{1}{2}\frac{\partial(1/\varepsilon_{ij})}{\partial X_{kl}}. \tag{7.6}$$

That is, the electrostrictive coefficient reflects the change in the dielectric permittivity caused by the mechanical stress.

For an isotropic polymer under an electric field along 3-direction with a polarization response P along 3-direction,

$$x_{33} = Q_{3333}P^2 = M_{3333}E^2 \quad \text{and} \quad x_{11} = Q_{1133}P^2 = M_{1133}E^2 \tag{7.7}$$

or

$$x_3 = Q_{33}P^2 = M_{33}E^2 \quad \text{and} \quad x_{11} = Q_{13}P^2 = M_{13}E^2, \tag{7.8}$$

where x_3 and x_1 are the strains along and perpendicular to the P direction, known as the longitudinal and transverse strains, respectively. For an isotropic polymer, it is experimentally and theoretically indicated that Q_{33} $(M_{33})<0$ and Q_{13} $(M_{13})>0$ (Kinase and Itoh 1977; Shkel and Klingenberg 1998). Hence, for a polymer, an increase in E or P will result in a contraction along the polarization direction.

For a linear dielectric polymer, the P is related to the permittivity as $P=(\varepsilon-\varepsilon_0)\,E$ so that the M is directly and only related to the Q. However, it should be noted that most polymers exhibit nonlinear dielectric behavior. Therefore, the $x \sim E$ plots do not exhibit a quadratic nature. In other words, the M is dependent on the strength of electric field. For an electrostrictive polymer, if a

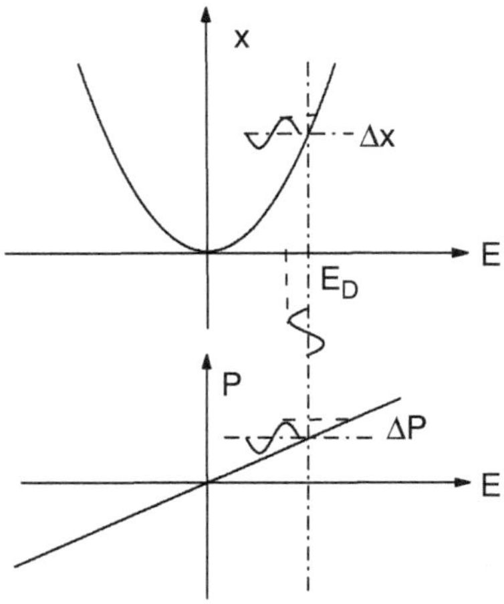

Fig. 7.1 A schematic view of an electrostrictive polymer, with a small AC electric field superimposed on a dc bias field

dc electric bias field with a small superimposed ac field is applied, an apparent piezoelectric effect can be observed as shown in Fig. 7.1.

For the field along 3-direction, the dominating ac strain response term is (from (7.5))

$$\Delta x_p = 2Q_{p3}P_D\Delta P, \quad p = 1-6, \tag{7.9}$$

where P_D is the P induced by the dc bias field and ΔP is the P change induced by the ac field as illustrated in Fig. 7.1. Under weak ac field, $\Delta P = (\varepsilon - \varepsilon_0)\,\Delta E$, which leads to

$$\Delta x_p = 2Q_{p3}(\varepsilon - \varepsilon_0)\Delta E. \tag{7.10}$$

Equation (7.10) is valid for both linear and nonlinear dielectric polymers as long as ΔE is small. The linear $x \sim E$ relationship in (7.10) resembles that in (7.1a). One can introduce an effective piezoelectric coefficient for dc field-biased electrostrictive polymers:

$$d = 2QP_D(\varepsilon - \varepsilon_0). \tag{7.11}$$

For an isotropic polymer with a P_D along 3-direction, (7.11) leads to

$$d_{33} = 2Q_{33}P_D(\varepsilon - \varepsilon_0) \quad \text{and} \quad d_{31} = 2Q_{13}P_D(\varepsilon - \varepsilon_0), \tag{7.12}$$

where d_{33} and d_{31} are the piezoelectric coefficients along and perpendicular to the induced polarization.

It should be noticed that the ε in **(7.10)–(7.12) are strongly dependent on the dc bias. Usually, the ε increases with the dc bias initially and then decreases with the dc bias. For electromechanical applications, electromechanical coupling factor k, which measures the efficiency of the material in converting energy between the electric and mechanical forms, is one of the most important parameters (IEEE Standard on Piezoelectricity 1988),

$$k^2 = \frac{\text{Convertedmechanicalenergy}}{\text{Inputelectricenergy}} = \frac{\text{Convertedmechanicalenergy}}{\text{Inputelectricenergy}}. \tag{7.13}$$

The k depends on the direction of the electric field and the directions associated with strain/stress. Hence, there are many coupling factors. For polymer actuators made with the electric field along 3-direction, if the actuation is along the 3-direction, the coupling factor is k_{33}, while if the actuation is along the 1-direction, the coupling factor is k_{31} as

$$k_{33}^2 = d_{33}^2/(\varepsilon_{33}^{\mathrm{X}} s_{33}^{\mathrm{E}}) \tag{7.14a}$$

and

$$k_{31}^2 = d_{31}^2/(\varepsilon_{33}^{\mathrm{X}} s_{11}^{\mathrm{E}}). \tag{7.14b}$$

For electrostrictive effect, the coupling factor has been derived based on the consideration of electrical and mechanical energies generated in the material under the external field (Hom et al. 1994):

$$k_{3p}^2 = \frac{g x_p^2}{s_{pp}^{\mathrm{D}} \left[P_E \ln\left(\frac{P_S+P_E}{P_S-P_E}\right) + P_S \ln\left(1-\left(\frac{P_E}{P_S}\right)^2\right)\right]}, \tag{7.15}$$

where $i = 1$ or 3 correspond to the transverse or longitudinal direction, s_{pp}^{D} is the elastic compliance under constant polarization, and x_p and P_E are the strain and polarization responses, respectively, for the material under an electric field E. The coupling factor depends on E. In (7.15), it is assumed that the polarization–field $(P-E)$ relationship follows approximately, $|P_E| = P_{\mathrm{S}} \tan h(g|E|)$, where P_{S} is the saturation polarization and g is a constant. The k can be related to the coefficients measured under different conditions. For example, the elastic compliance s_{33}^{E} is related to s_{33}^{D} (measured under constant field or charge condition) as $s_{33}^{\mathrm{D}} = (1-k_{33}^2)s_{33}^{\mathrm{E}}$. Therefore, a polymer with a large k will see a large difference in the elastic compliance when used under different external electric boundary conditions, which may be utilized to tune the elastic modulus of the polymeric material by varying the electric conditions.

7.3 PVDF-based Ferroelectric Polymers

A FE polymer can be in a single crystal form or, as in most cases encountered, a semicrystalline form in which the FE crystallites are embedded in an amorphous matrix. Examples of FE polymers include PVDF, copolymers of VDF with trifluoroethylene (TrFE) or tetrafluoroethylene (TFE), and odd-numbered nylons. This section will discuss PVDF homopolymer and P(VDF–TrFE) copolymers in detail.

7.3.1 Phases and Phase Transitions

PVDF is polymorphic and has at least four major crystalline phases (Lovinger 1982; Lovinger et al. 1982). Two of them, the form I (β-phase) and form II (α-phase), are the most relevant phases for practical ferro-/piezo-electric applications. In form I, two chains in an all-*trans* planar zigzag conformation are packed into individual orthorhombic as shown in Fig. 7.2 (Hasegawa et al. 1972), where a, b and c are lattice dimensions of the unit cell. It is noted that in the all-*trans* conformation, the fluorine atoms are positioned on one side of the unit cell, resulting in a net dipole moment. The form I unit cell is quite polar, having a net dipole of 2.1 D ($=7\times10^{-30}$ Cm), which results in a spontaneous polarization of $130\,\mathrm{mC/m^2}$. PVDF and P(VDF–TrFE) at β-phase is FE. In form II, the chain conformations are represented as a sequence of alternating *trans* and *gauche* sequences, or $\mathrm{TGT}_{\bar{G}}$ as shown in Fig. 7.3 (Takahashi et al. 1983), where each unit cell has two VDF repeat units. In form II, adjacent chains are packed so that the dipole moments of the individual carbon–fluorine bonds are aligned perpendicular to the chain direction, canceling one another out. The directions of the chains consist of a statistical average of up–up and up–down orientations.

For PVDF homopolymer, the first principle calculation results show that the α-phase is the most stable phase (Su et al. 2004). When prepared using the melt crystallization or solution cast, in most cases PVDF forms α-phase. Mechanical stretching is often used in order to convert the α-phase to the FE β-phase. On the other

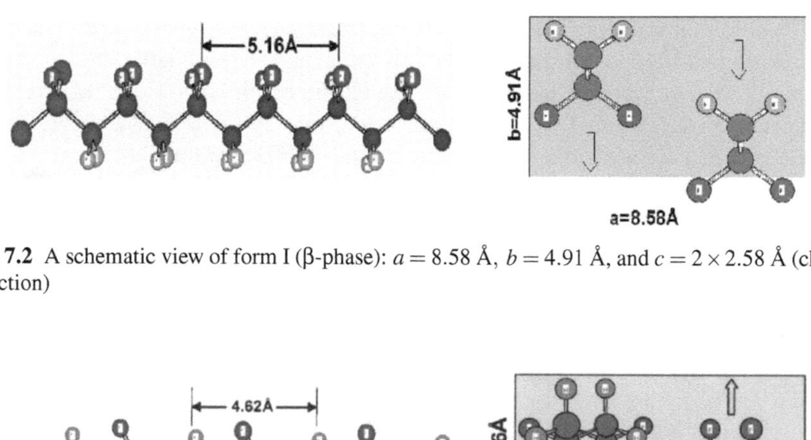

Fig. 7.2 A schematic view of form I (β-phase): $a=8.58$ Å, $b=4.91$ Å, and $c=2\times2.58$ Å (chain direction)

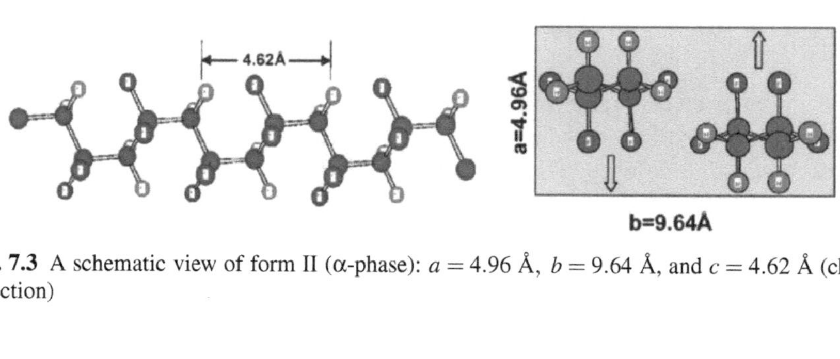

Fig. 7.3 A schematic view of form II (α-phase): $a=4.96$ Å, $b=9.64$ Å, and $c=4.62$ Å (chain direction)

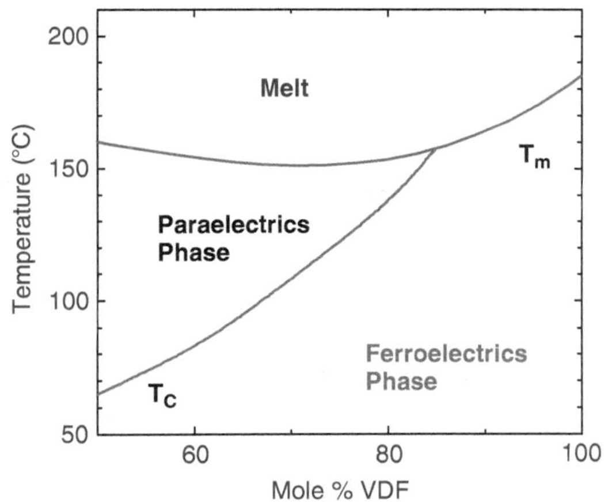

Fig. 7.4 Phase diagram of P(VDF–TrFE) polymers showing a ferroelectric–paraelectric transition

hand, for P(VDF–TrFE) copolymers with VDF content less than ~85 mol%, the β-phase will be formed directly. In the β-phase crystallites there are FE domains, which are polar but nonetheless orientated in all crystallographically allowed directions. Furthermore, in the semicrystalline polymer, these crystallites are randomly oriented within the sample. This accounts for the absence of any piezoelectric activity unless the sample is poled. To obtain piezoelectricity, the domains must be oriented in a strong electric field called the "poling field." Poling can be accomplished by electroding the polymer surfaces with a metal, followed by application of a strong electric field. An alternative method of poling is the use of a corona discharge where a corona charge is injected into the polymer from a needle electrode. In corona poling, no electroding is required.

Figure 7.4 shows the phase diagram for P(VDF–TrFE) system, where a FE to paraelectric (PE) transition that signals a change from a FE (polar) phase to a PE (nonpolar) phase is observed (Lovinger et al. 1983). The FE–PE transition temperature (T_c) increases with VDF mole fraction. Below the T_c, the crystal is best represented as an ordered form I structure with long sequences of all-*trans* bonds. As the temperature rises and goes through the FE–PE transition, an increasing number of *gauche* bonds are introduced into the ordered all-*trans* structure. As a result, the polarization in the crystal regions tends to take a disordered state, leading to the formation of the PE phase containing a random mixing of TG, T$\bar{G}$, TTTG, and TTT$\bar{G}$. Eventually, at higher temperatures the PE phase passes through the melt transition (Lovinger 1981; Takahashi and Tadokoro 1980).

One should note from Fig. 7.4 that P(VDF–TrFE) copolymers with high VDF concentrations do not appear to possess distinct FE–PE transitions; rather, it appears that melting takes place before a FE–PE transition. However, it must be mentioned that even in the FE phase, conformational defects can be introduced as the temperature of the polymers is raised. These defects are introduced so subtly that they may not be apparent in thermal studies such as DSC. In P(VDF–TrFE), in addition to a low-temperature (LT) phase where the chain conformation is predominantly all-*trans*, a cooled (CL) phase has been identified. Structural analysis has indicated that the most probable structure of the so-called CL phase is a mixture of two disordered crystalline phases, one transplanar and the other 3/1 helical (frozen-in high temperature phase) (Lovinger 1982; Lovinger et al. 1982; Davis et al. 1982; Fernandez et al. 1987; Tashiro and Kobayashi 1986). Because of this "frozen-in" disorder, copolymers with VDF contents below 50 mol% lose ferroelectricity and do not exhibit a clear phase transition signal.

7.3.2 Ferroelectric Responses

Figure 7.5a shows the *P–E* loop measured on a stretched film of P(VDF–TrFE) 68/32 mol% copolymer for which, at low cyclic electric fields (<55 MV/m), the loop is nearly linear with very little hysteresis. At field amplitudes exceeding 70 MV/m, a well-defined *P–E* loop appears. The coercive field E_c and remnant polarization P_r do not change appreciably with the applied field amplitude. In contrast, unstretched 50/50 mol% copolymer displays little hysteresis at fields below 25 MV/m. As the field amplitude increases beyond that, the loop gradually expands and both E_c and P_r increase with the applied cyclic field amplitude, as shown in Fig. 7.5b. Eventually, when the field amplitude exceeds 100 MV/m, E_c and P_r are saturated. Moreover, even at fields far below E_c measured from the saturated loop, polarization is often switchable. Such behavior is related to the nucleation process in the polarization switching (Lines and Glass 1977; Jona and Shirane 1993). Because of the semi-crystalline nature, the magnitudes of P_r and E_c are particularly sensitive to sample preparation conditions. For instance, Fig. 7.6 illustrates that both crystallinity and remnant polarization for stretched 68/32 mol% copolymer film increase with the sample annealing temperature.

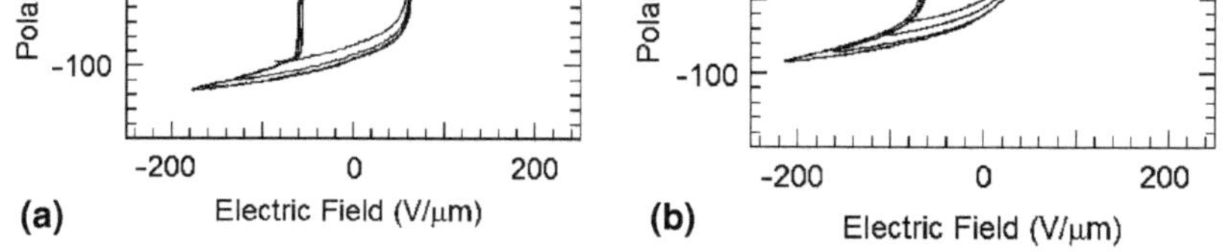

Fig. 7.5 *P–E* loops under different fields: (**a**) stretched P(VDF–TrFE) 68/32 mol% film; (**b**) unstretched of P(VDF–TrFE) 50/50 mol% film

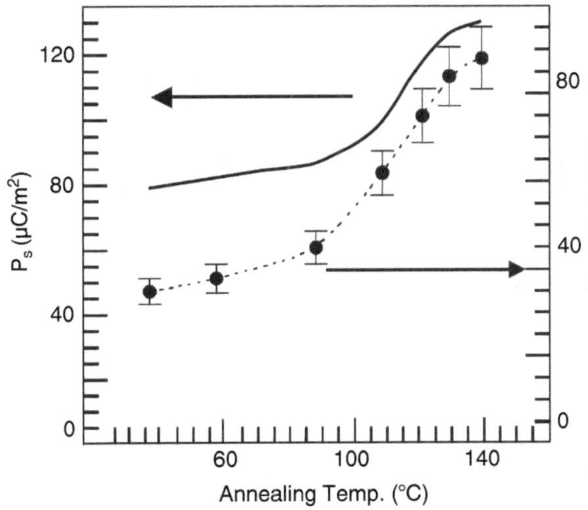

Fig. 7.6 The effect of annealing temperature on the polarization and crystallinity

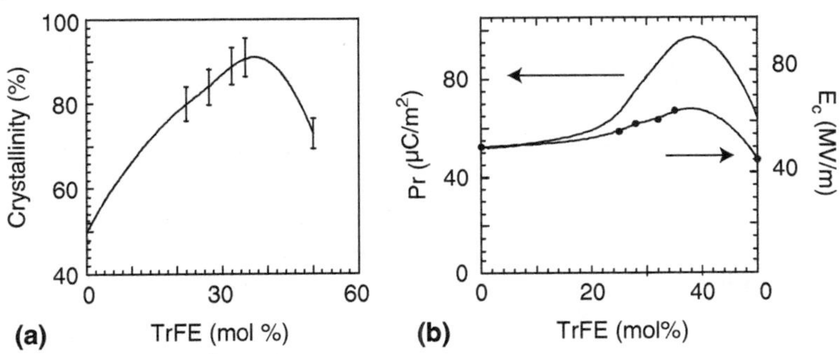

Fig. 7.7 (**a**) The dependence of the crystallinity on TrFE and (**b**) the dependence of the remnant polarization (P_r) and coercive field (E_c) on TrFE for P(VDF–TrFE) copolymers

The crystallization process in the copolymer is also influenced by the mobility as well as the stereoirregularity of the polymer chains (Sharples 1966). For PVDF homopolymer, the crystallinity is at about 50% (Scheinbeim et al. 1979). With increasing TrFE content, because of the larger size of the TrFE monomer unit, the lattice spacing between the polymer chains expands and thus facilitates and enhances the crystallization process (Fig. 7.7). Hence, in the copolymer of 75/25 mol%, the degree of crystallinity can reach 90%. In contrast, in polymer with higher TrFE content, the stereoirregularity introduced by TrFE reduces the stability of the crystalline phase with respect to the amorphous and leads to a lowering of the crystallinity. The dependence of the crystallinity with VDF content is depicted in Fig. 7.7a. Similarly, the dependence of the P_r and E_c on VDF content is illustrated in Fig. 7.7b.

Besides the crystallinity and P_r, the T_c for a given P(VDF–TrFE) copolymer can also be varied by conditions used in sample processing (Stack and Ting 1988; Green et al. 1986; Tanaka et al. 1999). For instance, it has been demonstrated that for P(VDF–TrFE) 60/40 mol% copolymer, the T_c can range from 65 °C to 85 °C (Green et al. 1986). This behavior is attributed to variations in the TrFE concentration in the crystalline phase for samples prepared under different conditions. Such variations lead to different T_c. However, it is also quite possible that the polar ordering in the crystalline region and the morphology of the polymer is affected by these differing processing conditions.

For example, the crystallites in the samples prepared using different methods may have different sizes or shapes with different interfacial layers. Based on Figs. 7.2 and 7.3, one can find that the α- and β-phase have a great difference in the dimension of their unit cell. That is, the FE–PE transition is associated with a large lattice strain. In the 65/35 mol% copolymer, a strain as high as 11% was observed (Tashiro et al. 1990). The corresponding lattice strain change through the transition perpendicular to the polymer chain is presented in Fig. 7.8.

For PVDF and its copolymers, the strong coupling between the P and x is electrostrictive in nature (Sundar and Newnham 1992). However, in the polarization switching process, it has been observed that (7.7) does not adequately describe the strain–polarization relationship. This is primarily due to the fact that the polarization switching is through the domain-wall motion. For example, the P–E loop and the corresponding strain change for 65/35 mol% copolymer are shown in Fig. 7.9. As can be seen, there is very little strain change as the polarization switches from "B" to "A" (Furukawa and Seo 1990).

Early Fourier transform infrared (FTIR) and X-ray studies demonstrated that the switching is primarily through successive 60° domain-wall motions. Owing to the pseudo-hexagonal symmetry of the unit cell in the directions perpendicular to the polymer chain, 60° domain-wall motions are not expected to generate very high strains (Furukawa and Seo 1990; Dvey-Aharon et al. 1980; Guy and Unworth 1988; Takahashi et al. 1980; Takahashi and Odajima 1984). Therefore, to achieve a

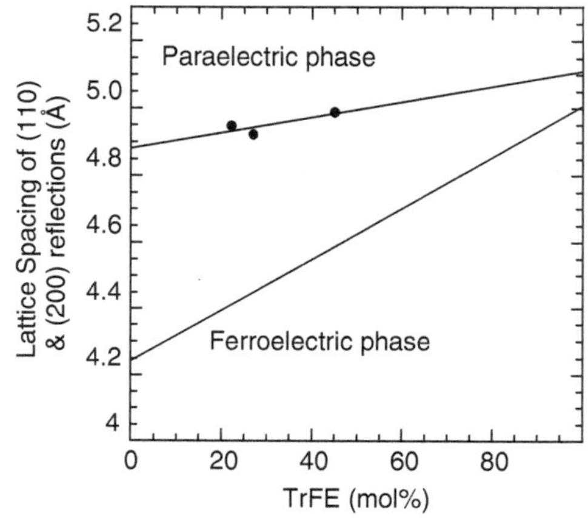

Fig. 7.8 The lattice spacing perpendicular to the polymer chain vs. the VDF content for P(VDF–TrFE) polymers

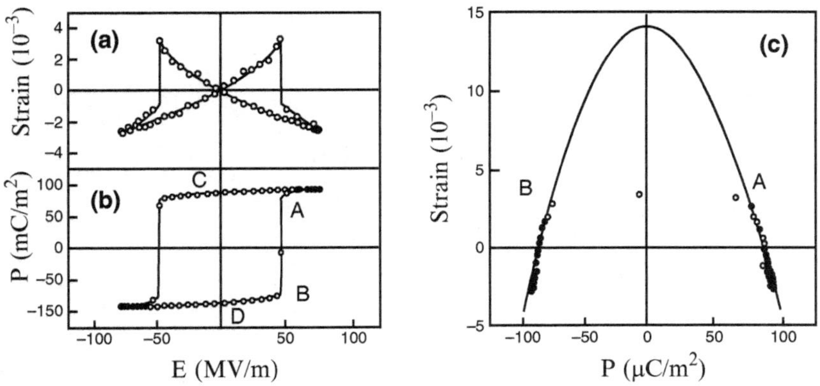

Fig. 7.9 A linear relationship of strain vs. P^2 is observed in the nonswitching part of the polarization response (Furukawa and Seo 1990)

Table 7.1 Piezoelectric, dielectric, and elastic properties of PVDF and P(VDF–TrFE) 75/25 mol%

Property	*PVDF*	P(VDF–TrFE) 75/25	
		Real	Imaginary
d_{31} (pC/N)	28	10.7	0.18
d_{32} (pC/N)	4	10.1	0.19
d_{33} (pC/N)	−35	−33.5	−0.65
s_{11} $(10^{-10}\,\mathrm{Pa}^{-1})$	3.65	3.32	0.1
s_{22} $(10^{-10}\,\mathrm{Pa}^{-1})$	4.24	3.24	0.07
s_{33} $(10^{-10}\,\mathrm{Pa}^{-1})$	4.72	3.00	0.07
s_{12} $(10^{-10}\,\mathrm{Pa}^{-1})$	−1.10	−1.44	−0.036
s_{13} $(10^{-10}\,\mathrm{Pa}^{-1})$	−2.09	−0.89	−0.022
s_{23} $(10^{-10}\,\mathrm{Pa}^{-1})$	−1.92	−0.86	−0.022
$\varepsilon_{33}/\varepsilon_0$	15	7.9	0.09
k_{33}		0.23	
k_{13}	0.13	0.07	
k_t	0.144	0.196	

high strain response in PVDF-based FE materials through the polarization switching mechanism, efforts should be expended to suppress the domain-wall motions that do not generate much strain response.

7.3.3 Electromechanical Properties

Various coefficients in the piezoelectric constitutive equations for PVDF and P(VDF–TrFE) 75/25 copolymer are listed in Table 7.1, where the data have been taken from several different sources (Wang et al. 1993a; Schewe 1982). In most

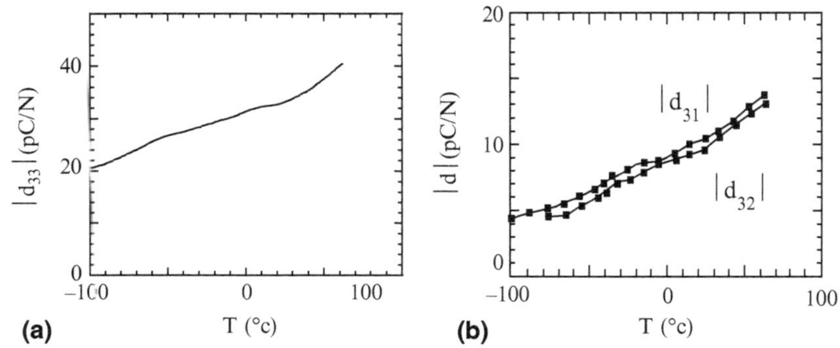

Fig. 7.10 The temperature dependence of the piezoelectric constant of P(VDF–TrFE) 75/25 mol% copolymer

polymers, the response to an external stimulus often shows strong relaxation behavior (McCrum et al. 1991). Hence, the coefficients in the piezoelectric constitutive equation are frequency dependent and complex (Holland 1967), which can be expressed as follows:

$$d^*_{ip} = d'_{ip} - jd''_{ip},\ s^*_{ip} = s'_{ip} - js''_{ip},\ \text{and } \varepsilon^*_{ij} = \varepsilon'_{ij} - j\varepsilon''_{ij}, \tag{7.16}$$

In Table 7.1, both the real and imaginary parts of these coefficients for 75/25 mol% copolymer are listed, which suggests that the imaginary part for the copolymer cannot be neglected. The temperature dependence of the piezoelectric constant of the copolymer is presented in Fig. 7.10. In semicrystalline copolymers, the piezoelectric properties can often be improved by raising the degree of crystallinity, since the piezoelectric response is mainly from the crystalline region. For example, using high-pressure crystallization to increase the crystallinity and improve the quality of the crystalline phase, Ohigashi et al. reported an improved thickness coupling factor for 80/20 copolymer to near 0.3 ($k_t = 0.3$) (Ohigashi and Hattori 1995).

More recently, a relatively large-sized single crystal of 75/25 mol% copolymer has been grown that has a room temperature $d_{33} = -38\,\text{pm/V}$ and $k_{33} = 0.33$, values which up to now represent the best piezoelectric performance of known piezoelectric polymers (Omote et al. 1997).

7.4 Electrostrictive Polymers

In most insulating polymers, the electrostriction is relatively weak. One reason for the small electrostrictive strain in polymers is the low dielectric constant. It was showed that the M for most polymers with the elastic modulus of 1 GPa is below $2 \times 10^{-20}\,\text{m}^2/\text{V}^2$ (Newnham et al. 1998). Hence, even under a field of 100 MV/m, the electrostrictive strain is below 2×10^{-4}. On the other hand, for polymers with low elastic modulus, the electrostrictive coefficient may become quite large (Newnham et al. 1998).

Recently, it has been shown that the electrostrictive response of FE polymers can be markedly improved by making use of the molecular conformation changes between the polar and nonpolar forms, which results in a large polarization change and consequently large electrostrictive strain and other electroactive response (Zhang et al. 1998). It is also found that the interfacial layer between the crystallite and amorphous regions, which is the key to understand some features of semicrystalline polymers, can generate large strain and polarization response (Li et al. 2004, 2006). This section will summarize recent progress for polymers with high electrostriction.

7.4.1 High Energy Electron Irradiated P(VDF–TrFE) Copolymers

Recently, it has been shown that in high energy electron irradiated (HEEI) P(VDF–TrFE) copolymers, the dielectric constant near 60 can be achieved at room temperature range and a high electrostrictive strain of about 5% under 150 MV/m field has been demonstrated, as shown in Fig. 7.11a (Zhang et al. 1998; Zhao et al. 1998; Cheng et al. 1999b).

The plot of strain vs. P^2 yields a straight line, indicating the response is electrostrictive in nature. For the HEEI copolymer, Q_{33} is found to be ranging from -4 to $-15\,\mathrm{m}^2/\mathrm{C}$ (the M is greater than 2×10^{-18}) (Cheng et al. 2000), depending on the sample processing conditions. The strain response does not change appreciably with temperature, as suggested by the plot in Fig. 7.11b.

The basis for such a large electrostriction in the HEEI copolymers is the large change in the lattice strain as the polymer goes through the transformation from the PE-like to FE-like phases. Besides the large strain response at FE–PE transition, there is another interesting feature associated with the FE–PE transition – the

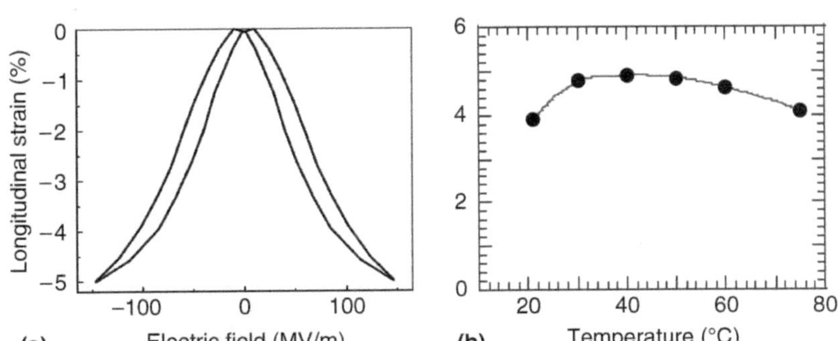

Fig. 7.11 A high electrostrictive strain, about −5% under 150 MV/m field from a high-energy electron irradiated P(VDF–TrFE) copolymer

possibility that a high coupling factor ($k \sim 1$) can be obtained near a first-order FE–PE transition (Zhang et al. 1995). Here, the irradiation introduces defects in the polymer, which lower the transformation temperature, eliminate the hysteresis, and broaden the transformation temperature region so that a high electrostriction can be obtained over a relatively broad temperature range.

Of special interest is the finding that in the HEEI copolymers, large anisotropy in the strain responses exists along and perpendicular to the chain direction. The transverse strain can be tuned over a large range by varying the film processing conditions. For unstretched films, the transverse strain is relatively small ($\sim +1\%$ at $\sim$100 MV/m) with an amplitude ratio of the transverse to longitudinal strain less than 0.33 (Cheng et al. 1999b). This feature is attractive for devices utilizing the longitudinal strain such as ultrasonic transducers in the thickness mode. On the other hand, for stretched films, a large transverse strain (x_1) along the stretching direction is obtained as shown in Fig. 7.12a, where a strain of about $+4.5\%$ was observed at 85 MV/m (Huang et al. 2004).

The coupling factors for the HEEI copolymers calculated using (7.16) are shown in Fig. 7.12b. At 80 MV/m, k_{33} reaches more than 0.3, which is comparable to that obtained in a single-crystal P(VDF–TrFE), while the k_{31} reaches 0.65 in a stretched copolymer (Cheng et al. 2001), which is much higher than that in piezopolymers

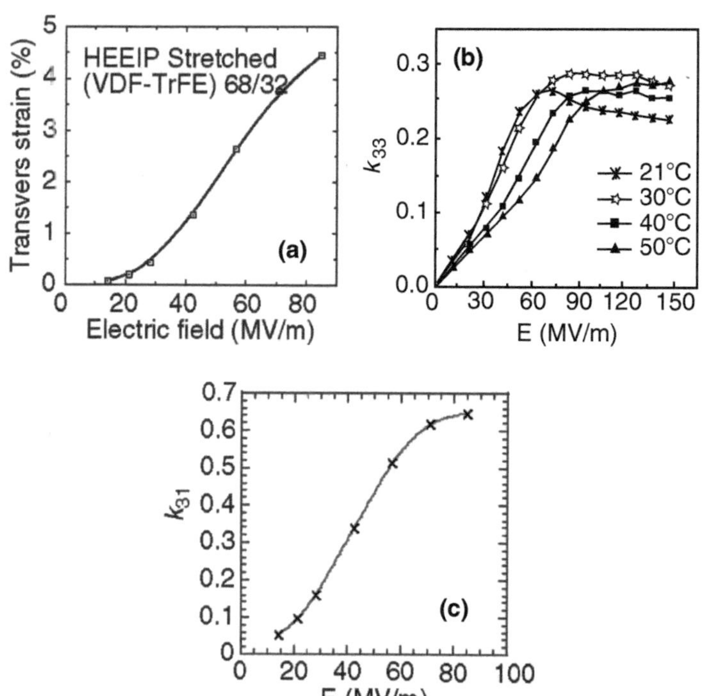

Fig. 7.12 Electromechanical response of HEEI Polymers: (**a**) transverse strain along stretching direction; (**b**) longitudinal coupling factor; (**c**) transverse coupling factor

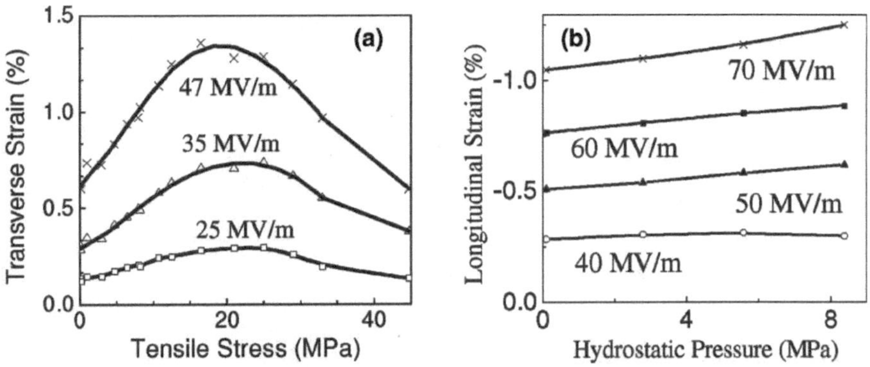

Fig. 7.13 (**a**) The transverse strain of stretched HEEI 65/35 copolymer under a tensile stress along the stretching direction and (**b**) the longitudinal strain of unstretched HEEI 65/35 copolymer under hydrostatic pressure

(Omote et al. 1997). These results are also verified by recent resonance studies in these polymers. In a polymer, there is always a concern about the electromechanical response under high mechanical load; that is, whether the material can maintain high strain levels when subject to high external stresses. Figure 7.13 depicts the transverse strain of a stretched HEEI copolymer under a tensile stress along the stretching direction and the longitudinal strain of unstretched HEEI copolymer under hydrostatic pressure (Bharti et al. 1999; Gross et al. 1999; Cheng et al. 1999a).

As can be seen from the figure, under a constant E the transverse strain increases initially with the load and reaches a maximum at the tensile stress of about 20 MPa. Upon a further increase of the load, the field-induced strain is reduced. One important feature revealed by the data is that even under a tensile stress of 45 MPa, the strain generated is still nearly the same as that without load, indicating that the material has a very high load capability.

The frequency response of the electrostrictive strain of the HEEI copolymer was studied and it was found that the high strain response can be maintained at least to 100 kHz, the upper limit of the study (Xu et al. 2002). In addition to the high electrostrictive strain, the HEEI copolymers also exhibit FE relaxor behavior, which is featured with high dielectric constant and strong dispersion as shown in Fig. 7.14a (Zhang et al. 1998; Cheng et al. 1998, 2002). The structure and thermal studies indicate that the broad dielectric constant maximum does not correspond to any phase transition. These behaviors are typical of FE relaxor, as observed in many inorganic materials (Cross 1994). It is noted that this is the first FE relaxor observed in polymers (Zhang et al. 1998).

Recently, high electrostrictive strain was also be obtained in the high energy proton irradiation on P(VDF–TrFE) 56/44 mol% copolymer (Guo et al. 2004). The M_{33} measured from the proton irradiated copolymer is about $-1.83\times10^{-18}\,\mathrm{m}^2/\mathrm{V}^2$, comparable to those in the HEEI copolymers. The irradiated copolymer also exhibits FE relaxor behavior. One principal effect of the irradiation is network formation (crosslinking) (Bharti et al. 2000). The 3D network of P(VDF–TrFE) with high

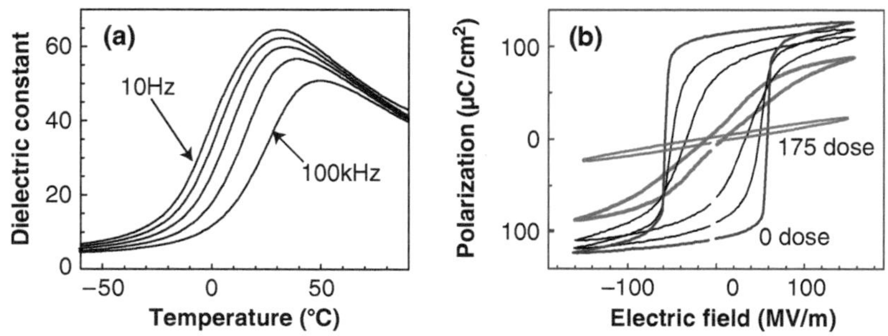

Fig. 7.14 (**a**) Dielectric constant at different frequencies vs. temperature for HEEI polymer; (**b**) *P–E* loops at room temperature for HEEI 68/32 polymer with different doses: 0, 35, 50, 75, and 175 Mrad, respectively

cross-link density was formed using a free-radical cross-linker and coagent. In such a P(VDF–TrFE) network, a thickness strain of more than -10% can be induced at 9 MV/m under static condition (Casalini and Roland 2001). By fitting the data using (7.8), the $M = -1.6\times10^{-15}\,\mathrm{m}^2/\mathrm{V}^2$, much higher than that in the HEEI copolymers. The authors suggested that the large strain is from the rotation of the small FE crystals, created because of the high density of cross-linking, which greatly reduces the crystallite sizes, under external field.

The HEEI copolymer also provides an interesting system to study FE relaxor phenomenon. Presented in Fig. 7.14b is the change in the *P–E* loop measured at room temperature for the P(VDF-TrFE) 68/32 mol% copolymer with the irradiation dose (Cheng et al. 2002b). With increased dosage, the near square *P–E* loop, characteristic for a normal FE material, is transformed to a slim *P–E* loop (at 75 Mrads). At very high dose (175 Mrad), the polymer becomes a linear dielectric in which the crystallinity is near zero.

7.4.2 P(VDF–TrFE)-Based Terpolymers

Although the irradiations can be used to convert P(VDF–TrFE) from the normal FE to a relaxor with high electrostriction, the irradiation also introduces many undesirable defects, such as the formation of radicals and chain scission (Bharti et al. 2000; Mabboux and Gleason 2002). From the FE response point of view, the defect modification can also be realized by introducing randomly in the polymer chain a third monomer, which is bulkier than VDF and TrFE. Furthermore, by a proper molecular design, the terpolymer can exhibit a higher electromechanical response than the HEEI copolymer. The high electrostrictive response has been observed in several fluoroterpolymers such as P(VDF–TrFE–CFE) (CFE, chlorofluoroethylene), P(VDF–TrFE–CTFE) (CTFE, chlorotrifluoroethylene), and P(VDF–TrFE–HFP) (HFP, hexafluoro-prolylene). To facilitate the discussion and comparison with the HEEI P(VDF–TrFE), the composition of the terpolymers (use P(VDF–TrFE–CFE) as an example) are labeled as VDF_x–$TrFE_{1-x}$–CFE_y, where the mole ratio of VDF/TrFE is $x/1-x$ and y is the mol% of CFE in the terpolymer.

Presented in Fig. 7.15a is the strain response along the thickness direction for a terpolymer. Under a field of 130 MV/m, a thickness strain of −4.5% is achieved, which is comparable to that observed in the HEEI copolymers. The high elastic modulus in this terpolymer (Y ∼ 1.1 GPa for unstretched films) results in a high elastic energy density, $YS^2/2$, ∼1.1 J/cm³. The k_{33} is presented in Fig. 7.15b. It is interesting to note that the k_{33} can reach more than 0.55 (Xia et al. 2002).

By increasing the ratio of VDF/TrFE in the terpolymer, the field-induced strain level can be raised because of the fact that the lattice strain between the polar conformation and nonpolar conformations increases with the VDF/TrFE ratio.

Presented in Fig. 7.16a is the thickness strain of the P(VDF–TrFE–CFE) terpolymer at a composition of 68/32/9 mol%, and a thickness strain of more than 7%

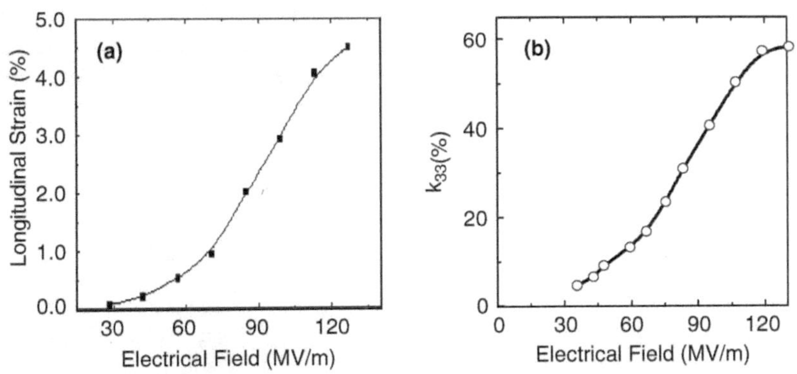

Fig. 7.15 (**a**) Longitudinal strain and (**b**) the k_{33} vs. applied field for the P(VDF–TrFE–CFE) 62/38/4 mol% terpolymer

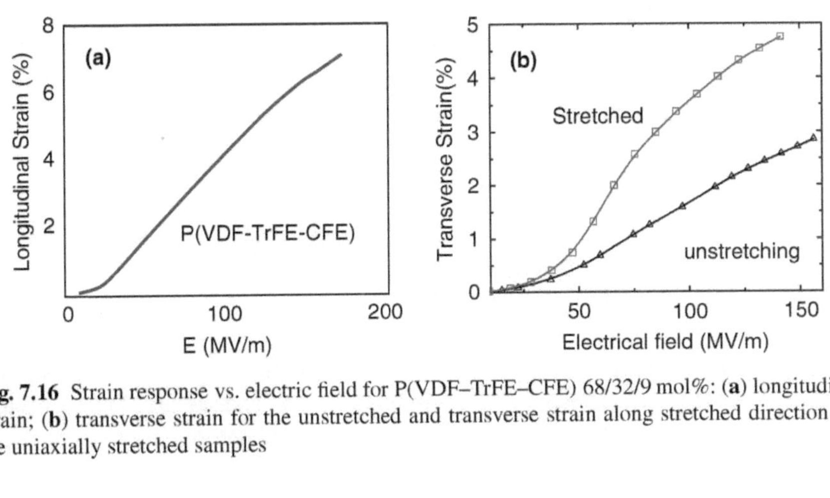

Fig. 7.16 Strain response vs. electric field for P(VDF–TrFE–CFE) 68/32/9 mol%: (**a**) longitudinal strain; (**b**) transverse strain for the unstretched and transverse strain along stretched direction for the uniaxially stretched samples

can be reached, which is by far the highest among all the HEEI copolymers and terpolymers investigated (Xia et al. 2002; Huang et al. 2004; Cheng et al 2001; Xu et al. 2001). The transverse strain response of this terpolymer is presented in Fig. 7.16b, where the transverse strain for uniaxially stretched film was measured along the stretching direction. For the uniaxially stretched films, the transverse strain perpendicular to the stretching direction is quite small and for stretched films, this strain is negative. Using the relationship $x_v = x_3 + x_1 + x_2$, where x_1 and x_2 are the transverse strains, the volume strain x_v can be determined. It was found that for the P(VDF–TrFE–CFE) terpolymer, the volume strain is very small (on the order of 10% of the thickness strain). This is different from the HEEI copolymers and suggests different strain generation mechanisms in these two classes of polymers.

In addition to P(VDF–TrFE–CFE) terpolymers, fluoropolymers with other termonomers such as CTFE, HFP, VC (vinyl chloride), and CDFE (chlorodifluoroethylene) (Xu et al. 2001; Chung and Petchsuk 2001; Chung and Petchsuk 2002; Garrett et al. 2003) have been studied. Among these terpolymers, P(VDF–TrFE–CTFE) was investigated in some detail regarding their electromechanical response and the results show that very large thickness strain was obtained at near static condition (strain of 1.2% under 20 MV/m, for example, corresponding an M of $3 \times 10^{-17}\,\mathrm{m^2/V^2}$). However, they exhibit a lower strain response at nonstatic condition than the P(VDF–TrFE–CFE) terpolymer (Xu et al. 2001; Buckley et al. 2002).

7.4.3 Computer Simulation of Nanoactuators in PVDF

The molecular origin of the large electrostrictive response in PVDF and related polymers makes it attractive for the nanoscale actuators and motors, which play central roles in nanotechnolgy. In PVDF, the most stable phase is the α-phase, which has TGTG′ modular conformation arranged in antiparallel manner. Unlike in macroscopic systems, where large energy barrier exists between different molecular conformations, in nano-crystals, reversible change between the α-phase and β-phase (all-*trans* conformation) can be induced electrically. Recently, Strachan and Goddard performed molecular dynamic simulation with a first principles-based force field (denoted MSXX) and show that large electrostrictive strain (∼5%) at extremely high frequencies ($>10^9$ Hz) can be achieved in PVDF nanocrystals (Strachan and Goddard 2005). The nanoactuator thickness is 2.3 nm.

7.4.4 Electrostrictive Graft Elastomers

Su et al. (2000) reported the observation of a high strain in a grafted elastomer. The graft elastomer consists of two components, a flexible backbone polymer and grafted crystalline polar groups. The flexible backbone polymer provides the amorphous chains for the formation of the three-dimensional network that is physically cross-linked by the grafted crystalline moieties. The grafted polar crystalline moieties also provide the electric field-responsive mechanism for the electric field-induced dimensional change, or electromechanical properties. The schematics in Figs. 7.17a, b show the structure and molecular morphology, respectively, of the grafted elastomers.

In the electrostrictive graft elastomer studied, the flexible backbone polymer is a copolymer of CTFE and TrFE, while the grafted polar crystalline polymer is a P(VDF–TrFE) copolymer. The thickness (longitudinal) strain response in the electrostrictive graft elastomer is shown in Fig. 7.18.

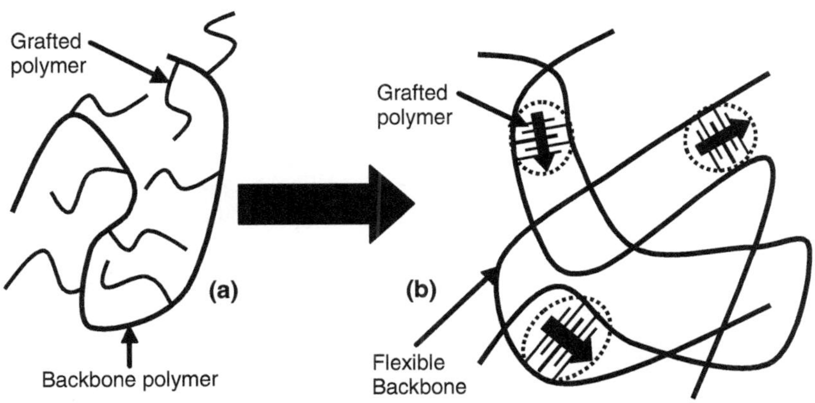

Fig. 7.17 Schematics of (**a**) structure and (**b**) morphology of the electrostrictive graft elastomers

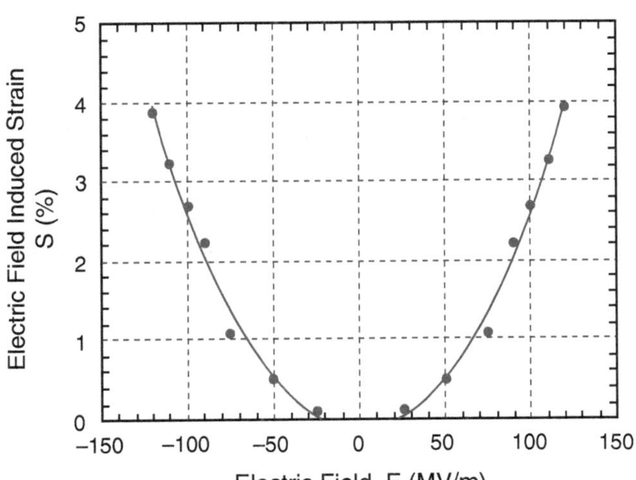

Fig. 7.18 Longitudinal strain vs. the electric field (Su et al. 2000)

In the case of polymeric elastomers exhibiting a large strain, two intrinsic mechanisms are considered as primary contributors: electrostriction and the Maxwell stress effect. For the newly developed electrostrictive graft elastomers, more than 95% of the strain response is contributed by the electrostriction mechanism, while the contribution from the Maxwell stress effect is less than 5%. Because of the relatively high elastic modulus (~0.55 GPa), the grafted elastomer also exhibits a high elastic energy density of $0.44\,\mathrm{MJ/m^3}$.

It is reported (Su et al. 2003) that the strain response of the elastomer can be controlled by thermal annealing treatment, which can increase the crystallinity of the grafted electric field responsive domains. The higher crystallinity is achieved, the higher strain is obtained. Since the mechanism of the electromechanical response in an electrostrictive graft elastomer is hypothesized to involve rotation of the polar crystal domains under an electric field, the overall strain should critically depend on the number or size of the polar domains; and therefore, the degree of crystallinity. Accordingly, the strain response of the elastomers significantly increases with annealing treatment. These trends are summarized in Fig. 7.19. In general, the strain responses increase with the crystallinity. At a given crystallinity, however, the elastomer containing more grafted polar domains gives a higher response. It is believed that the crystal size and distribution also affect the overall electromechanical response.

To understand the mechanisms of electrostriction in the graft elastomers, a model has been established (Wang et al. 1993). The model indicates that the mechanisms of the electric field-induced strain in the electrostrictive graft elastomer are attributed to the rotation of grafted polar crystal units and the reorientation of flexible backbone chains under an electric field.

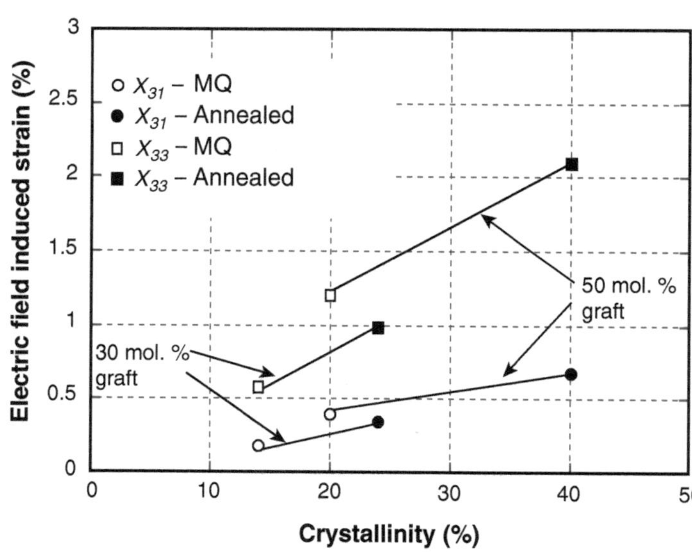

Fig. 7.19 Comparison of the crystallinity effects on the elastomer and the thermal treatment on the strain response (x_{31}, x_{33}) of the electrostrictive graft elastomers. (MQ: melt-pressed-quenched; annealed: annealed for 30 h at 120 °C)

7.4.5 PVDF-based Copolymers

To develop inexpensive electropolymers for mechanics, the commercially available PVDF-based copolymers have been investigated, inspired by the high electromechanical performance obtained in PVDF-based terpolymers. So far, most of the research in this area has been focused on P(VDF–HFP) and P(VDF–CTFE) copolymers, which are developed for other applications and have good stability and quality.

7.4.5.1 P(VDF–HFP) Copolymer

The electromechanical properties of P(VDF–HFP) copolymers were mainly investigated by two groups. In Dr. Gerhard-Muthaupt's group at University of Potsdam, the piezoelectric and pyroelectric properties were investigated. It is reported that the piezoelectric constant d_{31} can reach 30 pC/N in uniaxially stretched P(VDF–HFP) 85/15 copolymer film after poled with an electric field of more than 300 MV/m (Künstler et al. 2001). The piezoelectric constant d_{31} obtained in the polymer film is comparable or higher than that in other PVDF-based piezo-polymers. Further research on the copolymer film indicates that piezoelectricity originates from its ferroelectricity (Wegener et al. 2002). A remnant polarization of $24\,\mathrm{mC/m^2}$ and a coercive field of 100 MV/m were obtained. It seems that the HFP monomer as a comonomer in PVDF polymer decreases the switchable dipole moment.

In Dr. Scheinbeimn's group at Rutgers University, the electrostrictive performance of P(VDF–HFP) copolymer films under different thermal treatments was investigated (Lu et al. 2000; Jayasuriya et al. 2001). Although the copolymer exhibits a large polarization hysteresis loop, it is found that the quenched copolymer sample is at electrostrictive state with a relatively high strain response. The strain measurements were carried out at near static condition and the mechanism responsible for such a large strain response was not known.

7.4.5.2 P(VDF–CTFE) Copolymer

More recently, the electrostrictive behavior of P(VDF–CTFE) copolymers was investigated in Cheng's groups at Auburn University (Li 2004; Li et al. 2004, 2006). A high electromechanical response was obtained. As in others, the processing condition plays a very critical role on its properties. For a well-stretched and annealed P(VDF–CTFE) 88/12 copolymer film, a longitudinal electrostrictive strain reaches 5.5% as shown in Fig. 7.20a. It is found that the interfacial layer between the crystallite and amorphous regions plays an important role in the electrostrictive effect observed in P(VDF–CTFE). A linear $x_{33} \sim \mathrm{P}^2$ relationship was observed, which indicates the electrostrictive nature ($M_{33} = -1.23 \pm 0.02 \times 10^{-18}\,\mathrm{m^2/V^2}$) of the electromechanical response.

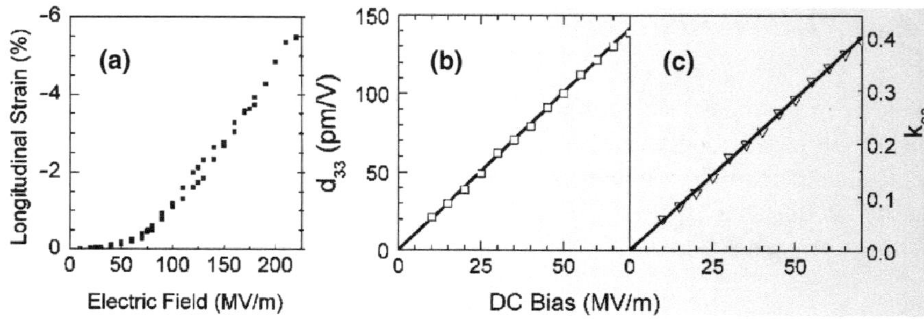

Fig. 7.20 Electromechanical response of P(VDF-CTFE) 88/12 copolymer at room temperature: (**a**) strain vs. electric field; (**b**) apparent piezoelectric constant d_{33}; and (**c**) corresponding k_{33}

The apparent piezoelectric response was determined by using a small ac field (fixed at 10 MV/m) imposing on a dc field and is shown in Fig. 7.20b. The piezoelectric constant d_{33} reaches 140 pm/V in the copolymer under 70 MV/m at room temperature. The elastic modulus and dielectric constant of the films were determined at different frequencies and temperatures. At room temperature, it is found that the dielectric constant is 12 and the Young's modulus is 820 MPa. Using (7.14a), the k_{33} for the film at room temperature is obtained as shown in Fig. 7.20c. It is very interesting to notice that at room temperature and under an electric field of 70 MV/m, k_{33} of 0.39, which is higher than that obtained in a single-crystal P(VDF–TrFE) and in the HEEI copolymers (Cheng et al. 2001), is obtained.

7.5 Ferroelectric Odd-Numbered Polyamides (Nylons)

Odd-numbered polyamides (nylon) is another class of FE polymer. In nylon, the $-$NH and $-$C $=$ O groups provide essential polar units for ferroelectricity and piezoelectricity. The piezoelectric properties of nylon 11 was reported in early 1980s (Mathur et al. 1988; Newman et al. 1980; Scheinbeim et al. 1986, 1991). Its ferroelectricity was predicted based on the chemical and crystal structure study. However, the ferroelectricity in nylons was not confirmed until 1990 when Scheinbeim et al. reported a typical FE *P–E* loop in a specially-treated, melt-quenched, and cold-stretching nylon 11 films (Lee et al. 1991a, 1991b). These films exhibit not only the FE switching mechanism but also a significantly enhanced piezoelectric property. The discovery has led to the discovery of the ferroelectricity and piezoelectricity of odd-numbered nylons, including nylon 9, nylon 7, and nylon 5.

7.5.1 Molecular Structures and Ferroelectricity

The α-phase structure of nylon 11 suggested by Slichter in 1959 is polar. The molecular conformation of the proposed structure is all *trans*, and for an odd-numbered

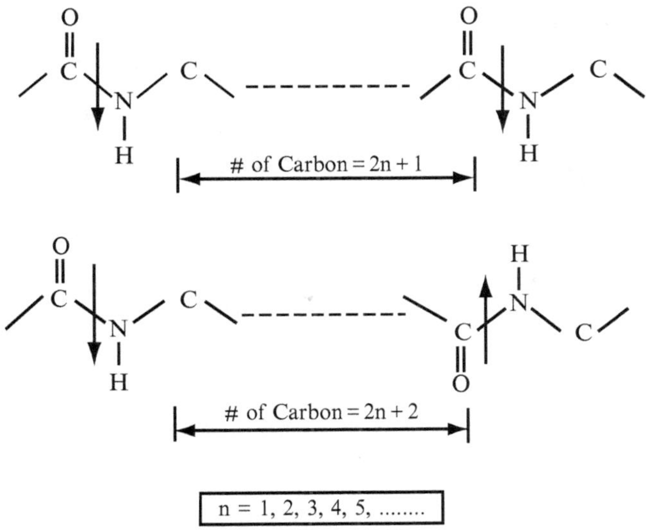

Fig. 7.21 Schematics of molecular structures of odd-numbered nylons and even-numbered nylons

nylon, this entails a net dipole moment per chain. Figure 7.21 shows the molecular structures of odd- and even-numbered nylons (Su et al. 1995). As can be seen, in odd-numbered nylons, the electric dipoles formed by amide groups (H–N–C = O) with a dipole moment of 3.7 D are sequenced in a way that all the dipoles are in the same direction synergistically, which results a net dipole moment. While in even-numbered nylons, the amide group dipoles are in two opposite directions, alternately. This results in an intrinsic cancellation of the dipole moments, as demonstrated schematically in the figure. For the polarization switching, a 90°-then-180° mechanism for the ferroelectricity of nylon 11 (Lee et al. 1991a, 1991b), the mechanism was confirmed by X-ray diffraction (XRD) and FTIR studies. This mechanism was also observed in other odd-numbered nylons, including nylon 9, nylon 7, and nylon 5. The investigation also discovered that the cold-stretching following the melt-quenching is a very critical step in obtaining the polyamide chains in the parallel form needed for the FE polarization switching and piezoelectricity. For instance, when a nylon 11 film is prepared using a melt-quenched and then uniaxially cold draw technique, a doubly-oriented structure was formed, with dipole orientation in the sheet plane parallel to the surface of the film while the molecular chains are oriented in the drawing direction.

There are two possible molecular arrangements: parallel (progressive) chain packing and antiparallel (staggered) chain packing, which are known as the molecular bases of the α-form and β-form crystals, respectively. In both cases, hydrogen-bonded sheets are formed. When an E is applied in the direction perpendicular to the film surface, the dipoles rotated to the field direction by 90° from a doubly-oriented state to be parallel to the applied electric field. In the following polarization reversals, the dipoles switched by 180° when the applied electric field

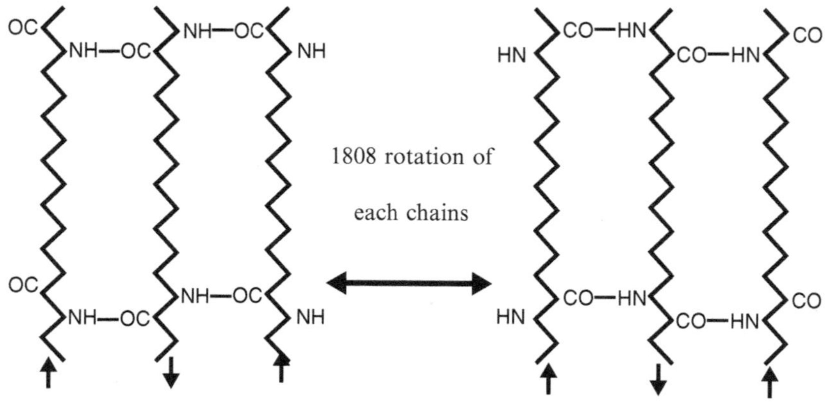

Fig. 7.22 A schematic representation of the 90°-then-180° polarization switching in the nylon 11 (redrawn and adopted from Lee et al. 1991a, 1991b)

Table 7.2 Structure and property of odd-numbered nylons

	Molecular weight of repeat unit	Dipole density (D/100A^3)	P_r (mC/m^2)
N-3	71.1	4.3	180[a]
N-5	99.1	2.92	125
N-7	127.2	2.12	86
N-9	155.2	1.65	68
N-11	183.3	1.4	56

[a]Predicted.

reversed direction (Scheinbeim and Newman 1993). A diagram of the simplified switching mechanism is shown in Fig. 7.22 and the arrows represent the orientation of the dipoles.

Studies indicate that the hysteresis loop of nylon 11 exhibit the following characteristics: It is found that the P_r increases but the E_c decreases with temperature. This is attributed to the increases of the mobility of the dipole moments. Therefore, the switching of the dipoles becomes easier and more dipoles can be switched. The P_r of the FE odd-numbered nylons is a function of the number, which is decisive to the electric dipole density. The smaller the number is, the higher the dipole density is, since the number represents the number of carbons between amide groups (H–N–C = O). The interrelationship between the number and the P_r is tabulated in Table 7.2 as reported by Mei et al. (1993).

7.5.2 *Piezoelectricity*

Similar to the PVDF, odd-numbered nylons are semicrystalline polymers and their piezoelectric properties are determined by the P_r and the mechanical properties. The

P_r depends on the crystallinity and the content of polar crystals and the alignment of the polar crystal domains. Thermal annealing is the most popular method to increase the crystallinity. For odd-numbered nylons, the annealing temperature should be significantly above the glass transition temperature and below the melting temperature. In the effective temperature range, the higher the temperature, the better the annealing effect. The method to generate the good alignment of polar crystal domains is electrically poling. Three primary factors are important for poling the nylons to achieve high P_r: poling temperature, field, and time. Generally, higher poling field and longer poling time result in higher P_r. For a well-poled (at room temperature) molten-quenched and uniaxially stretched (3.5 times) nylon 11 film, a P_r of $50\,mC/m^2$ and a d_{31} of 2.8 pC/N are obtained at room temperature. It should be noticed that the d_{31} is dependent on the temperature. For nylon 11, its glass transition temperature T_g ($\sim$70°C) is higher than room temperature. Therefore, a high piezoelectric coefficient is expected for nylon at higher temperature (Su et al. 1995; Takase et al. 1991). It is also demonstrated that the annealing treatment of poled nylon 11 and nylon 7 could enhance the piezoelectric strain coefficient and improved piezoelectric stability (Takase et al. 1991), which makes the odd-numbered nylons possess high and stable piezoelectric properties at high temperatures (up to 200 °C), which is a significant advantage over PVF_2, which will melt at the temperature.

7.5.3 Ferroelectric Polymer–Polymer Composites

Interfacial polarization and space charges may enhance polarization. Using two FE polymer constituents, PVDF and nylon 11, which have different dielectric constants, two typical polymer–polymer composite systems have been developed: (1) nylon 11-PVDF bilaminates (Su et al. 1995) and (2) nylon 11-PVDF blends (Gao et al. 1999). Both of them exhibit significantly enhanced P_r and piezoelectric properties. The enhancement should be attributed to the introduced interface between the two constituents (Su 1992; Gao and Scheinbeim 2000). The polarization, the piezoelectric properties, and the thermal stability of the composite systems depend on the relative composition of each constituent. In the case of the bilaminates, the interface exists between the layers of the two different constituents (oriented in the plane direction), while in the case of the blends, the interface exists between the domains of the two different constituents (randomly distributed). The difference of the interface types results in the difference of the enhanced FE and piezoelectric properties.

Acknowledgements Z.-Y. Cheng appreciates the financial support by USDA under Grant No. 2005-3560315817, USGS under grant G00003001, and 1 3M Earlier Career award. Mario El Tahchi gratefully acknowledges the support of Fulbright commission as a visiting scholar at the Pennsylvania State University. Q. Zhang thanks the financial support of this work by NIH under Grant No. R21EY016799-02, and ONR under Grant Nos. N00014-05-1-0455 and N00014-05-1-0541 (MURI).

References

Bharti V, Cheng ZY, Gross S, Xu TB, Zhang QM (1999) High electrostrictive strain under high mechanical stress in electron irradiated poly(vinylidene fluoride–trifluorethylene) copolymer. Appl Phys Lett 75: 2653.

Bharti V et al. (2000) Polarization and structural properties of high energy electron irradiated P(VDF–TrFE) copolymer films. J Appl Phys 87: 452–461.

Buckley GS et al. (2002) Electrostrictive properties of poly(vinylidenefluoride–trifluoroethylene–chlorotrifluoroethylene). Chem Mater 14 (6): 2590–2593.

Casalini R, Roland M (2001) Highly electrostrictive poly(vinylidene fluoride–trifluoroethylene) networks. Appl Phys Lett 79: 2627.

Cheng ZY, Katiyar RS, Yao X, Bhalla AS (1998) Temperature dependence of dielectric constant of relaxor ferroelectrics. Phys Rev B57: 8166–8177.

Cheng ZY et al. (1999a) Transverse strain responses in electrostrictive poly(vinylidene fluoride–trifluoroethylene) films and development of a dilatometer for the measurement. J Appl Phys 86: 2208.

Cheng ZY et al. (1999b) Transverse strain responses in the electrostrictive P(VDF–TrFE) copolymer. Appl Phys Lett 74: 1901–1903.

Cheng ZY et al. (2000) Effect of high energy electron irradiation on the electromechanical properties of poly (vinylidene fluoride–trifluorethylene) 50/50 and 65/35 copolymers. IEEE Trans Ultrason Ferroelectrics Freq Contr 47: 1296.

Cheng ZY et al. (2001) Electrostrictive poly(vinylidene fluoride–trifluoroethylene) copolymers. Sensor Actuator A: Phys 90: 138.

Cheng ZY, Zhang QM, Bateman FB (2002a) Dielectric relaxation behavior and its relation to microstructure in relaxor ferroelectric polymers – high-energy electron irradiated P(VDF–TrFE) copolymer. J Appl Phys 92: 6749–6755.

Cheng ZY et al. (2002b) Structural changes and transitional behavior studied from both micro- and macroscale in the high-energy electron-irradiated P(VDF–TrFE) copolymer. Macromolecules 35: 664.

Chung T, Petchsuk A (2001) Ferroelectric polymers with large electrostriction; based on semicrystalline VDF/TrFE/CTFE terpolymers. Ferroelectrics Lett 28: 135.

Chung T, Petchsuk A (2002) Synthesis and properties of ferroelectric fluoroterpolymers with Curie transition at ambient temperature. Macromolecules 35: 7678.

Cross LE (1994) Relaxor ferroelectrics. Ferroelectrics 151: 305.

Davis GT et al. (1982) Structural and dielectric investigation on the nature of the transition in a copolymer of vinylidene fluoride and trifluoroethylene (52/48 mol%). Macromolecules 15: 329.

Dvey-Aharon H et al. (1980) Kink propagation as a model for poling in poly(vinylidene fluoride). Phys Rev B: Condens Matter 21: 3700.

Fernandez MV et al. (1987) Study of annealing effects on the structure of VDF–TrFE copolymers using WAXS and SAXS. Macromolecules 20: 1806.

Furukawa T, Seo N (1990) Electrostriction as the origin of piezoelectricity in ferroelectric polymers. Jpn J Appl Phys 29 (4): 675.

Gao Q, Scheinbeim JI (2000) Dipolar intermolecular interactions, structure development, and electromechanical properties in ferroelectric polymer blends of Nylon-11 and poly(vinylidene fluoride). Macromolecules 33: 7564.

Gao Q, Scheinbeim JI, Newman BA (1999) Ferroelectric properties of Nylon 11 and poly (vinylidene fluoride) blends. J Polym Sci Part B: Polym Phys 37: 3217.

Garrett JT et al. (2003) Electrostrictive behavior of poly(vinylidene fluoride–trifluoroethylene-chlorotrifluoroethylene). Appl Phys Lett 83: 1190.

Green JB, Farmer BI, Rabolt JF (1986) Effect of thermal and solution history on the curie point of VF2–trFE random copolymers. J Appl Phys 60 (8): 2690.

Gross SJ, Cheng ZY, Bharti V, Zhang QM (1999) Mechanical load effects on the electrostrictive strain of P(VDF–TrFE) copolymer and the development of a high-resolution hydrostatic-pressure dilatometer. Proc IEEE 1999 Int Symp Ultrason: 1019–1024.
Guo S et al. (2004) High electrostriction and relaxor ferroelectric behavior in proton-irradiated P(VDF–TrFE) copolymer. Appl Phys Lett 84: 3349.
Guy IL, Unworth J (1988) Observation of a change in the form of polarization reversal in a P(VDF–TrFE) copolymer. Appl Phys Lett 52: 532.
Hasegawa R, Takahashi Y, Chatani Y, Tadokoro H (1972) Crystal structures of three crystalline forms of poly(vinylidene fluoride). Polym J 3: 600.
Holland R (1967) Representation of dielectric, elastic, and piezoelectric losses by complex coefficients. IEEE Trans Son Ultrason Su-14 (1): 18.
Hom C et al. (1994) Calculation of quasi-static electromechanical coupling coefficients for electrostrictive ceramic materials. IEEE Trans Ultrason Ferroelectrics Freq Contr 41: 542–551.
Huang C et al. (2004) P(VDF–TrFE–CFE) based high performance electroactive polymers. IEEE Transactions on Dielectrics and Electrical Insulation 11 (2): 299–311.
IEEE Standard on Piezoelectricity, ANSI/IEEE Std 176–1987, IEEE, New York (1988).
Jayasuriya AC et al. (2001) Crystal-structure dependence of electroactive properties in differently prepared P(VDF–HFP) copolymer films. J Polym Sci Part B: Polym Phys 39: 2793–2799.
Jona F, Shirane G (1993) Ferroelectric Crystals. Dover, New York, p 138.
Kinase W, Itoh M (1977) Theory of electrostriction of NaCl crystal. J Phys Soc Jpn 42 (3): 895–902.
Künstler W et al. (2001) Preparation and assessment of piezo- and pyroelectric P(VDF–HFP) copolymer films. Appl Phys A: Mater Sci Process 73 (5): 641–645.
Lee JW, Takase Y, Newman BA, Scheinbeim JI (1991a) Ferroelectric polarization switching in Nylon-11. J Polym Sci B: Polym Phys 29: 273.
Lee JW et al. (1991b) Effect of annealing on the ferroelectric behavior of Nylon-11 and Nylon-7. J Polym Sci: Part B: Polym Phys 29: 279.
Li ZM (2004) Novel electroactive poly(vinylidene fluoride)-based polymer systems and their applications. PhD Thesis, Auburn University.
Li ZM et al. (2004) Recrystalliation study of high-energy electron-irradiated P(VDF–TrFE) 65/35 copolymer. Macromolecules 37: 79–85.
Li ZM et al. (2006) Electromechanical properties of poly(vinyledene-fluoride–chlorotrifluoroethylene) copolymer. Appl Phys Lett 88: 062904.
Lines ME, Glass AM (1977) Principles and Applications of Ferroelectrics and Related Materials. Clarendon Press, Oxford.
Lovinger AJ (1981) Unit cell of the β phase of poly(vinylidene fluoride). Macromolecules 14: 322.
Lovinger AJ (1982) Poly(vinylidene fluoride). In: Bassett DC (Ed.) Developments in Crystalline Polymers-1. Applied Science Publishers, London, p. 195.
Lovinger AJ et al. (1982) Crystalline forms in a copolymer of vinylidene fluoride and trifluoroethylene (52/48 mol %) Macromolecules 15: 323.
Lovinger AJ, Furukawa T, Davis GT, Broadhurst MG (1983) Curie transitions in copolymers of vinylidene fluoride. Ferroelectrics 50: 227.
Lu XY et al. (2000) Giant electrostrictive response in P(VDF–HFP) copolymers. IEEE Trans Ultrason Ferroelectrics Freq Contr 47 (6): 1291–1295.
Mabboux P, Gleason K (2002) F-19 NMR characterization of electron beam irradiated vinylidene fluoride–trifluoroethylene copolymers. J Fluorine Chem 113: 27.
Mathur SC et al. (1988) Pizeoelectricity in uniaxially stretched and plasticized nylon 11 films. J Polym Sci B: Polym Phys 26: 447.
McCrum NG, Read BE, Williams G (1991) Anelastic and Dielectric Effects in Polymeric Solids, Chapter 4. Dover Publications, New York.
Mei BZ, Scheinbeim JI, Newman BA (1993) The ferroelectric behavior of odd-numbered Nylons. Ferroelectrics 144: 51.
Newman BA et al. (1980) Piezoelectricity in Nylon 11. J Appl Phys 51: 5161.

Newnham RE, Sundar V, Yimmirun R, Su J, Zhang QM (1998) Electrostriction in dielectric materials. Ceram Trans 88: 15–39.

Nye JF (1987) Physical Properties of Crystals. Clarendon Press, Oxford.

Ohigashi H, Hattori T (1995) Improvement of piezoelectric properties of PVDF and its copolymers by crystallization under high pressure. Ferroelectrics 171: 11.

Omote K, Ohigashi H, Koga K (1997) Temperature dependence of elastic, dielectric, and piezoelectric properties of "single crystalline" films of vinylidene fluoride trifluoroethylene copolymer. J Appl Phys 81 (6): 2760.

Scheinbeim JI, Newman BA (1993) Electric field induced changes in odd-numbered nylons. Trends in Polym Sci 1: 384.

Scheinbeim JI, Nakafuku C, Newman BA, Pae KD (1979) High-pressure crystallization of poly(vinylidene fluoride). J Appl Phys 50: 4399.

Scheinbeim JI, Mathur SC, Newman BA (1986) Field induced dipole reorientation and piezoelectricity in heavily plasticized Nylon 11 films. J Polym Sci B: Polym Phys 24: 1791.

Scheinbeim JI, Lee JW, Newman BA (1991) Ferroelectric polarization mechanisms in Nylon 11. Macromolecules 25: 3729.

Schewe H (1982) Piezoelectricity of uniaxially oriented polyvinylidene fluoride. Ultrason Symp Proc, Vol 1, IEEE, New York.

Sharples A (1966) Introduction to Polymer Crystallization. St Martin's Press, New York.

Shkel YM, Klingenberg DJ (1998) Electrostriction of polarizable materials: Comparison of models with experimental data. J Appl Phys 83: 415.

Stack GM, Ting RY (1988) Thermodynamic and morphological-studies of the solid state transition in copolymers of vinylidene fluoride and trifluoroethylene. J Polym Sci B: Polym Phys 26: 55.

Strachan A, Goddard WA III (2005) Large electrostrictive strain at gigahertz frequencies in a polymer nanoactuator: Computational device design. Appl Phys Lett 86: 083103.

Su J (1995) Ferroelectric and piezoelectric properties of ferroelectric polymer composite systems. PhD Thesis, Rutgers University.

Su J et al. (1995) Ferroelectric and piezoelectric properties of Nylon 11/poly(vinylidene fluoride) bilaminate films. J Polym Sci B: Polym Phys 33: 85.

Su J et al. (2000) Novel polymeric elastomers for actuation. Proc IEEE Int Symp Appl Ferroelectrics: 811.

Su J et al. (2003) Composition and annealing effects on the response of electrostrictive graft elastomers. Proc SPIE Smart Struc Mater 5051: 191.

Su HB, Strachan A, Goddard III WA (2004) Density functional theory and molecular dynamics studies of the energetics and kinetics of electroactive polymers: PVDF and P(VDF–TrFE). Phys Rev B: Condens Matter 70: 064101.

Sundar V, Newnham RE (1992) Electrostriction and polarization. Ferroelectrics 135: 431.

Takahashi Y, Tadokoro H (1980) Crystal structure of form III of poly(vinylidene fluoride). Macromolecules 13: 1317.

Takahashi N, Odajima A (1984) On the structure of poly(vinylidene fluoride) under a high electric field. Ferroelectrics 57: 221.

Takahashi T, Dale M, Fukada E (1980) Dielectric hysteresis and rotation of dipoles in polyvinylidene fluoride. Appl Phys Lett 37 (9): 791.

Takahashi Y, Matsubara Y, Tadokoro H (1983) Crystal structure of form II of PVDF. Macromolecules 16: 1588.

Takase Y et al. (1991) High-temperature characteristics of Nylon-11 and Nylon-7 piezoelectrics. Macromolecules 24: 6644.

Tanaka R et al. (1999) Annealing effect on the ferroelectric phase transition behavior and domain structure of P(VDF–TrFE): A comparison between uniaxially oriented VDF 73 and 65% copolymers. Polymer 40: 3855.

Tashiro K, Kobayashi M (1986) Ferroelectric phase transition and specific volume change in P(VDF–TrFE). Rep Progr Polym Phys Jpn 29: 169.

Tashiro K, Nishimura S, Kobayashi M (1990) Thermal contraction and ferroelectric phase transition in vinylidene fluoride–trifluoroethylene copolymers. II. An effect of tensile stress applied in the direction perpendicular to the chain axis. Macromolecules 23: 2802.

Uchino K, Nomura S, Vedam K, Newnham RE, Cross LE (1984) Pressure dependence of the refractive index and dielectric constant in a fluoroperovskite, $KMgF_3$. Phys Rev B: Condens Matter 29 (12): 6921–6925.

Wang H et al. (1993) Piezoelectric dielectric, and elastic properties of polyvinylidene fluoride/trifluoroethylene. J Appl Phys 74: 3394.

Wegener M et al. (2002) Ferroelectric polarization in stretched piezo- and pyroelectric P(VDF–HFP) copolymer films. J Appl Phys 92 (12): 7442.

Xia F et al. (2002) High electromechanical responses in a P(VDF–TrFE–CFE) terpolymer. Adv Mater 14: 1574.

Xu H et al. (2001) Ferroelectric and electromechanical properties of P(VDF–TrFE–CTFE) terpolymer. Appl Phys Lett 78: 2360.

Xu TB et al. (2002) High-performance micromachined unimorph actuators based on electrostrictive P(VDF–TrFE) copolymer. Appl Phys Lett 80: 1082.

Zhang QM et al. (1995) Characteristics of the electromechanical response and polarization of electric field biased ferroelectrics. J Appl Phys 77: 2549.

Zhang QM et al. (1998) Giant electrostriction and relaxor ferroelectric behavior in electron irradiated P(VDF–TrFE). Science 280: 2101.

Zhao X et al. (1998) Electromechanical properties of electrostrictive P(VDF–TrFE) copolymer. Appl Phys Lett 73: 2054.

Chapter 8
Low-Attenuation Acoustic Silicone Lens for Medical Ultrasonic Array Probes

Y. Yamashita,* Y. Hosono, and K. Itsumi

8.1 Introduction

Medical ultrasonic diagnostic apparatus have been used as a noninvasive method for diagnosing the human body since the 1970s. Figure 8.1 shows various types of commercialized medical pulse-echo ultrasound probes: (1) cardiac probe, (2) abdomen probe, and (3) high-frequency linear probe for diagnostic applications.

As shown in Fig. 8.2, the medical ultrasonic array probe consists of four basic materials: backing, piezoelectric, acoustic matching layers, and acoustic lens. A piezoelectric material such as lead zirconate titanate (PZT) or relaxor single crystals is the only active material used for the transducer, which transmits and receives ultrasound. On the other hand, acoustic matching layers are installed on the transducer in order to raise the transmission efficiency of the ultrasound, because the acoustic impedance (Z), which is expressed as sound velocity (c) $\times$ density(ρ), of PZT ceramic and that of human tissue are mismatched otherwise; i.e., $Z_{\mathrm{PZT}} = 35 \times 10^6\,\mathrm{kg/m^2\,s}$ vs. $Z_{\mathrm{tissue}} = 1.55 \times 10^6\,\mathrm{kg/m^2\,s}$. In general, the ultrasound is reflected at the boundary with different Z.

Moreover, a convex-shaped acoustic lens is attached on top of the acoustic matching layers to focus the ultrasound beam and to ensure good contact to the human body. Specifically, the acoustic lens is used for focusing of the short axis; the elevation dimension determines the effective slice thickness of the image plane. The ultrasound transmission is attenuated in acoustic matching layers, acoustic lens, and human tissue. The acoustic attenuation (α) of the acoustic lens is very important because it is usually thicker than the matching layers. Moreover, α becomes large with increasing frequency; the α of the acoustic lens has a great effect on the sensitivity of a medical ultrasonic probe as well.

Silicone rubber has been the predominant acoustic lens material for external focusing since 1974 (Iinuma and Hashiguchi 1974, 1978) because it is safe to use

---

Y. Yamashita
Toshiba Research Consulting Corporation, Kawasaki 212-8582, Japan

A. Safari, E.K. Akdoğan (eds.) *Piezoelectric and Acoustic Materials for Transducer Applications.*

Fig. 8.1 Various type of echo ultrasound probes: (I) cardiac probe, (II) abdomen probe, (III) high-frequency linear probe (Courtesy of Toshiba Medical Systems Corp.)

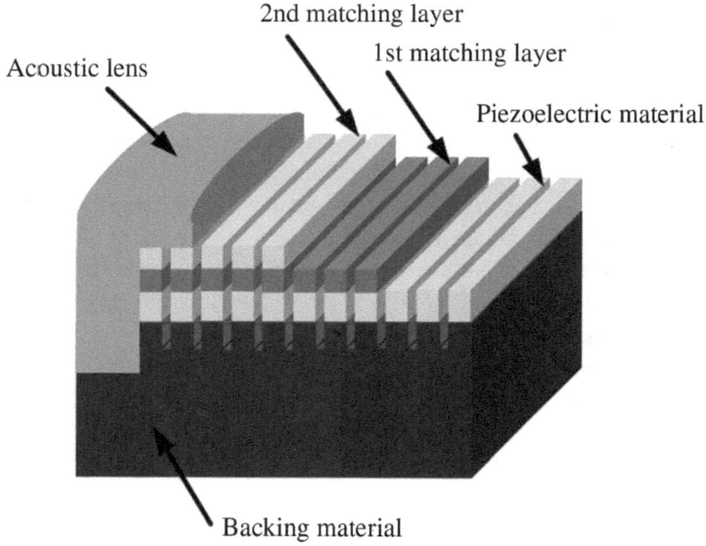

Fig. 8.2 Structure of medical ultrasound array probe

with the human body, it has a suitable sound velocity ($c \sim 1{,}000\,\mathrm{m/s}$), and it has reasonably low attenuation ($\alpha \sim 0.3\,\mathrm{dB/mm/MHz}$). However, the density (ρ) of undoped silicone rubber is $\sim 0.97 \times 10^3\,\mathrm{kg/m^3}$. Therefore, Z of undoped silicone rubber is $\sim 0.97 \times 10^6\,\mathrm{kg/m^2\,s}$, which is considerably different from Z of human tissue ($Z_{\text{tissue}} = 1.55 \times 10^6\,\mathrm{kg/m^2\,s}$). The difference in Z between the acoustic lens and the human tissue interface reduces the transmission efficiency of ultrasound. Therefore, many approaches have been developed to solve such problems in the last 30 years.

8.2 History of Acoustic Lens Development

Listed in Table 8.1 are the research and development milestones for acoustic lens materials (Iinuma and Hashiguchi 1974, 1978; Kikuchi 1979; Matsuo 1989; Kawabuchi et al. 1987; Shimamoto and Handa 1993; Sato and Shimamoto 1996; Solomon et al. 1996; Hayasaka et al. 1994; Fujii et al. 1995; Irioka et al. 2000; Saito and Fukase 2002; Yamashita et al. 2005, 2006; Hosono et al. 2007). The first report on silicone rubber-based acoustic lens was by Iinuma (Iinuma and Hashiguchi 1974, 1978). Kikuchi (1979) first reported that the *c* of silicone rubber can be changed by

Table 8.1 Historical research milestones of acoustic lens materials development

Authors	Year	Affirmation	Materials
Iinuma and Hashigushi	1974	Toshiba Corp., Japan	Silicone rubber doped with SiO_2
Kikuchi	1979	National Defense Academy, Japan	Silicone rubber doped with nylon, Al, Fe, Zn, and W metals
Matsuo	1989	Matsushita Elect. Indus. Co., Japan	Silicone rubber doped with TiO_2
Kawabuchi et al.	1987	Matsushita Elect. Indus. Co., Japan	4-Methylpentene
Shimamoto et al.	1993	Shin-Etsu Chemical Co., Japan	Silicone rubber doped with Al_2O_3, TiO_2, and plastic powder
Sato et al.	1996	Shin-Etsu Chemical Co., Japan	Silicone rubber doped with SiO_2 and polyether
Solomon et al.	1996	Hewlett Packard Co., USA	Hard plastics: PS, MMA, TRIS
Hayasaka et al.	1994	Nihonkai Medical, Japan	Silicone rubber doped with butadiene rubber
Fujii et al.	1995	Hitachi, Ltd., Japan	20 types of rubbers
Irioka et al.	2000	Matsushita Elect. Indus. Co., Japan	PS, TPX, silicone rubber
Saito et al.	2002	Matsushita Elect. Indus. Co., Japan	Silicone rubber doped with 40–50 wt% of SiO_2
Yamashita et al.	2005	Toshiba Corp., Japan	Silicone rubber doped with Pt metal powder
Hosono et al.	2007	Toshiba Corp., Japan	Silicone rubber doped with Yb_2O_3 powder

adding heavy-ρ materials such as Fe, Zn, and W metal powders. To bring Z of the silicone rubber close to that of human tissue, many inorganic materials such as SiO_2, Al_2O_3, and TiO_2 have been doped into the silicone rubber (Iinuma and Hashiguchi 1974, 1978; Kikuchi 1979; Matsuo 1989; Shimamoto and Handa 1993; Sato and Shimamoto 1996; Saito and Fukase 2002). Matsuo (1989) reported attenuation properties of silicone rubber doped with TiO_2 powder having a size of 80–200 nm. Since these inorganic materials have relatively low densities ranging from 2.2 to $4.0 \times 10^3\,\mathrm{kg/m^3}$, dopants in the amount of 44–65 wt% (16–46 vol%) are necessary for obtaining silicone rubber with $\rho > 1.55 \times 10^3\,\mathrm{kg/m^3}$. As a result, silicone rubber shows large $\alpha(>0.9\,\mathrm{dB/mm/MHz})$.

Kawabuchi (Kawabuchi et al. 1987) and Solomon (Solomon et al. 1996) investigated hard plastic materials, such as methylpentene (TPX), polystyrene (PS), methylmethacrylate (MMA), or trimethylsiloxy metaacryloxypropylsilane (TRIS), which has a c of 1,500–2,500 m/s. However, using these high-c plastics, it is impossible to fabricate an external focusing lens of desired design specifications (McKeighen 1998).

Fujii (Fujii et al. 1995) reported acoustic properties of 20 types of rubber materials. They found that fluorosilicone rubber, tetrafluroethylene polypropylene copolymer and phosphazene rubber have low c (1,021–1,150 m/s), which is suitable for external focusing lens applications. However, these materials have very high α(>2 dB/mm/MHz), which is unacceptable for use in medical probes. In fact, no practical material has yet been identified, which can possibly replace silicone rubber in the acoustic lens application.

8.3 Low-Attenuation Lens Development Concept

Generally, the formula for determining the requisite radius of curvature (R_c) for an acoustic lens is:

$$R_C = F_{\mathrm{eff}}\{(V_H - V_L)/V_L\}, \tag{8.1}$$

where F_{eff} is the desired effective geometric focal point, V_H is the sound velocity of human tissue, and V_L is the sound velocity of acoustic lens material ($V_L < V_H$ in this study).

Moreover, the thickness of the acoustic lens (T_L) is expressed by the following equation:

$$T_L = R_C - \{(R_C)^2 - (D/2)2\}^{1/2}, \tag{8.2}$$

where R_C is the radius of curvature and D is the aperture length. These equations indicate the following. First, the radius of curvature for the acoustic lens becomes large as the sound velocity of the acoustic lens becomes small. Second, the thickness of the acoustic lens becomes thin as the radius of the lens becomes large for constant aperture size (McKeighen 1998; Kossoff 1979). Figure 8.3 shows a relationship among thickness, aperture, radius of curvature, and collar of acoustic lens.

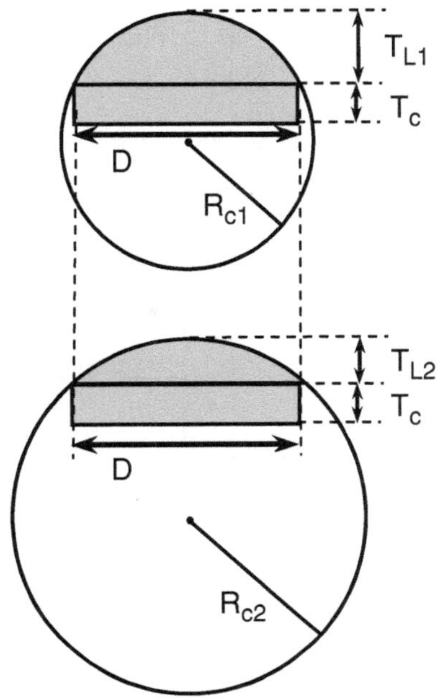

Fig. 8.3 Acoustic lens design, thickness, aperture, and radius of curvature of medical ultrasound probe

In general, the collar of the acoustic lens is added under the acoustic lens with a curvature in order to obtain proper hardness by increasing the thickness of the acoustic lens (the thickness of the collar is defined as T_C).

Therefore, total thickness of the acoustic lens is expressed as $T_L + T_C$. An acoustic lens material with low sound velocity is important for obtaining a thinner acoustic lens as shown in Fig. 8.3. Total α of an acoustic lens is determined by the following equation:

$$\text{Total}\,\alpha = \text{material} \quad \alpha \times \text{frequency} \times \text{thickness} \times 2(\text{transmit and receive}). \quad (8.3)$$

If a medical array probe has a center frequency of 8 MHz, then the lens thickness is 1.0 mm, which is typical for 30-mm focal point and material α of 1.0 dB/mm/MHz, resulting in a total α of 16 dB, which is very large. As we notice from (8.3), the thickness of the acoustic lens has a certain effect on the reduction of the total acoustic α as well as the material α of the plastic lens.

Figure 8.4 shows dopant densities vs. theoretical weight and volume % of silicone rubber to attain a silicone rubber density of $1.55 \times 10^3\,\text{kg/m}^3$, when base silicone rubber with ρ of $1.00 \times 10^3\,\text{kg/m}^3$ is used. High-ρ dopants, Pt, and Au require only 2.5–2.8 vol%, whereas mid-ρ dopants, TiO_2, Al_2O_3, $BaSO_4$, Fe_2O_3, and ZrO_2 require 10.5–17.3 vol%, and a low-ρ dopant, SiO_2, requires 46 vol% to attain ρ of $1.55 \times 10^3\,\text{kg/m}^3$. Powders with a size of less than 100 nm have large surface area (SA) and volume. Therefore, in the case of 10 vol% nanopowder charging of silicone rubber, fabrication becomes easy because of the lower viscosity of the liquid

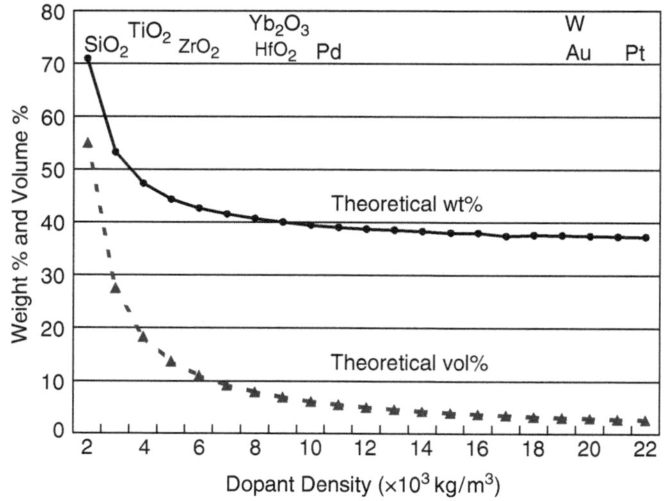

Fig. 8.4 Dopant densities vs. theoretical weight and volume % of silicone rubber to attain rubber density is $1.55 \times 10^3\,\text{kg/m}^3$

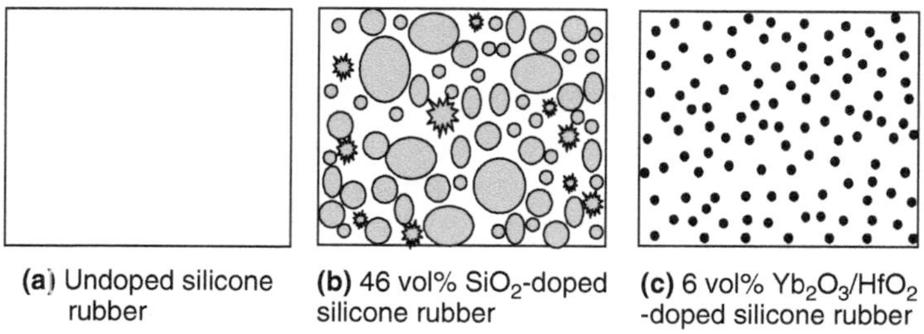

Fig. 8.5 Development concept of low-attenuation silicone-rubber-based lens materials: (**a**) Plain silicone rubber with $\rho = 0.97 \times 10^3\,\text{kg/m}^3$, $c = 980\,\text{m/s}$, $Z = 0.96 \times 10^6\,\text{kg/m}^2\,\text{s}$; (**b**) 46 vol% of SiO_2-doped silicone rubber with $\rho = 1.55 \times 10^3\,\text{kg/m}^3$, $c = 1{,}000\,\text{m/s}$, $Z = 1.55 \times 10^6\,\text{kg/m}^2\,\text{s}$; and (**c**) 6 vol% of Yb_2O_3/HfO_2-doped silicone rubber with $\rho = 1.55 \times 10^3\,\text{kg/m}^3$, $c = 880\,\text{m/s}$, $Z = 1.36 \times 10^6\,\text{kg/m}^2\,\text{s}$. Note: (**b**) and (**c**) have the same rubber density

room-temperature vulcanization (RTV) silicone rubber. However, in the case of charging with >10 vol% of nanopowder into the silicone rubber, fabrication is difficult since the viscosity becomes very high, and de-airing is impossible.

Figure 8.5 shows conceptual lens materials based on low-α silicone rubber. Undoped silicone rubber with a small amount of SiO_2 dopant has ρ of $0.97 \times 10^3\,\text{kg/m}^3$, c of 976 m/s, $Z = 0.96 \times 10^6\,\text{kg/m}^2\,\text{s}$, and reasonably low α as shown in Fig. 8.5a. Conventional acoustic silicone lens with 46 vol% of SiO_2 ($\rho = 2.2 \times 10^3\,\text{kg/m}^3$) dopant has ρ of $1.55 \times 10^3\,\text{kg/m}^3$, c of 1,000 m/s, Z of $1.55 \times 10^6\,\text{kg/m}^2\,\text{s}$, and a high $\alpha > 1.0\,\text{dB/mm/MHz}$ as shown in Fig. 8.5b. On the other hand, heavy metals or oxides such as Ni, Bi_2O_3, Yb_2O_3, Lu_2O_3, HfO_2, Ag, Pd, W, Au, and Pt, have high ρ of 8.9–21.5 $\times 10^3$ kg/m^3. Using these dopants, only 2.6–6.7 vol% of silicone rubber is necessary to realize a silicone rubber with ρ of 1.55×10^3 kg/m^3 as shown in Fig. 8.5c. It is well known that α is proportional to dopant volume concentration for a given particle size and ρ (Matsuo 1989). The acoustic α is also proportional to dopant ρ and particle size if the same volume quantity is added to silicone rubber. Therefore, doping a small amount of fine-particle-size heavy metal or oxide dopants into silicone rubber is one way of obtaining low-α silicone rubber. In addition, the high mechanical tear strength, low thermal conductivity, and reasonably high Shore-A hardness (>50) of silicone rubber along with a proper amount of dopants facilitates the fabrication of a thin acoustic lens with a complex cap-space shape. In general, silicone rubber doped with a large amount of coarse dopants shows poor mechanical strength and high thermal conductivity.

8.4 Sample Preparation and Measurements

RTV of liquid silicone rubber (TSE-3032, Momentive Performance Materials Japan Co.) was used as the base material in this study. The six metal and seven ceramic powders listed in Table 8.2 were added to the base silicone rubber. The dopant content was set such that $\rho \sim 1.5 \times 10^3$ kg/m^3 in this study ultimately. The average

Table 8.2 Physical properties of silicone rubbers doped with metal and ceramic powders

Dopant type	Dopant density ρ (10^3 kg/m^3)	Dopant particle size (nm)	Dopant content (wt%)	Dopant content (vol%)	Rubber density ρ (10^3 kg/m^3)
None	–	–	–	–	1.02
Ni	8.9	300	37.7	6.5	1.52
Ag	10.5	300	37	5.4	1.53
Pd	12	30	36.7	4.7	1.54
W	19.3	700	32.7	2.8	1.48
Au	19.3	500	35.8	2.8	1.53
Pt	21.5	10	34.9	2.5	1.52
Pt	21.5	35	45	2.5	1.53
Pt	21.5	94	45	2.5	1.53
SiO_2	2.2	2,000	45	27.8	1.35
SiO_2	2.2	2,000	64.8	46	1.54
TiO_2	3.89	100	44.4	17.3	1.52
Al_2O_3	3.97	100	43.8	16.7	1.51
$BaSO_4$	4.5	100	43.8	15	1.55
Fe_2O_3	5.24	30	41.9	12.3	1.54
ZrO_2	5.68	30	39.5	10.5	1.51
CeO_2	7.7	10	38.6	7.7	1.44
Bi_2O_3	8.9	100	37.8	6.5	1.54
Yb_2O_3	9.2	16	37.4	6.2	1.52
Lu_2O_3	9.4	16	37.4	6.1	1.54
HfO_2	9.7	7.3	37.4	5.9	1.54

primary particle size (D_p) of a powder was calculated using specific SA (m^2/g) measured by the Brunauer-Emmett-Teller (BET) method, assuming each particle to be spherical and of density ρ. The equation used for the calculation is:

$$D_p(\text{nm}) = 6{,}000/\rho\text{SA}. \quad (8.4)$$

These powders and liquid silicone rubber materials were put into a $150 - \text{cm}^3$ polyethylene cup and mixed using a high-speed rotation mixer at 1,000 rpm for 2 min and de-aired in a vacuum chamber for 10 min. Then, a 1.0-mm-thick silicone rubber sheet was fabricated using an applicator on a flat glass plate.

The ρ was measured by the water-immersion method. Moreover, Shore-A hardness was measured at 25°C by using a type A durometer (GS-719N, Teclock Corp.). Sound velocity and acoustic α (of longitudinal waves) were measured by a reflection method in water using a 5-MHz transducer with an element diameter of 15.8 mm (A309S, Panametrics Corp.) at 37°C ± 1°C. A 10-MHz pulser/receiver (Model 5800, Panametrics Corp.) were used to generate a pulse with an energy of 12.5 μ J and a damping value of 50 Ω. The output waveform from the transducer was sampled by a digital oscilloscope (TDS420, Tektronix, Inc.) through a 50-Ω coaxial cable with a length of 0.5 m. Attenuation figure of merit (α-FOM) for acoustic lens material was set and calculated by multiplying attenuation (dB/mm/MHz) and sound velocity c (m/s) to compare lens material property. Microstructure of the silicone rubber surface was observed by field emission scanning electron microscopy (FE-SEM) with energy dispersive X-ray spectroscopy (EDX).

8.5 Results and Discussion

8.5.1 Fine Metal Powder Dopants

Figure 8.6 shows FE-SEM images of 500-nm Au, 10-nm Pt and 16-nm Yb_2O_3 powders. Three Pt powders of different particle size, namely 10, 35, and 94 nm, were used to investigate the particle size effect for the experiment. The 10-nm Pt is very

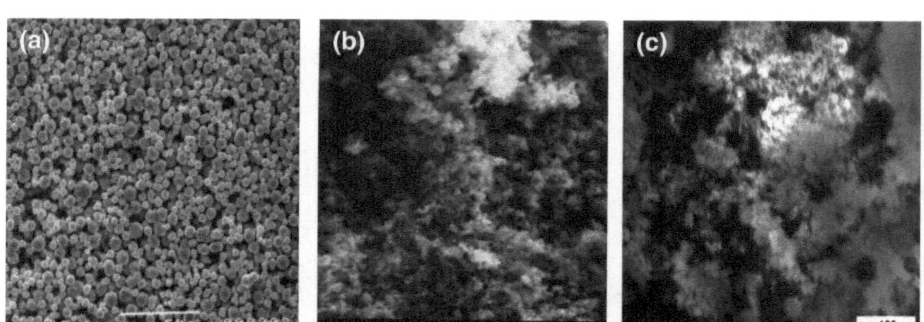

Fig. 8.6 FE-SEM images of (**a**) 500-nm Au, (**b**) 10-nm Pt, and (**c**) 16-nm Yb_2O_3 powders with 5,000, 10,000 and 100,000 magnifications

Table 8.3 Acoustic properties of silicone rubbers doped with metal and ceramic powders

Dopant type	Sound velocity, c (m/s)	Acoustic impedance, Z ($10^6\,kg/m^2\,s$)	Acoustic attenuation, α (dB/mm/MHz)	Attenuation figure of merit (α-FOM)	Shore-A hardness
None	976	1.00	0.43	420	35
Ni	896	1.36	1.03	922	50
Ag	885	1.35	1.89	1,672	45
Pd	847	1.30	0.81	686	40
W	983	1.45	3.2	3,140	34
Au	821	1.26	3.1	2,545	33
Pt	825	1.25	0.71	585	37
Pt	824	1.25	1.38	1,137	36
Pt	824	1.25	1.93	1,590	35
SiO_2	976	1.31	1.01	986	50
SiO_2	1,000	1.54	1.3	1,470	67
TiO_2	884	1.34	1.05	928	60
Al_2O_3	899	1.37	0.97	872	58
$BaSO_4$	923	1.42	1.84	1,698	58
Fe_2O_3	957	1.47	2.36	2,259	56
ZrO_2	887	1.34	0.92	816	54
CeO_2	922	1.33	1.33	1,229	60
Bi_2O_3	894	1.38	2.49	2,223	55
Yb_2O_3	882	1.36	0.93	820	55
Lu_2O_3	905	1.37	1.49	1,348	50
HfO_2	904	1.39	1.22	1,103	50

small compared with the other metal powders. Such a fine Pt powder is manufactured for application as catalyst for batteries. The Tables 8.2 and 8.3 show physical and acoustic properties of silicone rubber with different metal and ceramic powder dopants.

Figures 8.7 and 8.8 show the sound velocity vs. metal powder dopant density and attenuation α–FOM vs. acoustic impedance Z, respectively. Materials with lower α–FOMs with Z of $\sim 1.5 \times 10^6\,\mathrm{kg/m^2\,s}$ are better for the acoustic lens applications. High-ρ metal dopants decrease c to less than 900 m/s, except 700 nm W. Although Pt, Au, and W have similar bulk densities ($\rho = 19.3$–$21.5 \times 10^3\,\mathrm{kg/m^3}$) and thus similar volume fractions in the base silicone rubber, silicone rubber doped with 10-nm Pt shows the lowest α. On the other hand, silicone rubber doped with Au and W show the two highest α although their sound velocities are quite different. This result indicates α is not directly dependent on the volume of dopant but on the type of metal particles. In addition, large particle size, 94 nm, of Pt powder showed high α of 1.93 dB/mm/MHz, which indicates there is a size effect for α. It is well known that a smaller-particle-size dopant in the silicone rubber or epoxy resin shows lower α (Kikuchi 1979; Grewe et al. 1990). One of the reasons for lower α of silicone rubber doped with Pt is the smaller particle size of Pt (10 nm) compared with Au (500 nm) and W (700 nm). Another possible cause is a catalytic effect of metal powder. As shown in Table 8.3 and Fig. 8.8, silicone rubber doped with Pd

Fig. 8.7 Sound velocity of metal-doped RTV rubber vs. metal dopant density

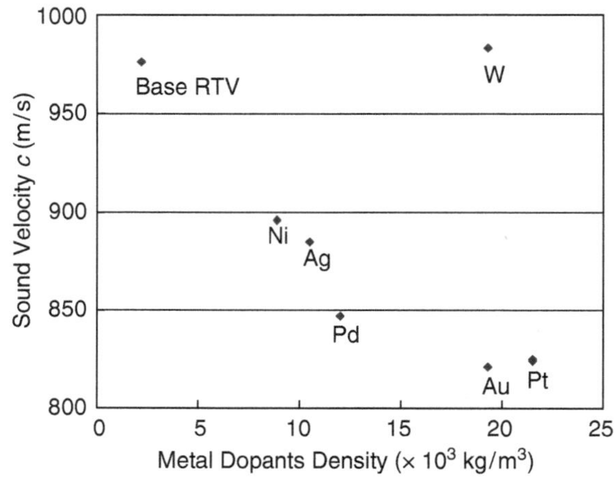

Fig. 8.8 Attenuation α-FOM vs. Z of metal powder doped RTV rubber

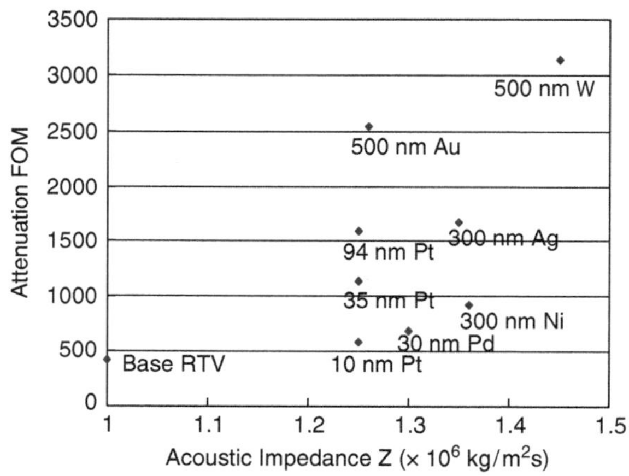

shows reasonably low α compared with those of Ag and Ni; all three metals have the same ρ ranging from 8.9 to $12 \times 10^3\,\mathrm{kg/m^3}$. Fine metals of Pt and Pd showed low α and also have a catalytic effect of vulcanization on silicone rubber.

Another important property of silicone rubber doped with Pt is its low c of 824 m/s. A low c of lens material makes the acoustic lens thinner and results in a probe with higher sensitivity. As mentioned earlier, it is possible to fabricate a thinner acoustic lens with the same focal point and lens width if the lens material has a low c as indicated by Hosono et al. (2007) and Kossoff (1979). Therefore, when 5–10-MHz array probe is manufactured by utilizing silicone rubber doped with Pt, it is possible to obtain a higher sensitivity compared with that of ordinary SiO_2-doped silicone rubber lens with the same Z. Hence, silicone rubber doped with Pt shows the best α-FOM as indicated in Table 8.3 and in Fig. 8.8, whereas silicone rubber doped with 10-nm Pt shows the lowest α-FOM with the same Z.

8.5.2 Fine Ceramic Powder Dopants

In the previous section, acoustic α properties of silicone rubber doped with heavy-ρ metal powders were reported (Yamashita et al. 2006, 2007). In particular, silicone rubber doped with 10-nm Pt powder showed good acoustic properties. However, the Pt-doped silicone rubber still has several disadvantages that require to be eliminated or overcomed. First, the Shore-A hardness of the Pt-doped silicone rubber is low. The low hardness induces the shape change of the acoustic lens when it is pressed to the human body, changing the focal point during a diagnosis thereby. Second, Pt is a precious metal, and therefore, very expensive. In addition, it has the propensity to catch fire easily when exposed to air. And at last, the color of silicone rubber doped with Pt is typically dark black, which is not preferable for users and designers of medical probes.

In what follows, the effects of nanometer-size ceramic powders, such as TiO_2, Al_2O_3, $BaSO_4$, Fe_2O_3, ZrO_2, CeO_2, Bi_2O_3, Yb_2O_3, Lu_2O_3, and HfO_2, on the acoustic α of silicone rubber will be discussed with the purpose of identifying the trends involved in developing a low-α silicone lens material with reasonable mechanical hardness and cost for medical probe applications.

Table 8.3 lists the physical and acoustic properties of silicone rubber with various dopants. Figure 8.9 shows c of silicone rubber doped with nanopowder vs. dopant densities of all samples. Silicone rubbers doped with ceramic TiO_2 and ZrO_2 showed low c of <900 m/s, whereas silicone rubber doped with Fe_2O_3 and $BaSO_4$ showed high c of more than 920 m/s, although all powder densities are similar ($\rho = 3.89$–$5.24 \times 10^3\,\mathrm{kg/m^3}$). High-$\rho$ oxide dopants, namely, Al_2O_3, TiO_2, ZrO_2, and Yb_2O_3, decreased c to less than 900 m/s, which is around 10% lower than that of undoped silicone rubber. However, low-ρ-dopant, SiO_2-doped silicone rubber showed high c of 976–1,000 m/s. Similar data was reported previously by Kikuchi (1979) and (Matsuo 1989).

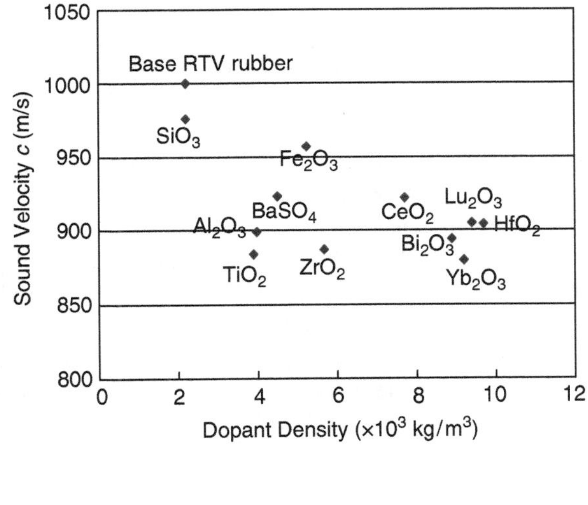

Fig. 8.9 Sound velocity of ceramic doped RTV rubber vs. ceramic dopant density

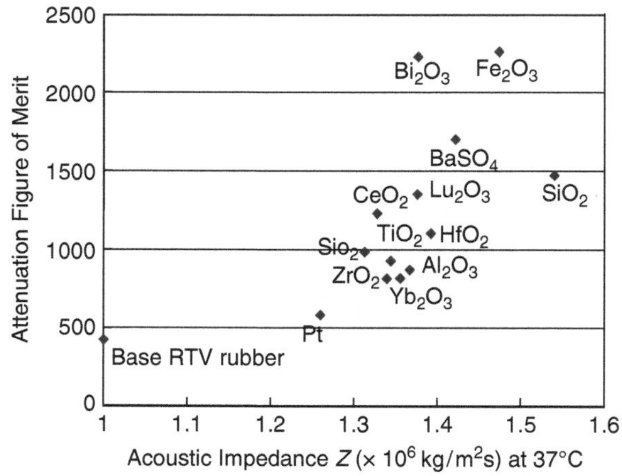

Fig. 8.10 Attenuation α-FOM vs. Z of ceramic powder doped RTV rubber

Figure 8.10 shows the α–FOM of each sample presented in Fig. 8.9. Materials with lower α–FOMs when $Z = 1.5 \times 10^6\,\text{kg/m}^2\,\text{s}$ are better for the acoustic lens application. Although these six dopants, namely TiO_2, Al_2O_3, $BaSO_4$, Fe_2O_3, ZrO_2, and Yb_2O_3, have similar bulk densities ($\rho = 3.9$–$9.2 \times 10^3\,\text{kg/m}^3$), and thus similar volume percentages (6.2–17.3 vol%) in the base RTV silicone rubber, acoustic α properties of the silicone rubber differ considerably.

As shown in Fig. 8.10, the RTV silicone rubber doped with Yb_2O_3 showed the lowest α-FOMs. It is interesting to note that, although Fe_2O_3 and ZrO_2 have similar bulk densities, $\rho = 5.24 \times 10^3\,\text{kg/m}^3$ and $5.68 \times 10^3\,\text{kg/m}^3$, and primary particle size of 30 nm, silicone rubber doped with ZrO_2 shows one of the lowest α, 0.92 dB/mm/MHz, whereas RTV silicone rubber doped with Fe_2O_3 shows the highest α, 2.36 dB/mm/MHz. This result indicates that α does not depend on the volume of dopants in the silicone rubber but on the characteristics of ceramic nanopowders. Therefore, microstructure observation was carried out using an FE-SEM and an EDX to investigate the mechanism of acoustic properties of silicone rubber doped with nanopowder.

Figure 8.11 shows FE-SEM photos of silicone rubbers doped with Pt, Yb_2O_3, and Fe_2O_3. RTV silicone rubber samples doped with metal Pt and ceramic Yb_2O_3 showed uniform surface. However, the sample doped with Fe_2O_3 showed large agglomeration (5–50 μm) in the silicone rubber matrix.

Figure 8.12 shows FE-SEM and EDX photos with magnification (×5,000) of silicone rubbers doped with Pt, Yb_2O_3, and Fe_2O_3. Metal Pt distribution in the RTV silicone rubber was very uniform than that of Yb_2O_3 powder, which showed agglomeration of 1–3 μm in the RTV silicone rubber. However, the Fe_2O_3 powders were located as agglomerated large particles, 5–50 μm, in the RTV silicone rubber. One reason for the agglomeration of nanopowder is considered to be the viscosity of the RTV silicone rubber liquids with different powders. When 2.5 vol% of 10-nm Pt powders was charged into liquid silicone rubber, the liquid had a low viscosity. However, when 12.3 vol% of 30-nm Fe_2O_3 powder was charged into the silicone

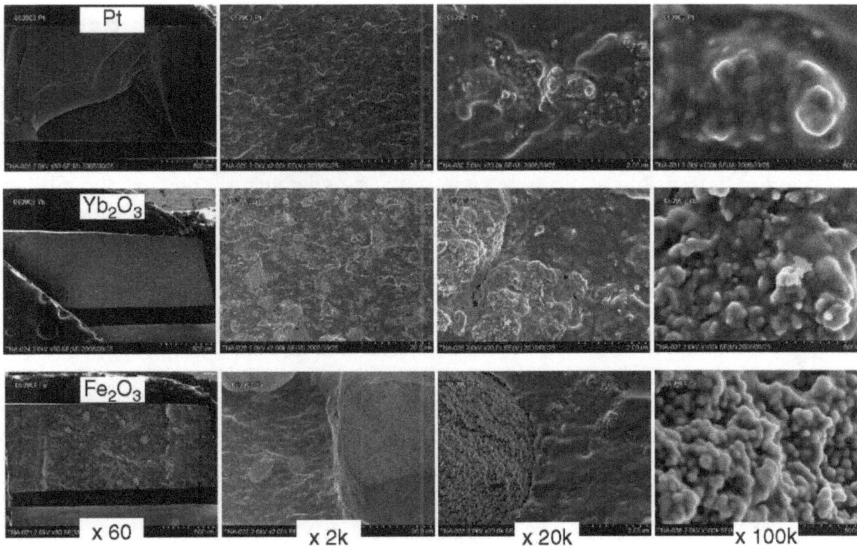

Fig. 8.11 FE-SEM images of silicone rubber doped with Pt, Yb_2O_3, and Fe_2O_3 nanopowders

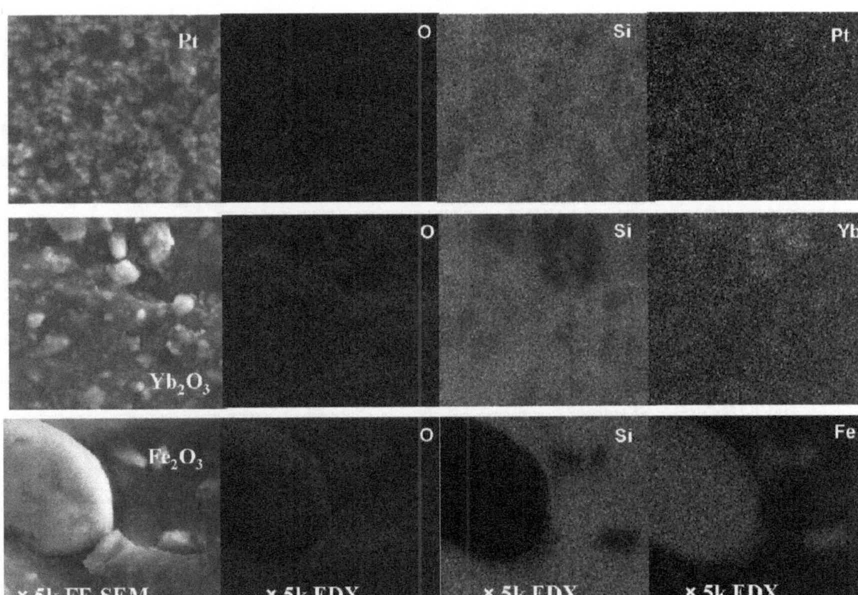

Fig. 8.12 FE-SEM and EDX images of silicone rubber doped with Pt, Yb_2O_3, and Fe_2O_3 nanopowders

rubber, the viscosity was very high and it was difficult to disperse the nanopowder into silicone rubber without any agglomeration. In the case of 16-nm Yb_2O_3 powder, the viscosity was between Pt and Fe_2O_3.

Some other researchers reported that silicone rubber with a smaller-particle-size dopant showed lower α (Yamashita et al. 2005; Grewe et al. 1990). The effects of Pt particle sizes ranging from 10 to 94 nm on α of RTV silicone rubber were also reported (Yamashita et al. 2006). In particular, it was shown that as the particle size of the powder became smaller, strong agglomeration induced by the Van der Waals forces and easy trapping of air and water on the powder surface occurred in silicone rubber, causing large-sized agglomerates in the RTV matrix. The experimental results indicated that α of silicone rubber doped with Pt, Yb_2O_3, and Fe_2O_3 powders does not depend on the primary particle size, but on the secondary particle size (agglomerate size) of powders in the silicone rubber. The said agglomerates (5–50 μm) were proposed as the root cause of the large-α of silicone rubber doped with Fe_2O_3 powder (primary particle size of 30 nm).

Therefore, to reduce α and c of RTV silicone rubber, one needs to take into consideration the agglomerate size in addition to the primary particle size, as well as agglomeration in the silicone matrix. Silicone rubber doped with Yb_2O_3 powder not only shows low α–FOM but also low c, as indicated in Table 8.3 and Fig. 8.10. The charging of high-ρ Yb_2O_3 ceramic powder ($\rho = 9.2 \times 10^3\,\mathrm{kg/m^3}$) to liquid RTV silicone rubbers is easy compared with the charging of other low-ρ nanopowders because of the lower volume content (<10 vol%) of nanopowder in the silicone rubber matrix as shown in Fig. 8.12. An acoustic lens material made out of silicone rubber doped with Yb_2O_3 powder provides for an increased sensitivity and bandwidth as it allows for a thinner acoustic lens than is conventionally used. Moreover, in such a lens, the focal point is not changed when the acoustic lens is pressed onto the human body since its Shore-A hardness is over 50. By virtue of its low c and low α, the silicone rubber doped with the high-ρ powder provides a better α–FOM for acoustic lens material applications than silicone rubber doped with conventional inorganic powder of SiO_2 does.

8.5.3 Temperature Dependence of Acoustic Properties of Silicone Rubber

Another important issue is the measurement condition of the acoustic properties. Figures 8.13a and 8.13b show temperature dependence of acoustic properties of base silicone rubber and RTV silicone rubber doped with Yb_2O_3. Acoustic attenuation α and sound velocity c change considerably by temperature in the 25–50°C interval. Since the transition temperature (T_g) of silicone rubber is around –150°C, the change in physical and acoustic properties is attributed to the changes in the structure on the silicone matrix with temperature, even around room temperature. Therefore, measurement of acoustic properties should be carried out in a temperature-controlled condition for the purpose of comparison with acoustic data reported by other researchers.

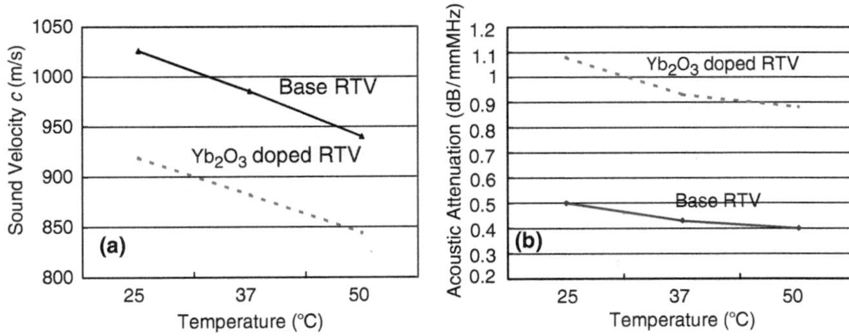

Fig. 8.13 (**a**) Temperature dependence of sound velocity of base RTV silicone rubber and Yb_2O_3-doped rubbers; (**b**) Temperature dependence of attenuation of base RTV silicone rubber and Yb_2O_3-doped rubbers

8.6 Future Prospect and Tasks

In the immediate future, we expect the c and α of the silicone rubber doped with Yb_2O_3 powder to be decreased by utilizing better nanopowder dispersion procedures, preventing agglomeration in RTV silicone rubber. Furthermore, we expect considerable improvement in silicone-based lens' properties such as low c, low α, low thermal conductivity, and moderate mechanical hardness, resulting in even higher $\alpha-$FOM. In medical array probe applications, increased sensitivity and bandwidth can be obtained by utilizing a thinner lens than what is conventionally used. Hence, future research is expected to also include studies on size, size distribution, stiffness, shape, surface treatment, dispersion, and agglomeration of the nanopowders on the acoustic property of the silicone rubber. Furthermore, development of new plastic materials, which meet all requirements for medical acoustic lens applications and not requiring dopants constitutes a major challenge for material scientists.

8.7 Conclusion

The effects of metal and ceramic nanoparticle powders with densities (ρ) 2.2–21.5 $\times$ $10^3\,\text{kg/m}^3$ and particle size 16–2,000 nm on the acoustic properties of RTV silicone rubber lens were investigated and the results obtained are summarized as follows.

Silicone rubber doped with fine metal particles, namely, 10-nm Pt powder and 30-nm Pd powder, showed the two lowest acoustic attenuations (α). There is a strong dependence between particle size of Pt dopant, ranging from 10 to 94 nm, and α of RTV silicone rubber. Silicone rubber doped with 10-nm Pt showed the following properties: $\rho = 1.54 \times 10^3\,\text{kg/m}^3$, sound velocity ($c$) $= 825\,\text{m/s}$, acoustic impedance $Z = 1.25 \times 10^6\,\text{kg/m}^2\,\text{s}$, and $\alpha = 0.71\,\text{dB/mm/MHz}$.

The mechanism of low α of silicone rubber doped with Pt is attributable to a small particle size of Pt powder. Another possible reason may be a catalytic effect of Pt and/or Pd powders in silicone rubber. The silicone rubber doped with high-ρ ($\rho = 9.2 \times 10^3\,\mathrm{kg/m^3}$) 16-nm Yb_2O_3 ceramic powder showed a low c and low α. The silicone rubber showed the following physical and acoustic properties: $\rho = 1.54 \times 10^3\,\mathrm{kg/m^3}$, $c = 880\,\mathrm{m/s}$, $Z = 1.36 \times 10^6\,\mathrm{kg/m^2\,s}$, $\alpha = 0.93\,\mathrm{dB/mm/MHz}$, and α-FOM of 822.

The silicone rubber doped with 30-nm Fe_2O_3 ($\rho = 5.24 \times 10^3\,\mathrm{kg/m^3}$) powder showed a high α of 2.4 dB/mm/MHz, whereas ZrO_2 ($\rho = 5.68 \times 10^3\,\mathrm{kg/m^3}$) having a similar particle size and ρ dopant showed a reasonably low α of 0.92 dB/mm/MHz. Microstructure observation by FE-SEM and EDX revealed that the high-α Fe_2O_3-doped silicone rubber has much more agglomerated powder of large size, 15–50 μm, in the silicone rubber matrix than does the low-α RTV silicone rubber doped with Pt or Yb_2O_3. Thus, α of the silicone rubber is not determined by the volume quantity and primary particle size of the dopant but by the secondary particle size in the RTV silicone rubber matrix.

High-ρ oxide dopants, namely, Al_2O_3, TiO_2, ZrO_2, and Yb_2O_3 ($\rho = 3.9$–$9.2 \times 10^3\,\mathrm{kg/m^3}$), decreased the c of the silicone rubber to below 900 m/s, which is around 10% lower than that of undoped silicone rubber. Acoustic properties of RTV silicone rubber have large temperature dependence. Therefore, the measurement should be carried out in a temperature-controlled condition for the purpose of comparison with other data. The discovery of the process parameter, namely, the better dispersion, to reduce c and α of the silicone rubber has important practical consequences in addition to being of scientific interest.

References

Fujii H, Nakaya C, Takeuchi H, Kondo T, Ishikawa Y (1995) Acoustic properties of lens materials for ultrasonic probes. Jpn. J. Appl. Phys. 34:312–315.

Grewe MG, Gururaja TR, Shrout TT, Newnham RE (1990) Acoustic properties of particle/polymer composites for ultrasonic transducer backing application. IEEE Trans. Ultrason. Ferroelect. Freq. Contr. 37:506–514.

Hayasaka T, Hayasaka H, Hosaka H (1994) Acoustic lens for medical diagnostic equipment. Japanese Published Patent Application No: H08–00615.

Hosono Y, Yamashita Y, Itsumi K (2007) Effects of metal oxide particle dopant on the acoustic properties of silicone rubber lens for medical echo probe. IEEE Trans. Ultrason. Ferroelectrics Freq. Contr. 54 (8):1589–1595.

Iinuma K, Hashiguchi M (1974) Ultrasonic transducer with silicone rubber lens. Jpn. J. Med. Ultrason. 28 (Suppl. 26):119–120 [in Japanese].

Iinuma K, Hashiguchi M (1978) Ultrasonic transducer with silicone rubber lens, Japanese Patent No. 1,191,390.

Irioka K, Tokunaga I, Fujii K (2000) Ultrasonic probe. Published Japanese Patent Application No. 2000–189417.

Kawabuchi M, Matsuo K, Muramatsu M, Fukumoto A, Saito K (1987) Ultrasonic transducer assembly for medical diagnostic examinations. United States Patent Application No. 4,699,150.

Kikuchi K (1979) Control of sound velocity in RTV silicone rubber by compounding various powders. Nihon Onkyou Gakkaishi 35:29–35 [in Japanese].

Kossoff G (1979) Analysis of focusing action of spherically curved transducers. Ultrasound Med. Biol. 5:359–365.

Matsuo K (1989) Ultrasonic probe lens. Japanese Patent No. 1,551,320.

McKeighen RE (1998) Design guideline for medical ultrasonic array. Proc. SPIE, 3341:2–18.

Saito K, Fukase K (2002) Ultrasonic diagnostic apparatus, probe acoustic lens and their production method. Published Japanese Patent Application No. 2002–95081

Sato M, Shimamoto N (1996) Acoustic lens composition. Japanese Patent No. 3,105,151.

Shimamoto N, Handa R (1993) Acoustic lens composition. Japanese Patent No. 1,799,240.

Solomon RJ, Vogel GG, King R.W, Williams S (1996) Interface element for medical ultrasound transducer. United States Patent No. 5,505,205.

Yamashita Y, Hosono Y, Itsumi K (2005) Effects of metal particle dopant on acoustic attenuation properties of silicone rubber lens for medical echo array probe. Jpn. J. Appl. Phys. 44:4558–4560.

Yamashita Y, Hosono Y, Itsumi K (2006) Effects of sub-100 nm platinum metal particle on the acoustic attenuation properties of silicone rubber lens for medical array probe. Jpn. J. Appl. Phys. 45:4684–4686.

Yamashita Y, Hosono Y, Itsumi K (2007) Effects of ceramic nanopowder dopants on the acoustic properties of silicone rubber lens for medical echo probe. Jpn. J. Appl. Phys. 46:4784–4789.

Chapter 9
Carbon-Fiber Composite Materials for Medical Transducers

Toshio Kondo

A composite, consisting of two or more materials, was developed, wherein optimum acoustic properties were obtained by the judicious choice of its constituents. Specifically, unidirectional carbon-fiber composite materials have a high acoustic impedance and a light weight. Therefore, it is possible to formulate a composite with broad-range acoustic impedance. By the approach described in this chapter, the applications utilizing such carbon-fiber composite in medical transducers are possible. In what follows, the acoustic properties of the carbon-fiber composite materials are presented and a method of fabricating the composite materials to form backing layers will be elaborated on. The characteristics of a prototype transducer, which uses this new composite, will be discussed.

9.1 Introduction

Most medical diagnostic and nondestructive testing systems use transducers composed of the piezoelectric lead–zirconate–titanate (PZT) ceramic as an active element. While the transducer is ringing, long pulse durations degrade the axial resolution. The acoustic wave can be damped by attaching an acoustically absorbing material to the back of the transducer, thereby absorb the ultrasonic energy from the active element. Ideally, this backing material should provide high attenuation, and it should match the acoustic impedance of the element to achieve efficient coupling. If the acoustic impedance of the backing is close to that of the element, most of the energy will be coupled into the backing. As a result, the transducer will have a shorter pulse. However, the use of a backing with higher acoustic impedance causes lower sensitivity because energy is absorbed into the backing. Hence, a compromise must be made in choosing the acoustic impedance of the backing layer (Auld 1990).

T. Kondo
Faculty of Engineering, Tokushima Bunri University, Sanuki-shi, 1314, Kagawa 769-2101, Japan

A. Safari, E.K. Akdoğan (eds.) *Piezoelectric and Acoustic Materials for Transducer Applications.*

The backing materials are generally made of composites of tungsten powder and epoxy resin or rubber. The specific acoustic impedance of ceramics such as the polycrystalline PZT family is about 30 MRayl. To match the impedance of PZT, a higher content of tungsten is required in the backing material. However, with tungsten and epoxy composites, it is difficult to achieve acoustic velocities above 2,000 m/s, uniform properties and light weight (Grewe et al. 1990). On the other hand, the acoustic velocity in the longitudinal direction of carbon-epoxy composites ranges from a very high value of ~10,000 m/s to a value of 6,000 m/s (Daniel and Ishai 1994). By varying the content of tungsten powder, the acoustic impedance of carbon-epoxy composites can be adjusted to suit the particular purpose for which it is to be used. Considering such composite material features, this material would allow the realization of a backing material with high specific acoustic impedance and a light weight. In this study, a method of fabricating such a carbon-fiber composite loaded with tungsten metal powder as backing material with controllable characteristics is presented. The acoustic impedance of the backing ranging from 14 to 34 MRayl has been achieved by employing the composite materials.

9.2 Unidirectional Carbon-Fiber Composite Materials

9.2.1 Theoretical Acoustic Properties

We discuss the general mechanical properties of composite materials in order to obtain a suitable acoustic impedance of the backing of transducers. The following consideration (Fukuda and Ben 1989) is based on the model of a unidirectional composite material shown in Fig. 9.1.

The properties of the composite material related to loading in the fiber direction L, E_L, and ν_{LT} are dominated by fibers that are usually stiffer and have a lower ultimate strain. The longitudinal Young's modulus is given by

$$E_{\mathrm{L}} = V_{\mathrm{f}}E_{\mathrm{f}} + V_{\mathrm{m}}E_{\mathrm{m}}, \tag{9.1}$$

where E_{f} and E_{m} are the longitudinal fiber and matrix Young's moduli, and V_{f} and V_{m} are the fiber and matrix volume ratios, respectively.

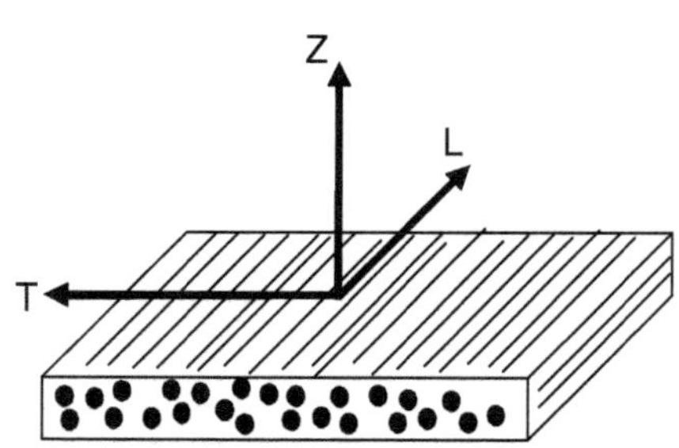

Fig. 9.1 Model of unidirectional carbon-fiber composite material

For composites with very stiff fibers, i.e., $E_f \gg E_m$ and a practical V_f of 0.45, the above relation can be simplified to

$$E_L \cong V_f E_f. \tag{9.2}$$

The rule of mixture prediction for the longitudinal Poisson's ratio is given by a similar relation

$$\nu_{LT} = \nu_{Lf} V_f + \nu_m V_m, \tag{9.3}$$

where ν_{Lf} is the longitudinal Poisson's ratio of the fiber and ν_m is Poisson's ratio of the matrix.

The composite is represented by a series model of matrix and fiber elements yielding the following relation for transverse Young's modulus.

$$E_T = E_{Tf} E_m / (V_f E_m + V_m E_{Tf}). \tag{9.4}$$

Here, E_{Tf} is the transverse Young's modulus of the fiber. E_{Tf} is generally not provided by the manufacturers of carbon fibers.

The transverse Poisson's ratio ν_{TL} of the composite is given by (Daniel and Ishai 1994)

$$\nu_{TL} = \nu_{LT}(E_T/E_L). \tag{9.5}$$

On the other hand, the acoustic velocity C in a solid rod is given by (Daniel and Ishai 1994)

$$C = (E/\sigma)^{1/2}, \tag{9.6}$$

where E and σ are Young's modulus and density, respectively (Kumamoto 1985).

Therefore, for the unidirectional composite material, the longitudinal C_L and transverse C_T velocities can be respectively derived from (9.1), (9.4), and (9.6) as

$$C_L = (E_L/\sigma)^{1/2} = [(V_f E_f + V_m E_m)/\sigma]^{1/2}, \tag{9.7}$$

$$C_T = (E_T/\sigma)^{1/2} = [E_{Tf} E_m / (V_f E_m + V_m E_{Tf})\sigma]^{1/2}, \tag{9.8}$$

$$\nu_{TL} = \nu_{LT}(C_T/C_L)^2. \tag{9.9}$$

The following equations can be derived from (9.6) and (9.7).

$$Z_L = \sigma C_L = [(V_f E_f + V_m E_m)\sigma]^{1/2}, \tag{9.10}$$

where Z_L is the transverse acoustic impedance.

Typical Young's modulus of a carbon-fiber E_f is typically 54–900 GPa and that of resin E_m is not greater than 40.

Therefore,

$$Z_L \cong (V_f E_f \sigma)^{1/2}. \tag{9.11}$$

The Z_L value of 30 MRayl is attainable with a carbon-fiber composite.

9.2.2 Processing and Fabrication

Typical physical properties of the carbon fiber are shown in Table 9.1. (Nippon Graphite Fiber Corporation 2007). A wide variety of carbon fibers having Young's modulus ranging from 54 to 900 GPa have been developed (Kondo et al. 2004, 2006).

Consequently, it is possible to make backing materials with appropriate acoustic impedances by selecting the grade of the carbon fibers. In Fig. 9.2, the structure of a typical unidirectionally woven carbon cloth is shown, which is used as the

Table 9.1 Typical physical properties of carbon fibers

Grade	Tensile Young's Modulus (GPa)	Density (g/cm^3)	Diameter (μm)
XN-05	54	1.65	10
XN-10	110	1.70	10
XN-15	155	1.85	10
YT-50A-04H	400	2.08	–
YSH-50A-04S	520	2.10	–
CN-60	620	2.12	–
YS-80A	785	2.17	7
YS-90A	880	2.18	7
YS-90A	900	2.19	7

Courtesy of Nippon Graphite Fiber Corporation

Fig. 9.2 Photograph of unidirectional woven cloth

starting material. To form homogenous composite materials, the said cloth is directly impregnated with liquid thermostat resin.

The processing steps for the fabrication of composites laden with carbon-fiber and tungsten powder are as follows:

1. Tungsten powder, epoxy resin, and their hardening agent are weighed in required proportions according to the desired volume fraction of each constituent in the final composite.
2. The tungsten powder, epoxy resin, and their hardening agent are put in a container, which is then placed in a spinner. The container simultaneously rotates and revolves under atmospheric pressure. The centrifugal force reaches $\sim 400 \times g$ at the highest speed, ensuring homogenous mixing thereby.
3. The carbon-fiber cloth is then impregnated with the homogenously mixed tungsten laden resin using screen-printing technology.
4. The impregnated carbon-fiber cloth is placed in a vacuum chamber to remove air from the interstices, i.e., degassing step.
5. The impregnated and degassed carbon-fiber cloth is cured in an oven, thereby forming a carbon-fiber composite.
6. The composite formed in step 5, is then covered with pristine carbon-fiber cloth, and then re-impregnated, degassed, and cured, resulting in the final composite.

The cross-sectional view of the composite, which was fabricated by the process, is schematically depicted in Fig. 9.3. By this method, a uniform distribution of the fibers is possible.

In Fig. 9.4, the cross-sectional view of the composite, which were subjected to a bending test, is depicted under high and low resolution. It was observed that the fibers exhibited fiber-pull-out in addition to fracture.

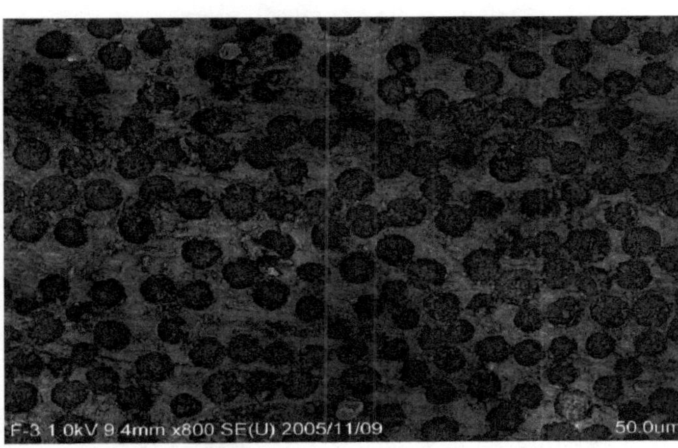

Fig. 9.3 Sectional view of carbon fibers in the composite. Micron bar represents 50 μm in linear dimension

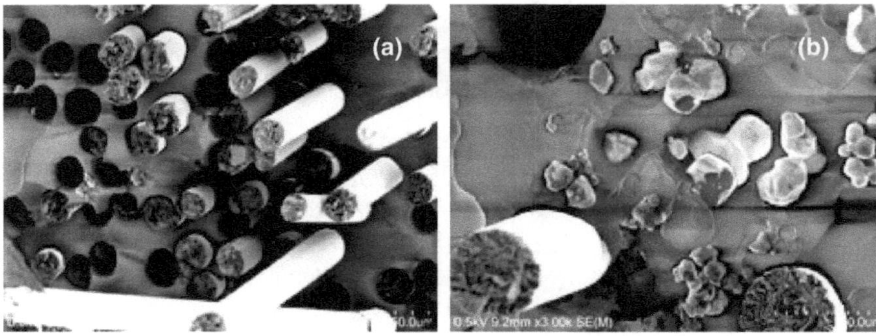

Fig. 9.4 Scanning electron micrographs showing fracture surface of the composite due to failure by bending (low magnification). (**b**) The same as (**a**), but higher magnification

9.2.3 Measurement of Acoustic Properties

Samples of the composites with thickness ranging from 3 to 10 mm were prepared. The samples used in measurements were cut to rectangular sections. The thickness of the sample was measured with a micrometer and the fractional thickness variation over the sample surface was confirmed to be less than $3 \sim 4$ μm, and the length of the sides was measured to an accuracy of 0.05 mm with a slide caliper. The sound velocity of the samples was measured by sing-around method. The time travel delay due to the sample of a 2-MHz burst wave was measured at room temperature using the SING-AROUND Sound Velocity Measurement Device UVM-2 made by Ultrasonic Engineering Co., Ltd., and converted to a longitudinal sound velocity, utilizing the thickness of the sample determined previously. Densities were determined by the Archimedes method.

The physical properties of the matrixes of the composites are shown in Table 9.2. It is of considerable interest to observe the decrease in sound velocity in epoxy resin mixed with tungsten powder. The effect prevents high acoustic impedance of a backing made of tungsten–epoxy composite, and mixing of large quantity of tungsten powder in the composites allows the realization of high-acoustic-impedance materials. However, it leads to heavy weight of the transducer. Typical experimental results of the acoustic properties of tungsten–vinyl composites were shown in early publications (Lees et al. 1973).

The measured values of the density, sound velocity, and the derived acoustic impedance for the composites fabricated in this chapter are given in Table 9.3. (Kondo et al. 2006). The increases in attenuation and acoustic impedance ar0e brought about by the addition of tungsten powder to the composites. As seen from the data shown in Table 9.3 (Kondo and Kitatuji 2004), heavy backing materials with any acoustic impedance can be easily manufactured with the appropriate volume of carbon fiber. The adoption of the composite as a backing material can lead to consistency in the acoustic impedance of the backing and of PZT, the acoustic impedance of which is 28 MRayl.

Table 9.2 Physical properties of the matrixes used for the composites

Materials	Supplier and specification	Sound velocity (m/s)	Density (g/cm^3)
Epoxy resin	Japan Epoxy Resin Co. Epikoto 807-TEPA	2,670	1.22
Epoxy resin mixed with tungsten powder	Tungsten powder (67 wt%)	1,980	2.93

Table 9.3 Physical parameters of the carbon-fiber composites

Sample number	Sound velocity (m/s)	Attenuation (dB/cm)	Acoustic impedance (MRayl)	Carbon fiber (vol%)	Tungsten powder (wt%)	Diameter of tungsten powder (μm)
1	11,400	2	22.7	71	0	–
2	11,400	36	28.1	57	60	3.1
3	11,300	32	33.2	57	70	3.1
4	10,900	21	38.3	64	80	3.1
5	10,900	27	34.2	54	70	15.2
6	11,500	24	39.2	63	80	15.2

9.3 The Performance of Transducer with Composite Backing

9.3.1 Computer Simulation

Computer simulation based on the KLM model (Krimholtz et al. 1970) is widely used to design various kinds of transducers. The model is physically intuitive. Because the objective is to estimate the effect of the backing material, it is discussed on the basis of the model, that frequency response and transient response of the transducer are governed by acoustic impedance Z_P of the piezoelectric plate and acoustic impedance Z_B of the backing material. The dependences of the frequency and transient responses on Z_B and Z_P have been obtained by computer simulation. The simulation results are shown in Fig. 9.5a, b.

9.3.2 Fabrication of Transducers

A simple ultrasonic transducer, as shown in Fig. 9.6, has been fabricated for experiments. It is a single-element piston transducer without matching layers. The transducer has a PZT piezoelectric ceramic with a flat surface. Consequently, it is not focused. The simple structure is effective for estimating the damping effect of the backing.

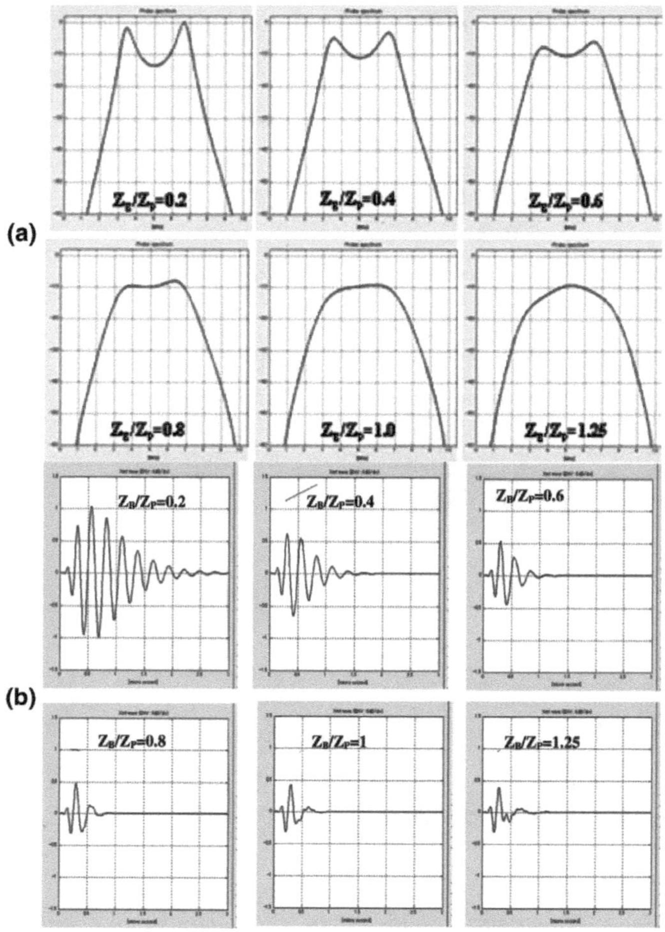

Fig. 9.5 Frequency response of the transducer. $Z_P = 28$ MRayl; drive voltage is 10 cycles sine burst wave of backing. (**b**) Transient response of the transducer. $Z_P = 28$ MRayl; drive voltage, 1 cycle sine wave

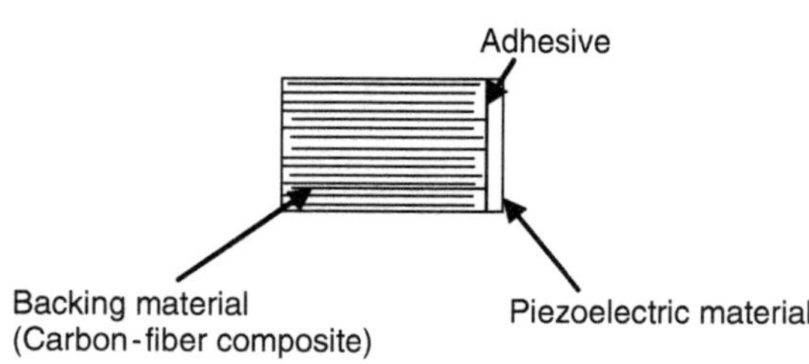

Fig. 9.6 Sectional view of the testing system used in experiments

The increase in bandwidth when using a high acoustic impedance backing has been examined experimentally. Figure 9.7 shows the experimental arrangement for measuring the frequency response. The transducer is excited with a 5.0-MHz

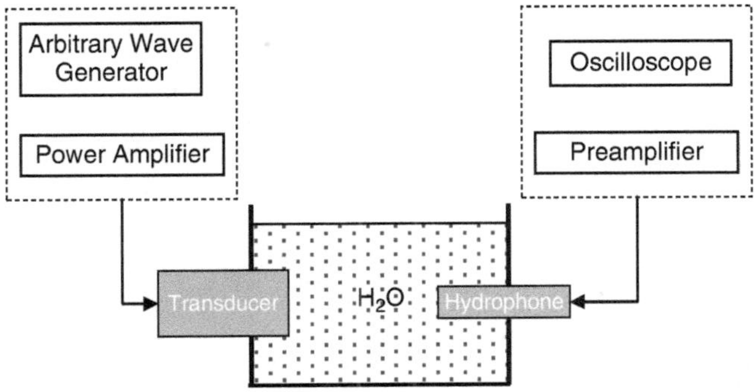

Fig. 9.7 Block diagram of the experimental arrangement for measuring frequency and transient responses

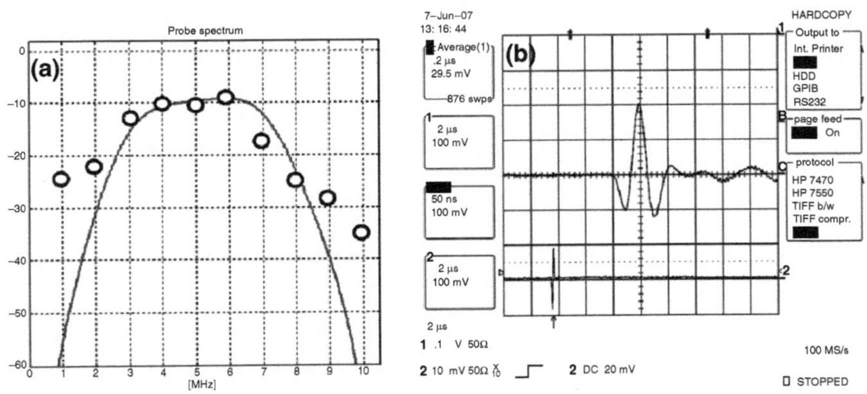

Fig. 9.8 Frequency response of the transducer (*x* axis spans from 0 to 10 MHz, with 5 MHz as center frequency). (**b**) Transient response of the transducer. C: receiving signal; 2: driving 1-cycle pulse

10-cycle sinusoidal tone burst wave. The experimental result is shown in Fig. 9.8a. The predicted broad-band characteristic of the transducer is proved. The experimental values dispute the computer simulation results obtained with the KLM model. As the beam behavior of the transducer is affected by the emitted sound frequency, the receiving signal should be compensated to obtain the actual frequency response. Another consideration in the transducer characteristics is the transient response. The transient response of the transducer is measured using the same experimental system as shown in Fig. 9.7. The emitted sound from the transducer is received by a wide-band hydrophone, and the signal is accumulated 1,000 times in order to eliminate noise. Here, a 5.0-MHz, 1-cycle sinusoid excitation pulse is used. Figure 9.8b shows the experimental result of the transient response.

9.4 Conclusion

A unidirectional carbon-fiber composite as backing material for transducers has been newly developed. Its use enables us to tune the transducer to arbitrary acoustic impedance. Even though the composite is lightweight low dense), which makes it convenient to use, a high acoustic impedance of 39 MRayl can be attained.

References

Auld BA (1990), Acoustic Fields and Waves in Solids, 2nd ed., Vol. 1. Krieger Publishing, Florida.

Daniel IM and Ishai O (1994), Engineering Mechanics of Composite Materials, Oxford University Press, New York.

Fukuda H and Ben G (1989), Introductory Mechanics of Composite Materials [in Japanese], Kokon Shoion, Tokyo.

Grewe MG, Gururaja TR, Shrout TR and Newnham RE (1990), Acoustic Properties of Particle/Polymer Composites for Ultrasonic Transducer Backing Applications, IEEE Trans, Ultrason., Ferroelect., and Freq. Cont., vol. 37, no. 6, pp. 506–514.

Kondo T and Kitatuji M (2004), Composite Materials and Measurement of Their Acoustic Properties, Japanese Journal of Applied Physics, 42–5B:2914–2915.

Kondo T, Izumi M, and Kanda H (2006), Broad Band Ultrasonic Transducer Using Carbon Fiber Composites for a Backing Material, IEEE Ultrasonic Symposium Proceedings, pp. 1537–1540.

Krimholtz R, Leedom D, and Matthaei G (1970), New Equivalent Circuit for Elementary Piezoelectric Transducers, Electronics Letters, 6:398.

Kumamoto O (1985) Ultrasonic Handbook [in Japanese], Nikkan Kogyo Shinbunsha, Tokyo.

Lees S, Gilmore RS and Kranz PR (1973), Acoustic Properties of Tungsten-Vinyl Composites, IEEE Trans, Sonics Ultrason., vol. SU-20, no. 1, pp. 1–2.

Nippon Graphite Fiber Corporation (2007), Technical Data, Granoc Yarn XN, YT, YSH and YS, NGF Information, Tokyo, Japan.

Thinky Technology (2007), Revolution in Mixing Technology , pp. 1–2, THINKY Corporation, Tokyo.

Part III
Transducer Design and Principles

Chapter 10
Piezoelectric Transducer Design for Medical Diagnosis and NDE

Marc Lethiecq, Franck Levassort, Dominique Certon, and Louis Pascal Tran-Huu-Hue

10.1 Introduction

The number one technique for medical imaging and non-destructive evaluation (NDE) is ultrasound. This is due to its non-ionizing character, low cost and to the fact that images and measurements contain data linked to several physical and structural parameters of the explored media.

The overall performance of an ultrasonic system is mainly determined by the transducer characteristics. Consequently, each application having its specific requirements, very different transducers need to be designed. Furthermore, the measurement techniques and imaging modalities are in constant evolution, requiring higher performance and versatility of the transducers. Not only must frequency bandwidth and sensitivity be increased, but transducers must also be able to operate in various modes such as pulse-echo (classical A, B or C modes), burst (Doppler or other velocity measurements) or harmonic reception (non-linear acoustics). Innovations such as ultrasound stimulated elastography and combination of different techniques such as ultrasound and MRI or ultrasound therapy and imaging are only possible if specific transducers are developed.

The structure of a single-element ultrasonic transducer based on piezoelectric effect is first described, with particular focus on the influence of the constitutive materials on transducer performance. More complex transducers such as annular, linear, curved, phased, and 2D arrays are described and most important design issues discussed. Piezoelectric material issues relative to transducer applications are then addressed. Methods to characterize and models to predict transducer behaviour are presented, and several comparisons are shown to illustrate achieved performance. Finally, transducers designed for high-resolution imaging are presented, and other current developments briefly described. Some particularly interesting future trends are highlighted in the conclusion.

M. Lethiecq, F. Levassort, D. Certon, and L.P. Tran-Huu-Hue
Ultrasonics Group (LUSSI), Francois Rabelais University, Tours, France

A. Safari, E.K. Akdoğan (eds.) *Piezoelectric and Acoustic Materials for Transducer Applications.*

10.2 Performance of Ultrasonic Transducers

The performance of an ultrasonic transducer is a multi-parameter issue (Kossof 1966; Desilets et al. 1978). Section 10.5 deals with methods to predict, optimize and measure these parameters. Here, simple definitions are given to justify the choices one has to make when designing a transducer. These design rules will be addressed in Sect. 10.2, 10.3 and 10.4 for typical transducer configurations.

Transducer performance can be described by considering three main issues: (a) electrical input impedance, (b) electroacoustic response and (c) radiation pattern. It must be noted that many other issues exist, among others electromagnetic compatibility (EMC), operation temperature and homogeneity of elements in arrays. These are outside the scope of this chapter, and will be addressed only in the context of specific cases in the ensuing sections.

The geometry and size of the piezoelectric element are of great importance, since they influence all transducer properties. In particular, the main vibration mode of the structure is selected by the geometry and the operating frequency by the size. It must be noted that different types of materials perform better in some vibration modes than in others. This will often govern the choice of the material type. In most cases, PZT-based compositions or PZT–polymer composites offer the best compromise for the vibration modes used in medical diagnosis and NDE applications. This will be addressed for specific designs in the following sections.

The electrical input impedance of a transducer is linked to its size and geometry as well as to the properties of its constitutive materials. Generally the transducer will be connected to its driver electronics and to its receiver electronics by a coaxial cable. Commercially available cables have characteristic impedance between 50 and 80 Ω. Consequently, to maximize electrical energy transfers, the transducer input impedance must be close to the cable characteristic impedance. This is taken into account in design procedures, and the main variable one can use to adjust the input impedance is the piezoelectric material dielectric constant. Indeed, most other parameters are imposed by the electroacoustic and radiation requirements. In some cases, electrical elements such as transformers and/or inductors are used to improve electrical matching.

The most-studied electroacoustic response of a transducer is its pulse-echo response: It represents the signal (or echo) received following a short electrical pulse excitation of the transducer with a perfect reflector placed in front of it, at a relatively short distance. The amplitude of the echo is representative of the transducer sensitivity and will impose signal to noise ratios and depth of penetration. It is generally measured as the ratio in dB of the amplitude of the echo to that of an ideal transducer, or insertion loss as defined in electronics. The time duration of the echo, measured at half its amplitude, defines the axial resolution. In imaging applications, this resolution represents the inverse of the smallest distance between two point-targets placed one behind another that will be seen as two separate reflectors. For medical imaging and NDE, this distance typically ranges from one wavelength, the theoretical maximum resolution, to three wavelengths. This is in the mm range for most common applications using 2–10 MHz frequencies.

The radiation pattern of a transducer can be measured in transmit mode, which gives the spatial distribution of acoustic pressure. It is often measured in pulse-echo mode, which is a configuration closer to the actual use of most transducers. A small target is scanned in the propagation medium, and the echo amplitude is plotted as a function of the target position. It allows parameters such as lateral resolution depth of field and level of acoustic noise to be quantified. The lateral resolution has the same definition as the axial resolution, but in a direction perpendicular to that of the wave propagation. It is determined by the width of the acoustic beam at half its maximum value. Lateral resolution changes with depth, i.e. distance to transducer surface. It is typically lower than the axial resolution; especially at distances far from the transducer focus. The depth of field, which is similar to that of an optical camera, is defined between the shortest and longest distances at which the echoes of a target along the transducer axis have amplitude at least half of the maximum value. This is one of the main parameters defining the distance range of the transducer. Finally, the amplitude of echoes produced by a target, placed off-axis, defines the acoustic noise of the transducer. Ideally, a transducer should have electrical input impedance around $50\,\Omega$, a high sensitivity, high axial and lateral resolutions, an extended depth of field and low acoustic noise. Unfortunately, several of these requirements are in contradiction with others. Hence, trade-offs will have to be resolved, making transducer design a complex process. For example, if a very high bandwidth is required, the transducer resonance will be highly damped, which will reduce its sensitivity. In the following sections the focus will be on electro-acoustic responses of transducers, since it is on these aspects that the properties of the piezoelectric element have most influence.

10.3 Basic Transducer Design

10.3.1 Operation of a Single-Element Transducer

The classical single-element transducer (Fig. 10.1) is based on a piezoelectric plate or disc poled along the thickness direction and used in its thickness mode, so its thickness defines the resonance frequency of the device through the equation: thickness = half wavelength. When an electrical impulse is applied to the plate, an acoustical resonance is induced, which will lead to the generation of pressure waves both in the front and back directions. Since the plate, typically a piezoceramic, has high acoustic impedance ($\sim$33 MRa) compared with that of biological tissues or water (1.5 MRa), the duration of the acoustic resonance will be very high. Thus, a piezoceramic plate alone would lead to a very poor axial resolution. Consequently, other layers are added to the active layer (Kossof 1966).

On the rear face of the active element, a thick layer is usually added. It is referred to as the backing. It serves as a mechanical support for the active element, but it also allows acoustic energy to flow by the rear face and thus induces damping of the transducer resonance. The closer its acoustical impedance is to that of the active

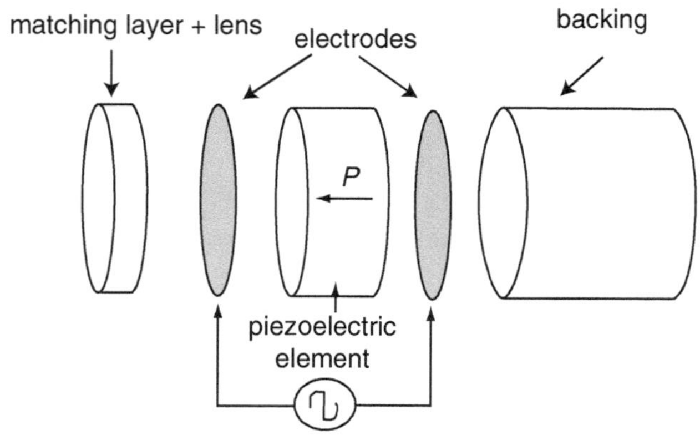

Fig. 10.1 Diagram of a classical single-element transducer

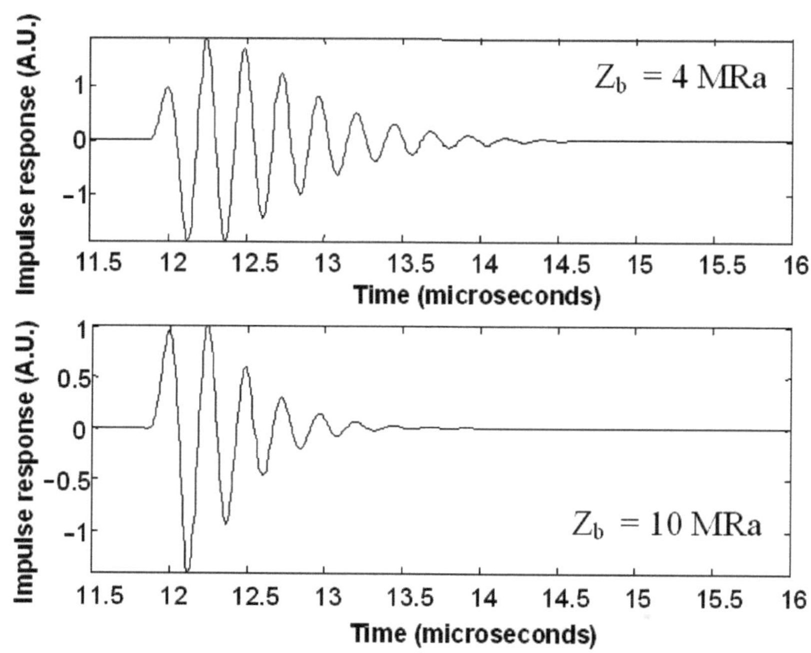

Fig. 10.2 Impulse responses of a light-backed ($Z = 4$ MRa) and a heavy-backed ($Z = 10$ MRa) piezoceramic disc

layer, the more energy is lost. The consequence is a lower sensitivity but a higher axial resolution. Thus, a trade-off has to be found for each application. The attenuation coefficient and the thickness of the backing layer must be sufficient so that no energy can be radiated back to the active layer, which would produce parasitic echoes. Figure 10.2 shows the response of a simple transducer made of a piezoceramic disc and a backing with an acoustical impedance of 4 MRa (typical backing), and another one with an acoustical impedance of 10 MRa (heavy backing).

Between the piezoceramic and the propagation medium, one or several matching layers are used. Their design is optimized in order to increase the transfer of energy from the active layer to the propagation medium (Desilets et al. 1978). Typically, the thickness of these matching layers is around a quarter wavelength at the resonance frequency of the piezoceramic element, and their acoustical impedance is intermediate between those of the piezoceramic and propagation medium. The use of matching layers thus improves the sensitivity of the transducer. Moreover, since the acoustical energy can better flow towards the propagation medium, the duration of acoustical resonance in the active layer is decreased. Consequently, the matching layers also improve axial resolution. Figure 10.3 shows a comparison of the response of a transducer made of a piezoceramic disc on a backing with an acoustical impedance of 4 MRa with that of the same transducer on which a matching layer has been added. The use of multiple matching layers (Inoue et al. 1987) produces further improvement of transducer performance, but cost and technical complications often limit their number to only two.

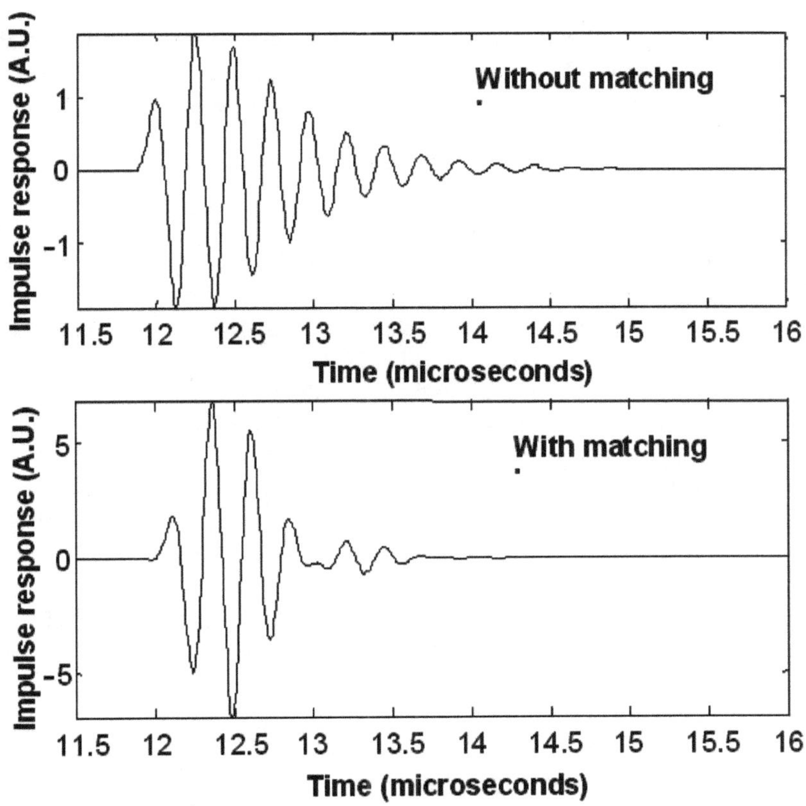

Fig. 10.3 Impulse responses of a light-backed ($Z = 4\,\mathrm{MRa}$) piezoceramic disc without and with matching layer

10.3.2 Material Aspects

The active layer is typically a PZT-based material, while matching layers and backings are polymer-based. The available piezoelectric materials – ceramic–polymer composites that are widely used in modern transducers – are discussed in Sect. 10.4, wherein their properties are considered.

High loss polymers are preferred for the backing such as high viscoelastic attenuation epoxy resins, and different particles are added in order both to adjust the acoustic impedance and to increase the attenuation through scattering. Particle sizes must be much lower than the wavelength of the acoustic radiation used to avoid parasitic echoes. To increase acoustic impedance (the so-called heavy backing), metal particles such as tungsten are currently used (Nguyen et al. 1996). On the contrary, hollow plastic spheres are used to obtain light backings. To achieve very high attenuation without large variations of acoustic impedance, mixtures of several different types of particles are used. The thickness of the backing layer can then be reduced, which allows one to achieve miniaturization of the transducer.

For matching layers, particle-loaded low-loss polymers are used. Here, attenuation must be kept as low as possible to avoid losses that would reduce transducer sensitivity. Consequently, a single type of particle is added in proportions determined in order to obtain the desired acoustic impedance (Inoue et al. 1987). Hard setting epoxy resins are preferred here (Nguyen et al. 1996).

10.3.3 Simple Transducer Configurations and Design Rules

In medical applications single- and dual-element transducers are mainly used for Doppler blood flow measurements (Fig. 10.4). In the continuous wave Doppler method, transducers are designed so that the radiation patterns of the two elements overlap at a given depth, allowing the selection of blood vessels around this depth. For pulsed Doppler measurements, a single element is used together with a focusing lens, which allows increasing the sensitivity in a given range of depths. For these applications, transducer sensitivity is optimized while resolution and bandwidth are not critical. Consequently, light backings and even air-backing are used in certain cases.

Single-element transducers can also be used for imaging applications, the scanning being obtained by a motor. In this case, the sensitivity-resolution trade-off is shifted to higher resolution, so that sensitivity is lower than that of Doppler transducers. These types of transducers, which are designed for high-resolution medical imaging, will be presented in Sect. 10.4.

For NDE applications, single-element transducers without focusing lenses are used for contact measurements on flat surfaces, while a focusing lens is generally added for immersion testing. Dual-element transducers are used for thickness measurements in order to allow a reduction of the dead zone produced by the transmit pulse when the same element is used both in transmit and receive modes.

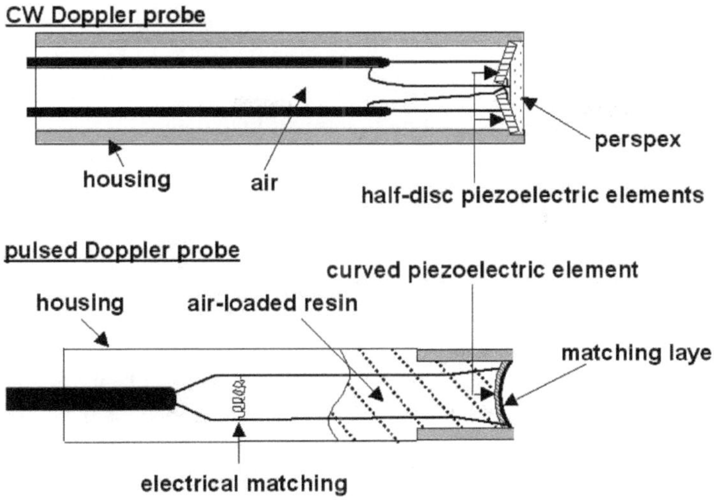

Fig. 10.4 Diagram of Doppler probes

The radiation pattern of single-element transducers is defined by the size of the active element and the lens curvature. Lateral resolution, which is linked to the lateral dimensions of the acoustic beam, is optimum at the focal distance, but becomes very poor when shorter or longer distances are observed. In most cases the piezoelectric element is a disc of diameter D. If a lens is added with a radius of curvature chosen to obtain a focusing distance F, the radiation characteristics (Fig. 10.5) can be approximated by the following formulas in which λ is the wavelength (ratio of speed of sound to resonance frequency): lateral resolution $\approx \lambda$ F/D at the focal distance and depth of field $\approx 7\lambda(F/D)^2$. A particular case is a transducer without a lens, for which the "natural focus" is $N = D^2/4\lambda$, and called its Fresnel distance. Its lateral resolution is then $D/4$.

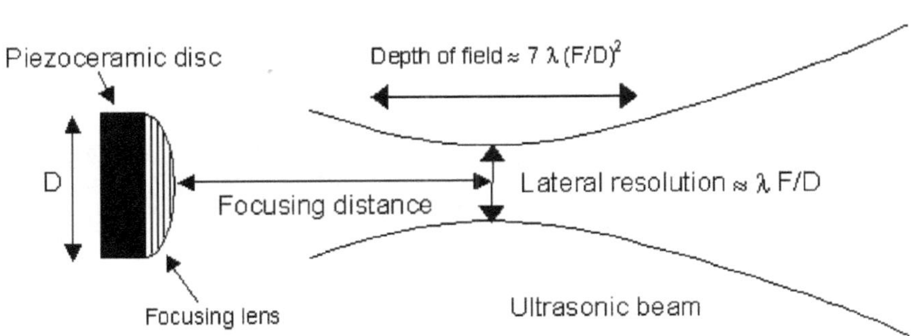

Fig. 10.5 Radiation pattern of single-element transducer with a focusing lens

10.4 Transducer Arrays

As pointed out in the previous section, single-element transducers can be focused to improve their lateral resolution, but their depth of field is then reduced, so they can only be used in a very limited range of distances. In particular, only the middle of the images will be exploitable for imaging applications. All structures situated at distances either significantly lower or significantly higher than the focus will produce a blurry image.

To overcome this problem, annular arrays have been designed (Powers 1980) (Fig. 10.6). Here, focusing is obtained by applying delays to the rings in such a way that the acoustic pressure generated by each ring arrives at the focal point at the same time. The focal distance can thus be adapted to the region of interest. Moreover, when receiving the echoes, the delays can be continuously adapted to focus on the current depth at which echoes are produced. Thanks to this procedure, which is referred to as dynamic focusing, much larger ranges of depths can be imaged with good lateral resolution as compared with imaging with single-element transducers. Annular arrays have been used in many commercial imaging systems during the 1980s. Today there are mainly replaced by linear, curved or phased arrays, except for low-cost systems or for niche applications. One of the main reasons for this loss of popularity of annular arrays is due to the need for a mechanical device to scan the ultrasonic beam.

In most of the current imaging systems, the scanning of the acoustic beams is obtained by electronic means, thus avoiding mechanical devices that are not as reliable. Two different transducer configurations can be used: linear arrays or phased arrays.

When the imaging window is large, which provides easy access to the explored media, linear arrays can be used. They can either be flat or curved, which gives either rectangular- or conical-shaped images (Fig. 10.7). The number of elements is typically around 200, each element having a width (along the largest dimension of the array) around 1–2 wavelengths and a length (elevation direction) of several tens of wavelengths. Scanning of the beam is obtained through multiplexing. For a given line of the image, only a given number of neighbouring elements are selected

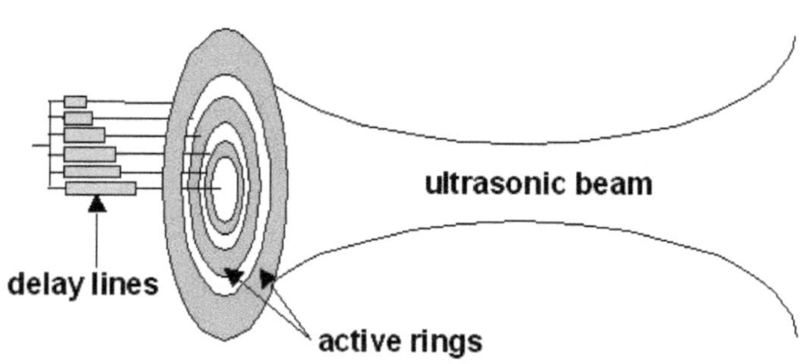

Fig. 10.6 Operation of an annular array

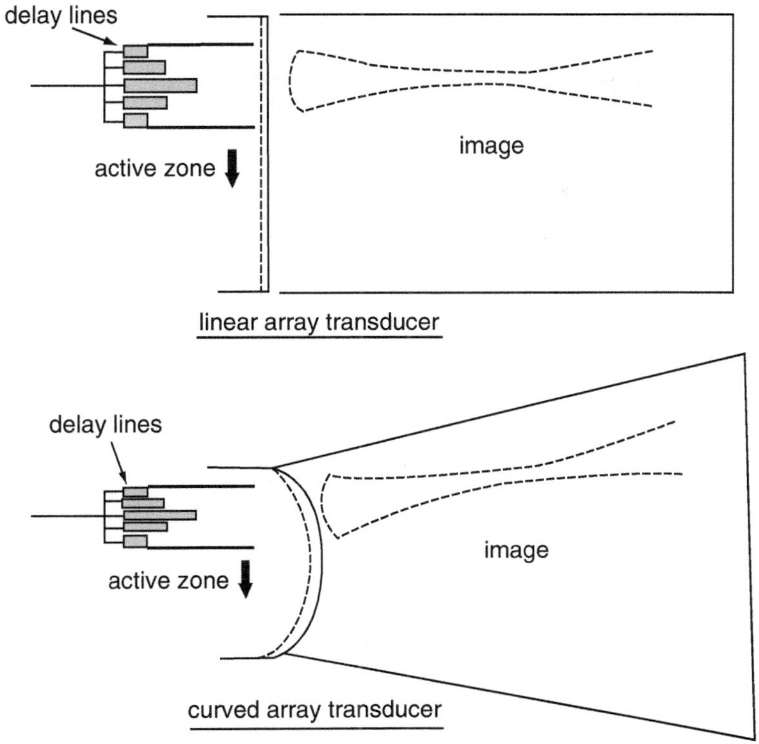

Fig. 10.7 Operation of a planar and curved array transducers

to produce a radiating aperture. For the following line, multiplexing circuits simply shift this aperture one pitch; i.e. distance between the centres of two adjacent elements. The only limit to image frame rate is the time needed for the ultrasound waves to propagate from the transducer to the deepest structures of interest and return back from where it originated. Focusing is obtained by electronic delay lines in the scanning direction and by a lens in the elevation direction. Dynamic focusing, as described for annular arrays, is also implemented in the scanning direction. The internal structure of such arrays is described in Fig. 10.8. In the case of arrays based on bulk piezoelectric ceramics, each array element must be cut into a few sub-elements in order to guarantee the absence of parasitic vibration modes. These sub-elements are nevertheless electrically connected, and so the 'electrical pitch" of the array is higher than its mechanical pitch. A deep dicing cut is performed between neighbouring elements and filled with an attenuative material in order to limit inter-element coupling. If piezocomposites with 1–3 connectivity are used, the sub-dicing is no longer required, and in some cases, even the dicing between elements can be avoided without prohibitive inter-element coupling.

When the acoustic window is small, such large arrays cannot be used. For example, to obtain images of the heart, the acoustic beam must fit between the ribs. In this

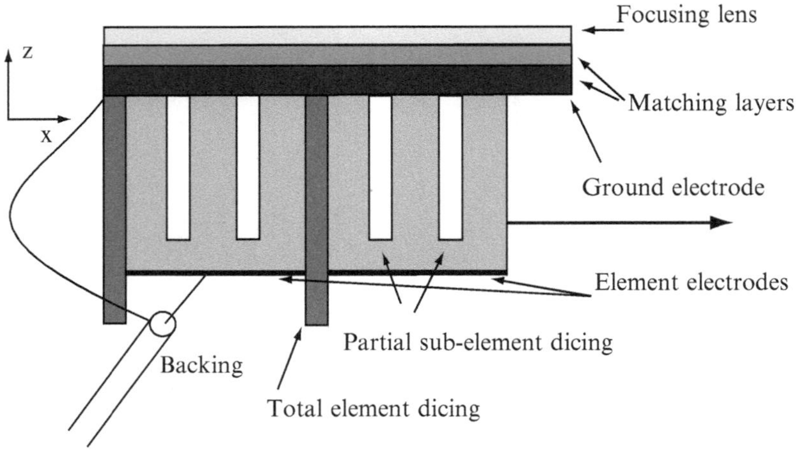

Fig. 10.8 Internal structure of planar and curved array transducers

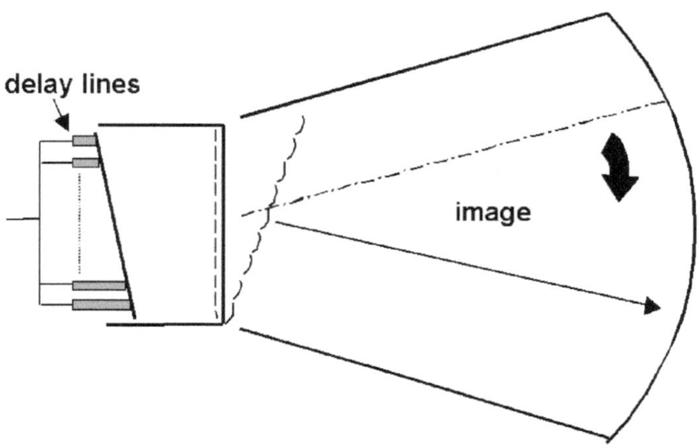

Fig. 10.9 Operation of a phased-array transducer

case, electronic scanning by multiplexing is no longer used. Consequently, a scanning method based on electronic delays is deployed. For this purpose, phased-array transducers are designed (Fig. 10.9). Here, all elements of the array are used for each line of the image and the delays allow steering of the beam (typically $+/-45^{\circ}$) as well as focusing. The width of phased-array elements must be lower than that of classical linear arrays, since each element must be able to radiate energy at large steering angles. A typical design criterion is a width equal to half a wavelength, which ensures in principle a radiation pattern without parasitic grating lobes. The internal structure of a typical phased array is shown in Fig. 10.10.

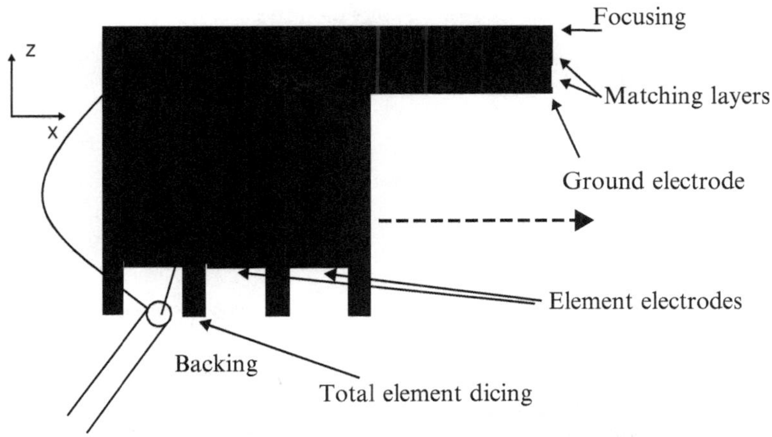

Fig. 10.10 Internal structure of a phased-array transducer

10.5 Piezoelectric Materials for Medical Diagnosis and NDE Applications

10.5.1 General and Specific Requirements

A large number of piezoelectric material properties must be taken into account in transducer design. These are: (a) elastic, (b) dielectric, (c) piezoelectric constants, and (d) machinability as well as cost. Two of the most important material parameters for transducer applications are the effective electromechanical coupling coefficient k_{eff} of the main vibration mode and the acoustic impedance Z_{ac}.

The k_{eff} factor represents the magnitude of the piezoelectric effect of the material for the vibration mode of interest. In other words the capability of the material to convert electrical energy into acoustical energy (or vice-versa) in a short time is quantitavely described by k_{eff}. Its value ranges from 0 (no piezoelectric activity) to 100%, and it should be as high as possible. It should be noted that k_{eff} depends not only on the material properties but also on the geometry of the active element. In medical imaging applications, all vibration modes are longitudinal, i.e. the displacements are in the poling direction defined by the thickness dimension. For large plates or discs in which the thickness is much smaller than the lateral dimension(s), the thickness coupling factor k_t is used. For bars or pillars (thickness higher than lateral dimensions), the factor of interest is k_{33}. For intermediate cases (one small and one large lateral dimensions with a thickness value between them), the k'_{33} factor is used (IEEE 1987) (Fig. 10.11).

The acoustic impedance, defined as the ratio between acoustical pressure and particle velocity in a given media, is obtained by multiplying the density by the acoustic wave velocity: $Z_{ac} = p/v = \rho c$. It should be as close as possible to that of

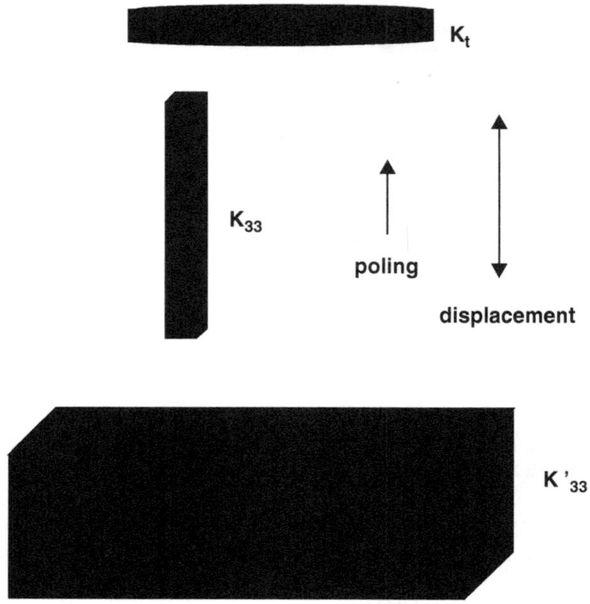

Fig. 10.11 Vibration modes and coupling factor

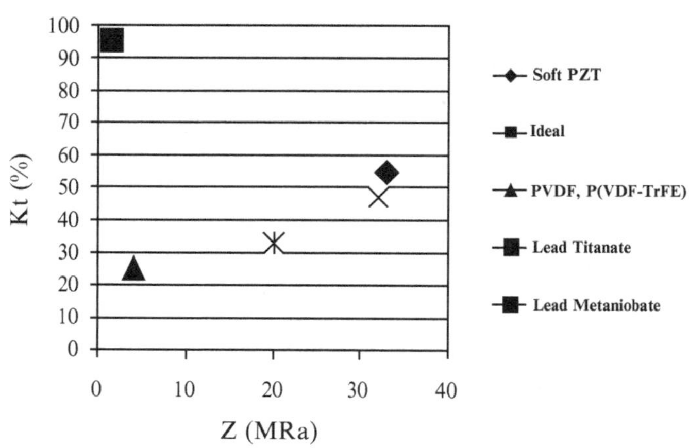

Fig. 10.12 Electromechanical coupling factor in thickness mode (k_t) vs. the acoustical impedance (Z) of different piezoelectric materials

the propagation medium, i.e. biological tissues or often water in NDE; both have an acoustic impedance of ~1.5 MRa.

Figure 10.12 shows a plot of k_t vs. Z_{ac} for several piezoelectric materials. It can be observed that no material allows both high coupling and low acoustic impedance. Consequently, the material chosen is the one which offers the best trade-off. For almost all medical and NDE transducer applications, PZT-based piezoceramics are used because of their high coupling factors (k_t around 50% and k_{33} around 70%). Their high acoustic impedance can be compensated for by using acoustic matching layers in the transducer structures, as described in Sect. 10.2.

There are many types of PZT-based piezoceramics on the market, which have different properties due to doping by additives in varying proportions and specific fabrication processes. For a given application, properties such as dielectric constant and grain size will allow to choose a specific material reference. Most piezoceramic manufacturers offer a selection of references, from relatively low (a few hundred) to very high (a few thousand) relative dielectric constants, and with grain sizes from 1 to 10 μm.

For large area devices, such as single- or dual-element transducers, a moderate dielectric constant allows good electrical matching to the electronics, while array elements require much higher dielectric constants. For high-frequency devices or for arrays (whenever small dimensions are needed), fine-grained materials will be required. Loss factors are usually not a critical parameter because transducer losses do not only appear in the piezoelectric layer, but also in the backing and matching layers. Consequently, moderate mechanical and dielectric losses can be accepted, and a certain degree of mechanical losses can even be useful to damp internal parasitic resonances in the active layer.

10.5.2 Commonly Used Piezoceramics

Considering these requirements, soft PZT materials such as PZT-5A or Pz27 are mostly used (Ferroperm Piezoceramics 2007), with relative dielectric constants in the order of 1,000. If higher dielectric constants are required, PZT-5H or Pz29 (Ferroperm Piezoceramics 2007) will be preferred. Piezoceramic manufacturers are constantly improving the properties of their materials, and so new references often appear on the market.

The trends in the past few years have been on one hand to increase dielectric constants while maintaining high coupling factors (k_t over 50%), and on the other hand to decrease grain size and porosity. This is achieved by developing ceramics based on PZT but with relatively large proportions of additives, leading to compositions such as PNNZT or PLZT, and by optimizing fabrication processes. Ferroperm Pz21 is an example of this type of high-performance piezoceramic (Bove et al. 2001).

10.5.3 Other Available Piezoelectric Materials

Many other types of piezoceramics are available on the market. Hard PZTs are seldom used because their coupling factors ($k_t \sim 45\%$) and dielectric constants (a few 100) are lower than those of soft PZT, and their very low losses are of no use for high-bandwidth transducers. Lead titanate compositions can be useful for relatively large area devices such as single-element transducers, because of their low planar and lateral coupling factors, which allow parasitic vibration modes to be eliminated. However, they are rarely used in arrays because of their moderate k_{33} factor and low dielectric constant. Lead metaniobate, which is widely used in NDE applications, suffers from low thickness coupling, which is only partly compensated by its low acoustic impedance. It is only used in applications where very high bandwidth is required while sensitivity is of minor importance; for example in high-resolution imaging of superficial structures. Porous PZT has also been studied for many years, and recently, several compositions have been put on the market (Levassort et al. 2007). They have coupling coefficients equal to those of bulk materials of similar chemical composition, with the advantage of significantly lower acoustic impedance. They allow wide-band transducers to be developed without numerous matching layers.

Non-ceramic piezoelectric materials are also available. Piezopolymers such as PVDF or copolymers such as P(VDF–TrFE) can be purchased as films or sheets that are poled (Kawai 1969). Their coupling factors are modest (20–30%), their dielectric constant is very low (<10), but their acoustic impedance ($\sim$4–5 MRa) is close to that of water so that acoustic matching is not required. Their properties tend to decrease when the thickness of the films increase because of poling difficulties. Therefore, they are used in high-frequency applications only. Unfortunately, the overall transducer performance of polymer-based transducers is lower than that of piezoceramic-based ones. Several single crystals such as quartz or lithium niobate have interesting piezoelectric properties: They are used in resonators and surface acoustic wave filters. Thickness coupling is very low in quartz, while it is close to that of PZT ceramics in lithium niobate. These materials have very low dielectric constants, are quite expensive and their machining requires specific tools and methods. They do not have internal grain structure such as polycrystals, and so they are well adapted to obtain very thin plates or discs. Consequently, they tend to be limited to specific applications such as very high frequency single-element transducers. Lead-based single crystals with very high coupling coefficients have recently been developed. They will be introduced in Sect. 10.7.2.

10.5.4 Piezocomposites

It had been observed earlier that no available material could combine a high coupling and a low acoustic impedance. At the end of the 1970s, several publications (Newnham et al. 1978; Newnham 1986) showed that by combining a high-coupling piezoceramic (such as soft PZT) and a low acoustic impedance polymer (such as an epoxy resin), it was possible to obtain performance trade-offs that no other material could achieve. Different arrangements between the active phase and the passive phase were investigated, and it appeared that one arrangement was of great interest for ultrasonic transducer applications: The 1–3 connectivity piezocomposite corresponding

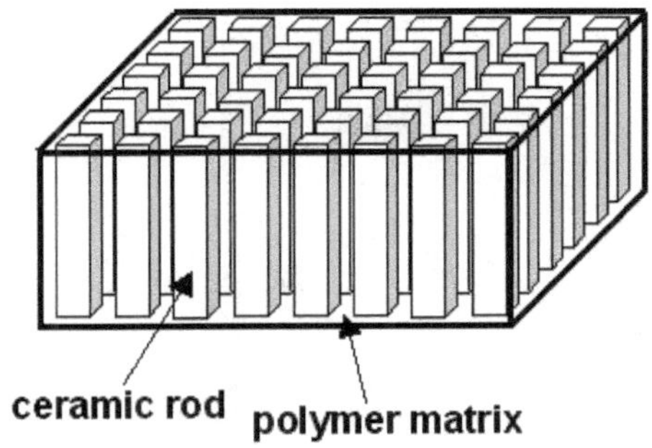

Fig. 10.13 Schematic representation of a 1–3 piezocomposite

to ceramic pillars in a polymer matrix (Fig. 10.13) (Savakus et al. 1981). The electromechanical properties of such a composite can be obtained by several models, for example using a unit-cell approach (Banno 1983; Rittenmeyer et al. 1982; Levassort et al. 1998). A high thickness coupling factor k_t, close to the value of k_{33} in the ceramic alone, is obtained even for relatively low ceramic contents, for which the acoustic impedance is around 3–4 times lower than that of pure ceramics (PZT ceramics are used because they exhibit values of k_{33} around 70% while their value of k_t is only around 50%). These properties allow both sensitivity and bandwidth to be increased in transducers, in comparison with classical piezoceramic devices. It is interesting to notice that the 2–2 connectivity piezocomposite, corresponding to alternating layers of piezoceramic and polymer, also exhibits high performance. This configuration was, however, already known and used before the concept of piezocomposite was introduced as linear and phased arrays are constructed this way.

10.5.5 Systems and Methods to Characterize Functional Material Parameters

Because of the piezoelectric effect, the electrical impedance of an active layer depends not only on the geometry and dielectric properties of the material, but also on its mechanical and electromechanical constants. Consequently, effective properties can be derived from complex impedance vs. frequency curves. The measurements are performed using an impedance (or network) analyzer, in such a way that the sample under test is in free resonator conditions, i.e. with no mechanical load. The measured curves are usually sent to a computer to be processed and memorized. In some cases the exploitation of the curves can be performed manually using the IEEE standard on piezoelectricity (IEEE 1987). For other cases, if high precision is required for some materials such as composites, phenomenological models are developed (Tran-Huu-Hue et al. 1997, 2000; Kwok et al. 1997).

A typical set of complex impedance (Z)/admittance (Y) curves is shown in Fig. 10.14. The resonant frequency (f_r) corresponds to maximum conductance component, while the anti-resonant frequency (f_a) corresponds to the maximum of resistance component. High electromechanical coupling corresponds to a large

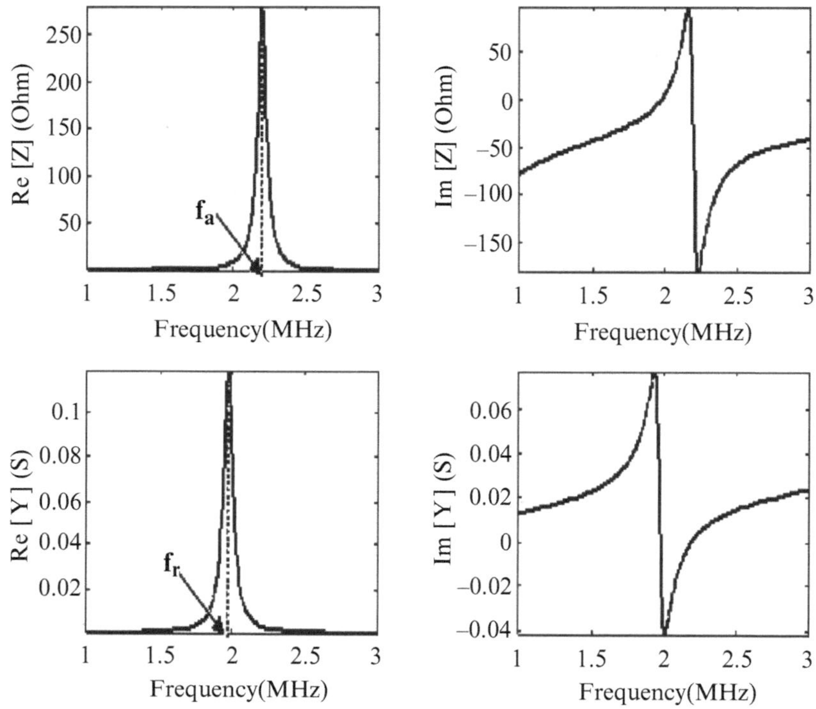

Fig. 10.14 Typical curves of the complex electrical impedance/admittance of free resonator

value of the frequency gap $f_a - f_r$. For example, if the vibration is a thickness mode, its coupling factor can be calculated by

$$k_t^2 = \frac{\pi}{2}\frac{f_r}{f_a}\cot\left(\frac{\pi}{2}\frac{f_r}{f_a}\right) \tag{10.1}$$

Different formulas have been derived for other vibration modes. The value of f_a is linked to layer thickness and acoustic wave velocity. Combined with a density measurement, acoustic impedance can be obtained. The dielectric constant is obtained through the capacitive susceptance at frequencies where no resonances occur. A typical measurement frequency is twice the anti-resonant frequency, for which no thickness resonance can appear. Dielectric losses appear as an offset on the conductance component curves. Measurements can be performed at $2f_a$ frequency or, if dielectric losses change with frequency, at multiples of this frequency. Finally, mechanical losses are evaluated through the width of anti-resonance (resistance) peak, typically by measuring the width Δf at half the maximum resistance. The mechanical quality factor is evaluated as $Q_m = f_a \Delta f^{-1}$. These methods can be adapted to higher order harmonics of the vibration mode, which allows material properties to be plotted vs. frequency (Tran-Huu-Hue et al. 2000).

10.6 Modeling and Characterization of Transducers

10.6.1 Modeling of Transducer Performance

Transducer performance can be divided into two groups: electroacoustic and radiation. Electroacoustic behaviour, which includes (a) electrical input impedance governing energy transfer between the imaging system and the transducer and (b) pulse-echo response whose amplitude is linked to sensitivity and duration to axial resolution from which frequency response can be deduced by the Fourier transform. The centre frequency and bandwidth can be read directly therefrom. All these curves can be predicted with the assumption of a single-mode vibration, and by using equivalent electrical circuits such as the one proposed by Krimholtz, Leedom and Matthei (KLM) (Krimholtz et al. 1970) (Fig. 10.15). All transducer layers

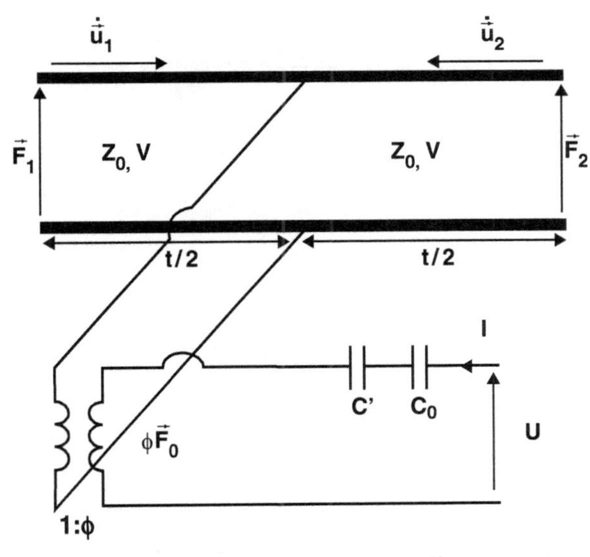

A: area; t: thickness; e: piezoelectric coefficient

$$C_0 = \frac{\varepsilon^S A}{t};\; C' = \frac{-C_0}{k_t^2 \sin c\,(kt)};\; \phi = \frac{2e \sin\left(\frac{kt}{2}\right)}{A\varepsilon^S Z_0 \omega}$$

Fig. 10.15 Krimholtz, Leedom and Matthei (KLM) scheme for thickness mode vibration of a piezoelectric plate (thickness t and area A). The force (pressure) at each face of the piezoelectric material is considered through an equivalent voltage while the derivative of displacement (particle velocity) is considered through an equivalent current. The required material parameters are ultrasonic wave velocity V, acoustical impedance Z_0, dielectric constant ε^S at constant strain and electromechanical coupling coefficient k_t. The piezoelectric coefficient e and the wave number $k = \omega/V$, with ω the angular frequency, are deduced from these parameters, as well as all other components of the circuit are taken into account as well as electrical elements such as matching impedances and cables. Numerical methods such as finite element methods are also available (Assaad et al. 1991) in cases where several acoustic modes coexist.

On the other hand, acoustic radiation patterns allow defining the following: (a) lateral resolution, which is high when lateral beam dimensions are low, and (b) acoustic noise level, which is the off-axis power level measured in dB using the on-axis power level as a reference. The appearance of side lobes, which exist whenever there are amplitude steps on the surface defined by the transducer front face, or grating lobes, which are due to space sampling in arrays, tends to increase acoustic noise level. Lateral resolution is responsible for image sharpness and low parasitic lobe levels ensure the absence of artifacts in the lateral direction. Radiation pattern calculations can be performed by applying Huygen's principle, by assuming that the transducer surface is a series of point-sources, and the pressure radiated at any point is the sum of pressures radiated by each of the point-sources (Kino 1987). For simple geometries (circular, annular or rectangular transducers), analytical results have been derived using impulse diffraction theories, which allow more efficient calculations (Penttinen and Luukkala 1976, Selfridge et al. 1980, Arditi et al. 1981). A code has been developed by Jensen and made available on the Internet (Jensen 2007).

10.6.2 Characterisation of Transducer Performance

Electrical input impedance can be measured directly using a network or an impedance analyzer, the front face of the transducer being in contact with water. Typical frequencies range from one-tenth of the resonance frequency of the piezoelectric layer to over twice of the same. Pulse-echo response is obtained by exciting the transducer by an electrical pulse of duration less than half a period of the active element resonance, and by measuring the echo reflected by a large acoustic mirror immersed in water (Fig. 10.16). The waveform is usually digitized, which allows fast-Fourier transform (FFT) to be used to obtain frequency response curves.

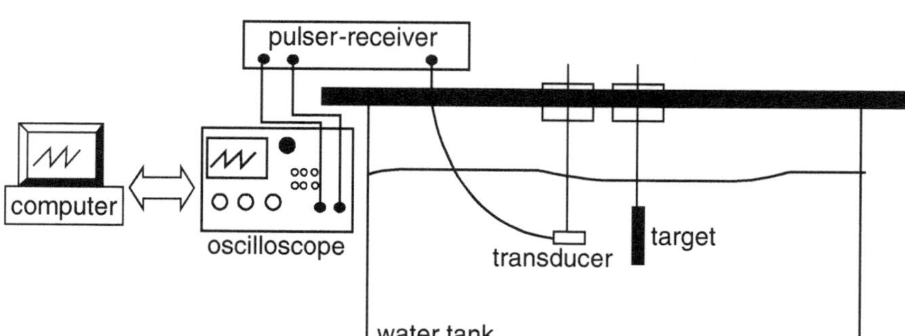

Fig. 10.16 Set-up for pulse-echo response measurement

Radiation patterns can be measured in the transmit mode using a point hydrophone receiver, which is scanned in the half space of the transducer's front side. In the pulse-echo mode, a point-target is scanned, and the amplitude of the echo is plotted as a function of target position.

10.7 High-Frequency Transducers and Other Future Trends

10.7.1 New Piezoceramic Compositions and Structures

Both piezoceramic manufacturers and research institutes are currently developing new piezoceramics with improved properties. Several groups are working on relaxor-ferroelectric compositions and on lead-free materials. More details are given in other chapters of this book; hence no discussion on these topics will be presented herein.

Fabrication process optimization is also a field of intense research. The objectives of these studies are to improve reproducibility, to reduce porosity and grain size. Technologies such as hot isostatic pressing (Millar et al. 1994) and hydrothermal powder synthesis (Millar et al. 1992) have been implemented to this end.

For high-frequency imaging (>20 MHz), piezoelectric elements with a few tens of micrometers thickness are required. Lateral dimensions also need to be controlled precisely. This is traditionally accomplished either by machining bulk piezoceramic or single crystals using diamond tooling and high-precision cutting and lapping systems, or by using piezopolymers. The machining operations are costly and samples often suffer from chipping, or even breaking, giving rise to low yields. The sensitivity is low with piezopolymers.

To avoid these difficulties, thin plates or thick films are an alternative. The typical thickness of such elements is in the 10–100-μm range. Several fabrication processes have been developed for thick films: screen printing (Vechembre et al. 1999), spin coating (Lukacs et al. 1999), dip coating (Gentry et al. 2000) or spray technique (Kobayashi et al. 2000) for high-frequency transducer applications. Generally, the electromechanical performance obtained is lower than the same compositions in bulk form. However, optimized processes such as screen printing, and more recently pad-printing (Levassort et al. 2006), which allows focusing through curved films, have produced samples with properties almost as good as those of bulk PZT.

Thin piezoceramic discs (or plates) are now available through the tape casting route, which allows for the fabrication of self-supporting structures, with different high dielectric constant compositions. Multilayer structures with internal electrodes can also be produced by this process (Sung 1984; Omote et al. 1994; Goldberg and Smith 1994; Desmare et al. 1999). Disks with diameters ranging from 1 to 15 mm and thickness from 60 to 90 μm have been fabricated. The corresponding frequency range is around 25–40 MHz (Levassort et al. 2000). Figure 10.17

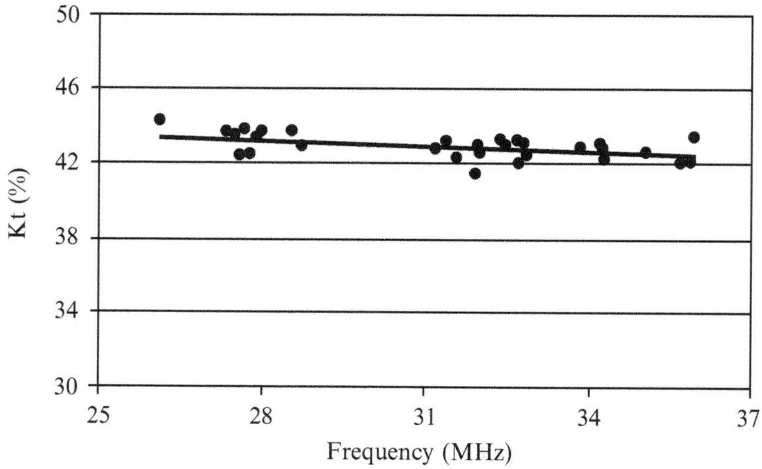

Fig. 10.17 Thickness mode coupling factor (k_t) as a function of frequency for Pz21 ceramics

Fig. 10.18 SEM micrographs of cross-sectioned samples of PZT layer on gold-coated PZT substrate

shows the frequency dependence of the thickness mode coupling factor measured at each fundamental resonance of 32 Ferroperm Pz21 samples. The k_t values obtained are satisfactory to produce miniature single-element transducers with good performance.

For thick films deposited on a substrate, maintaining a k_t value over 40% is often difficult. A PZT composition with the addition of lead germanate as a low-temperature sintering aid has been used to obtain films with 35–50-μm thickness by screen printing on alumina and porous PZT substrates (Tran-Huu-Hue et al. 2001; Levassort et al. 2001) (Fig. 10.18). The determination of the thickness mode parameters of the piezoelectric layer is challenging because of the acoustic resonances that appear in the substrate as well as in the bottom electrode whose thickness is generally not negligible. Electromechanical parameters can be obtained through fitting with a KLM model. Satisfactory performance has been observed, with k_t values comparable to that of bulk ceramics.

The relatively low dielectric constant is well adapted to high-frequency single-element transducer applications. Moreover, thick films on porous PZT substrate are an interesting route for an integrated device. That is so because the relatively high attenuation of the substrate limits the size required for the backing layer. Furthermore, its high acoustical impedance, which is relatively close to that of the PZT thick film, leads to high-transducer bandwidth.

10.7.2 High Coupling Single Crystals

In the beginning of the 1980s and more recently in the early 1990s, several publications (Kuwata et al. 1981, 1982; Shrout et al. 1990; Saitoh et al. 1999a,b) on the PZN–PT solid solution system have reported very high longitudinal coupling coefficients (k_{33}), around 90%.

Currently, two families are being investigated for transducer applications: PZN–PT and PMN–PT, both with a tetragonal macroscopic symmetry. The materials are discussed in other chapters of this book and a review of their use in ultrasonic transducers has been published by Oakley (Oakley and Zipparo 2000).

Compared with those of soft PZT ceramics, the longitudinal wave velocity of these lead-based single crystals is lower but their density is slightly higher, resulting in significantly smaller acoustic impedance. The transmission of acoustic energy to the human body will thus be easier. The very high values of k'_{33} and k_{33} are of interest for 2–2 connectivity (classical linear array technology) and 1–3 connectivity piezocomposites. However, using the dice-and-fill technique, risks of breakage and chipping of single-crystal materials are increased. In the case of piezocomposites with 1–3 connectivity, using a homogenization model (Hashimoto and Yamaguchi 1986; Levassort et al. 1997), the effective thickness coupling coefficient can be calculated.

Figure 10.19 shows the results for three active materials (PMN–33%PT, PZN–8%PT and Navy type IV PZT) and two polymers. One can observe that higher coupling coefficients are indeed obtained with single-crystal-based composites.

Lead-free single crystals have also been developed (Najamura and Kawamura 2000). For instance, $KNbO_3$ displays k_{33} values close to that of PZN–PT, and values of k_t higher than that of both PZN–PT and PMN–PT compositions. Operating temperatures are also increased in $KNbO_3$.

Several transducer applications of single crystals have been published (Saitoh et al. 1998). Single-element transducers using 1–3 composites, operating around 4 MHz (Ritter et al. 2000), have shown high sensitivities and large bandwidths.

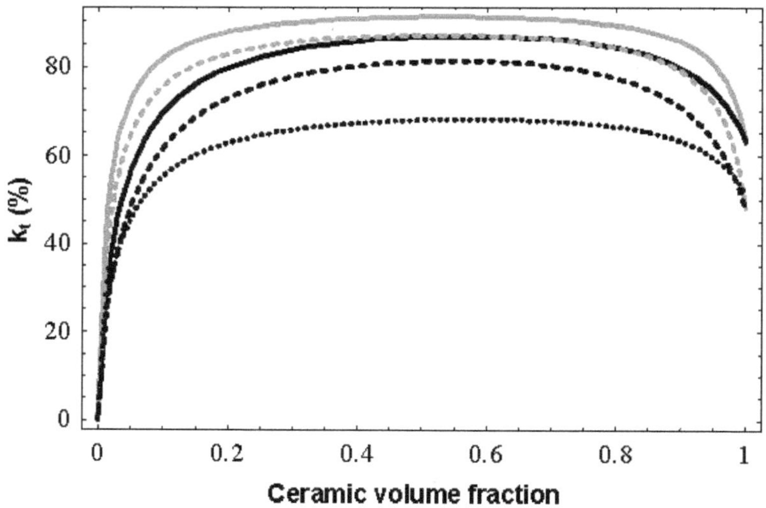

Fig. 10.19 Theoretical thickness coupling factor for 1–3 piezocomposites as a function of the active-phase volume fraction (*black dotted line*, PZT (Navy IV)/epoxy resin; *dashed lines*, with PZN–PT single crystal; solid lines, with PMN–PT single crystal; *grey lines*, with polyethylene; *black lines*, with epoxy resin)

In array configurations, an increase in sensitivity by 5 dB and a 25% wider bandwidth have been obtained in comparison with a PZT-5H-based array. The left ventricle of an adult has been imaged in B-mode by using a 128 element PZN–PT array operating at 3.5 MHz (Saitoh et al. 1999a,b).

10.8 Conclusion

New modalities such as non-linear or high-frequency imaging demand improved transducer performance, which in turn requires more efficient piezoelectric materials. For instance, harmonic imaging requires very high transducer bandwidth to cover more than an octave frequency range, while high-frequency transducers require fine-grained materials. Combined ultrasonic therapy and imaging or different imaging methods make new transducer requirements very stringent and challenging.

For medical imaging and NDE transducers, PZT piezoceramics remain the de facto standard material, and are often combined with a polymer phase to obtain a piezocomposite. Arrays using relaxor-ferroelectric compositions, with low porosity and high dielectric constants have been produced, and on-going research targets further improvements of properties. Fabrication processes are also being developed to improve material properties and to obtain integrated structures such as films on metallized substrates for high-resolution imaging applications. Piezoelectric single crystals with very high electromechanical coupling have also been produced. These new materials appear as promising for array transducers, even though their use in industrial production has a series of obstacles to surmount, including purity, reproducibility, brittleness and cost.

Design of 2D arrays for volumetric imaging applications are pushing towards new structures such as multilayer piezoelectric elements and alternative techniques such as capacitive micro-machined ultrasonic transducers (CMUTs), or their piezoelectric counterparts (PMUTs). These technologies are expected to open new possibilities for transducer integration in the future, such as disposable arrays, but piezoceramic-based transducers should continue to be the workhorse of the industry for many years to come.

References

Arditi M, Foster FS, Hunt JW (1981) Transient fields of concave annular arrays. Ultrasonics Imaging, vol. 3, pp. 37–61.

Assaad J, Dubus B, Hamonic B, Decarpigny JN, Debus JC (1991) Finite element modelling of ultrasonic transducers using the ATILA code. Proceedings of Ultrasonics International, pp. 371–374.

Banno H (1983) Recent development of piezoelectric ceramic products and composites of synthetic rubber and piezoelectric ceramic particles. Ferroelectrics, vol. 50, pp. 329–338.

Bove T, Wolny W, Ringaard E, Pedersen A (2001) New piezoceramic PZT–PNN material for medical diagnosis applications. Journal of the European Ceramic Society, vol. 21, pp. 1469–1472.

Desilets CS, Fraser JD, Kino GS (1978) The design of efficient broadband piezoelectric transducers. Ultrasonics, vol. 25, pp. 115–125.

Desmare R, Tran-Huu-Hue LP, Levassort F, Lethiecq M (1999) Modeling of multilayer piezoelectic structures. Ferroelectrics, vol. 224, pp. 623–630.

Ferroperm Piezoceramics (2007) www.ferroperm-piezo.com.

Gentry KL, Zara JM, Bu S, Eom C, Smith SW (2000) Thick film sol PZT transducer using dip coating. IEEE Ultrasonics Symposium Proceedings, vol. 2, pp. 977–980.

Goldberg RL, Smith SW (1994) Multilayer piezoelectric ceramics for 2D array transducers. IEEE Transactions on Ultrasonics, Ferroelectrics, and Frequency Control, vol. 41, pp. 761–771.

Hashimoto KY, Yamaguchi M (1986) Elastic, piezoelectric and dielectric properties of composite materials. IEEE Ultrasonics Symposium Proceedings, pp. 697–702.

IEEE (1987) Standard on Piezoelectricity ANSI/IEEE Std.

Inoue T, Ohta M, Takahashi S (1987) Design of ultrasonic transducers with multiple acoustic matching layers for medical application. IEEE Transactions on Ultrasonics, Ferroelectrics, and Frequency Control, vol. 34, pp. 8–16.

Jensen JA (2007) www.es.oersted.dtu.dk/staff/jaj/field.

Kawai M (1969) The piezoelectricity of polyvinilydene fluoride. Japanese Journal of Applied Physics, vol. 8, pp. 975–976.

Kino GS (1987) Acoustic waves: devices imaging and analog signal processing: processing. Prentice-Hall, Englewood Cliffs, New Jersey.

Kobayashi M, Olding TR, Zou L, Sayer M, Jen CK, Rehman AU (2000) Piezoelectric thick film ultrasonic transducers fabricated by spray technique. IEEE Ultrasonics Symposium Proceedings, vol. 2, pp. 985–989.

Kossof G. (1966) The effects of backing and matching on the performance of piezoelectric ceramic transducers. IEEE Transactions on Sonics and Ultrasonics, vol. 13, pp. 20–30.

Krimholtz R, Leedom DA, Matthei GL (1970) New equivalent circuit for elementary piezoelectric transducers. Electronic letters, vol. 38, pp. 338–339.

Kuwata J, Uchino K, Nomura S (1981) Phase transitions in the Pb(Zn1/3Nb2/3)O3–0.09PbTiO3 system. Ferroelectrics, vol. 37, pp. 579–582.

Kuwata J, Uchino K, Nomura S (1982) Dielectric and piezoelectric properties of 0.91Pb(Zn1/3Nb2/3)O3–0.09PbTiO3 single crystals. Japanese Journal of Applied Physics, vol. 21, pp. 1298–1302.

Kwok KW, Chan HLW, Choy CL (1997) Evaluation of the material parameters of piezoelectric materials by various method. IEEE Transactions on Ultrasonics, Ferroelectrics, and Frequency Control, vol. 44, pp. 733–742.

Levassort F, Lethiecq M, Certon D, Patat F (1997) A matrix method for modeling electroelastic moduli of 0–3 piezo-composite. IEEE Transactions on Ultrasonics, Ferroelectrics, and Frequency Control, vol. 44, pp. 445–452.

Levassort F, Lethiecq M, Millar CE, Pourcelot L (1998) Modeling of highly loaded 0–3 piezoelectric composites using a matrix method. IEEE Transactions on Ultrasonics, Ferroelectrics, and Frequency Control, vol. 45, pp. 1497–1505.

Levassort F, Tran-Huu-Hue LP, Lethiecq M, Bove T, Wolny W (2000) New piezoceramics films for high resolution medical imaging applications. IEEE Ultrasonics Symposium Proceedings, vol. 2, pp. 1125–1128.

Levassort F, Tran-Huu-Hue LP, Holc J, Bove T, Kosec M, Lethiecq M (2001) High performance piezoceramic films on substrates for high frequency imaging. IEEE Ultrasonics Symposium Proceedings.

Levassort F, Filoux E, Lou-Møller R, Ringgaard E, Lethiecq M, Nowicki A (2006) Curved piezoelectric thick films for high resolution medical imaging. Proceedings of IEEE International Ultrasonics Symposium, pp. 2361–2364.

Levassort F, Holc J, Ringgaard E, Bove T, Kosec M, Lethiecq M (2007) Fabrication, modelling and use of porous ceramics for ultrasonic transducer applications. Journal of Electroceramics, vol. 19, pp. 125–137.

Lukacs M, Olding T, Sayer M, Tasker R, Sherrit S (1999) Thickness mode material constants of a supported piezoelectric film. Journal of Applied Physics, vol. 85, pp. 2835–2843.

Millar CE, Wolny WW, Pardo L (1992) Field dependence of the electromechanical properties of fine grained hydrothermally process lead titanate ceramics. IEEE ISAF Proceedings, pp. 59.

Millar E, Pedersen L, Pardo L, Ricote J, Alemany C, Jimenez B, Feuillard G, Lethiecq M (1994) Effect of processing on surface acoustic wave properties of modified lead titanate ceramic. IEEE ISAF Proceedings, pp. 138–141.

Najamura K, Kawamura Y (2000) Orientation dependance of electromechanical coupling factors in KnbO3. IEEE Transactions on Ultrasonics, Ferroelectrics, and Frequency Control, vol. 47, pp. 750–755.

Newnham RE (1986) Composite electroceramics. Ferroelectrics, vol. 68, pp. 1–32.

Newnham RE, Skinner DP, Cross LE (1978) Connectivity and piezoelectric–pyroelectric composites. Mat. Res. Bull., vol. 13, pp. 525–536.

Nguyen TN, Lethiecq M, Levassort F, Pourcelot L (1996) Experimental verification of elastic properties using scattering approximation in (0–3) connectivity composite materials. IEEE Transactions on Ultrasonics, Ferroelectrics, and Frequency Control, vol. 43, pp. 640–645.

Oakley CG, Zipparo MJ (2000) Single crystal piezoelectrics: a revolutionary development for transducers. IEEE Ultrasonics Symposium Proceedings, vol. 1, pp. 1157–1167.

Omote K, Park KS, Li G, Ohigashi H (1994) Performance of multilayered ultrasonics transducers comprising vinylidene fluoride and trifluorethylene copolymer films and ferroelectric ceramic plates. Japanese Journal of Applied Physics, vol. 33, pp. 2966–2971.

Penttinen A, Luukkala M (1976) The impulse response and pressure nearfield of a curved ultrasonics radiators. Journal of Physics D: Applied Physics, vol. 9, pp. 1547–1557.

Powers JE (1980) An ultrasonic annular array based on quadrature sampling. Ph. D. Thesis, Washington University, Saint Louis, MO.

Rittenmeyer K, Shrout TR, Schulze WA, Newnham RE (1982) Piezoelectric 3–3 composites. Ferroelectrics, vol. 41, pp. 289–295.

Ritter T, Shung KK, Geng X, Lopath PD, Park SE, Shrout TR (2000) Single crystal PZN–PT-polymer composites for ultrasound transducer applications. IEEE Transactions on Ultrasonics, Ferroelectrics, and Frequency Control, vol. 47, pp. 792–800.

Saitoh S, Kobayashi T, Harada K, Shimanuki S, Yamashita Y (1998) A 20 MHz single-element ultrasonic probe using 0.91Pb(Zn1/3Nb2/3)O3–0.09PbTiO3 single crystal. IEEE Transactions on Ultrasonics, Ferroelectrics, and Frequency Control, vol. 45, pp. 1071–1076.

Saitoh S, Imuzi M, Shimanuki S, Hashimoto S, Yamashita Y (1999a) US Patent 5,295,487.

Saitoh S, Takeuchi T, Kobayashi T, Harada K, Shimanuki S, Yamashita Y (1999b) Forty-channel phased array ultrasonic probe using 0.91Pb(Zn1/3Nb2/3)O3–0.09PbTiO3 single crystal. IEEE Transactions on Ultrasonics, Ferroelectrics, and Frequency Control, vol. 46, pp. 152–157.

Savakus HP, Klicker KA, Newnham RE (1981) PZT-epoxy piezoelectric transducers: a simpliflied fabrication procedure. Mat. Res. Bull., vol. 16, pp. 677–680.

Selfridge AR, Kino GS, Khury-Yakub BT (1980) A theory for the radiation pattern of a narrow strip acoustic transducer. Applied Physics Letters, vol. 37, pp. 35–36.

Shrout TR, Chang ZP, Kim N, Markgraf S (1990) Dielectric behavior of single crystals near the (1-x)Pb(Zn1/3Nb2/3)O3–(x)PbTiO3 morphotropic phase boundary. Ferroelectrics Letters Section, vol. 12, pp. 63–69.

Sung KM (1984) Piezoelectric multilayer transducers for ultrasonic pulse compression. Ultrasonics, pp. 61–68.

Tran-Huu-Hue LP, Levassort F, Lethiecq M, Certon D, Patat F (1997) Characterization of the piezoelectric and dielectric relaxation parameters of 0–3 composite and PVDF materials in thickness mode. Ultrasonics, vol. 34, pp. 317–324.

Tran-Huu-Hue LP, Levassort F, Felix N, Damjanovic D, Wolny W, Lethiecq M (2000) Comparison of several methods to characterize the high frequency behavior of piezoelectric ceramics for transducer applications. Ultrasonics, vol. 38, pp. 219–223.

Tran-Huu-Hue LP, Levassort F, Vander Meulen F, Holc J, Kosec M, Lethiecq M (2001) Preparation and electromechanical properties of PZT/PGO thick films on alumina substrate. Journal of the European Ceramic Society, vol. 21, pp. 1445–1449.

Vechembre J, Sagalowicz L, Setter N (1999) Screen printed PZT layer-fabrication and properties. Ferroelectrics, vol. 224, pp. 145–152.

Chapter 11
Piezoelectric Transducer Designs for Sonar Applications

James F. Tressler

11.1 Introduction

An electroacoustic transducer is a device that converts acoustic energy (sound) into electrical energy (voltage or current) or vice versa. When the transducer is used to generate sound, it is called a projector, transmitter, or source. When it is used to detect sound, it is called a receiver. Furthermore, when the receiver is employed underwater, it is referred to as a hydrophone. An underwater sonar system consists of projectors, hydrophones, and associated electronics such as amplifiers and data acquisition systems. This chapter, however, will only cover the description and operational principles of the projector and hydrophone components. More specifically, the chapter will focus on piezoelectric ceramic-based transducer designs intended for underwater use that operate in the frequency band spanning from 1 kHz to 1 MHz. This span covers weapons sonar (1–100 kHz) and imaging sonar (100 kHz to 1 MHz) applications.

Section 11.2 describes the fundamental parameters and measurement techniques necessary to characterize transducers both in air and in water. The piezoelectric ceramic materials that are most commonly used in sonar transducers and their relevant materials properties are discussed in Sect. 11.3. The most common piezoelectric ceramic-based projector designs for sonar applications are described in Sect. 11.4, where they are grouped according to their operational frequency range. A single transducer cannot operate efficiently over the entire 1-kHz to 1-MHz frequency band – primarily due to impedance matching issues with the associated amplifiers. Consequently, a variety of transducer designs have been developed and optimized for different frequency ranges. The chapter concludes with hydrophone characterization, designs, and some performance results in Sect. 11.5.

J.F. Tressler
Naval Research Laboratory, Washington, DC 20375, USA

A. Safari, E.K. Akdoğan (eds.) *Piezoelectric and Acoustic Materials for Transducer Applications.*

11.2 Transducer Characterization

11.2.1 In-Air Characterization

Piezoelectric transducers designed for underwater use are initially characterized in air, preferably before being either potted in polyurethane or filled with oil. With modern impedance analyzers, electrical immitance (i.e., impedance or admittance) as a function of frequency, both in terms of real and imaginary components, is a simple measurement to perform and provide a wealth of information to the transducer designer.

A very informative way to plot admittance data is shown in Fig. 11.1. Here, the in-air admittance is plotted as a vector in the conductance–susceptance plane, where frequency increases along the loop as shown in the figure. The circle describes the admittance behavior in the neighborhood of the transducer fundamental resonance frequency. It should be noted that this particular plot is representative of a lightly-loaded (such as air) transducer with a strong resonance.

The noteworthy frequencies obtained from this plot are as follows:

- f_s is the series resonance frequency and is the frequency of maximum conductance. It is the frequency at which the motional power is a maximum for a constant applied voltage (Stansfield 1991).
- f_p is the parallel resonance frequency and is the frequency of maximum motional resistance.
- f_m is the frequency of maximum admittance.
- f_n is the frequency of minimum admittance.
- f_r is the electrical resonance frequency (susceptance is zero).
- f_a is the electrical antiresonance frequency (susceptance is also zero).

The "sharpness" of the resonance of a transducer is known as the *mechanical quality factor*, or simply Q_m. It is found from the electrical conductance vs. frequency curve as

$$Q_m = \frac{f_s}{f_2 - f_1}, \tag{11.1}$$

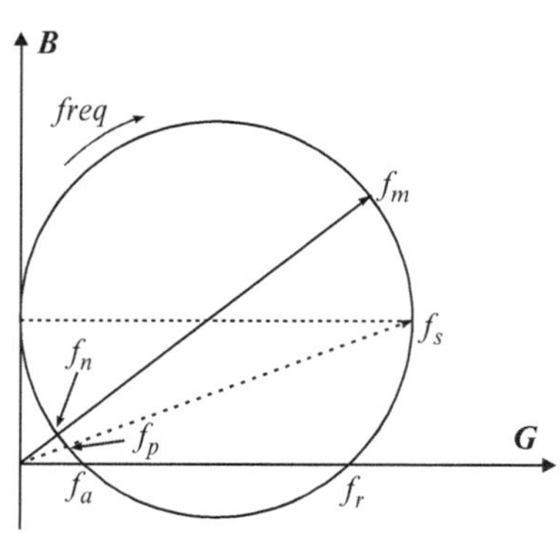

Fig. 11.1 Admittance circle of a lightly-loaded piezoelectric transducer

where f_1 and f_2 are the frequencies on either side of the resonance where the conductance value is one half its maximum value.

The *electromechanical coupling factor*, k, is a measure of the effectiveness with which a piezoelectric material or device converts the input electrical energy into mechanical energy, or vice versa (Moulson and Herbert 1992). It will also be shown to be a measure of the maximum achievable bandwidth of the transducer. When referring to coupling factor, one needs to differentiate between the material coupling factor of the piezoelectric ($k_{31}, k_{33}, k_{15}, k_t$, or k_p) and the effective coupling factor (k_{eff}) of the transducer itself. Typically k_{eff} is calculated from admittance measurements taken in air rather than in water. The magnitude of the mechanical quality factor determines the formula for calculating k_{eff} (Moffett and Marshall 1994).

$$k_{\text{eff}} = \sqrt{1 - \left(\frac{f_{\text{s}}}{f_{\text{p}}}\right)^2} \quad \text{for} \quad Q_{\text{m}} > 50 \tag{11.2a}$$

$$k_{\text{eff}} = \sqrt{\frac{1}{1 + Q_{\text{m}} Q_{\text{e}}}} \quad \text{if} \quad 10 < Q_{\text{m}} < 50. \tag{11.2b}$$

Here Q_{m} is calculated from (11.1) and the electrical quality factor Q_{e} is

$$Q_{\text{e}} = \frac{B_{\text{s}}}{G_{\max}}, \tag{11.3}$$

where $G_{\max}$ and B_{s} are the respective conductance and susceptance values measured at the series resonance frequency.

Along with electrical admittance, one can also easily measure the parallel capacitance, C_{p}, and dielectric loss, $\tan \delta$. Below the resonance frequency, the parallel capacitance is nearly constant and this value, which is usually measured at either 100 Hz or 1 kHz, is the one that is reported.

11.2.2 In-Water Characterization

When submerged under water, the effect of the water loading on the transducer must be taken into account. Three new parameters are the radiation resistance, R_{r}, radiation reactance, X_{r}, and the radiation mass, M_{r}. The power dissipated in R_{r} represents the far-field acoustic output power from the projector and the energy stored in M_{r} is equal to the stored near-field acoustic energy, which cyclically returns to the transducer (Grantner et al. 1976). Taken together, the radiation resistance and radiation mass constitute the radiation impedance as

$$Z_{\text{r}} = R_{\text{r}} + \text{j}X_{\text{r}} = R_{\text{r}} + \text{j}\,\omega M_{\text{r}} \tag{11.4}$$

For the common cases of a radiating sphere and of a circular piston (both mounted in an infinite rigid baffle), the radiation impedances can be expressed as

$$\text{Sphere}: \quad Z_r = 4\pi a^2 \rho c \left[\frac{(k_\lambda a)^2}{1+(k_\lambda a)^2} + \mathrm{j}\frac{k_\lambda a}{1+(k_\lambda a)^2} \right] \tag{11.5a}$$

and

$$\text{Piston}: \quad Z_r = \rho c \pi a^2 [R_1(2k_\lambda a) + \mathrm{j}X_1(2k_\lambda a)], \tag{11.5b}$$

where a is the piston or sphere (outer) radius, k_λ is the wavenumber and equals ω/c or $2\pi/\lambda$ (λ being wavelength, ω radial frequency, and c sound speed), R_1 is the radiation resistance function, X_1 is the radiation reactance function, ρc is known as the specific acoustic impedance of the fluid and is the product of its density ρ and sound speed c. The normalized resistance and reactance functions of a sphere and a piston in an infinite rigid baffle are shown in Fig. 11.2.

Returning to the case of the piston – whereas the radiation resistance dissipates itself as acoustic energy, the radiation reactance manifests itself as an added mass to the transducer of magnitude

$$M_r = \frac{X_r}{\omega} = \pi a^2 \rho \left[\frac{X_1(2k_\lambda a)}{k_\lambda} \right]. \tag{11.6}$$

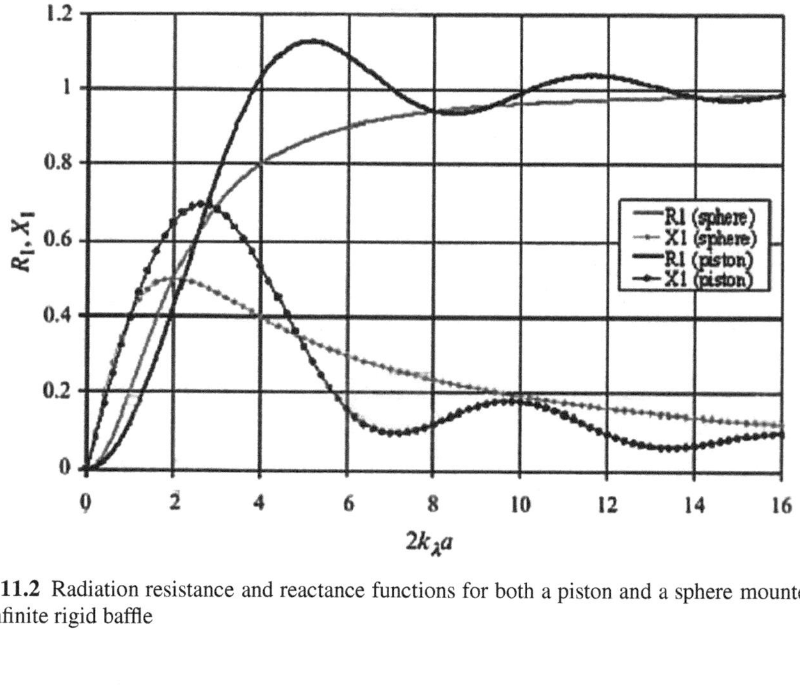

Fig. 11.2 Radiation resistance and reactance functions for both a piston and a sphere mounted in an infinite rigid baffle

When the parameter $2k_{\lambda}a$ is large, R_1 tends to a value of 1 and X_1 approaches zero. In this instance, the transducer is said to be ρc-loaded and radiation efficiency into the water is maximized. If the piston is very small or operated at low frequencies (i.e., $2k_{\lambda}a < 1$), then the piston-added mass term in (11.6) approaches a constant value of $(8/3)\rho a^3$. This is equivalent to the mass of an imaginary cylinder of the acoustic medium with the same radius as the piston and a height of $(8a)/3\pi$. In light acoustic media such as air, this added mass is negligible. However, when in water, this added mass will markedly reduce the resonance frequency.

The acoustic output power (P_a) from the transducer is related to the radiation resistance and transducer velocity u (and subsequently transducer displacement x) as

$$P_{\mathrm{a}} = \frac{1}{2}u^2 R_{\mathrm{r}} = \frac{1}{2}\omega^2 x^2 R_{\mathrm{r}}. \quad (11.7)$$

When $2k_{\lambda}a$ is small, that is, for small pistons or at very low frequencies, R_1 in Fig. 11.2 goes as $(k_{\lambda}a)^2/2$. Hence R_{r} is proportional to the square of the piston area as well as the square of frequency. This shows that it is quite difficult to produce a large acoustic output from a small size source at low frequency unless very large displacements can be achieved. This is noteworthy because monolithic piezoelectric ceramic elements have inherently small displacements. The displacement necessary to produce a desired output power for a known transducer size at any frequency can be calculated from (11.7).

A representative admittance loop for a transducer measured in water is shown in Fig. 11.3. The mechanical Q in water (Q_{w}) is calculated from (11.1) but with f_{s} replaced by $f_{\mathrm{s,w}}$ (w subscript denotes measurement taken in water) and $f_{1,\mathrm{s}}$ and $f_{2,\mathrm{s}}$ replaced by $f_{1,\mathrm{w}}$ an $f_{2,\mathrm{w}}$, respectively. Here $f_{1,\mathrm{w}}$ and $f_{2,\mathrm{w}}$ are the frequencies on the admittance loop located on the line that bisects D_{w}, where D_{w} is the length of the horizontal line extending from $G_{\mathrm{w}} = f_{\mathrm{s,w}}$ to its intersection with the $Y_{\mathrm{b,w}}$-line.

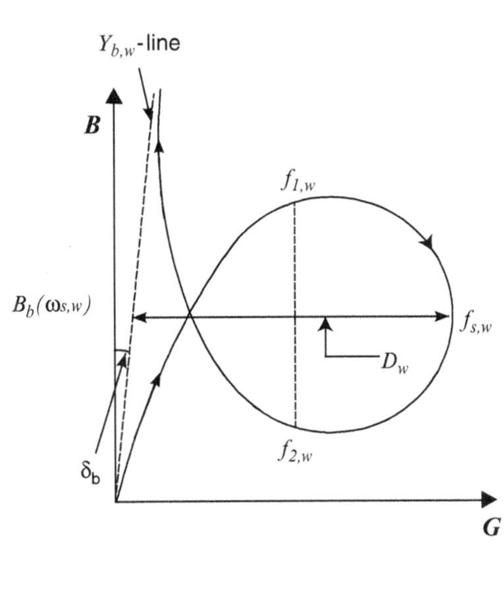

Fig. 11.3 Admittance circle of a water-loaded piezoelectric transducer (Grantner et al. 1976)

The in-water electrical $Q_{e,\mathrm{w}}$ is now calculated as:

$$Q_{\mathrm{e,w}} = \left[\tan\delta_{\mathrm{b}} + Q_{\mathrm{w}}\left(\frac{k^2_{\mathrm{eff,w}}}{1-k^2_{\mathrm{eff,w}}}\right)\right]^{-1}. \tag{11.8}$$

Note that the dielectric loss in (11.8) is for blocked conditions. Normally, the tan δ for a material is measured under free conditions. The two are related as tan $\delta_{(\mathrm{free})} = (1-k^2_{\mathrm{eff,w}})$ tan δ_b (Grantner et al. 1976).

The electromechanical coupling factor in water is:

$$k^2_{\mathrm{eff,w}} = \left[1 + \frac{Q_{\mathrm{w}} B_{\mathrm{b}}(\omega_{\mathrm{s,w}})}{D_{\mathrm{w}}}\right]^{-1}, \tag{11.9}$$

where the value $B_b(\omega_{\mathrm{s,w}})$ denotes the value of susceptance at the series resonance frequency.

Electroacoustic efficiency (η) is defined as the ratio of the output-radiated power to the total electrical input power (Hughes 1998). This is a computed parameter that is best derived from direct in-water acoustic measurements (as per (11.11) or (11.12)). However, it is also possible to calculate its value at the transducer resonance frequency by comparing the diameters of the in-air and in-water admittance circles (Bobber 1988) as

$$\eta = \frac{D_{\mathrm{w}}(D_{\mathrm{a}} - D_{\mathrm{w}})}{G_{\mathrm{max}} D_{\mathrm{a}}}, \tag{11.10}$$

where D_{a} is the diameter of the in-air admittance circle from Fig. 11.1, D_{w} is the diameter of the in-water admittance circle from Fig. 11.3, and G_{max} is the value of maximum conductance measured in air.

11.2.3 In-Water Acoustic Calibration Measurements

The characterization of an underwater transducer is not complete until its performance as both a sound projector and sound receiver has been measured. This procedure is known as calibration. Described below are the most important parameters that need to be determined. For more detailed information and other parameters, the reader is encouraged to review the textbooks by Bobber (1988), Stansfield (1991), or Wilson (1988).

The *directivity* (aka *beam*) *pattern* of a transducer shows the relative acoustic pressure (generated for a projector, received for a hydrophone) as a function of angle relative to the face of the transducer. From the beam pattern, one can locate the main or primary beam on the *acoustic axis*, which is defined as the direction where the acoustic pressure has its maximum value. The beam pattern of a transducer is the same regardless of whether it is operating as a projector or as a receiver. The *beam width* is the angular width (in degrees) of the primary beam of the transducer. Its measure is taken between the two points 3 dB lower than the maximum point on the main beam.

The *directivity index* (DI) is the decibel representation of the *directivity factor*, which is a number that reveals the relative degree of directionality of a transducer at a given frequency (Beranek 1993). In general, it depends upon the aperture area (and to a lesser extent its shape) of the transducer as well as its mounting conditions. The directivity factor/index for a number of transducer configurations can be found in the texts by Beranek (1993), Bobber (1988), and Stansfield (1991).

The *transmitting voltage response* (TVR) of a projector is a measure of its acoustic output referenced to a distance 1 m from the face of the transducer for an input of 1 V_{rms}. It is reported in terms of decibels as dB re: 1 μPa$\,$V^{-1} at 1 m. The TVR is related to the conductance, DI, and efficiency as:

$$\mathrm{TVR} = 10\log(G_{\mathrm{p}}) + \mathrm{DI} + \eta_{(\mathrm{dB})} + 170.8. \tag{11.11}$$

Transmitting current response (TCR) of a projector is analogous to the TVR, but with the acoustic output referenced at 1 m to an input current of 1 A_{rms}. It is thus reported as dB re: 1 μPa$\,$A^{-1} at 1 m. It is related to the electrical series resistance of the transducer, DI, and efficiency as

$$\mathrm{TCR} = 10\log(R_{\mathrm{s}}) + \mathrm{DI} + \eta_{(\mathrm{dB})} + 170.8. \tag{11.12}$$

The TCR of a projector can be calculated from the TVR via the transducer impedance as

$$\mathrm{TCR} = \mathrm{TVR} + 20\log|Z|. \tag{11.13}$$

The idealized transmitting voltage and current responses for a piezoelectric transmitter are illustrated in Fig. 11.4. The TVR increases at +12 dB per octave below resonance and then drops at −12 dB per octave above resonance before flattening out. On the other hand, the TCR curve increases at +6 dB per octave below resonance and then decreases by −6 dB per octave above resonance.

In practice, a projector will be driven at levels greater than 1 V$_{\mathrm{rms}}$. The drive level is dictated by the cavitation threshold and the materials properties of the piezoelectric ceramic elements that constitute the transducer. The *source level* (SL) of a projector is a measure of the acoustic output (again referenced to 1 m) for any given

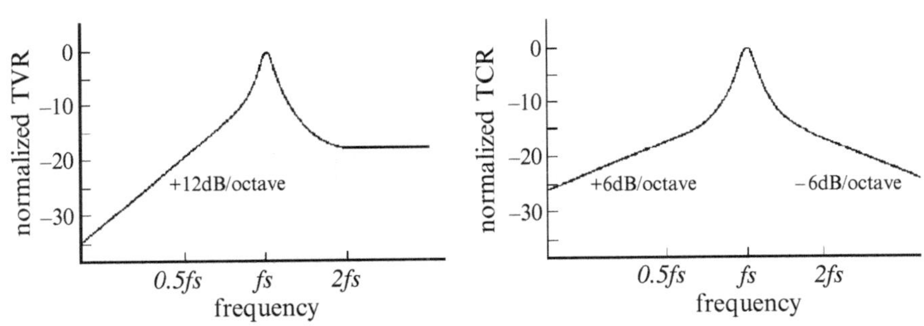

Fig. 11.4 Idealized transmitting voltage response (TVR) and transmitting current response (TCR) of a piezoelectric transducer

applied rms voltage as:

$$\mathrm{SL} = 10\log(P_\mathrm{a}) + \mathrm{DI} + 170.8 \quad (\mathrm{dB\ re:1\ \mu Pa\ at\ 1\ m}). \tag{11.14}$$

The SL can be related to the TVR as:

$$\mathrm{SL} = \mathrm{TVR} + 20\log(V) = \mathrm{TPR} + 10\log(P_\mathrm{in}). \tag{11.15}$$

Here TPR is the *transmitting power response*. It is equal to $\mathrm{DI} + \eta + 170.8$ and is a measure of the acoustic output of the projector referenced to a distance of 1 m from the face of the transducer for an electrical input of 1 W. The input power is designated by P_in and is equal to V^2G_p.

The *bandwidth* is the frequency span over which the projector can effectively operate. Unfortunately, there is no universally agreed upon definition for bandwidth. Quite often it is reported as being equal to $1/Q_\mathrm{m}$. And more often than not, it is taken to be the frequency span from the TVR plot, which lies within 3 dB of its maximum value. However, these definitions are really only valid if the power source can deliver a constant voltage across the band. More precisely, the power amplifier and projector properties combine to determine practical bandwidth limitations (Moffett et al. 2007). The so-called Stansfield criteria for bandwidth is to have no greater than a 2:1 variation in transducer electrical admittance magnitude over the band as well as no more than a ±37° change in phase angle (Stansfield 1991).

Stansfield (1991) further states that in order to optimize bandwidth, the transducer should be designed so that its $Q_\mathrm{m} \cdot k_\mathrm{eff}$ product is approximately equal to 1.2. This condition provides the best matching between the transducer and power source. The effective coupling factor is related to the maximum attainable bandwidth. However, to reach this best-case scenario, an appropriate Q_m must also be used. Figure 11.5 shows how bandwidth is affected by Q_m for transducers with ef-

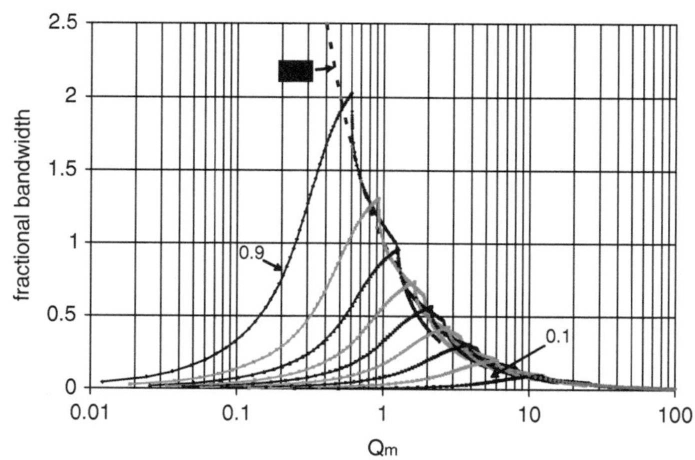

Fig. 11.5 Achievable fractional bandwidth as a function of mechanical Q_m for transducers with k_eff ranging from 0.9 to 0.1 (in decreasing increments of 0.1 from left to right)

fective coupling coefficients ranging from 0.1 to 0.9 (Moffett et al. 2007). This result further shows the fundamental importance of the coupling coefficient in determining the bandwidth of a transducer, since wide bandwidth can only be obtained by means of a transducer with a high effective coupling coefficient (Stansfield 1991).

11.3 Transducer Materials

The two types of piezoelectric ceramic most commonly used in sonar transducers are those from the lead zirconate titanate (aka PZT) and barium titanate ($BaTiO_3$) family of compositions. Commercial manufacturers have developed a general nomenclature based on US Military Standard (1995) specifications to differentiate these piezoceramic materials. Table 11.1 lists some of the relevant properties of these transducer materials.

The materials properties provided in product literature, such as from Channel Industries (2006) or Morgan Electro Ceramics(2006), are typically for low-electrical drive levels. Under high-electrical drive conditions, such as in the case of projector operation, these properties can change considerably. For example, Fig. 11.6 shows how the dielectric constant and dielectric loss change as a function of electric field for various piezoelectric ceramic types (Grantner et al. 1976). A typical drive level for a projector is nominally 2 kV cm^{-1}. At this level, a good projector material should have its $\tan\,\delta < 0.02$ in order to prevent the material from overheating. A good rule of thumb is never to exceed 75% of the Curie temperature during any phase of projector fabrication or operation in order to prevent the piezoelectric from depoling. In addition, the linear relationship between strain (displacement) and electric field only holds over a limited range of electric field. The extent of the linearity depends on the coercive field of the ceramic and hence is dependent upon the ceramic type. Nonlinearity results in harmonic distortion of the acoustic output as well as lower efficiency and heating via dielectric loss.

As seen in Fig. 11.7, large static compressive stresses (and consequently operating depth) also strongly affect the materials properties of piezoelectric ceramics. The data shown in the figure were culled from the work of Krueger (1967a) for compressive stresses parallel to the poling direction. Stresses perpendicular to the poling direction and hydrostatic stresses affect the materials as well, albeit differently (Krueger 1967b). It should also be noted that the materials properties are also dependent on the number of stress cycles as well as exposure time (Krueger 1967a, b; Krueger and Berlincourt 1961). The change in dielectric constant under stress seen in Fig. 11.7 (as well as under high electric field as in Fig. 11.6) consequently means a change in the material capacitance and hence a change in its electrical impedance. The end result being that the electrical impedance match between the transducer and the amplifier is no longer optimized. Likewise, the change in d_{33} is associated with a change in displacement, which when coupled with the variation in permittivity, results in a change in sensitivity. The generally accepted safe compressive stress limits for the ceramic materials are provided in Table 11.1. Because of their inherently

Table 11.1 Room temperature low electric field properties of common piezoelectric materials used in sonar transducers (Grantner et al. 1976; Stansfield 1991; Channel Industries 2006; Morgan Electro Ceramics 2006)

Military specification	Commercial specification	k_{33}	k_{31}	d_{33} (pm V^{-1})	d_{31} (pm V^{-1})	g_{33} (V m N^{-1})	g_{31} (V m N^{-1})	K_{33}^{T}	K_{31}^{T}	tan δ	Field limit at fs (kV cm^{-1})	Maximum compressive stress (MPa)	Curie temperature (°C)
Navy type I	PZT-4	0.71	−0.36	300	−135	0.0261	−0.0117	1,300	1,475	0.004	2.8	100	300
Navy type II	PZT-5A	0.73	−0.37	400	−185	0.0258	−0.0119	1,750	1,775	0.02	–	35	350
Navy type III	PZT-8	0.66	−0.32	240	−105	0.0258	−0.0113	1,050	1,340	0.004	4	100	300
Navy type IV	$BaTiO_3$	0.45	−0.18	145	−56	0.0122	−0.0047	1,350	1,450	0.008	2–3	70	115
Navy type V	PZT-5J	0.73	−0.36	505	−225	0.0220	−0.0098	2,600	2,400	0.02	–	<35	240
Navy type VI	PZT-5H	0.72	−0.37	550	−250	0.0194	−0.0088	3,200	2,950	0.02	1.6	<35	190

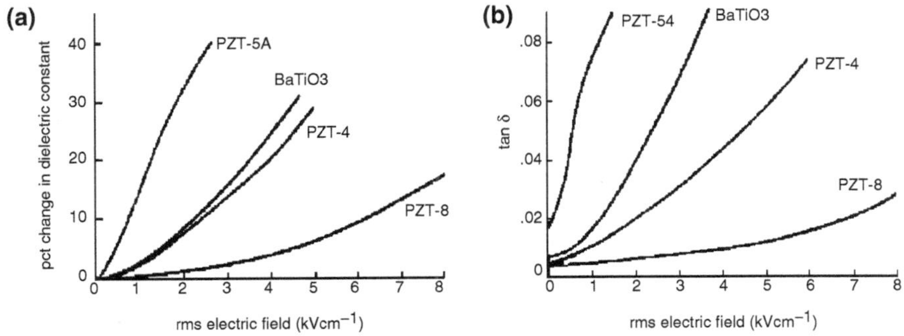

Fig. 11.6 Change in (**a**) dielectric constant and (**b**) dielectric loss as a function of drive field for different piezoelectric ceramic materials (Grantner et al. 1976)

better stability under high drive levels, projectors usually incorporate PZT-4 or PZT-8 as the active material. Hydrophones, on the other hand, utilize the higher sensitivity and dielectric constant of the PZT-5 compositions.

11.4 Projector Designs

Underwater projectors are ideally operated at or very near their resonance frequency so as to optimize efficiency and sound output into the water. In the literature, underwater projectors are often grouped into three categories: low-frequency, mid-frequency, and high-frequency designs. The terms low, mid, and high are somewhat arbitrary. But in general, low frequency refers to <5 kHz and high frequency pertains to frequencies above 100 kHz. Mid-frequency covers the band in between.

11.4.1 Low-Frequency Projector Designs

Low-frequency projector designs have historically been the most challenging for transducer engineers. As alluded to in Sect. 11.2.2, to achieve a high sound output at low frequency, it is necessary to generate very large particle velocities (and hence displacements) or else push a large volume of water. As an illustration, assume one has a 5-cm diameter piston and the desired SL is 180 dB. The displacement required for a given frequency can be calculated using (11.14), (11.7), and the real part of (11.5b). At 10 kHz, a 5-cm diameter piston requires a displacement of 1.7 μm to produce 180 dB. However, down at 1 kHz, $\sim$155 μm is needed. Alternatively, given a 1.7 $-$ μm displacement at 1 kHz, the piston aperture needs to be nearly 31 cm in diameter to generate 180 dB. Thus, at low frequency, the use of monolithic piezoelectric ceramic elements by themselves becomes impractical. Piezoelectric ceramics are inherently mechanically stiff (resulting in a high resonance frequency) and

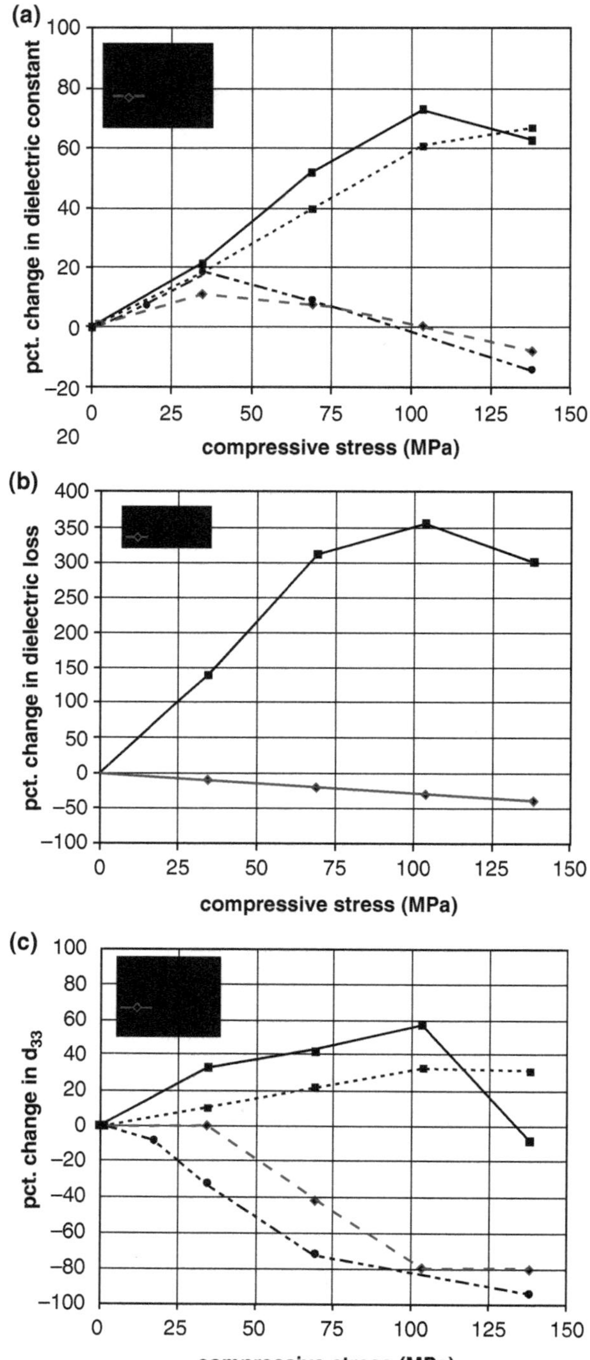

Fig. 11.7 Percent change in (**a**) dielectric constant, (**b**) dielectric loss, and (**c**) d_{33} as a function of compressive stress parallel to the poling direction for different piezoelectric ceramic materials (from Krueger 1967a)

generate very small displacements on their own. To utilize them in low-frequency applications, it is necessary to couple them with a displacement amplification mechanism. This also acts to lower the stiffness of the device, which in turn lowers the resonance frequency so that it appears in the band of interest.

11.4.1.1 Ring Projectors

Piezoelectric ceramic ring projectors are actually an exception to the rule stated above in that they do not utilize a displacement amplification mechanism. Instead, this design can be employed when large projector size is not an issue. In this projector design, acoustic energy is generated by the radial expansion and contraction of the ring. Consequently, sound is radiated from both the outer and inner surface (as well as the ends). The ring can be operated either as free-flooded or pressure release. In the free-flooded case, water is free to flow over the entire volume of the ring. This configuration has the advantage of being a simple resonator whose acoustic performance is virtually independent of hydrostatic pressure. Unfortunately, the acoustic radiation from the inner and end surfaces opposes that from the outer surface and sound cancellation occurs. However, through the appropriate selection of dimensions, namely ring height, it is possible to minimize this cancellation effect.

The ring can also be manufactured with a pressure-release material, such as cellular rubber, on its inner surface. This attenuates the inward radiation from the ring so that the acoustic energy flows only in the outward direction from the outer ring surface. Use of a pressure-release material, however, greatly increases the pressure dependence of the acoustic performance.

A ring projector can be fabricated from a single piece of ceramic, electroded on its inner and outer surfaces, and poled in the radial direction. From a manufacturing perspective, however, it is very difficult to make a monolithic ring of large enough diameter to resonate at a frequency less than ~10 kHz. Larger diameter rings can be made, though, by gluing together wedge-shaped pieces to form a segmented ring as shown in Fig. 11.8. In this case, the electrodes are on the abutting surfaces of the segments and poled through the wedge thickness where the poling direction is indicated by the arrows in the figure. This segmented construction has the added advantage of operating in the k_{33} mode rather than the k_{31} mode as in the case of the radially poled monolithic ring. Since these projectors are typically operated at high

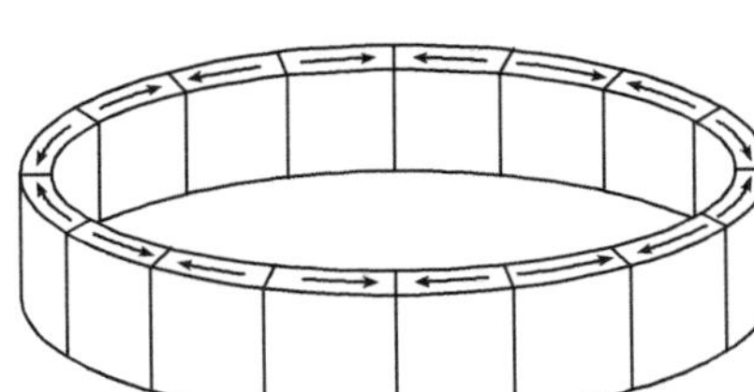

Fig. 11.8 The ring-stave projector. *Arrows* show the poling direction in each ceramic segment

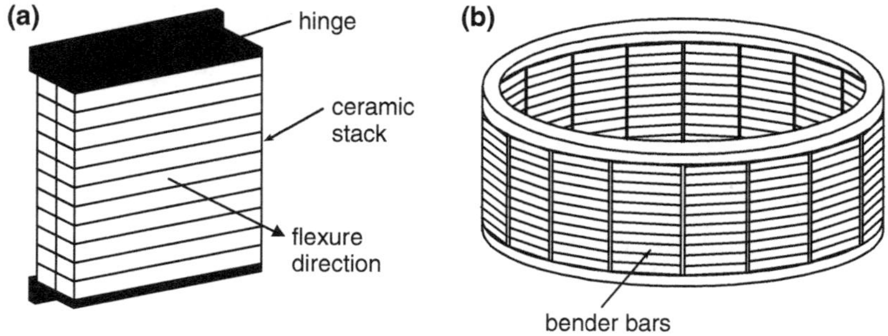

Fig. 11.9 The bender bar projector showing (a) an individual bender bar element and (b) a completed assembly. Reprinted with permission from Blue and Van Buren (1997). © 1997 Wiley

power, the rings are prestressed in compression during the fabrication stage using fiberglass wrapping. This prevents tensile failure during high-power operation.

11.4.1.2 Bender Bars

A bender bar element consists of two stacks of piezoelectric ceramic bars paired together with the stack being hinged at each end (Fig. 11.9a). Each pair of prestressed bars is oppositely poled so that when driven by an ac electric field one bar expands and the other contracts. This gives rise to bending or flexing motion, which generates acoustic pressure waves (Blue and Van Buren 1997).

The individual bender bars are subsequently assembled into a barrel stave arrangement as shown in Fig. 11.9b. Sizes for use at very low frequency (~1 kHz) can be on the order of 0.5 m in diameter and 0.5–1 m in length with the mass exceeding 100 kg. As in the case of the ring transducer described above, sound radiates both outwardly as well as inwardly toward the central cavity. A compliant tube pressure-release system is therefore typically incorporated into the central cavity in order to cancel the inward radiation. Utilizing this pressure-release system thus limits the operational depth to nominally 750 m (Grantner et al. 1976).

Projector construction is completed by placing the barrel stave assembly inside an acoustically transparent rubber boot, sealing the top and bottom with metal caps, and filling the unit with oil. The resonance frequency of the projector is a function of the length of the ceramic bender bars as well as the transducer configuration – namely the resonance is changed by altering the number of compliant tubes in the pressure-release system.

11.4.1.3 Flexural Disks

There are a variety of flexural disk design configurations. In its simplest form, known as a bilaminar disk, a piezoelectric ceramic disk is adhered to a flat metal disk whose annular edge is fastened to a baffle structure. Another common design is the trilaminar disk. It consists of two piezoelectric ceramic disks that are cemented

Fig. 11.10 The flexural disk transducer

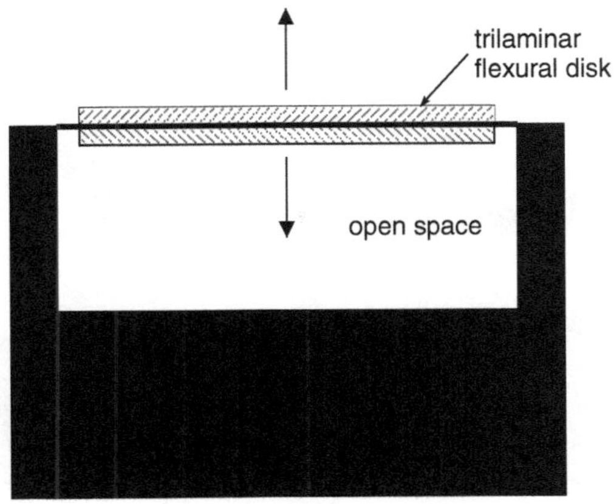

to each side of a central metal disk that is likewise mounted to a baffle (Fig. 11.10). The two piezoelectric disks can be either poled in opposite directions and wired electrically in series or poled in the same direction but wired in parallel. The choice depends on the desired capacitance. When a voltage is applied to the disks, they want to vibrate in the radial direction. However, since they are coupled to the stiff metal, the radial vibration is converted to a bending motion with the two disks vibrating in phase. Consequently, sound is radiated from both the front and back surfaces.

The purpose of the baffle is to mitigate the inward direction radiation so that the transducer delivers sound in just one direction. The baffle open space can be either air or oil filled. Oil backing allows for deeper water operation. An air-backed projector can also be used to deep depths if a pressure compensation system is employed. This will equilibrate the inside air pressure with the external water pressure.

A very thorough description of the physics of flexural disk transducers can be found in Woollett (1960).

11.4.1.4 Flextensional Transducers

Flextensional transducers consist of a piezoelectric ceramic drive element that is mechanically coupled to a metal shell. The metal shell acts as a mechanical impedance transformer. The small velocity imparted at the ceramic is converted into a larger velocity at the major faces of the shell. So as opposed to the aforementioned flexural disk (where the ceramic element bends), in flextensional transducers, it is the attached metal shell that flexes.

A sufficient number of flextensional geometries have been developed to merit their grouping into classifications (Royster 1970; Rynne 1993). To date, there are seven primary classes. A schematic of the design configurations is shown in Fig. 11.11.

The Class I transducer consists of a cylindrical ceramic stack held in compression by a metal shell along its major axis. This football-like transducer design, however,

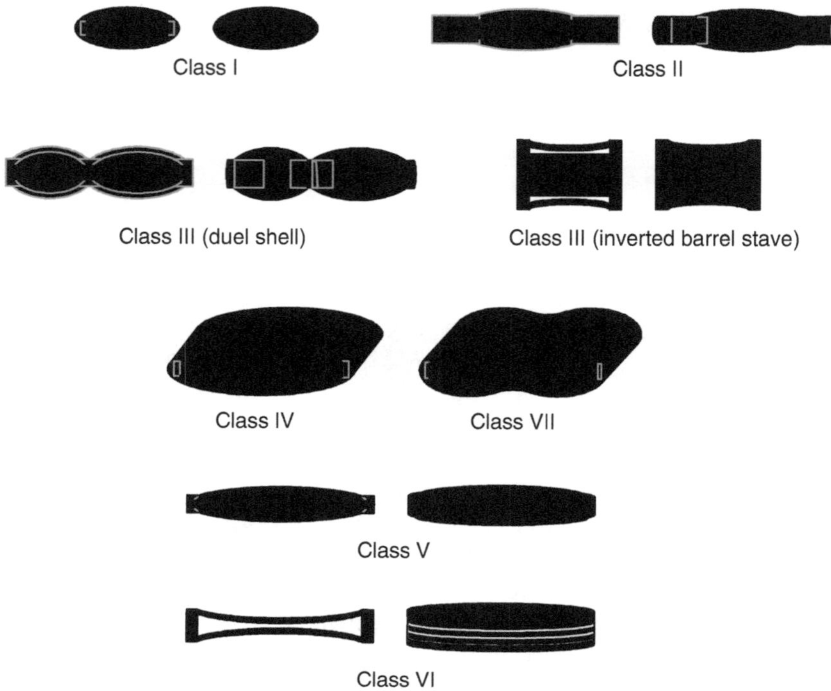

Fig. 11.11 Cross-sectional (and three-dimensional) perspectives of the seven flextensional transducer classes

is limited in the amount of ceramic it can hold and hence is power limited. The Class II is a second-generation Class I, now with extensions at the ends, which allow for the accommodation of more ceramic material. This additional ceramic boosts the power capability without markedly changing the resonance characteristics of the shell. There are two Class III designs. The first is a double shell design consisting of two slightly different sized Class I transducers joined together to form a double resonance shell assembly. The objective here is to broaden the acoustic bandwidth by utilizing two closely spaced resonance modes. The second type is known as an inverse barrel stave. It consists of a piezoelectric ceramic ring stack that drives concave metal plates that are arranged in the layout of a barrel.

The Class IV transducer is the most well known, understood, and widely used of the flextensional configurations. It consists of a piezoelectric ceramic bar stack held in compression by a thick elliptical convex shell. Unlike the previous three designs, all of which exhibit circular symmetry, the Class IV has a tubular configuration.

The Class V "ring-shell" design has a thin piezoelectric disk sandwiched between and bonded to two shallow convex spherical shells. The radial expansion and contraction of the piezoelectric disk causes the curved shells to execute an amplified axial motion.

The Class VI and VII (aka dogbone) designs are modifications of the Class V and IV transducers, respectively, but having concave radiating surfaces, which allow the shell motion to all be in phase (Rynne 1993).

11.4.2 Mid-Frequency Projector Designs

The most common transducer design for the so-called mid-frequency band is the tonpilz projector, which is discussed in further detail below. Other mid-frequency transducers can be built by modifying low-frequency designs via scaling their geometry or from high-frequency designs by adding mass to lower the resonance frequency. An example of the former is ceramic rings, which can easily be reduced in diameter to provide a resonance frequency in this band, albeit at the price of lower acoustic output.

11.4.2.1 Tonpilz Projectors

Tonpilz transducers are easily recognized by their distinctive "mushroom" shape as seen in Fig. 11.12. The flared head mass provides a means of obtaining a better impedance match between water and the active transducer material as well as generating an increased power output for a given strain in the ceramic (Wilson 1988). The maximum radiating area of the head mass, however, is limited by the need to avoid spurious resonances in the head of the projector.

The basic design is quite simple. A stack of longitudinally poled piezoelectric ceramic rings is held in compression by a stress bolt between two metal masses.

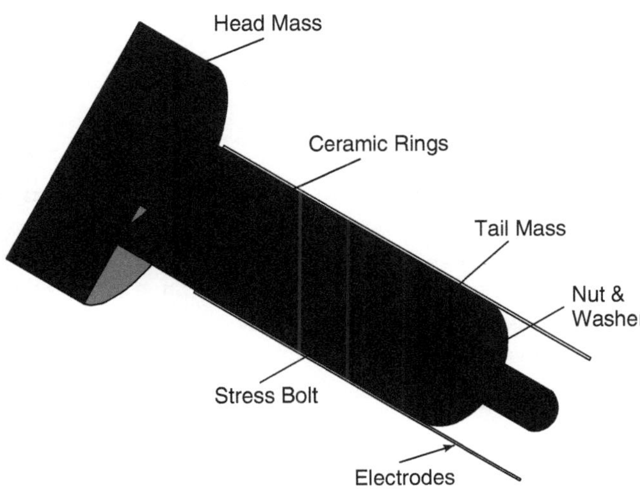

Fig. 11.12 Three-dimensional cross-sectional view of a tonpilz transducer

The ceramic elements are stacked with the polarization directions antiparallel so that they may be connected electrically in parallel while remaining mechanically in series. The purpose of the stress bolt is to apply a compressive load to the ceramic stack so that the ceramic elements avoid experiencing a tensile stress during high-power operation. A good rule of thumb is to use a bolt material whose stiffness is $<20\%$ of the stack so as not to restrain the motion of the metal masses (Stansfield 1991). The head mass is made from a lightweight material such as aluminum or magnesium and its thickness is determined by setting its flexural resonance at ~ 2 times the fundamental resonance of the transducer. The tail mass is typically 3–5 times more massive than the radiating head mass (Butler et al. 1996) in order to minimize the acoustic radiation from the back of the transducer. Tungsten, brass, and stainless steel are good candidate materials. The assembly can be easily incorporated into spherical, cylindrical, or planar arrays.

Stansfield (1991) devotes two chapters in his textbook to a thorough description of the construction procedure and design equations required to build a practical device.

11.4.3 High-Frequency Projector Designs

At frequencies above 100 kHz, it is possible to simply use a monolithic ceramic element in its pure vibration mode. The resonance frequency is obtained from the frequency constant of the material divided by its relevant dimension.

Piezoelectric ceramic–polymer composites, especially those with 1–3-type connectivity, are beginning to see more widespread use as high-frequency projectors. Advantages to using 1–3 piezoelectric composites include the ability to fabricate large area conformal architectures.

11.5 Hydrophone Designs

It is generally preferred to use hydrophones below their resonance frequency so that the received response remains flat across a wide frequency band. As a result, hydrophones are designed to have their mechanical resonance frequency above the desired operational frequency band. This is best accomplished by making the overall dimensions of the hydrophone small when compared with a wavelength in water.

The receiving sensitivity (M) of a piezoelectric element is equal to the open circuit voltage (V_i) that it generates because of an applied stress (σ_j), or

$$\mathrm{M} = \frac{V_i}{\sigma_j} = g_{ij}h, \tag{11.16}$$

where g_{ij} is the piezoelectric voltage coefficient and h is the element thickness.

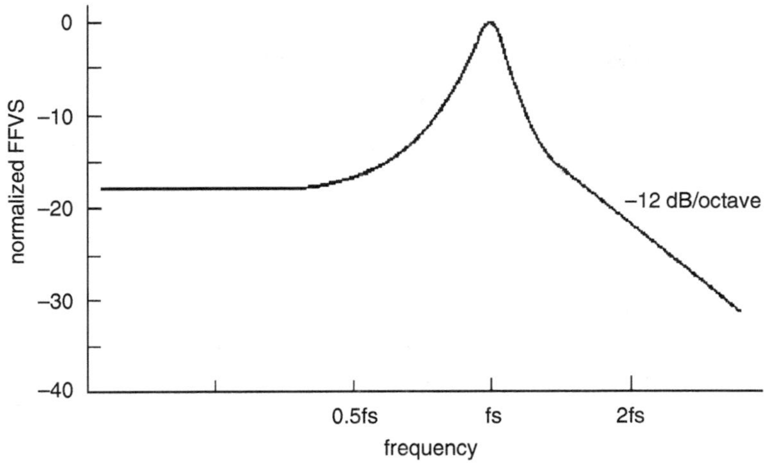

Fig. 11.13 Idealized free field voltage sensitivity of a piezoelectric hydrophone

The acoustic performance of a hydrophone is usually expressed in terms of *free field voltage sensitivity* (FFVS). It is the rms voltage measured at the open circuit terminals of the hydrophone referenced to an incident pressure of 1 μPa. It is reported in terms of decibels as dB re: $1\,\mathrm{V}\,\mu\,\mathrm{Pa}^{-1}$ at 1 m and is related to the receiving sensitivity as:

$$\mathrm{FFVS} = 20\log(M \times 10^{-6}). \tag{11.17}$$

The FFVS can also be calculated from the TVR as:

$$\mathrm{FFVS} = \mathrm{TVR} + 20\log|Z| - 354 - 20\log(f_{\mathrm{kHz}}). \tag{11.18}$$

The idealized (i.e., ρc-loaded) FFVS of a piezoelectric hydrophone is shown in Fig. 11.13. The response is flat until the resonance is reached and then it drops off at −12 dB per octave. It is generally desirable to have the FFVS greater than −200 dB over the flat region.

11.5.1 Common Hydrophone Configurations

The most common piezoelectric ceramic hydrophone designs used in sonar applications are discussed below.

11.5.1.1 Rectangular Elements

The simplest hydrophone configuration is that of a rectangular block of piezoelectric material fully electroded on its top and bottom surfaces and poled in the thickness direction. The sides are typically shielded from the incoming pressure waves and hence the acoustic energy is incident only on the top and bottom faces of the material.

11.5.1.2 Cylinders

Cylinders are a basic shape used for hydrophone applications in the audio and low ultrasonic frequency range (Burdic 1984). There are three possible poling configurations for cylinders. They can be poled either through their wall thickness (i.e., radially), along the length of the cylinder (longitudinally), or divided into segments via segmented electrodes and poled tangentially (akin to Fig. 11.8).

The FFVS of piezoelectric ceramic cylinders poled radially, longitudinally, and tangentially are shown graphically in Fig. 11.14. The data were calculated from equations provided by Wilder (1977). As can be seen, for certain ratios of inner diameter to outer diameter, the sensitivity of the cylinder can become negligible. So the transducer designer must be careful to avoid these dimensions.

Radial poling generally exhibits lower sensitivity than either longitudinal or tangential poling. Nevertheless, radially poled cylinders are often preferred in some applications, namely when driving a long cable, because of their higher capacitance.

11.5.1.3 Spheres

The sensitivity of a spherical hydrophone for various piezoelectric materials as a function of wall thickness to outer diameter ratio is shown in Fig. 11.15. The data were calculated via equations given in Stansfield (1991). As in the case of cylindrical hydrophones, there are certain wall thicknesses for which the sensitivity goes to zero and must be avoided.

Spherical hydrophones can be constructed from one piece of material or from two hemispheres glued together. The sensitivity can be doubled by wiring two hemispheres in electrical series, but at the cost of reducing the capacitance by a factor of 4.

11.5.1.4 Piezoelectric Ceramic–Polymer Composites

There has been a longtime interest in developing piezoelectric ceramic-polymer composites (aka piezocomposites) for underwater sensor applications because of their high hydrostatic sensitivity, good acoustic impedance match to water, and high pressure tolerance. So-called piezocomposite hydrophones are predominantly of 1–3-type connectivity. However, there is also rather extensive literature available on 0–3-type composites as well.

Piezoelectric-polymer composites with 1–3 connectivity consist of piezoelectric ceramic (generally PZT) rods or posts imbedded within a polymer host matrix

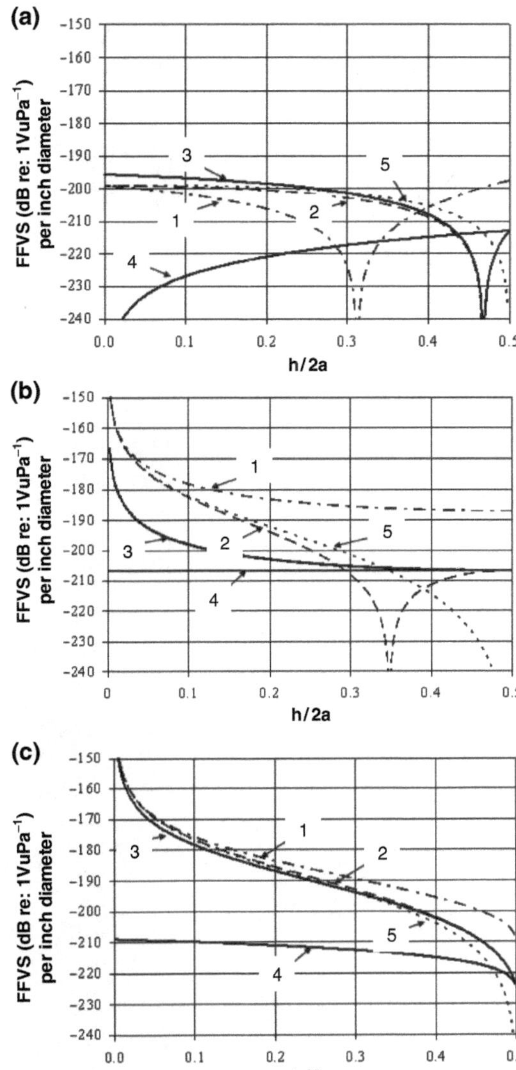

(d)

Case	Inside of cylinder	Outside of cylinder	Ends of cylinder
1	shielded	exposed	shielded
2	shielded	exposed	exposed
3	shielded	exposed	capped
4	shielded	exposed	exposed
5	exposed	shielded	shielded

Fig. 11.14 FFVS (below resonance) of piezoelectric ceramic cylinders poled (**a**) radially, (**b**) longitudinally, and (**c**) tangentially. The table in (**d**) denotes the cylinder boundary conditions for curves 1–5. Also note that case (**c**) is for a cylinder with four segments

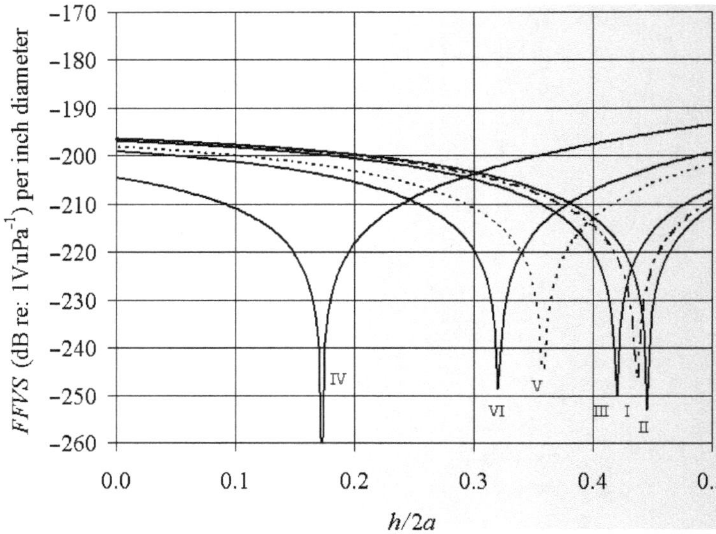

Fig. 11.15 The FFVS (below resonance) of piezoelectric ceramic spheres poled with polarization direction through the ceramic wall. The roman numerals I through VI represent the Navy classification scheme as per Table 12.1

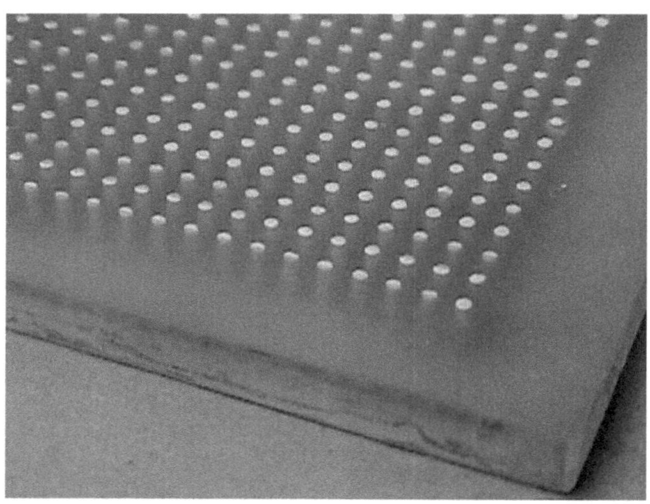

Fig. 11.16 Photograph of a 1–3 piezoelectric ceramic-polymer composite showing ceramic rods embedded in a polymer matrix

(Fig. 11.16). This composite design is intended to operate in its pure thickness mode. The 1–3-type piezocomposites have been used effectively as sensors at frequencies spanning from Hz to MHz. When used solely as a hydrophone, the volume fraction of PZT in the composite is generally between 15 and 25%. When used for both transmit as well as receive, the optimum PZT volume fraction is between 30 and 50%, and when used solely as a transmitter, PZT volume fractions can be upwards of 80%.

References

Beranek LL (1993) Acoustics. American Institute of Physics, Woodbury, NY

Blue JE, Van Buren AL (1997) Transducers. In: Crocker MJ (ed), Encyclopedia of acoustics, vol 1. Wiley, New York, pp 597–606

Bobber RJ (1988) Underwater electroacoustics measurements. Peninsula, Los Altos, CA

Burdic WS (1984) Underwater acoustic system analysis. Prentice-Hall, Eaglewood Cliffs, NJ

Butler SC, Lindberg JF, Clark AE (1996) Hybrid magnetostrictive/piezoelectric tonpilz transducer. Ferroelectrics 187: 163–174

Channel Industries, Inc. (2006) Piezoelectric ceramics. Channel, Santa Barbara, CA

Grantner L, Kubick F, Williams R (1976) Handbook of acoustic projector technology, vol 1. Naval Command, Control and Ocean Surveillance Center, San Diego

Hughes WJ (1998) Transducers, underwater acoustic. In: Trigg GL (ed), Encyclopedia of applied acoustics, vol 22. VCH, New York, pp 67–84

Krueger HHA (1967a) Stress sensitivity of piezoelectric ceramics, part 1: sensitivity to compressive stress parallel to the polar axis. J Acoust Soc Am 42: 636–645

Krueger HHA (1967b) Stress sensitivity of piezoelectric ceramics, part 3: sensitivity to compressive stress perpendicular to the polar axis. J Acoust Soc Am 43: 583–591

Krueger HHA, Berlincourt D (1961) Effects of high static stress on the piezoelectric properties of transducer materials. J Acoust Soc Am 33: 1339–1344

Military Standard (1995) Piezoelectric ceramic material and measurements guidelines for sonar transducers. MIL–STD–1376B (SH)

Moffett MB, Marshall WJ (1994) The importance of coupling factor for underwater acoustic projectors. NUWC_NPT technical document 10,691, Naval Undersea Warfare Center, Newport, RI

Moffett MB, Robinson HC, Powers JM, Baird PD (2007) Single-crystal lead magnesium niobate–lead titanate (PMN/PT) as a broadband high power transduction material. J Acoust Soc Am 121: 2591–2599

Morgan Electro Ceramics (2006) Piezoelectric ceramics. Morgan Electroceramics, Inc., Bedford, OH

Moulson AJ, Herbert JM (1992) Electroceramics. Chapman & Hall, New York

Royster LH (1970) The flextensional concept: a new approach to the design of underwater acoustic transducers. Appl Acoust 3: 117–126

Rynne EF (1993) Innovative approaches for generating high power, low frequency sound. In: McCollum MD, Hamonic BF, Wilson OB (eds), Transducers for sonics and ultrasonics. Technomic, Lancaster, PA, pp 38–49

Stansfield D (1991) Underwater electroacoustic transducers. Bath University Press, Bath, UK

Wilder WD (1977) Electroacoustic sensitivity of ceramic cylinders. J Acoust Soc Am 62: 769–771

Wilson OB (1988) Introduction to theory and design of sonar transducers. Peninsula, Los Altos, CA

Woollett RS (1960) Theory of the piezoelectric flexural disk transducer with applications to underwater sound. USL Research Report No 490, S–F001 03 04–1, US Navy Underwater Sound Laboratory, New London, CT

Chapter 12
Finite Element Analysis of Piezoelectric Transducers

Anne-Christine Hladky-Hennion* and Bertrand Dubus

12.1 Introduction of the Finite Element Method

Electromechanical transducers, first used in sonar systems, convert electrical energy into mechanical energy, thanks to the piezoelectric effect, then to acoustic energy, with the generation and the radiation of an acoustic wave in a fluid. To design new transducers and to understand their physical behaviour, several physical mechanisms have to be described (piezoelectric–elastic–fluid-structure coupling–acoustic radiation). To solve such problems, analytical and semi-analytical approaches often rely upon simplifying hypotheses, in terms of geometry of the transducers, behaviour of the piezoelectric part of the device, radiation condition or frequency range of interest. Among these approaches, the equivalent electrical scheme (Beranek 1954) is based on a classical lumped constant representation with masses and springs, and the transfer matrix (Neppiras 1973) uses plane wave approximations. With the help of analytical expressions, these approaches become operational tools, leading to the realization of numerous transducers.

Nevertheless, to overcome restrictive hypotheses, more advanced numerical techniques have rapidly been developed. Numerical modelling activities have been first initiated by the U.S. Navy Laboratories in the late sixties. The radiating and scattering of waves problems were solved using Helmholtz integrals (Schenk 1968), whereas, from the structural point of view, variational principles have been applied to these problems (Allik and Hughes 1970). First devoted to naval studies, the finite element method is now being used for many other applications: ultrasonic arrays for medical imaging, directional transducers for the oil industry, integrated transducers for health monitoring, transducers for non-destructive testing and evaluation – commonly referred to as NDT or NDE.

This chapter first presents how an electromechanical problem is described with the finite element method: basic physical partial differential equations, the

A.-C. Hladky-Hennion
IEMN, ISEN Department, UMR CNRS 8520, 59046 Lille, France

A. Safari, E.K. Akdoğan (eds.) *Piezoelectric and Acoustic Materials for Transducer Applications.*

variational principle, discretisation techniques, and the final system of equations are presented. Throughout the text, the advantages and limitations of the classical finite element method as well as the model specific aspects according to applications are discussed. An example is presented, for which a complete analysis is carried out. Finally, current needs and future trends of these numerical tools are discussed.

12.2 Analysis of an Electromechanical Problem with the Finite Element Method

12.2.1 The Equations of the Problem

In this section, a heterogeneous structure made out of several elastic materials as well as piezoelectric materials is considered without any geometrical restriction. Only elastic and piezoelectric effects are taken into account, and the model excludes nonlinearities due to the behavioural law or to the large strains as well as dissipative effects due to material losses or contact friction. The equations are developed in the case of a piezoelectric material. The case of an elastic material can be deduced by withdrawing physical variables and constants associated to electrical effects. Under these conditions, the basic equations of state for converse and direct piezoelectric effects are expressed as

$$\begin{cases} T_{ij} = c^{\mathrm{E}}_{ijkl} S_{kl} - e_{kij} E_k \\ D_i = e_{ikl} S_{kl} + \varepsilon^{\mathrm{S}}_{ij} E_j \end{cases}, \tag{12.1}$$

where $[T]$ is the stress tensor, $[S]$ the strain tensor, $\boldsymbol{E}$ is the electric field vector, $\boldsymbol{D}$ is the electric displacement vector, $[c^E]$ is the elastic stiffness tensor at constant electric field, $[e]$ is the piezoelectric stress tensor, $[\varepsilon^S]$ is the dielectric permittivity tensor at constant strain, with i, j, k and l varying from 1 to 3 and being related to the coordinate axis. Einstein notation is systematically used throughout the text. Considering small strain hypothesis and theory of electrostatics, the strain tensor is written as:

$$S_{kl} = \frac{1}{2}\left(\frac{\partial u_k}{\partial x_l} + \frac{\partial u_l}{\partial x_k}\right), \tag{12.2}$$

where $\boldsymbol{u}$ is the displacement vector and the electrical potential φ is related to the electric field vector $\boldsymbol{E}$ by:

$$E_i = -\frac{\partial \varphi}{\partial x_i}. \tag{12.3}$$

In the whole domain, denoted Ω (Fig. 12.1), Newton's law is written as:

$$\rho \frac{\partial^2 u_i}{\partial t^2} = \frac{\partial T_{ij}}{\partial x_j}, \tag{12.4}$$

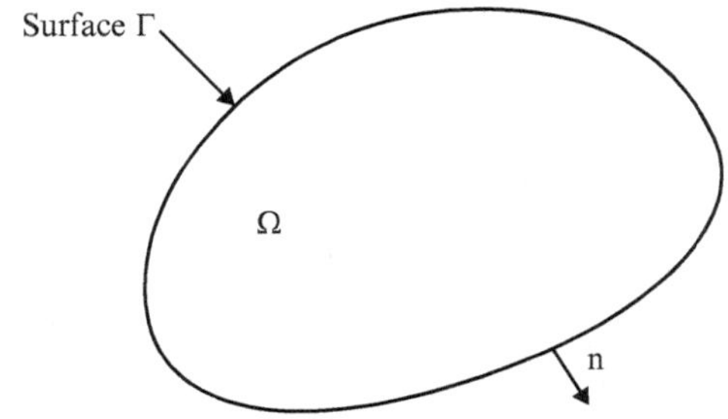

Fig. 12.1 Schematic representation of the Ω domain with the corresponding Γ surface

where ρ is the mass density of the material, and Gauss's equation is expressed:

$$\frac{\partial D_i}{\partial x_i} = 0 \tag{12.5}$$

because no macroscopic charges are present in the medium. If all the variables are considered with an $e^{+j\omega t}$ time dependency, ω being the angular frequency, (12.1)–(12.5) are combined to give:

$$\begin{cases} -\rho\omega^2 u_i = \dfrac{\partial}{\partial x_j}\left(c^{\mathrm{E}}_{ijkl} S_{kl} - e_{kij} E_k\right) \\ \dfrac{\partial}{\partial x_i}\left(e_{ikl} S_{kl} + \varepsilon^{\mathrm{S}}_{ij} E_j\right) = 0 \end{cases} . \tag{12.6}$$

Several types of mechanical and electrical boundary conditions are considered:

- B1: The displacement u_i is known on a surface, denoted (S_u). This condition is a Dirichlet condition, and is written as $u_i = u_i'$, where u_i' is known.
- B2: The stress field is related to an applied force on a surface, denoted (S_T). This condition is a Neumann condition given by

$$T_{ij} n_j = f_i, \tag{12.7}$$

where $\boldsymbol{n}$ is the normal vector to the surface, outward oriented, and $\boldsymbol{f}$ is a known vector of applied forces.

- B3: On surface denoted (S_σ), surface charges vanish as piezoelectric materials are insulators and

$$D_i n_i = 0, \tag{12.8}$$

where $\boldsymbol{n}$ is the normal vector to the surface. This condition can be used for non-electroded boundaries of piezoelectric materials having high dielectric permittivity, as, in this case, the electric displacement can be neglected outside the Ω domain.

- B4: On electroded surfaces, denoted (S_p), the electrical potential is uniform and $\varphi = \varphi_p$,
- B5: The corresponding electric charge is

$$-\iint_{S_p} D_i n_i dS_p = Q_p. \tag{12.9}$$

Either the electrical potential φ_p or charge Q_p must be prescribed, the remaining variable being unknown. For $M + 1$ separate electrodes, $M + 1$ independent conditions are required. Finally, it is necessary to set a reference point for the electrical potential, usually by grounding one electrode and denoted it by the subscript 0 ($\varphi_0 = 0$).

12.2.2 Variational Principle

The solution of the defined problem can be obtained either by solving directly the aforementioned partial differential equations with associated boundary conditions defining the behaviour of the continuous problem or by using a variational approach. The essence of the variational approach is to invoke the stationarity of a variational quantity L, also called the *functional*, of the system, it means $\delta L = 0$, with respect to a small change of the variables (Zienkiewicz 1977). Thus, for solid submitted to conservative forces, the functional is the potential energy for static problems and time integral of the Lagrangian function for dynamical problems. One of the advantages of this approach is that some boundary conditions can be taken into account when using a variational approach. How can the appropriate functional corresponding to a given problem be found? Mathematical manipulations have to be used but a number of valuable theorems exist, which provide further guidelines in establishing an appropriate functional (Segerlind 1976). One example is the Ritz theorem, which will be used as follows.

Let us find function $f(x)$, solution of the differential equation

$$\frac{\mathrm{d}^2 f}{\mathrm{d}x^2} = -g(x), \tag{12.10}$$

with the boundary conditions $f(a) = A$ and $f(b) = B.g(x)$ is supposed to be continuous on $[a,b]$. Then, the problem is equivalent to the minimization of the functional

$$L = \int_a^b \left[\frac{1}{2} \left(\frac{\mathrm{d}f}{\mathrm{d}x} \right)^2 - g(x) f(x) \right] \mathrm{d}x. \tag{12.11}$$

The starting point is generally the functional L and, by invoking the stationarity condition on L, the governing differential equations and the corresponding boundary conditions are derived. *Therefore, the method can be used if a functional can be found that yields to the differential equations and boundary conditions.* When the variational principle is used, the solution of a continuous problem can be approximated in an integral form that is suitable for the finite element analysis.

The weighted residual approach is an alternate method that leads to an integral form, one of the procedures being Galerkin's method (Zienkiewicz 1977). Contrary to the variational methods, Galerkin's method *starts* with the differential equations and uses a trial solution. The goal is to minimize the error or residual between the approximate solution and the exact solution. The application of Galerkin's method yields to the same discrete integral form than variational methods.

12.2.2.1 Functional Related to the Problem

Equation (12.6) on the one hand, (12.7)–(12.9) on the other hand are respectively the Euler equations (Morse and Fesbach 1953) and the boundary conditions associated to a stationary quantity L defined by (Allik and Hughes 1970, Decarpigny 1984)

$$\begin{aligned} L = & \iiint_{\Omega} \frac{1}{2}\left(S_{ij} c^{\mathrm{E}}_{ijkl} S_{kl} - \rho\omega^2 u_i^2\right) d\Omega \\ & - \iint_{(S_u)} (u_i - u_i') n_j \left(c^{\mathrm{E}}_{ijkl} S_{kl} - e_{kij} E_k\right) \mathrm{d}S_u \\ & - \iint_{(S_T)} f_i u_i dS_T - \iiint_{\Omega} \frac{1}{2}\left(2 S_{kl} e_{ikl} E_i + E_i \varepsilon^{\mathrm{S}}_{ij} E_j\right) d\Omega \\ & - \sum_{p=0}^{M} \iint_{(S_p)} (\varphi - \varphi_p) n_i \left(e_{ikl} S_{kl} + \varepsilon^{\mathrm{S}}_{ij} E_j\right) \mathrm{d}S_p + \sum_{p=0}^{M} \varphi_p Q_p. \end{aligned} \tag{12.12}$$

The first term in the right-hand side of (12.12) corresponds to the Lagrangian of the mechanical problem. Satisfying the stationary condition of L implies that (12.6) and the boundary conditions (B1) to (B5) are satisfied.

12.2.3 *Application of the Finite Element Method*

12.2.3.1 Discretization

In a finite element analysis, the whole domain Ω in which the displacement field $\boldsymbol{u}$ and the electrical potential φ have to be determined, is fictively cut into small volumes called *elements* (denoted Ω_{e}) interconnected at a finite number of points of their boundaries, denoted *nodes*. The whole set of elements is *the mesh* (Fig. 12.2). Inside each element, the displacement field and the electrical potential are uniquely defined from their values at the nodes of the element, using *interpolation functions*. Because the domain subdivision into elements is arbitrary, many different meshes can be used for a given problem. Contrary to the Ω domain, the geometrical shape of each elementary domain Ω_{e} is simple and homogeneous. For domain Ω, a trial

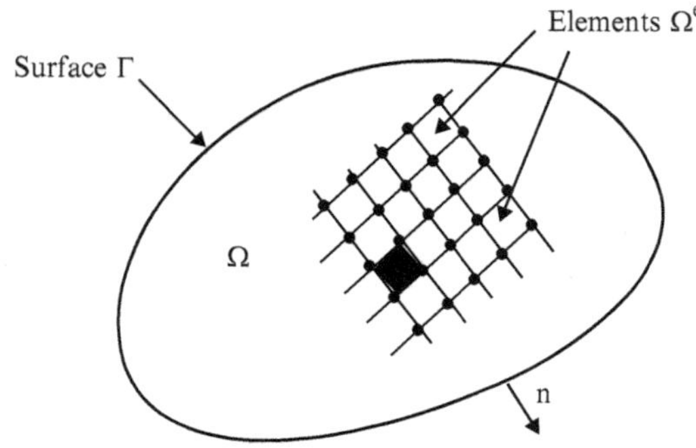

Fig. 12.2 Discretisation of the Ω domain

function is obtained, which only depends on the nodal values of $\boldsymbol{u}$ and φ and is defined by pieces between two adjacent elements. This function is combined with the expression of L to obtain, after minimization with respect to each nodal value, a linear system of equations, where the unknowns are the nodal values.

12.2.3.2 Interpolation Functions

For the sake of simplicity, the first part of this section corresponds to a two-dimensional mesh, but the theory can be easily extended to three-dimensional structures. An elementary domain is defined by m nodes and rectilinear lines. The coordinates of node i of the mesh are denoted x_i and y_i. For a triangular element $m = 3$ and for a quadrilateral element $m = 4$, when only nodes at the corner of the element are considered. For quadratic elements having additional nodes in the middle of each side, $m = 6$ for a triangular element and $m = 8$ for a quadrilateral element.

A scalar field as the electrical potential is defined within the elementary domain with the help of the interpolation function N_i and of the nodal values Φ_i of the electrical potential

$$\varphi(x,y) = \sum_{i=1}^{m} N_i^e(x,y)\Phi_i^e. \tag{12.13}$$

The interpolation functions have a polynomial form, depending on the x and y variables, and they have the following properties:

$$\begin{aligned} N_i^e(x,y) &= 1 \quad \text{if } x = x_i \text{ and } y = y_i, \\ N_i^e(x,y) &= 0 \quad \text{if } x = x_j \text{ and } y = y_j, \text{ with } i \neq j. \end{aligned} \tag{12.14}$$

When the nodal coordinates are reported in the right-hand side of (12.13), the left-hand side matches the nodal electrical potential. Interpolation functions are linear for $m = 3$ triangular elements and $m = 4$ quadrilateral elements (two nodes on each element side), whereas interpolation functions are quadratic for $m = 6$ triangular elements and $m = 8$ quadrilateral elements (three nodes on each element side) as shown in Fig. 12.3.

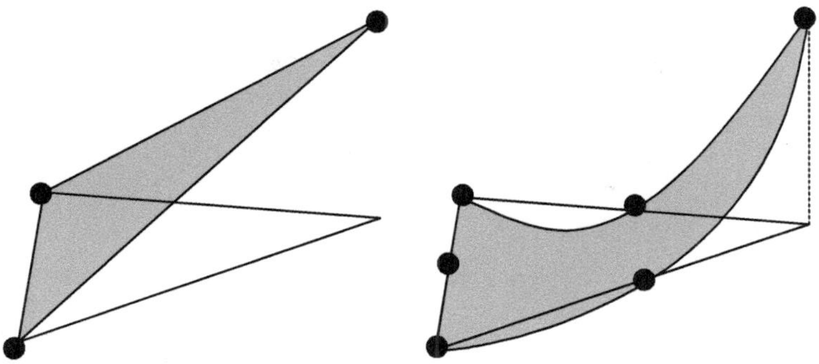

Fig. 12.3 Interpolation functions for a triangular element. *Left*: linear – three nodes; *Right*: quadratic – six nodes

As the displacement field $\boldsymbol{u}$ is a vector field having two (u_x and u_y) or three (u_x, u_y and u_z) components, for two-dimensional or three-dimensional meshes respectively, $2m$ or $3m$ interpolation functions are used on each element. Thus, the displacement field is written as

$$\boldsymbol{u}(x,y) = \sum_{i=1}^{m} [N_i^e]\boldsymbol{U}_i^e, \tag{12.15}$$

with

$$[N_i^e] = N_i^e(x,y)\begin{bmatrix}1 & 0 & 0\\ 0 & 1 & 0\\ 0 & 0 & 1\end{bmatrix}, \tag{12.16}$$

when the displacement field has three components. It is then possible to generalize this result to a higher order of interpolation functions, with more nodes on each element side. The interpolation functions must prescribe the continuity of the geometry between elements. Relations (12.15) and (12.16) also prescribe the continuity of the displacement field and of the electrical potential at the interface between two elements. The continuity of the spatial derivatives is not enforced in this case. Moreover, the order of the interpolation functions, i.e. the degree of the polynomial functions in (12.13) and (12.15) has to be the same for each sub-domain of the structure (elastic, piezoelectric etc.). Nodal values of electrical potential and displacement components are also called *degrees of freedom* of the finite element problem.

12.2.3.3 Tensors

In this section, (12.13) and (12.15) are derived to get the electric field vector $\boldsymbol{E}$ and strain tensor $[S]$. Then, the stress tensor $[T]$ and the electric displacement field vector $\boldsymbol{D}$ can be deduced. The derivation is presented in a three-dimensional case, but can be easily reduced in the case of a two-dimensional problem. The relationship between the electrical field vector $\boldsymbol{E}$ and the electrical potential φ is

$$E = -\overrightarrow{\text{grad}}\,\varphi = -\left[\overrightarrow{\text{grad}} N_1^e \quad \overrightarrow{\text{grad}} N_2^e \quad ...\overrightarrow{\text{grad}} N_m^e\right] \boldsymbol{\Phi}^e, \tag{12.17}$$

where m is the number of nodes in the element. Equation (12.17) is written as:

$$E = -\left[B_\varphi^e\right] \boldsymbol{\Phi}^e, \tag{12.18}$$

with

$$\left[B_\varphi^e\right] = \left[B_{\varphi 1}^e, B_{\varphi 2}^e, \ldots\ldots, B_{\varphi m}^e\right] \text{ and } \left[B_{\varphi i}^e\right] = \begin{bmatrix} \frac{\partial N_i^e}{\partial x} \\ \frac{\partial N_i^e}{\partial y} \\ \frac{\partial N_i^e}{\partial z} \end{bmatrix}, \tag{12.19}$$

where $\boldsymbol{\Phi}^e$ is the vector containing the nodal values of the electrical potential. The $B_{\varphi i}^e$ terms are the first spatial derivative of the interpolation functions. In the same way, for the strain, a condensed notation is used, with six components

$$S = \left[B^e\right] U^e, \tag{12.20}$$

with

$$\left[B^e\right] = \left[B_1^e, B_2^e, \ldots, B_m^e\right] \text{ and } \left[B_i^e\right] = \begin{bmatrix} \frac{\partial N_i^e}{\partial x} & 0 & 0 \\ 0 & \frac{\partial N_i^e}{\partial y} & 0 \\ 0 & 0 & \frac{\partial N_i^e}{\partial z} \\ 0 & \frac{\partial N_i^e}{\partial z} & \frac{\partial N_i^e}{\partial y} \\ \frac{\partial N_i^e}{\partial z} & 0 & \frac{\partial N_i^e}{\partial x} \\ \frac{\partial N_i^e}{\partial y} & \frac{\partial N_i^e}{\partial x} & 0 \end{bmatrix}. \tag{12.21}$$

In (12.20), $\boldsymbol{U}^e$ is the vector containing the nodal values of the displacement field. As a consequence, (12.1) is written as:

$$\begin{cases} \mathbf{T} = \left[c^{\mathrm{E}}\right]\left[B^e\right]\mathbf{U} + \left[e\right]^{\mathrm{T}}\left[B_\varphi^e\right]\boldsymbol{\Phi} \\ \mathbf{D} = \left[e\right]\left[B^e\right]\mathbf{U} - \left[\varepsilon^{\mathrm{S}}\right]\left[B_\varphi^e\right]\boldsymbol{\Phi} \end{cases}, \tag{12.22}$$

where $[c^E]$ is a 6×6 matrix, obtained by a condensation of the c_{ijkl}^{E} tensor, and $[e]$ is a 3×6 matrix, obtained by a condensation of the e_{ikl} tensor.

12.2.3.4 Application of the Variational Principle

The expressions given above are combined with the functional L expressed in (12.12). Therefore, L is decomposed into a sum of integrals on each element e of the mesh. The volume of the element e is noted $\boldsymbol{\Omega}_e$ and the corresponding surfaces are S_{ue}, S_{Te}, S_{pe}, if they belong to element e. Then, L becomes

$$
\begin{aligned}
L = \sum_e \Bigg\{ & \frac{1}{2} \iiint_{\Omega^e} \mathbf{U}^{e\mathbf{T}} [B^e]^{\mathrm{T}} \left[c^{\mathrm{E}}\right] [B^e] \mathbf{U}^e \mathrm{d}\Omega_e \\
& - \frac{1}{2}\omega^2 \iiint_{\Omega^e} \rho \mathbf{U}^{e\mathbf{T}} [N^e]^{\mathrm{T}} [N^e] \mathbf{U}^e \mathrm{d}\Omega_e \\
& + \iiint_{\Omega^e} \mathbf{U}^{e\mathbf{T}} [B^e]^{\mathrm{T}} [e] \left[B_\varphi^e\right] \boldsymbol{\Phi}^e d\Omega_e \\
& - \frac{1}{2} \iiint_{\Omega^e} \boldsymbol{\Phi}^{e\mathrm{T}} \left[B_\varphi^e\right]^{\mathrm{T}} \left[\varepsilon^S\right] \left[B_\varphi^e\right] \boldsymbol{\Phi}^e \mathrm{d}\Omega_e \\
& - \iint_{S_{Te}} \mathbf{U}^{e\mathbf{T}} [N^e]^{\mathrm{T}} \mathbf{f} \mathrm{d}S_{Te} + \sum_{p=0}^{M} \varphi_p Q_p \Bigg\}
\end{aligned}
\tag{12.23}
$$

where $\boldsymbol{U}^e$ and $\boldsymbol{\Phi}^e$ are the vectors of the nodal values of the displacement field and of the electrical potential in element e, respectively.$[N^e]$ is the interpolation matrix of element e. $[B^e]$ and $\left[B_\varphi^e\right]$ are the first spatial derivatives of the interpolation matrices of element $e$for the displacement field and for the electrical potential respectively. The superscript T denotes the transposition.

In (12.23), the integrals on the S_u and S_p surfaces are withdrawn. In fact, because of the boundary conditions, the displacements or the electrical potentials are known on given nodes and the corresponding integrals in (12.12) vanish.

In (12.23), $\boldsymbol{U}^e$ and $\boldsymbol{\Phi}^e$ do not depend on the integration variables, and thus L can be written as

$$
\begin{aligned}
L = \sum_e \Bigg\{ & \frac{1}{2} \mathbf{U}^{e\mathrm{T}} \left[K_{uu}^e\right] \mathbf{U}^e - \frac{1}{2}\omega^2 \mathbf{U}^{e\mathrm{T}} [M^e] \mathbf{U}^e + \mathbf{U}^{e\mathrm{T}} \left[K_{u\varphi}^e\right] \boldsymbol{\Phi}^e \\
& + \frac{1}{2} \boldsymbol{\Phi}^{e\mathrm{T}} \left[K_{\varphi\varphi}^e\right] \boldsymbol{\Phi}^e - \mathbf{U}^{e\mathrm{T}} \mathbf{F}^e \Bigg\} + \sum_{p=0}^{M} \varphi_p Q_p,
\end{aligned}
\tag{12.24}
$$

with

$$
\begin{aligned}
\left[K_{uu}^e\right] &= \iiint_{\Omega^e} [B^e]^{\mathrm{T}} \left[c^{\mathrm{E}}\right] [B^e] \mathrm{d}\Omega_e \\
[M^e] &= \iiint_{\Omega^e} \rho [N^e]^{\mathrm{T}} [N^e] \mathrm{d}\Omega_e \\
\left[K_{u\varphi}^e\right] &= \iiint_{\Omega^e} [B^e]^{T} [e] \left[B_\varphi^e\right] \mathrm{d}\Omega_e \\
\left[K_{\varphi\varphi}^e\right] &= \iiint_{\Omega^e} \left[B_\varphi^e\right]^{\mathrm{T}} \left[\varepsilon^S\right] \left[B_\varphi^e\right] \mathrm{d}\Omega_e \\
\boldsymbol{F}^e &= \iint_{S_{Te}} [N^e]^{\mathrm{T}} \boldsymbol{f} \, \mathrm{d}S_{Te}.
\end{aligned}
\tag{12.25}
$$

The $[K_{uu}^e]$, $[K_{u\varphi}^e]$ and $[K_{\varphi\varphi}^e]$ matrices are the elastic, piezoelectric and dielectric elementary matrices, $[M^e]$ is the mass elementary matrix. As seen in (12.25), these matrices depend on the integration of the interpolation functions and their spatial derivatives. Therefore, linear interpolation functions lead to simple calculations of the elementary integrals, because the degree of $N_i(x,y)$ is equal to one and their spatial derivatives are constant. But a complicated geometry, with curved sides for instance, requires a large number of elements with rectilinear sides. Even if the calculation of each elementary matrix is fast, the computation time can increase rapidly because of the large number of elements. On the other hand, higher orders of interpolation functions lead to more complex calculations of the elementary integrals. In that case, Gauss method is often used for the numerical calculation of integrals of (12.25). A complex geometry often requires a reduced number of elements with parabolic sides (degree 2–3 nodes for each side) or cubic sides (degree 3–4 nodes for each side). *Therefore, a compromise has to be found between the degree of interpolation function and the number of the elements in the mesh.*

12.2.3.5 Assembly

The calculation of the elementary matrices given in (12.25) is performed for each element Ω_e. Depending upon the degree of interpolation, this calculation is either analytical (linear interpolation) or numerical (higher degrees of interpolation) based on a Gauss integration scheme (Zienkiewicz 1977). Then, the elementary matrices are rearranged and added in a global matrix. This procedure is called *assembly:*

$$\begin{gathered}[K_{uu}] = \sum_e [K_{uu}^e];\ [M] = \sum_e [M^e];\ [K_{u\varphi}] = \sum_e [K_{u\varphi}^e];\\ [K_{\varphi\varphi}] = \sum_e [K_{\varphi\varphi}^e];\ \boldsymbol{F} = \sum_e \boldsymbol{F}^e; \boldsymbol{U} = \sum_e \boldsymbol{U}^e;\ \boldsymbol{\Phi} = \sum_e \boldsymbol{\Phi}^e.\end{gathered} \tag{12.26}$$

The size of the global matrix is the number of degrees of freedom in the whole structure. In a piezoelectric domain, one electrical potential Φ_i and two ($U_{i,x}$ and $U_{i,y}$ for a two-dimensional structure) or three ($U_{i,x}$, $U_{i,y}$ and $U_{i,z}$ for a three-dimensional structure) components of the displacement field are attached to each node. Therefore, each node is characterized by three (two-dimensional structure) or four (three-dimensional structure) degrees of freedom. In the case of an elastic medium, the degrees of freedom related to the electrical potential are withdrawn.

Then, the application of the variational principle implies the minimization of the functional of (12.24), with respect to the nodal values i.e. the components of $\boldsymbol{U}$ and $\boldsymbol{\Phi}$

$$\frac{\partial L}{\partial U_{i\alpha}} = 0 \text{ and } \frac{\partial L}{\partial \Phi_i} = 0, \quad \forall i,\ \alpha = x \text{ or } y \text{ or } z, \tag{12.27}$$

and leads to the following system of equations:

$$\begin{bmatrix} [K_{uu}] - \omega^2 [M] & [K_{u\varphi}] \\ [K_{u\varphi}]^{\mathrm{T}} & [K_{\varphi\varphi}] \end{bmatrix} \begin{bmatrix} \mathbf{U} \\ \boldsymbol{\Phi} \end{bmatrix} = \begin{bmatrix} \mathbf{F} \\ -\mathbf{Q} \end{bmatrix}. \tag{12.28}$$

The matrix in the left-hand side of (12.28) is real and symmetric if losses are not taken into account. Mechanical, piezoelectric, or electrical losses associated to strain–stress, stress–electric field and electrical displacement–electric field hysteresis respectively can be described using complex physical constants (Holland 1967). In that case, system (12.28) is complex and symmetric.

The unknowns are $\boldsymbol{U}$ and $\boldsymbol{\Phi}$, the vectors of the nodal values of the displacement field and electrical potential respectively. If the structure is not submitted to an external force, then $\boldsymbol{F}$ is equal to $\boldsymbol{0}$. The components of the vector of nodal charges $\boldsymbol{Q}$ are equal to zero for all nodes that do not belong to an electrode (boundary condition B3, Sect. 12.2.1), whereas the sum of the charges for all the nodes belonging to the same electrode is related to φ_p (boundary condition B5, Sect. 12.2.1).

12.2.4 Resolution of the System

Resolution of the system (12.28) can be carried out for several cases of analysis.

12.2.4.1 Static Analysis

In a static analysis, the angular frequency ω is equal to zero. Electrical degrees of freedom are split into electroded (subscript p) and internal (subscript i) degrees of freedom, which are and are not connected to an electrode, respectively. Two specific cases are considered: on the one hand, if the structure is electrically isolated ($\boldsymbol{Q}_p = \boldsymbol{0}$) and submitted to an external force $\boldsymbol{F}$, then the system to be solved is

$$\begin{bmatrix} [K_{uu}] & [K_{u\varphi_i}] & [K_{u\varphi_p}] \\ [K_{u\varphi_i}]^{\mathrm{T}} & [K_{\varphi_i\varphi_i}] & [K_{\varphi_i\varphi_p}] \\ [K_{u\varphi_p}]^{\mathrm{T}} & [K_{\varphi_i\varphi_p}]^{\mathrm{T}} & [K_{\varphi_p\varphi_p}] \end{bmatrix} \begin{bmatrix} \mathbf{U} \\ \boldsymbol{\Phi}_i \\ \boldsymbol{\Phi}_p \end{bmatrix} = \begin{bmatrix} \mathbf{F} \\ \mathbf{0} \\ \mathbf{0} \end{bmatrix}. \tag{12.29}$$

The resolution of system (12.29) gives $\boldsymbol{U}$ and $\boldsymbol{\Phi}$ for each node of the structure. The electrical potentials are induced by the direct piezoelectric effect. On the other hand, if external electrical excitations are applied on the free structure ($\boldsymbol{Q}_p \neq \mathbf{0}$), the displacement field is induced by reverse piezoelectric effect and the set of equations is written as:

$$\begin{bmatrix} [K_{uu}] & [K_{u\varphi_i}] & [K_{u\varphi_p}] \\ [K_{u\varphi_i}]^{\mathrm{T}} & [K_{\varphi_i\varphi_i}] & [K_{\varphi_i\varphi_p}] \\ [K_{u\varphi_p}]^{\mathrm{T}} & [K_{\varphi_i\varphi_p}]^{\mathrm{T}} & [K_{\varphi_p\varphi_p}] \end{bmatrix} \begin{bmatrix} \mathbf{U} \\ \boldsymbol{\Phi}_i \\ \boldsymbol{\Phi}_p \end{bmatrix} = \begin{bmatrix} \mathbf{0} \\ \mathbf{0} \\ \mathbf{Q}_p \end{bmatrix}. \tag{12.30}$$

The first two lines of the set of equations (12.30) are rearranged into:

$$\begin{bmatrix} [K_{uu}] & [K_{u\varphi_i}] \\ [K_{u\varphi_i}]^{\mathrm{T}} & [K_{\varphi_i\varphi_i}] \end{bmatrix} \begin{bmatrix} \mathbf{U} \\ \boldsymbol{\Phi}_i \end{bmatrix} = \begin{bmatrix} -[K_{u\varphi_p}]\,\boldsymbol{\Phi}_p \\ -[K_{\varphi_i\varphi_p}]\,\boldsymbol{\Phi}_p \end{bmatrix}. \tag{12.31}$$

After solving the set of the equations (12.31), the last line of the set of (12.30) gives the nodal charges $\boldsymbol{Q}_p$ delivered by the external source. Any combination of the two previous cases is possible.

12.2.4.2 Harmonic Analysis

This analysis describes the response of the piezoelectric medium to an applied external load at angular frequency ω, mostly an external force or an electrical potential φ_p applied to the electrodes. The system (12.28) is used and its resolution is carried out using the same procedure as in the static case but with ω different from zero. In the former cases, the electrical current entering the electrodes is written as $I_p = j\omega Q_p$. Therefore, the impedance of each electrode is given by the ratio φ_p/I_p.

12.2.4.3 Modal Analysis

The modal analysis gives the eigenmodes of the *lossless* structure. The eigenmodes are obtained by cancelling the right-hand side term in the set of (12.28). For a piezoelectric structure, two cases are considered: short-circuited and open-circuited electrodes.

The *resonance* frequencies f_r are obtained when the electrodes are short-circuited. Therefore, the electroded electrical potentials $\boldsymbol{\Phi}_p$ in (12.28) vanish and the resulting system is:

$$\begin{bmatrix} [K_{uu}] - \omega^2 [M] & [K_{u\varphi_i}] \\ [K_{u\varphi_i}]^{\mathrm{T}} & [K_{\varphi_i\varphi_i}] \end{bmatrix} \begin{bmatrix} \mathbf{U} \\ \boldsymbol{\Phi}_i \end{bmatrix} = \begin{bmatrix} \mathbf{0} \\ \mathbf{0} \end{bmatrix}. \tag{12.32}$$

Delivered charges are calculated using

$$\mathbf{Q}_p = -\left[K_{u\varphi_p}\right]^{\mathrm{T}} \mathbf{U} - \left[K_{\varphi_i\varphi_p}\right]^{\mathrm{T}} \boldsymbol{\Phi}_\mathbf{i}. \tag{12.33}$$

The *anti-resonance* frequencies f_a are obtained when the electrodes are open-circuited. Therefore, the delivered electric charges $\boldsymbol{Q}_p$ in (12.28) vanish and the resulting system is:

$$\begin{bmatrix} [K_{uu}] - \omega^2 [M] & [K_{u\varphi_i}] & [K_{u\varphi_p}] \\ [K_{u\varphi_i}]^{\mathrm{T}} & [K_{\varphi_i\varphi_i}] & [K_{\varphi_i\varphi_p}] \\ [K_{u\varphi_p}]^{\mathrm{T}} & [K_{\varphi_i\varphi_p}]^{\mathrm{T}} & [K_{\varphi_p\varphi_p}] \end{bmatrix} \begin{bmatrix} \mathbf{U} \\ \boldsymbol{\Phi}_i \\ \boldsymbol{\Phi}_p \end{bmatrix} = \begin{bmatrix} \mathbf{0} \\ \mathbf{0} \\ \mathbf{0} \end{bmatrix}. \tag{12.34}$$

Several cases can be envisaged, if the structure contains several hot electrodes, that can be connected or not.

In these two calculations, the eigenvalues correspond to the resonance and anti-resonance frequencies, and the eigenvector $\boldsymbol{U}$ gives the displacement field of the mode.

These results are used for the calculation of the electromechanical coupling factor of the transducer:

$$k = \sqrt{1 - \frac{f_{\mathrm{r}}^2}{f_{\mathrm{a}}^2}}. \tag{12.35}$$

k^2 is related to the ratio between the energy transduced and the total energy input (Wilson 1988). Therefore k^2 appears as a *transduction efficiency*, but all dissipative mechanisms are not considered.

12.2.4.4 Transient Analysis

The transient analysis corresponds to an arbitrary time dependency of the physical variables. In that case, for the sake of simplicity, the system to be solved can be deduced from (12.28), by substituting each $j\omega$ factor by a time derivation. More rigorous demonstration can be found in (Dubus 2002). Equation (12.28) is then written as:

$$\begin{bmatrix} [K_{uu}] & [K_{u\varphi}] \\ [K_{u\varphi}]^{\mathrm{T}} & [K_{\varphi\varphi}] \end{bmatrix} \begin{bmatrix} \mathbf{U} \\ \boldsymbol{\Phi} \end{bmatrix} + \begin{bmatrix} [C_{uu}] & [C_{u\varphi}] \\ [C_{u\varphi}]^{\mathrm{T}} & [C_{\varphi\varphi}] \end{bmatrix} \begin{bmatrix} \dot{U} \\ \dot{\boldsymbol{\Phi}} \end{bmatrix} + \begin{bmatrix} [M] & [0] \\ [0] & [0] \end{bmatrix} \begin{bmatrix} \ddot{U} \\ \ddot{\boldsymbol{\Phi}} \end{bmatrix} = \begin{bmatrix} \mathbf{F} \\ -\mathbf{Q} \end{bmatrix}. \tag{12.36}$$

The dot above vectors designate time derivatives, the double dot designate the same twice repeated. The $[C]$ matrices correspond to intrinsic losses in the piezoelectric medium, which are taken into account by adding a viscous damping term. They are generally frequency dependent, thus they are associated to the first time derivative terms.

Time discretization is usually performed by considering a constant time increment. Classical marching-in-time finite difference schemes are used for solving (Zienkiewicz 1977). In some cases, the transient analysis is associated to a Fourier transformation to obtain problem solution in the frequency domain.

12.2.5 Discussion

The goal of this section is to address the main advantages and limits of the finite element method for the analysis of piezoelectric transducers.

The finite element method allows the modelling of any geometry as well as structures containing several materials, by simply building specific meshes, without any simplification assumptions for the structure or its displacement field. As a result, many cases have been considered over the years, and the accuracy of the numerical results has been demonstrated by comparison with analytical results or experiments. Nevertheless, the following points have to be particularly studied when starting an analysis with a finite element model:

- Complex structures with curved sides can be modelled using a reduced number of elements. Nevertheless, the best results are obtained when the elements have *a reasonable aspect ratio*, while distortions lead to inaccuracies because matrices will be badly conditioned.
- The previous point can be particularly tricky when a transducer contains parts having dimensions different by several orders of magnitude. In a first step, the structure has to be *idealized*, with some hypotheses for simplifications. This point is very important, because the accuracy of the results depends on the simplifying hypotheses. As an example, some parts of transducers are stuck. The thickness of the glue is often very thin (few μm) compared to the size of the transducer (few mm or cm). In a first approximation, the glue can be neglected but some studies have shown its role in the transmission of the mechanical motion and in the transducer efficiency. Taking into account the thin slice of glue requires a much larger mesh, to meet a reasonable element aspect ratio.
- As seen in the previous sections, the finite element method formulation relies upon linear or quadratic or cubic interpolation functions. The interpolation functions define the set of possible trial functions that have to describe at best the spatial variation of the real field. For propagating acoustic or elastic waves, *the mesh criterion* states that the largest length of each element in a given mesh has to be smaller than a tenth (for linear functions) or a quarter (for quadratic functions) of the wavelength in the material at the working frequency to reach a reasonable (a few percents) accuracy. The criterion has to be verified for the lowest value of wave velocity. It is often the transverse wave velocity, but, when plates are studied, the flexural wave velocity has to be considered, which is in some cases, lower than the previous one. The finite element method is often called "a low frequency method" because of these mesh criteria, for the design of transducers.
- It is sometimes difficult to model some mechanical, as well as electrical *boundary conditions*. As an example, on a free structure, no specific mechanical boundary conditions are applied. On a structure clamped on one side, the components of the displacement fields are equal to zero for the nodes where the structure is clamped. Sometimes, this idealized condition is not realistic, and experimentally a very small displacement is nevertheless authorized.
- Electrical boundary conditions B3 to B5 given in Sect. 12.2.1 on a piezoelectric material are also idealized. They can be reasonably used for an electroded surface connected to an ideal voltage generator (condition B4) or isolated (condition B5). To model a real generator having a non-zero output impedance, a new type of boundary condition is needed. For a non-electroded surface, the electrical charge density vanishes and the exact boundary conditions imply the continuity of the normal component of the electrical displacement and the continuity of the tangential displacement of the electric field through the surface. If the dielectric permittivity of the piezoelectric medium is much larger than the dielectric permittivity of the external medium, the normal component of the electrical displacement vanishes and condition B3 can be used. Otherwise, the electrical boundary conditions cannot be simplified and the external model must be meshed with electrostatic finite elements.

- The *physical properties* of the materials are introduced in the matrices used in the finite element method. Even if a large number of materials are well-known, the physical characteristics of some of them are not known with enough precision: for instance polymers such as glue often have properties that vary with frequency and temperature. Bulk properties are known, but because of the fabrication process, thin films (few μm) have not exactly the same properties as bulk materials. The determination of losses in the material is also a tricky point, because they vary with frequency and field amplitude. In piezoelectric materials, elastic and dielectric losses are often known at low level and sometimes at high level. Piezoelectric losses are not given by manufacturers and are difficult to measure.
- The computation of the finite element model gives the vector of the nodal values of the displacement field and of the electrical potential in the whole structure. If the structure is complex, too much information prevents from seeing where the main phenomenon appears.
- The finite element formulation presented above is a *linear model*, concerning the behaviour law. It also excludes hystereses. Non-linearities, for instance in the case of electrostrictive materials, can be taken into account, but necessitates new formulations. In the same way, *small deformations* are considered. This point is no more valid when using some materials such as electroactive polymers.
- Finally, piezoelectric transducers are often in contact with a fluid in which they radiate acoustic energy. For a long time, finite element methods have been developed for solid–fluid coupling. Additional matrices are introduced in the previous system (12.28), related to dynamic and cinematic conditions at the interface between the fluid and the solid. Moreover, the finite element mesh of the fluid is limited and a non-reflecting condition is added on its external surface. This condition is valid if the external surface is assumed to stand in the far field. In some cases, it requires large fluid domains. Therefore, it is difficult to analyze more general radiating problems with the finite element method. Many different methods have been proposed to deal with this problem: dampers, infinite elements, perfectly matched layers, Dirichlet to Neumann method, Boundary Element Method (BEM); this later method seems to be the best-suited approach (Zhang 2000).

12.3 Example: The Cymbal Transducer

All the results presented in this chapter have been obtained with the help of ATILA finite element code (ATILA 2002). As an illustration of the finite element method, the proposed example concerns the cymbal transducer, already presented in the chapter Piezoelectric Actuator Design.

The transducer is a miniaturized version of the class V flextensional transducer (Doğan et al. 1997), and, because of its similarity with the musical instrument, the "cymbal" name is often used for this sensor or actuator (Fig. 12.4). It consists of a piezoelectric disk, sandwiched between two metal end caps, with a small cavity

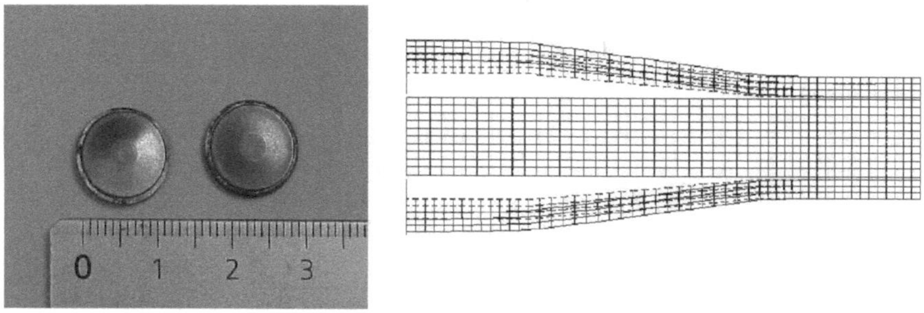

Fig. 12.4 Picture of the cymbal transducer and corresponding displacement field of the cymbal transducer at 32 kHz. *Dashed lines* correspond to the rest position. Amplitude is normalised

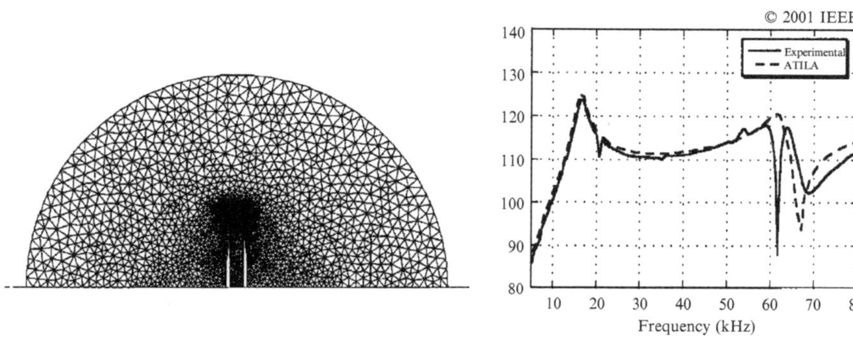

Fig. 12.5 Finite element mesh of the immersed cymbal transducer (*left*). Measured and calculated TVR as a function of frequency of the cymbal transducer (*right*) (Zhang et al. 2001)

between the disk and the caps. A lateral motion of the ceramic disk is converted into a bending motion of the cap, generating moderate force and displacement. In the finite element analysis, because of the symmetry of the cymbal, a 2D axi-symmetrical model is used. Typically, the diameter of the ceramic disk is 12.7 mm, its thickness is 1 mm and it is poled in the thickness direction. Hard PZT ceramic is used, with metal caps. The epoxy layer between the ceramic and the caps is taken into account and is assumed to be 40-μm thick.

A harmonic analysis is performed and gives the admittance spectrum vs. frequency, presenting several peaks. The first peak (32 kHz) corresponds to a bending mode, also called "umbrella mode", in which all parts of the caps move in phase, as presented in Fig. 12.4.

Immersed in water, the finite element mesh also contains fluid elements, with interface elements between the solid part and the fluid part. As shown in Fig. 12.5, the mesh is limited, with a non-reflecting condition on its external surface, as explained at the end of Sect. 12.2.5. Once again, a harmonic analysis is performed, leading to an admittance spectrum. Because of water loading, the flextensional resonance frequencies are shifted to lower frequencies, as the first bending mode of caps that shifts down from 32 to 16 kHz.

The calculation also provides the transmitting voltage response, which is the transmitted free-field pressure produced for a 1-V signal applied across the electrical terminals of the transducer and referenced at a distance of 1 m. Figure 12.5 presents the corresponding TVR, as a function of frequency. The first peak is related to the first bending mode of the caps. The experimental curve presents additional peaks, probably caused by the potting. Moreover, directivity beam patterns can be deduced from the pressure field in the fluid part.

12.4 Perspectives for the Modelling of Transducers

Numerical models allow the description of electromechanical transducers, fluid-structure coupling and acoustic radiation. With the development of new materials and the integration of piezoelectric transducers in more complex systems, the following trends can be identified for the future numerical models:

- The formalism described above concerns linear models, in terms of behaviour law. Modelling new materials, such as electrostrictive materials, magnetostrictive materials or single crystals, will necessitate taking nonlinear behaviour into account, with the description of hysteresis (Newnham 1997). Another trend concerning the active materials is the development of oriented or textured ceramics, made of grains, from different sizes and shapes. In a first approximation, homogenization models are used to determine the effective properties of these materials, but relying upon simplifying hypotheses, in terms of shape of the grains, percentage of grains. Therefore, to study these materials, micro–macro models have to be developed, which could provide a relation between their local micro-structural properties and their macroscopic properties.
- The transducer is often one part of a more general system. As presented above, the coupling with an external fluid medium is well described, but there is a lack in the description of the coupling with the electrical part of the system. Two levels can be considered for this coupling.
  - More realistic linear electrical boundary conditions corresponding to passive circuits could replace conditions B3 to B5 of Sect. 12.2.1.
  - Interaction with more complex electronic control system requires extracting a new transducer model that can be used in a circuit simulation model (VHDL-AMS type).
- In the same way, other phenomena could be taken into account for a good description of the whole device, such as thermal effects, that are frequently neglected in a first approximation. Therefore, multi-physic models are emerging.

References

Allik H, Hughes TJR (1970) Finite element method for piezoelectric vibration. Int. J. Num. Meth. Eng. 2, pp. 151–157.

ATILA (2002) Finite Element Code for Piezoelectric and Magnetostrictive Transducers Modeling, Version 5.2.1, User's Manual, ISEN, Acoustics Laboratory, Lille, France.

Beranek LL (1954) Acoustics. Mc Graw Hill, New York.

Decarpigny JN (1984) Application de la méthode des éléments finis à l'étude de transducteurs piézoélectriques. Doctoral thesis, ISEN, Lille, France.

Doğan A, Uchino K, Newnham RE (1997) Composite piezoelectric transducers with truncated conical endcaps 'cymbals'. IEEE Trans. Ultrason. Ferroelectrics Freq. Contr. Vol 44, no 3, pp. 597–605.

Dubus B (2002). Numerical modeling of sonic and ultrasonic devices. In: New acoustics. Selected topics. Ed. by C. Ranz-Guerra and J.A. Gallego-Juarez. pp. 37–73.

Holland R (1967) Representation of dielectric, elastic and piezoelectric losses by complex coefficients. IEEE Trans. Sonics Ultrason. SU-14, pp. 18–20.

Morse PM, Fesbach H (1953) Methods of theoretical physics. Mc Graw Hill, New York.

Neppiras EA (1973) The pre-stressed piezoelectric sandwich transducer. Ultrasonics International 1973 Conference Proceedings, pp. 295.

Newnham RE (1997) Molecular mechanisms in smart materials. MRS Bull., pp. 20–34.

Schenk HA (1968) Improved integral formulation for acoustic radiation problems. J. Acoust. Soc. Am. 44, pp. 41–58.

Segerlind LJ (1976) Applied finite element analysis, Wiley, ISBN: 0-471-77440-5.

Wilson OB (1988) Introduction to theory and design of sonar transducers. Peninsula Publishing, Los Altos, CA.

Zhang J (2000) Miniaturized flextensional transducers and arrays. Ph D thesis, The Pennsylvania State University.

Zhang J, Hladky-Hennion AC, Hughes WJ, Newnham RE (2001) Modeling and underwater characterization of cymbal transducers and arrays. IEEE Trans. Ultrason. Ferroelectrics Freq. Contr. Vol 48, no 2, pp 560–568.

Zienkiewicz OC (1977) The finite element method. Mc Graw Hill, London.

Part IV
Piezoelectric Transducer Fabrication Methods

Chapter 13
Piezoelectric Fiber Composite Fabrication

Andreas Schönecker

Abstract This chapter considers the status of piezoelectric fiber composite fabrication. The description is focused on three topics, the preparation of sol–gel-derived PZT fiber/polymer composites as the first exciting development phase of piezoelectric fiber composites, the soft-mold method with high achievement potential for preparing designed composites and understanding the structure–property relationships, and finally the preparation of powder suspension-derived PZT fiber/polymer composites as the technologically advanced and commercialized process. The use of piezoceramic fibers allows for the fabrication of high-quality fiber composites, covering performance data that cannot be achieved by the conventional dice and fill technique.

13.1 Introduction

Composites of parallel aligned piezoelectric rods embedded in a passive polymer matrix (Fig. 13.1) exhibit superior properties for ultrasound transduction over bulk piezoelectric ceramics because of the following (Newnham et al. 1978; Smith et al. 1989):

- Lower acoustic impedance and therefore good matching to the human body and water,
- Higher coupling coefficient k_t and therefore better conversion of electrical energy into mechanical energy and vice versa,
- Reduced lateral coupling due to acoustic isolation between neighboring active elements und therefore no need for sub dicing,
- Higher flexibility because of the matrix polymer and therefore easier fabrication of curved transducers.

A. Schönecker
Fraunhofer Institute for Ceramic Technologies and Systems, 01277 Dresden, Winterbergstrasse 28, Germany

A. Safari, E.K. Akdoğan (eds.) *Piezoelectric and Acoustic Materials for Transducer Applications*.

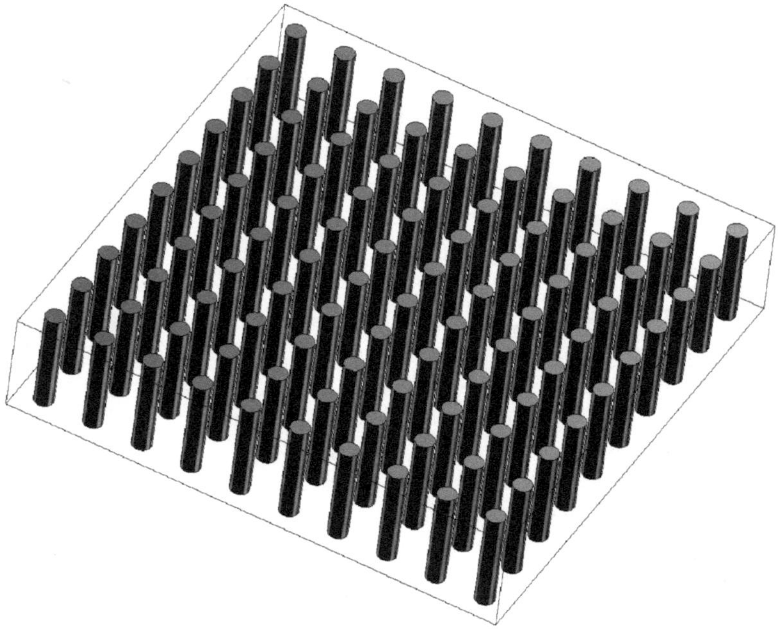

Fig. 13.1 Sketch of a 1–3 piezoelectric fiber composite material. Piezoceramic fibers are parallel arranged in an epoxy matrix. Fiber spacing, fiber diameters, and their variations are defined by the applied technology

These beneficial properties give reason for hydrophone applications, nondestructive testing and medical diagnostics. Usually, 1–3 composites are prepared using the dice-and-fill technique (Savakus et al. 1981). In this approach, parallel cuts in one direction and perpendicular to it are made into a sintered and poled piezoceramic block. The resulting grooves are backfilled with a polymer and the ceramic base is ground off. The dice-and-fill technique is time consuming, relatively expensive and limits the producible composite structures to regular and cube-shaped piezoceramic elements, which implies the excitation of spurious modes. To overcome these drawbacks, net shape technologies, like injection molding (Bowen and French 1992), the lost mold method (Lubitz et al. 1992; Janas and Safari 1995), the soft-mold method (Starke and Schönecker 1998; Gebhardt 2000), and fiber insertion methods (Safari et al. 1997) were introduced.

An essential motivation for the development of piezoceramic fibers arose from the concept of smart fiber-reinforced structural materials with integrated fibers for sensing and actuation (Hagood and Bent 1993). Respectively, tiny and flexible PZT fibers were looked for (Sporn and Schönecker 1999). In this respect the sol–gel process was seen as the most promising one, because only in this way thin and separated fibers sinterable at lower temperatures have been achieved (Yoshikawa et al. 1994; Glaubitt et al. 1995; Watzka et al. 1996).

At the same time, 1–3 composites with sol–gel-derived PZT fibers were prepared to characterize the elastic, dielectric, and piezoelectric properties of the fibers. This approach of reverse estimation of PZT fiber properties from measured effective data on composites was introduced in 1996 (Watzka et al. 1996). Fiber composites made from sol–gel PZT fibers were not commercialized, so far, although the potential for high-frequency transducer applications (>50MHz) exists.

Powder suspension based piezoceramic fibers have been developed to make high-performance, single fibers in the thickness range of 100–800 μm available. Suspension extrusion (Klicker et al. 1981) and suspension spinning (Cass 1991; French et al. 1997; Teager et al. 1998) were used as basic process routes. The latter one is considered as well suited for mass production because basically textile fabrication methods are used. The American company Advanced Cerametrics, Inc. (ACI), New Jersey, adopted the classical viscous suspension spinning technology for PZT fiber fabrication and sells PZT fiber materials commercially. The German research institute Thüringisches Institut für Textil- und Kunststoffforschung (TITK) modified the viscous suspension spinning using NMMNO as solvent and water for coagulation in view of environmental issues (ALCERU® process). A polysulfon spinning technology has recently been developed by Fraunhofer IKTS and TU Dresden for PZT fiber fabrication with the benefit of room temperature processing and the use of environmental harmless solvent NMP (*N*-methylpyrrolidon) in small quantities. These three mentioned methods allow for the production of high performance PZT fibers.

In 1999, Smart Material Corp. in co-operation with smartfiber AG, Germany, established the production of PZT fibers by the ALCERU® process and launched their 1–3 piezoelectric fiber composites on the market, allowing for fiber composite engineering and production. A variety of custom products have become available since that.

In view of understanding the static and dynamic properties of 1–3 composites, the soft-mold process initiated essential progress. The soft-mold process is basically a slip casting process. It uses a soft plastic template taken from a master mold prepared by micromachining, chemical, or plasma etching. This gives the opportunity to fabricate 1–3 composites with a high variety of structures, i.e., rod size, shape, spacing, and arrangement. Pilot production was achieved for high-frequency transducers with working frequencies in the range between 4 and 8 MHz.

Parallel to the technological achievements, characterization methods and modeling were developed to understand the performance of 1–3 composites. The analytical model of Smith and Auld (Smith and Auld 1991; Smith 1993) delivers accurate dielectric and piezoelectric materials data and can be used as an effective design tool for custom composite developments. Finite element analysis was applied for the detailed understanding especially of the dynamic properties, as vibration modes, the impedance spectra, the electroacoustic coupling factors, and further acoustical properties. Measurement of surface velocity and sound fields give insight in all excited vibration modes, the generated sound field and the correlation to the composite structure (Rödig et al. 2005).

This chapter is focused on three topics, the preparation of sol–gel-derived PZT fiber/polymer composites as the first exciting development phase of piezoelectric fiber composites, the soft-mold method with high achievement potential for preparing designed composites and understanding the structure–property relationships, and finally the preparation of powder suspension-derived PZT fiber/polymer composites as the technologically advanced and commercialized process.

13.2 Composites with Sol–Gel-Derived PZT Fibers

13.2.1 Technology Overview

Tiny sol–gel-derived PZT fibers with diameter ranging between 10 and 50 μm were developed in the mid-1990s by different groups with the intention to provide active fiber materials for ultrasound transducers and smart structures. The sol–gel process is favorable, because in this way fibers sinterable at lower temperatures with diameters <50 μm have been achieved (Seth and Schulze 1990; Boulton et al. 1992; Glaubitt et al. 1994; Shimono et al. 1993; Toyoma et al. 1995; Watzka et al. 1996; Meyer et al. 1998). Because of the comparably low sintering temperatures (<1,000°C), it is possible to fabricate fibers that can be handled in the sintered state as single filaments.

The process of piezoelectric fiber/polymer composite preparation with 1–3 connectivity starts with a "pick and place" handling to obtain carefully aligned fiber bundles (Fig. 13.2). These bundles are infiltrated with an epoxy resin and are cured under degassing conditions in order to avoid air bubbles in the matrix. The obtained composite blocks are sliced perpendicular to the fiber axis in the desired thickness. Finally, the samples are ground, metalized, and poled. At that point samples are ready to be used as piezoelectric transducer element. A detailed description of the technology was published in Watzka et al. (1996).

13.2.2 Fabrication of Sol–Gel-Derived PZT Fibers

The sol–gel technique of PZT fiber preparation relies on hydrolization and polycondensation of a metal alkoxide or metal salt precursor solution. These reactions are initiated by the addition of water, acid, and alcohol. The viscosity of the solution increases as the reaction proceeds. Fibers are produced by extruding the gel through a spinneret. Drying at room temperature and firing at 700–1,250°C complete fiber formation. The obtained diameters of sol–gel-derived PZT fibers range between 10 and 50 μm. Figure 13.3 shows the route delivering single PZT fibers, schematically.

Sol–gel-derived PZT fibers show an extraordinary high sintering activity. They can be completely densified at temperatures <950°C and can be handled as single fibers after sintering. The fiber microstructure is pore free and consists of grains between 2 and 4 μm in diameter (Fig. 13.4). The average tensile strength amounts to 100 MPa. Young's modulus was found to be 70 GPa.

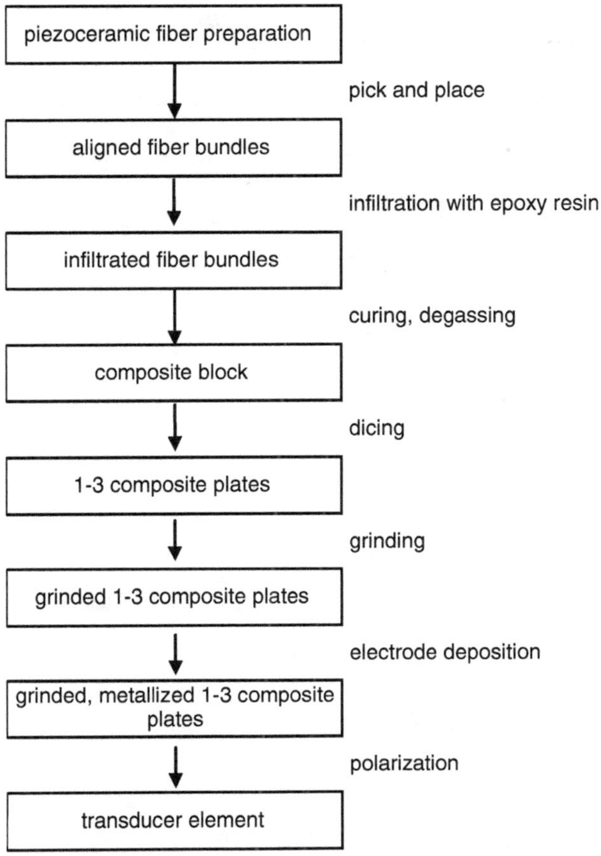

Fig. 13.2 Process chart of 1–3 piezoelectric fiber composite fabrication using sol–gel-derived PZT fibers

The poling conditions depend on the ferroelectric properties of the PZT material. In case of hard PZT fibers, typical poling conditions are temperature, 80°C, time, 2 h, and electric field strength, 55 kV/cm (Watzka et al. 1996). Fraunhofer ISC attained considerable progress of the fiber performance by tailoring the microstructure and the chemical composition. Solid solution formation of PZT with lead magnesium niobate (PZT–PMN) raised the permittivity and piezoelectric charge constant considerably.

The influence of the grain size reduction down to $<1\ \mu m$ of neodymium-doped PZT fibers on the electromechanical properties was studied in (Helbig 2006). It was concluded that the directional stress within the grains is responsible for the destabilization of the tetragonal phase, thus shifting the morphotropic phase boundary (MBP). As a consequence, the material properties of small grain size ceramics, such as PZT fibers, can be improved by adjustment of the chemical composition, e.g., the Zr/Ti ratio.

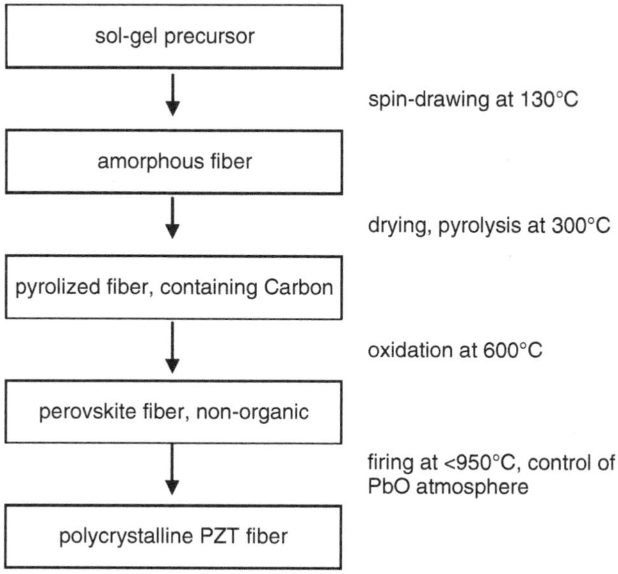

Fig. 13.3 PZT fiber preparation route based on the sol–gel process (Glaubitt et al. 1995)

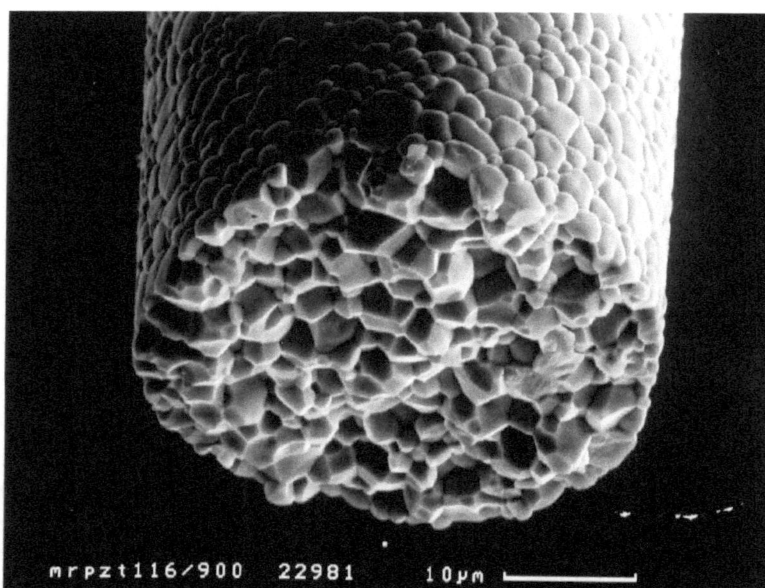

Fig. 13.4 SEM micrograph of a PZT fiber with the composition $Pb(Zr_{0,53}Ti_{0,47})O_3$ sintered at 925°C for 5 h

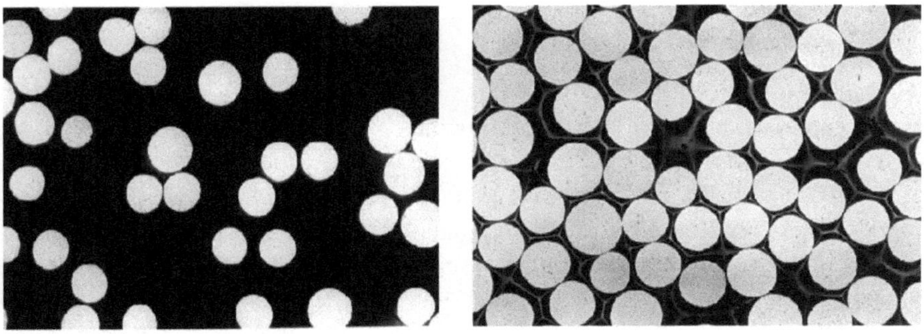

Fig. 13.5 Cross section of 1–3 piezoelectric fiber composites with 25 vol% (*left*) and 65 vol% (*right*) of PZT

13.2.3 Composite Structure

1–3 piezoelectric fiber composites are prepared by infiltration of aligned fiber bundles with an epoxy resin. Because of the bended and twisted shape of the sintered sol–gel-derived PZT fibers, random element spacing in the composite is obtained. By applying a static pressure during the curing step perpendicular to the fiber direction, the fiber volume fraction in the composite can be adjusted in the range between 25 and 65 vol%. Higher phase volume fractions due to applied external forces bear the risk of fiber cracking. This can clearly be controlled by permittivity measurements. From optical micrographs a statistical distribution of the fibers was found for composites with low fiber content of about 25 vol%, whereas a more regular arrangement of the fibers occur in samples with higher fiber contents, e.g., >50 vol% (Fig. 13.5).

The cross section of sol–gel-derived piezoelectric fiber composites show two characteristic features (Fig. 13.5). First, the PZT fibers are spaced randomly with a tendency of an ordered arrangement at higher volume content. Second, the diameters of fibers vary randomly, as well. This variation results from fiber processing.

13.2.4 Dielectric and Electromechanical Properties

Electromechanical properties of the piezoelectric fiber composites depend strongly on the phase volume fraction of PZT fibers and their arrangement. Therefore, it is necessary to understand the composite structure and property relationships. Results are summarized in this chapter.

Structure–Property Relationships

The effective material data of 1–3 piezocomposites containing piezoelectric fibers can be expressed by analytical approximations in terms of single phase data of the piezoceramic fibers and the polymer and their relative volume fractions (Smith and Auld 1991; Smith 1993). This allows either for the prediction of the effective properties of 1–3 piezocomposites from the data of the constituent phases, or the extraction of unknown fiber data from measured composite data. The complete set of equations is given in the Appendix. Equations (13.1) and (13.2) correlate, for example, the fiber data for the piezoelectric charge constant $d_{33,\mathrm{f}}$ and the permittivity $\varepsilon^{\mathrm{T}}_{33,\mathrm{eff}}$ with single phase data and the phase volume fraction v of the piezoceramic. The piezoelectric charge constant of the fibers $d_{33,\mathrm{f}}$ can be calculated from the effective composite value $d_{33,\mathrm{eff}}$, the elastic coefficients of the fibers $s^{\mathrm{E}}_{33,\mathrm{f}}$ and the polymer $s_{33,\mathrm{p}}$, and the piezoceramic volume fraction v

$$d_{33,\mathrm{f}} = d_{33,\mathrm{eff}}\left(1+\frac{(1-v)s^{\mathrm{E}}_{33,\mathrm{f}}}{v s_{33,\mathrm{p}}}\right) \tag{13.1}$$

The permittivity $\varepsilon^{\mathrm{T}}_{33,\mathrm{eff}}$ of the piezo fibers in the poled state is related to the effective permittivity of the composite $\varepsilon^{\mathrm{T}}_{33,\mathrm{eff}}$ by (13.2) (Smith and Auld 1991).

$$\varepsilon^{\mathrm{T}}_{33,\mathrm{eff}} = v\varepsilon^{\mathrm{T}}_{33,\mathrm{f}} - \left(\frac{\frac{v(d_{33,\mathrm{f}})^2}{s^{\mathrm{E}}_{33,\mathrm{f}}}}{1+\frac{v s_{33,\mathrm{p}}}{(1-v)s^{\mathrm{E}}_{33,\mathrm{f}}}}\right) + (1-v)\,\varepsilon_{\mathrm{p}}, \tag{13.2}$$

where $\varepsilon^{\mathrm{T}}_{33,\mathrm{eff}}$ and ε_p are the dielectric permittivity of the composite and the polymer matrix material, respectively. For calculations of the dielectric coefficient $\varepsilon_{\mathrm{r,f}}$ of the unpoled fibers, (13.2) reduces to a simple parallel circuit model, because the second term is equal to zero ($d_{33,\mathrm{f}} = 0$).

Reported Experimental Results

Table 13.1 summarizes the characteristic data of the sol–gel-derived PZT fibers. The electrical properties of the PZT fibers are partly known from measurements on 1–3 composites (Steinhausen et al. 1999). The elastic coefficients of PZT fibers were de-

Table 13.1 Properties of PZT fibers of various chemical compositions deduced by extrapolation of measured data on 1–3 composites

Fiber composition	Diameter (μm)	E_c (kV/cm)	P_r (μC/cm^2)	$\varepsilon^{\mathrm{T}}_{33}/\varepsilon_0$	Tanδ	d_{33} ($\times10^{-9}$C/N)	$-d_{31}$ ($\times10^{-9}$C/N)
Pb $(Zr_{0,53}Ti_{0,47})O_3$	30±10	25	<5	1,170	0.014	0.150	0.044
PZT–PMN	10,…,40	11	12	2,100	0.035	0.350	0.080

termined from the effective elastic coefficients of the 1–3 composite, which can be measured by electromechanical resonance experiments (Steinhausen et al. 2002). The method was applied to characterize the properties of undoped PZT fibers depending on the Zr content in the range between 48 and 58 mol%. The elastic coefficient s_{33}^{E} of the fiber has its maximum of $15 \times 10^{-12}\,m^2/N$ at a Zr/Ti ratio of 51/49. This result is comparable with undoped bulk PZT.

Impedance Spectrum

Conventional ceramic-polymer 1–3 composites for ultrasonic transducers have mostly a periodical inner structure due to the preparation technology. This causes spurious lateral resonances and can reduce the thickness coupling coefficient. Already first measurements of the impedance spectrum of piezoelectric fiber composites with nonregular arrangement of the piezoelectric elements in the composite showed the benefit of spurious mode suppression (Fig. 13.6).

Steinhausen studied the influence of the variations of the fiber spacing and the fiber diameters in composites by finite element modeling in detail (Steinhausen et al. 2004). The impedance spectra of fiber composites with a nonregular distribution of piezoelectric fibers were modeled by the harmonic response analysis of ANSYS. A number of 20 single unit cells with six randomly arranged fibers was generated using a random generator. Three of the unit cells are shown in Fig. 13.7.

The impedance spectra of each unit cell were calculated and added using a parallel equivalent circuit. In this first approximation, the mechanical interaction between the unit cells was neglected because of the definition of lateral clamping of every cell. Figure 13.8 shows the impedance amplitude around thickness resonance of all of the 20 unit cells and the total spectra according to the parallel equivalent circuit

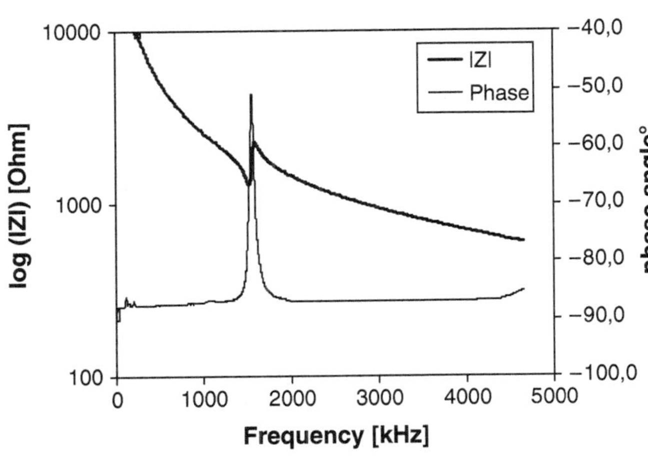

Fig. 13.6 Impedance spectrum of a 1–3 piezoelectric fiber composite with sol–gel-derived PZT fibers. Essentially, spurious modes are suppressed in the frequency range up to 5 MHz

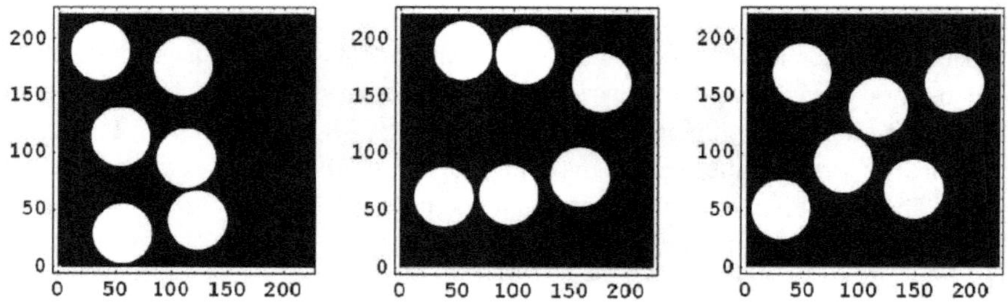

Fig. 13.7 Model unit cells with 6 fibers. The fiber positions are randomly generated. All unit cells are mechanically clamped. Used material data correspond to PZT ceramics PIC 151 (PI Ceramic, Germany) for the piezoelectric active part and epoxy resin Araldit (Ciba Geigy) for the polymer matrix. (Reprinted with permission from Steinhausen et al. 2004. Copyright 2004 IEEE)

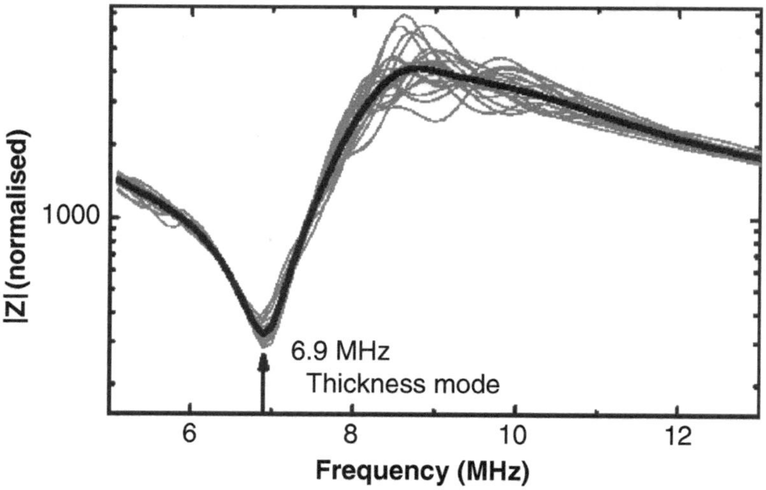

Fig. 13.8 Amplitude of the impedance of 20 unit cells each with 6 fibers and the resulting spectra. As a result of averaging (*black graph*), the spurious lateral resonances vanish completely. (Reprinted with permission from Steinhausen et al. 2004. Copyright 2004 IEEE)

approach. A sharp resonance of the thickness resonance mode occurs at 6.9 MHz with minor influence of the considered fiber position. In contrast, the influence of the fiber distribution around antiresonance frequency is evident. But, as a result of averaging, the spurious lateral resonances vanish completely. Thus, the modeled impedance spectra are in good agreement with experimentally obtained data of the 1–3 random spaced fiber composite (Fig. 13.6).

The technological process of the fiber preparation can cause a variation of the diameter of the ceramic fibers. It is conceivable that a statistical fluctuation of the fiber diameter has a positive influence on the dynamic behavior similar to the statistical distribution of fiber spacing described earlier.

The influence of the fiber diameter on the impedance spectrum was investigated by FEM using a unit cell with nine regularly arranged fibers embedded in the

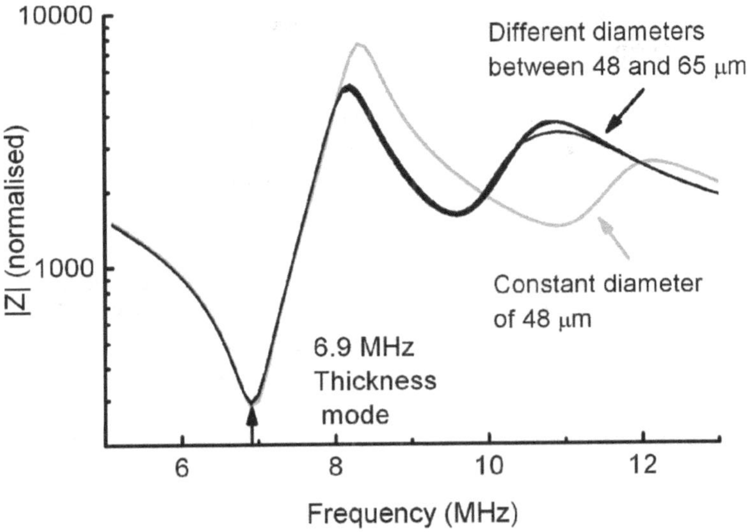

Fig. 13.9 Impedance amplitude of unit cells containing 9 fibers with different diameters between 48 and 65 μm, compared with the impedance amplitude of a unit cell with constant fiber diameter of 48 μm. (Reprinted with permission from Steinhausen et al. 2004. Copyright 2004 IEEE.)

polymer, each with a constant PZT fraction of 22 vol% (Steinhausen et al. 2004). The radius of each of the fibers was randomly generated within the range of $d = 56.5\,\mu\text{m} \pm 15\%(48,\ldots,65\,\mu\text{m})$, considering a standard deviation of $\sigma = 2.833\,\mu\text{m}$. Again 20 unit cells of different randomly generated fiber diameters were built. The impedance spectra were calculated using the same materials data and added assuming parallel circuit approach.

The obtained frequency dependence of the impedance amplitude is shown in Fig. 13.9. Again, the antiresonance vibration is effected. Lateral modes occur because of regular element spacing. Averaging influences the amplitude of the impedance, but does not suppress the spurious mode effect, like in nonregular spaced composites.

13.3 High-Frequency 1–3 Piezoelectric Fiber Composites Based on Soft Molding

13.3.1 Technology Overview

Soft-mold composites are manufactured using a patented technology that was invented at the Fraunhofer IKTS, Germany (Gebhardt 2000; Gebhardt et al. 2003). This process consists of copying a soft mold from a positive form of the final structure, filling the mold with a slip loaded with the calcined piezoceramic powder and subsequent firing of the element in PbO-controlled atmosphere. Once the piezoceramic pillars have been formed, the remaining spaces are filled with a polymer matrix material that protects the brittle fibers from accidental breakage. Next the base is removed by grinding. Metal electrodes are then bonded to the ends of the fibers and first used to polarize the piezoceramic at an elevated temperature and then to apply an electric field or collect developed charges from the material. At this point, the active elements are ready to be used as piezoelectric transducer element. The process chart is shown in Fig. 13.10.

Master molds are prepared by micromachining, chemical or plasma etching. First experiments were done using silicon (Si) master molds, which have been structured by cutting grooves with a diamond saw in one direction and perpendicular to it. For master molds with cylindrical rods, an anisotropic silicon etching (ASET) process is usually applied (Fig. 13.11).

From the Si master molds, numerous templates of a soft plastic can be taken and then filled with the ceramic slurry. Using a soft transfer mold and a ceramic slip

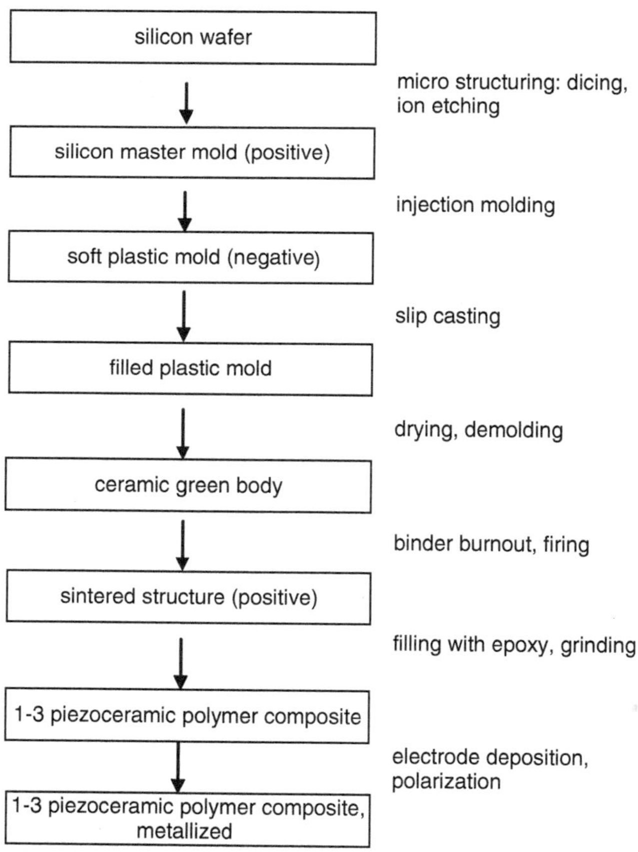

Fig. 13.10 Slip casting process using reusable soft plastic molds to manufacture 1–3 composites

Fig. 13.11 SEM micrograph of microstructures: *left*, master mold prepared by anisotropic silicon etching (ASET); *right*, ceramic replica of the master mold using the soft-mold process

Fig. 13.12 SEM micrograph of the as-sintered surface of PZT pillar prepared by the soft-mold process

with high green strength, the dried ceramic green body can be demolded with low amounts of defects.

The sintered PZT ceramic had grain sizes of 2–5 μm as shown in the SEM micrograph of the as-sintered surface of a PZT rod in Fig. 13.12. For fine-scale 1–3 composites made by the dice-and-fill technique, submicron grain sized materials are favored to get smooth cut surfaces and edges with fewer grain pullouts. Because here a slip casting technique is applied, smooth and undamaged as-sintered surfaces are already obtained.

This soft-mold process holds many advantages over conventional die-and-fill, injection molding or dicing techniques. First, the molds are reusable, thus allowing thousands of copies to be made of the original form with a single soft mold. While some configurations could be made by dicing the desired final shape from a monolithic piezoceramic block, the soft-molding process alleviates the need for

Table 13.2 Technological data of the soft-mold process, as developed at Fraunhofer IKTS

Key technological data of the soft-mold process	
Piezoceramic	Commercial PZT material (PIC 151, a solid solution of $PbTiO3$, $PbZrO3$, and $Pb(Ni1/3\ Sb2/3)O3$ at the morphotropic phase boundary, PI Ceramic GmbH, Lederhose, Germany) 97–98% of the theoretical density attained
Slip	Solvent, binder, plasticizer, and deflocculant
Binder burnout	2 K/min to 400°C/2h in air
Sintering	5 K/min to 1,150°C/2 h, PbO-controlled atmosphere
Epoxy	Araldit 2020 from Ciba Geigy
Polarization	2.2 kV/mm for 30 min at room temperature

Table 13.3 Geometrical data of prepared 1–3 composites by soft molding

Rod shape	Rod diameter (μm)	Spacing (μm)	PZT (vol%)
Cylinder	65	79	18
Cylinder	65	62	24
Cylinder	65	43	33
Cylinder	65; 48	Nonregular	22
Square	65	79	20
Square	65	62	26
Square	65	43	36
Square	81	36	48

such expensive cutting machinery. Finally, the soft-molding process is capable of producing elements with various types of fiber shapes, sizes, and pitches and offers the capacity to manufacture high volumes while maintaining superior quality at a reasonable price per unit. Experiments have shown that high-performance composites can be prepared. Details are published in Gebhardt et al. (2003). Table 13.2 gives the key data of the process.

13.3.2 Composite Structure

Rods with diameters as fine as 35 μm with an aspect ratio (height to width) of up to 6 were fabricated successfully. Attained geometrical data are shown in Table 13.3. The high regularity of the sintered PZT ceramic arrays can be seen in Fig. 13.13a. With a change of symmetry from a square to a hexagonal arrangement of rods by the same volume fraction, it is possible to shift lateral stop band resonance modes caused by the periodicity of the rods to higher frequencies. A nonregular arrangement of rods may help to suppress lateral stop band resonances.

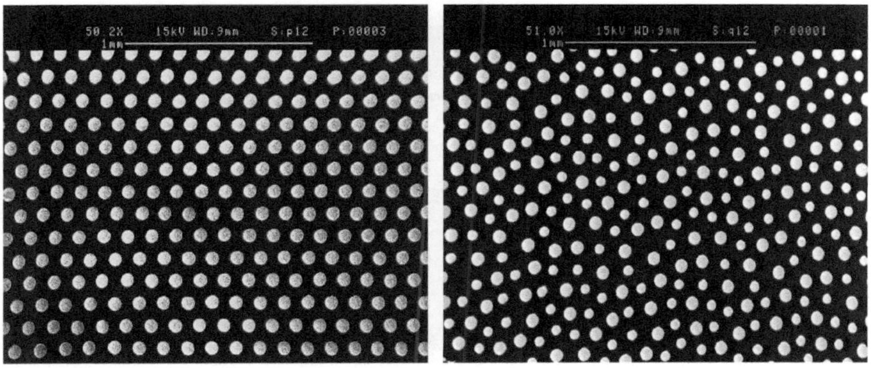

(a) (b)

Fig. 13.13 SEM micrograph of sintered PZT ceramic arrays (*top view*). *Left*, pillar with 64 μm in diameter, regular spacing of 43 μm, hexagonal symmetry; *right*, pillars with 48 and 65 μm, nonregular spacing

13.3.3 Dielectric and Electromechanical Properties

Analytical approximations (Smith and Auld 1991; Smith 1993) and finite element modeling (Steinhausen et al. 1999) were successfully applied to predict effective material parameters of the composite. The set of equations given in the Appendix allows for the calculation of composite data from materials data of constituent phases.

The required data set of the PZT material PIC 151 ($\varepsilon_{33}^{T} = 2{,}109$, $s_{33}^{E} = 1.9 \cdot 10^{-11}\,\mathrm{m^2/N}$, $d_{33} = 423\,\mathrm{pC/N}$) was given from the powder supplier (PI Ceramic, Lederhose, Germany). Young's modulus $Y = 3.89\,\mathrm{GPa}$ and Poisson ratio $\nu = 0.38$ of the polymer were determined from sound velocity measurements (Rabe, unpublished).

An excellent agreement between experimental data derived from soft-mold composites made of PIC 151 (Fig. 13.3) and the predicted data from modeling was found (Gebhardt 2000). Results concerning the piezoelectric coefficient of the composites $d_{33,\mathrm{eff}}$, the permittivity ε_{33}^{T}, the coupling factor k_t, and the acoustic impedance Z_a are given in the Fig. 13.14. The experimental data are deduced from the composites in Table 13.3 and the here-described measuring methods.

The piezoelectric coefficient of the composites $d_{33,\mathrm{eff}}$ was determined by strain measurements carried out at 130 Hz, using an equipment based on a capacitive detector. The relation between composite and fiber piezoelectric modules is given by (13.1).

The capacitance C was measured at 1 kHz, using a Hewlett Packard 4194A impedance analyzer, to determine the relative dielectric constant of the composites $\varepsilon_{33,\mathrm{eff}}^{T}/\varepsilon_0$.

The thickness mode coupling coefficient $k_{t,\mathrm{eff}}$ was calculated from minimum ($f_{\min}$) and maximum ($f_{\max}$) frequencies of the impedance curve, recorded by an

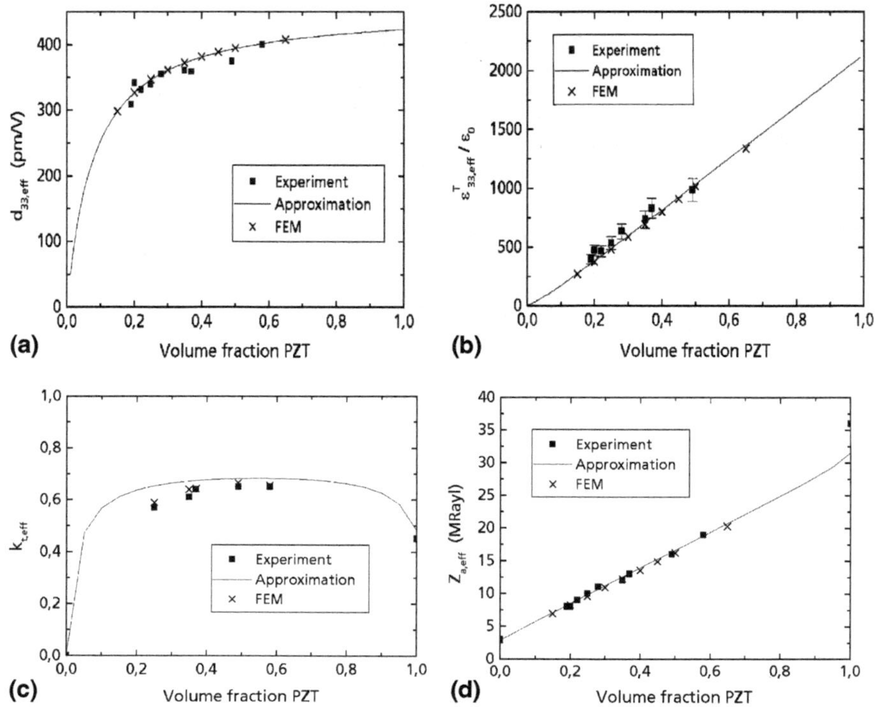

Fig. 13.14 Effective material data of soft-mold composites vs. PZT volume fraction: Comparison between experimental and analytical (Smith and Auld 1991; Smith 1993) and finite element (Steinhausen et al. 1999) modeling; (**a**) effective piezocharge constant; (**b**) effective permittivity; (**c**) thickness coupling factor; (**d**) effective acoustic impedance

Hewlett Packard 4194A impedance analyzer using the following equation:

$$k_t^2 = \frac{\pi}{2}\frac{f_{\min}}{f_{\max}}\tan\left(\frac{\pi}{2}\frac{f_{\max}-f_{\min}}{f_{\max}}\right). \tag{13.3}$$

The thickness mode coupling coefficient $k_{t,\mathrm{eff}}$ showed the expected behavior (Fig. 13.14), although $k_{t,\mathrm{eff}}$ attained slightly lower values in samples with a low content of PZT rods. This is caused by an overlapping of the first lateral stop band resonance with the thickness mode resonance for these samples.

The acoustic impedance $Z_{\mathrm{a,eff}}$ is given by:

$$Z_{\mathrm{a,eff}} = 2\rho_{\mathrm{eff}} t f_{\max}, \tag{13.4}$$

where t is the thickness and ρ_{eff} the density of the piezocomposite. The density was determined from sample dimension and weight.

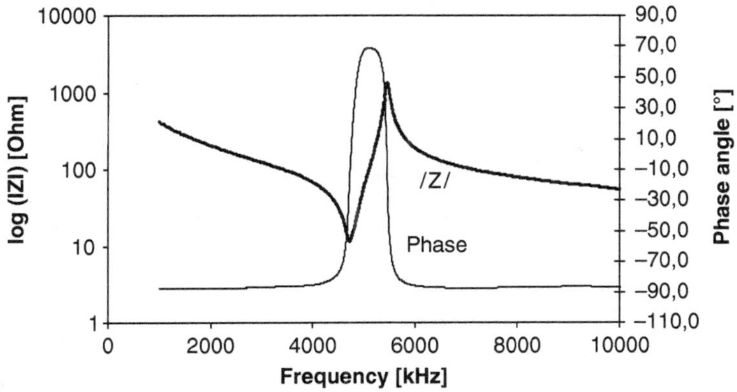

Fig. 13.15 Typical impedance spectrum of a 4-MHz transducer prepared by the soft-mold process

Vibration Modes

The vibrational characteristics of the composites are strongly influenced by the rod geometry and the rod arrangement. Using the soft-mold process, ultrasound transducers with resonance frequencies of 4–8 MHz with $k_t > 60\%$ could be prepared (Fig. 13.15).

13.4 Custom Piezo Fiber Composites by "Arrange & Fill"

13.4.1 Technology Overview

Piezoelectric 1–3 fiber composites using the Fraunhofer methodology are prepared by epoxy infiltration of fiber bundles or fixed fiber patterns and curing and dicing of the composite (Fig. 13.16). The obtained 1–3 fiber composite materials can be machined into virtually any shape or size transducer element. This "Arrange & Fill" process enables a quick, cost-efficient method of producing large quantities of composites. The piezoceramic fibers can be aligned either with or without spacer in a form and filled with polymer. Thus, regular or nonregular spacing patterns are obtainable.

Figure 13.17 shows an arranged fiber multitude and a cured composite ready for dicing. The fibers are 250 μm in diameter.

Especially know–how is required concerning the piezoceramic fiber fabrication, the fiber arrangement, epoxy selection, and the curing step. Originally developed at Fraunhofer IKTS, the process has been commercialized and up scaled by Smart Material Corp. (www.smart-material.com).

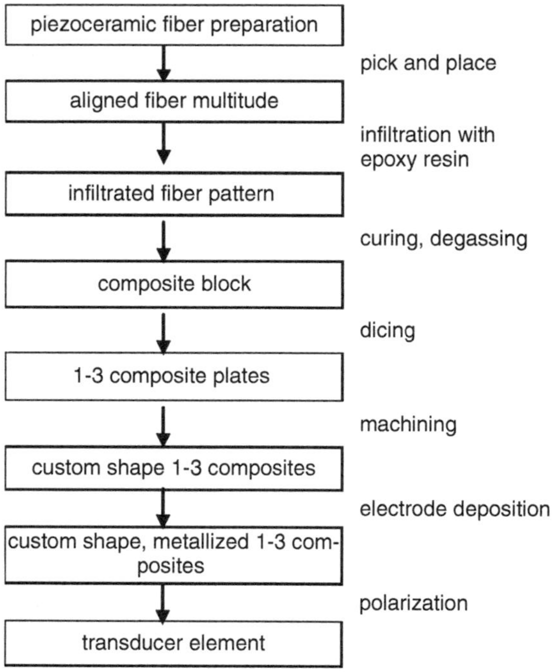

Fig. 13.16 Process chart of piezoelectric 1–3 fiber composite fabrication using powder-derived PZT fibers

13.4.2 High-Performance PZT Fibers – A Key to Quality Composites

One of the keys to manufacturing quality composites is the use of straight fibers with homogeneous consistency. Spinning processes are considered as well suited for mass production because basically well-established textile fiber fabrication methods have been adopted. Figure 13.18 shows a generalized flow chart of powder-derived PZT fiber fabrication. The Lyocell process (trade name "ALCERU® process") is basically a cellulose forming, suspension and spinning process (Teager et al. 1998). Essentially, commercial PZT powders are dispersed in a mixture consisting of cellulose NMMO (*N*-methylmorpholin-*N*-oxid-monohydrate) solution. The dispersion is pressed through a nozzle defining the fiber cross-section. The NMMO organic is replaced by water in the coagulation bath accompanied by network formation of the cellulose binder. As a result, filaments of ceramic green fibers are obtained. Subsequently, the green fibers are dried and fired.

The Lyocell process allows for the production of a variety of advanced ceramics on an industrial scale. Different filament shapes, hollow fibers, as well as bicomponent fibers have been manufactured successfully. Figure 13.19 shows sintered PZT fibers consisting of 250 μm in diameter and 150 mm in length. Table 13.4

Fig. 13.17 *Top*, Arrangement of the PZT fiber multitude; *bottom*, composite block ready for dicing. The obtained 1–3 composite shows nonregular element spacing

shows typical fiber data. The functional properties of the fibers are on a reduced performance level as compared with monolithic ceramics, which is attributed to the unusual high surface to volume ratio of fibers. Improvements are expected by compositional modifications.

ALCERU® PZT fibers have been commercialized by the smartfiber AG and Smart Material Corp. (Florida). They are made from Type II and Type VI (U.S. Navy designation standards) piezoceramics and are offered in the diameter range between 105 and 1,000 μm (see Table 13.5).

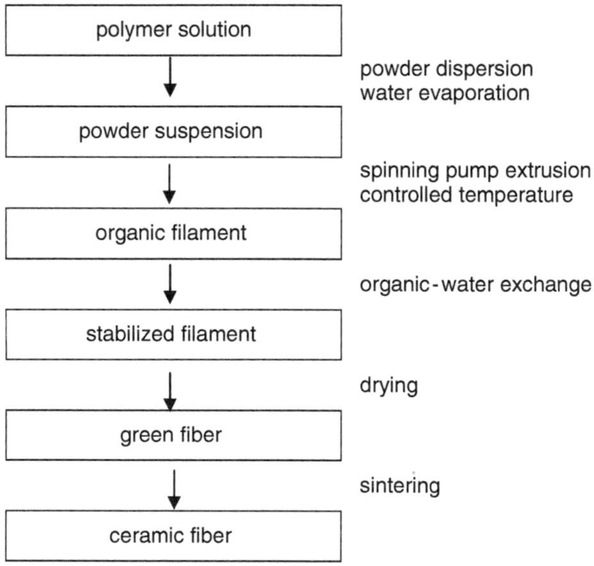

Fig. 13.18 General flow chart of powder-derived PZT fiber fabrication according to the Lyocell process

Fig. 13.19 Sintered PZT fibers ready for composite fabrication

13.4.3 Composite Structure

The PZT volume fraction of fiber composites can be varied from 25% to 65% depending upon the desired behavior characteristics. Piezoelectric 1–3 fiber composites are commercially available with random element spacing (Fig. 13.20) for a

Table 13.4 Properties of piezoceramic fibers

Piezoceramic	Diameter (μm)	$\varepsilon_{33,f}^{T}/\varepsilon_0$	Relative εt value (%)	$d_{33,f}$ (pC/N)	Relative d-value (%)
Navy Type IV	300	2,525	66	690	55
Navy Type II	250	1,300	62	470	51
Navy Type II	140	1,063	63	374	59

The relative values are deduced by relating the fiber data to the monolithic ceramic data of the same composition.

Table 13.5 Commercial PZT fibers and tubes offered by Smart Material

Fiber diameters	105–1,000 μm (other diameters on request)
Tube outer diameters	400 and 1,000 μm
Tube wall thickness	>100 μm
Fiber and tube length	<200 mm
Piezoceramic materials	Type: PZT 4, PZT 5A, PZT 5H

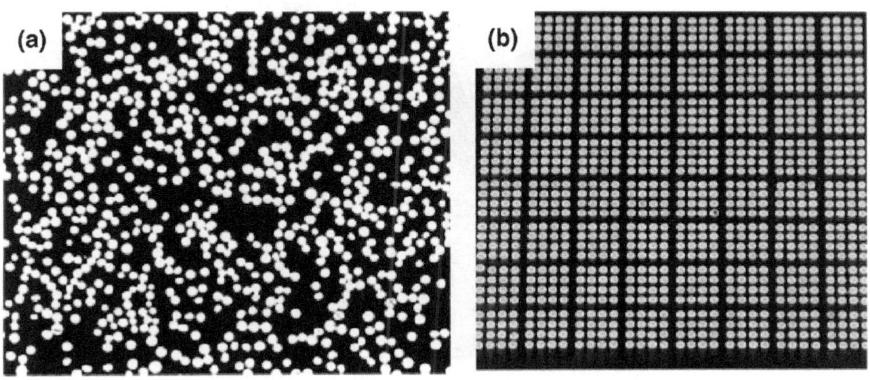

Fig. 13.20 Cross section of a piezoelectric 1–3 fiber composite with (**a**) nonregular element spacing (the fiber diameter is 250 μm) and (**b**) regular element spacing (the fiber diameter is 800 μm)

wide frequency range of 40 kHz up to 8 MHz. Fiber material, diameter, and spacing are subject for custom designs. Piezoelectric fiber composites with regular element spacing are available for frequencies <500 kHz (see Fig. 13.20). The properties of such fiber materials are summarized in Table 13.6.

13.4.4 Adaptable Shapes

Obtained piezocomposite blocks can be machined into virtually any shape or size transducer element (Fig. 13.21). Customers have reported further dicing or machining on their own to adapt them for other projects – an advantage not afforded by

Table 13.6 Specification of fiber composites materials offered by Smart Material

Sizes of samples	up to 80×80-mm^2 or 20×150-mm^2 square, round, or rectangular cross section
Fiber diameter	105, 250, 800 μm
Resonance frequency	40 kHz up to 8 MHz
Thickness	>250 μm up to 4 cm
Piezoceramic	PZT type 5A, 5H, 4
Polymer matrix	Epoxy, polyurethane
PZT volume fraction	25% up to 65%
Electrode material	Sputtered gold or silver, polymer silver
Fiber arrangement	Regular and nonregular element spacing
Custom shapes	concave, convex, biconcave

Fig. 13.21 Example of a custom-shaped piezoelectric 1–3 fiber composite. PZT fibers with 250 μm in diameters are used

conventional materials. Through the use of custom molds and precision machining, concave surfaces can be produced to create a broadband sensor. The presence of polymer creates a "cushion-effect" that protects the fibers and enables the structure to resist chipping and cracking during the machining process.

In guided wave inspections, where sheets or tubes type surfaces require inspection without extensive scanning, a broadband transducer is required. The piezoelectric fiber composite not only provides this desirable characteristic, but also allows for a high-pulse amplitude (low noise), conformability, and an excellent acoustic impedance match to the test piece.

13.4.5 Dielectric and Electromechanical Properties

Advanced piezoelectric fiber composites utilize fiber elements embedded in epoxy in a random or regular spacing pattern. These composites exhibit superior properties for ultrasound transduction because of lower acoustic impedance, broader bandwidth, and greater sensitivity and enhanced spurious mode suppression. Owing to their excellent performance characteristics, there is often no need for matching or backing layers. This can save on space, costs and decreases the weight of the overall transducer package. The effective properties of the here-considered 1–3 composites can be described by the equations given in the Appendix.

Impedance Spectrum

The thickness resonance frequency is defined by the frequency constant of the material of about 1,550 Hzm and the thickness of the sample. So far, samples in the wide resonance frequency from 50 kHz to 20 MHz have been prepared and tested. As seen in Fig. 13.22, spurious modes around the thickness vibration are completely suppressed.

The impedance spectra together with the surface velocity of fiber composites with regular elements spacing patterns were measured and compared to finite element modeling results (Rödig et al. 2005). ATILA finite element modeling and scanning vibrometer measurements proved to be well-suited design tools for the development of custom piezofiber composites. In case of regular spacing, ATILA allows the calculation of vibration modes, the resonance and antiresonance frequencies, the electro acoustic coupling factors, the admittance the displacement field, in

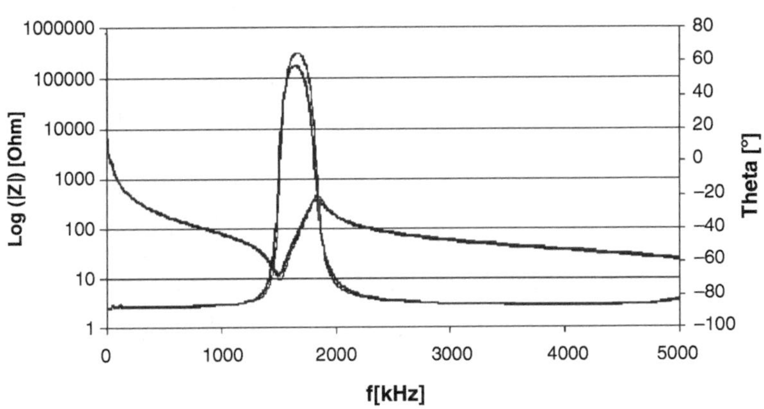

Fig. 13.22 Impedance plot IZI and phase angle theta as function of frequency measured on 1–3 fiber composites with random element spacing (fibers: PZT Navy Type II, 250 μm; composite: 65 vol%; sample size: 20×20 mm^2) showing only the thickness vibration mode at 1.5 MHz. No spurious modes occur

air or in water, as well as the free-field voltage sensitivity and the transmitting voltage response. For random spaced composites measurement of surface velocity and sound fields give insight in all excited vibration modes, the generated sound field and the correlation to the composite structure.

13.5 Conclusions

Piezoceramic fiber composites allow for a unique and cost-efficient piezo transducer technology. They incorporate a combination of piezoelectric fibers and polymers to enhance the ultrasound performance of the device. Resulting benefits include a high coupling coefficient, low acoustic impedance, minimized lateral mode coupling and an intermediate dielectric constant. Ultrasound sensor elements with suppressed spurious modes can be prepared for working frequencies between 40 and 6 MHz. The acoustic impedance can be adjusted to the needs of sonar applications, nondestructive testing, and biomedical diagnosis. Additionally, the transducer can be designed to match nonplanar surfaces, increasing its versatility when a difficult shape is involved or broadband sensor is required.

An outstanding feature of fiber composites with 1–3 connectivity, as prepared by the IKTS technology, results from the finding, that block material can be machined by conventional methods, such as dicing, grinding, and drilling. Thus, the possibility arises to get free-shaped active parts of mechanical structures.

A new technique of forming 1–3 composites for applications at frequencies $\geq$1 MHz which is called the *soft-mold process* was developed. The technology uses master molds, which have been structured by microsystems technologies. Because no extensive equipment and reusable transfer molds are used, the process is considered to be cost effective. In contrast to the fine-scale dice-and-fill technique conventional PZT powders with sintered grain sizes of 2–5 μm can be applied since no surface processing is needed. The soft-mold process allows for the preparation of fine-scale 1–3 composites with PZT rods of different size, shape, and spacing, which can be used as ultrasonic transducers for frequencies $\geq$4 MHz.

The advent of piezoelectric fiber composites will not entirely replace monolithics (bulk PZT ceramics). In fact, there is a need for both materials in the ultrasound transducer industry. Both types offer advantages under certain circumstances. For example, piezocomposites are ideal for applications requiring a resonant frequency between 40 and 6 MHz. Outside the scope of that range, monolithics are recommended. In general, piezocomposites cost more than monolithics, however the transducer manufacturer must weigh in the added benefits in performance and the possible material cost reduction from the elimination of matching/backing layers.

Acknowledgments The work on piezoelectric fiber composites was carried out in close cooperation with Fraunhofer ISC (Dieter Sporn), Martin Luther University Halle (Horst Beige, Ralf Steinhausen), and Smart Material Corp. (Thomas Daue). Essential contributions were made by Sylvia Gebhardt and Lutz Seffner, Fraunhofer IKTS.

Appendix

Correlation between single phase data (index "f" fiber, index "p" polymer) and effective composite data (index "eff") (Smith and Auld 1991; Smith 1993).

$$\rho_{\mathrm{eff}} = v\rho_{\mathrm{f}} + (1-v)\rho_{\mathrm{p}}$$

$$\varepsilon_{\mathrm{r,eff}} = v\varepsilon_{\mathrm{r,f}} + (1-v)\varepsilon_{\mathrm{r,p}}$$

$$\varepsilon^{\mathrm{T}}_{33,\mathrm{eff}} = v\varepsilon^{\mathrm{T}}_{33,\mathrm{f}} - \left(\frac{\frac{v(d_{33,\mathrm{f}})^2}{s^{\mathrm{E}}_{33,\mathrm{f}}}}{1+\frac{vs_{33,\mathrm{p}}}{(1-v)s^{\mathrm{E}}_{33,\mathrm{f}}}}\right) + (1-v)\varepsilon_{\mathrm{p}}$$

$$d_{33,\mathrm{eff}} = \frac{d_{33,\mathrm{f}}}{1+\frac{(1-v)s^{\mathrm{E}}_{33,\mathrm{f}}}{vs_{33,\mathrm{p}}}}$$

$$k_{t,\mathrm{eff}} = \frac{e_{33,\mathrm{eff}}}{\sqrt{c^{\mathrm{D}}_{33,\mathrm{eff}}\varepsilon^{\mathrm{S}}_{33,\mathrm{eff}}}}$$

$$Z_{\mathrm{a}} = \sqrt{c^{\mathrm{D}}_{33,\mathrm{eff}}\rho_{\mathrm{eff}}}$$

By using

$$e_{33,\mathrm{eff}} = v\left[e_{33,\mathrm{f}} - \frac{2(1-v)e_{31,\mathrm{f}}(c^{\mathrm{E}}_{13,\mathrm{f}} - c_{12,\mathrm{p}})}{v(c_{11,\mathrm{p}}+c_{12,\mathrm{p}}) + (1-v)(c^{\mathrm{E}}_{11,\mathrm{f}}+c^{\mathrm{E}}_{12,\mathrm{f}})}\right]$$

$$c^{\mathrm{E}}_{33,\mathrm{eff}} = v\left[c^{\mathrm{E}}_{33,\mathrm{f}} - \frac{2(1-v)(c^{\mathrm{E}}_{13,\mathrm{f}} - c_{12,\mathrm{p}})^2}{v(c_{11,\mathrm{p}}+c_{12,\mathrm{p}}) + (1-v)(c^{\mathrm{E}}_{11,\mathrm{f}}+c^{\mathrm{E}}_{12,\mathrm{f}})}\right] + (1-v)c_{11,\mathrm{p}}$$

$$\varepsilon^{\mathrm{S}}_{33,\mathrm{eff}} = v\left[\varepsilon^{\mathrm{S}}_{33,\mathrm{f}} + \frac{2(1-v)(e_{31,\mathrm{f}})^2}{v(c_{11,\mathrm{p}}+c_{12,\mathrm{p}}) + (1-v)\left(c^{\mathrm{E}}_{11,\mathrm{f}}+c^{\mathrm{E}}_{12,\mathrm{f}}\right)}\right] + (1-v)\varepsilon_{\mathrm{p}}$$

$$c^{\mathrm{D}}_{33,\mathrm{eff}} = c^{\mathrm{E}}_{33,\mathrm{eff}} + \frac{(e_{33,\mathrm{eff}})^2}{\varepsilon^{\mathrm{S}}_{33,\mathrm{eff}}}$$

$$v_{\mathrm{l,eff}} = \sqrt{\frac{c^{\mathrm{D}}_{33,\mathrm{eff}}}{\rho_{\mathrm{eff}}}}$$

with

- v – volume fraction of piezoceramic
- ρ – density
- ε – the permittivity
- $\varepsilon^{\mathrm{T}}_{33}$ – the permittivity after polarization
- d_{ij} – the piezoelectric charge constant

e_{ij} – the piezoelectric strain constant
s_{ij} – the elastic compliance
c_{ij} – the elastic stiffness
k_t – electromechanical coupling factor (thickness mode)
Z_a – acoustic impedance.

References

Boulton JM (1992) Sol–Gel Derived PZT Fibers. Mat Res Soc Symp Proc 271:517–522

Bowen LJ and French KW (1992) Fabrication of Piezoelectric Ceramic/Polymer Composites by Injection Molding. Proceedings of the 8th IEEE Int Symp on Appl of Ferroelectrics, ISAF' 92:160–163

Cass RB (1991) Fabrication of Continuous Ceramic fiber by the Viscous Suspension Spinning Process. Am Ceram Bull 70:424–429

French JD et al. (1997) Proc SPIE – Int Soc Opt Eng 3044:406–412

Gebhardt S (2000) Herstellung und Charakterisierung von feinskaligen 1–3 Piezo-kompositen für Ultraschallwandler. Thesis TU Bergakademie Freiberg

Glaubitt W et al. (1994) Formkörper auf der Basis von PZT, Bleizirkonat-Titanat, Verfahren und Zwischenprodukt zu deren Herstellung. Patent application DE 4332831 C1

Glaubitt W et al. (1995) Sol–Gel Processing of PZT Long Fibers. Adv Sci Technol 10: Intelligent Materials and Systems (Ed.: P Vinzenzini, Techna Srl):47–54

Hagood NW and Bent AA (1993) Development of Piezoelectric Fiber Composites for Structural Actuation. Proc 43th AIAA/ASME, paper 93–1717

JanasVF, Safari A (1995) Overview of Fine-Scale Piezoelectric Ceramic/Polymer Processing. J Am Ceram Soc 95:2945–2955

Kitaoka K et al. (1998) Preparation of lead lanthanum zirconate titanate (PLZT, $(Pb,La)(Zr,Ti)O_3$) fibers by sol–gel method. J Am Ceram Soc 81:1189–1196

Klicker KA et al. (1981) Composites of PZT and Epoxy für Hydrostatic Transducer Applications. J Am Soc 64:5–9

Lubitz K et al. (1992) New Piezoelectric Composites for Ultrasonic Transducers. Ferroelectrics 92:21–26

Meyer R et al. (1998) Lead Zirconate Titanate Fine Fibers Derived from Alkoxide-Based Sol–Gel Technology. J Am Ceram Soc 81(4):861–868

Newnham RE et al. (1991) Smart Electroceramics. J Am Ceram Soc 74:463–480

Newnham RE et al. (1978) Connectivity and Piezoelectric–Pyroelectric Composites. Mat Res Bull 13:525–536

Rabe U, IZFP Saarbrücken, Unpublished

Rödig T et al. (2005) Design and Characterization of 1–3 Ultrasonic Composites using ATILA and Ultra Fast Laser Measurements (20 MHz), Proceedings 1:353–356

Safari A et al. (1997) Development of Fin-Scale Piezoelectric Composites for Transducers. AIChE J 97:2849–2856

Savakus HP et al. (1981) PZT-Epoxy Piezoelectric Transducers: A Simplified Fabrication Procedure. Mat Res Bull 16:677–680

Seth VK and Schulze WA (1990) Fabrication and Characterization of Ferroelectric PLZT 7/65/35 Ceramic Thin Films and Fibers. Ferroelectrics 112:283–307

Shimono I et al. (1993) Preparation of $Pb(Zr,Ti)O_3$ Fibers by the Alginate Method. J Ceram Soc Jpn (Int Ed) 101:700–703

Smith WA (1993) Modeling 1–3 Composite Piezoelectrics: Hydrostatic Response. IEEE Trans Ultrason Ferroelectrics Freq Cntr 40(1):41–49

Smith WA and Auld A (1991) Modeling 1–3 Composite Piezoelectrics: Thickness Mode Oscillation. IEEE Trans Ultrason Ferroelectrics Freq Cntr 38(1):40–47

Smith WA et al. (1989) Design of Piezocomposites for Ultrasonic Transducers. Ferroelectrics 91:155–162

Sporn D and Schönecker A (1999) Composites with Piezoelectric Thin Fibers – First Evidence of Piezoelectric Behaviour. Mat Res Innovat 2:303–308

Starke S et al. (1998) Fine Scale Piezoelectric 1–3 Composites: A New Approach of Cost Effective Fabrication. Proc 11th IEEE International Symposium on the Applications of Ferroelectrics, ISAF'98 Vol. I: 393–396

Steinhausen R et al. (1999) Finescaled Piezoelectric 1–3 Composites: Properties and Modeling. J Europ Ceram Soc 19:1289–1293

Steinhausen R et al. (2002) A New Method for the Determination of Elastic Properties of Thin Piezoelectric PZT Fibers. Ferroelectrics 268:53–58

Steinhausen R et al. (2004) Finite Element Analysis of the Thickness Mode Resonance of Piezoelectric 1–3 Fibre Composites. 2004 IEEE Ultrason Symp Proc: 1678–1681

Teager E et al. (1998) Lyocell products with build-in functional properties. Chem Fibers Int 48:32–35

Toyoma M et al. (1995) Fabrication of $Pb(Zr,Ti)O_3$ Ceramic Fibers by Sol–Gel Processing II. Viscosity Change and Spinnability of Pb, Ti, Zr Complex Alkoxide Solutions and Hydrolysis and Morphology of $Pb(Zr,Ti)O_3$ Ceramic Fibers. Nippon Kagaku Kaishi: 150–155

Watzka et al. (1996) Dielectric and Ferroelectric Properties of 1–3 Composites Containing Thin PZT-Fibers. Proceedings of the 10th IEEE – Int Symp Appl Ferroelectrics, ISAF'96 Vol. II:569–572

Yoshikawa S et al. (1994) Ferroelectrics 154:325–330

Chapter 14
Composition Gradient Actuators

Ralf Steinhausen* **and Horst Beige**

Piezoelectric functionally graded materials (FGMs) are attractive alternatives to homogenous-single phase materials for actuator applications because of their reduced internal mechanical stresses and lower production costs. Furthermore, such FGM structures have increased band width if used as an ultrasonic transducer. One of the most effective ways to prepare piezoelectric and dielectric gradients based creating a gradient in chemical composition by powder processing prior to sintering. The sharp chemical interfaces between the green layers disappear because of diffusion during sintering. The chemical gradient is then transformed into a gradient in the piezoelectric properties by a poling process after sintering. Several models have been developed for the description of poling of layered systems, which is a formidable challenge. The ferroelectric properties, such as spontaneous polarization and coercive field strength, also depend on the local chemical composition. This causes an inhomogeneous electric field distribution, which is usually not stable in time because of conductive currents and space charges.

In this chapter, different types of composition gradients for bending actuators are described. Combinations of hard and soft piezoelectric ceramics and electrostrictive and electroconductive materials are presented. The theoretical results are compared with experimental data for lead-free systems based on barium titanate.

14.1 Introduction

Functionally graded materials (FGMs) have a one or more dimensional gradient of at least one material property. Usually, several material properties change their values continuously in the direction of the gradient. Thus, a gradient in piezoelectric coefficients is often combined with a gradient in dielectric or elastic coefficients.

R. Steinhausen
Institute of Physics, Martin-Luther-Universität Halle-Wittenberg, Friedemann-Bach-Platz 6, 06108 Halle, Germany

A. Safari, E.K. Akdoğan (eds.) *Piezoelectric and Acoustic Materials for Transducer Applications.*

Piezoelectric FGMs are suitable for applications where materials with different properties need to be mechanically connected. Thereby, residual mechanical stresses at interfaces between the different materials can be reduced. Piezoelectric FGMs with a gradient of the electromechanical properties can be used in micromechanical systems (MEMS), or as actuators for connecting piezoelectric active parts with substrates or other passive parts. In this way, reliability and lifetime of electromechanical devices can be improved. Furthermore, FGMs can be used to improve the bandwidth in ultrasonic applications (Sakamura et al. 2000).

Monolithic bending actuators are a special application of piezoelectric FGMs. The bending effect strongly depends on the difference of the lateral strain at the top and the bottom surface of the actuator. Usually, two layers with different piezoelectric properties or with an electrode in between the layers are bonded together. It was shown that piezoelectric FGMs can reduce the mechanical stresses at the interface (Hauke et al. 2000). A good performance of the bending actuator depends on the quality of the piezoelectric gradient.

Usually, the piezoelectric gradient in monolithic samples is obtained by creating a chemical composition gradient along a desired direction in the medium of interest. One of the most important preparation steps is the poling process, which transforms the chemical composition gradient to a piezoelectric one. Since ferroelectric properties such as spontaneous polarization and coercive field also depend on the chemical composition, poling is very complicated and is a time-dependent process. In this chapter, a model of the poling process is discussed and applied to different combinations of ferroelectric and nonferroelectric materials.

14.2 Preparation of a Graded Piezoelectric Material

The electromechanical response depends on the piezoelectric or electrostrictive coefficients as well as on the electric field strength, and hence, on the dielectric coefficient. Thus, a gradient of one or both coefficients can be used for preparing FGMs. Since the electromechanical and dielectric properties can be influenced by different material parameters, different preparation approaches are conceivable. The simplest and the most established method to date is the use of a chemical composition gradient. Other approaches are based on a gradient of microstructure; i.e., grain size dependence of ferroelectric and piezoelectric properties (Arlt et al. 1985). The increase in electromechanical response with increasing grain size was observed in $BaTiO_3$ ceramics (Arlt and Pertsev 1991) and other perovskite structures. For instance, the piezoelectric coefficient d_{33} of $Pb(Mg_{1/3}Nb_{2/3})O_3$–$PbTiO_3$ (PMN–PT) ceramics grows several times at the nm- and low μm-scale in grain size (Wang et al. 1992; Shvartsman et al. 2002). In most cases, the grain size in chemically identical materials is varied by the sintering temperature and time. Thus, a temperature gradient should be applied to the sample during the sintering process in order to prepare a grain size gradient in ceramics with homogeneous chemical composition. Such a localized temperature gradient can be induced by microwave sintering

(Yasuoka et al. 2006). However, the temperature change is high enough only at the mm scale. Thus, this technique is suitable only for thick bulk samples.

The most common method for preparing a piezoelectric gradient in monolithic ceramics starts with a chemical gradient in one or more dimensions of the sample. By adding dopants or changing the volume ratio in multicomponent systems, a series of parameters can be influenced. In ferroelectrics, the microstructure (grain sizes, porosity, and domain structure) depends not only on the sintering conditions but also on the chemical composition. Slight shifts in the volume fractions near the morphotropic phase boundary in a given ferroelectric-piezoelectrics system yield large differences in properties (spontaneous and remnant polarization, coercive field strength, among others). A well-known example are the piezoelectric, dielectric, and elastic properties of $Pb(Zr,Ti)O_3$ (PZT), which reach their maximum at the morphotropic phase boundary (Jaffe et al. 1971).

A popular monolithic actuator with a gradient in chemical composition is the so-called RAINBOW (reduced and internally biased oxide wafer) actuator, introduced by Haertling (1994). A piezoelectric ceramic disk (PZT or $(Pb, La)(Zr,Ti)O_3$) with a high content of lead is thermally reduced at one side. This reduced layer primarily consists of metallic lead, and hence is inactive piezoelectrically. However, it clamps the unreduced piezoelectric later, thereby acting as a strain amplifier.

The preparation of the chemical composition gradient is based on a multilayer structure in the green body by a stepwise preparation, using different powders or slurries with various chemical compositions. Depending upon the final thickness of the sample and the materials used, different technologies could be deployed. Various thin and thick film techniques could be used for thinner samples at the nano- to the micrometer scale. In the micrometer range, tape casting techniques appear to be most convenient. Another approach layered manufacturing technology such as fused deposition of multimaterials (FDMMs) used by Hall et al., who showed that it is a viable method in obtaining functional gradient electromechanical actuators using piezoelectric and electrostrictive ceramics in tandem (Hall et al. 2005).

In all methods, the chemical composition of each layer slightly differs from that one of the neighboring layer. The gradient itself can be varied by the number of different layers and the value of the difference of the chemical composition between the layers. Jeon et al. used up to 21 powder layers to prepare a graded barium–strontium titanate ceramic (Jeon et al. 2001). In our work, samples prepared from 2, 3, or 4 powder layers were investigated (see Fig. 14.1).

Different graded materials based on $BaTiO_3$ ceramics are described next. By adding different dopants, the ferroelectric and electromechanical properties can be changed over a wide range. Thus, combinations of piezoelectric ceramics with electrostrictive, electroconductive, and hard piezoelectric ceramics were investigated. The ceramic powders needed for the different layers were prepared by the conventional mixed-oxide technique. During the sintering process, samples consisting of layers with different chemical compositions bent because of the differential sintering and thermal expansion coefficient mismatch. It was observed that, the higher the number of pressed layers or the smaller the difference in chemical composition of neighboring layers, the smaller the deformation. Here, the chemical

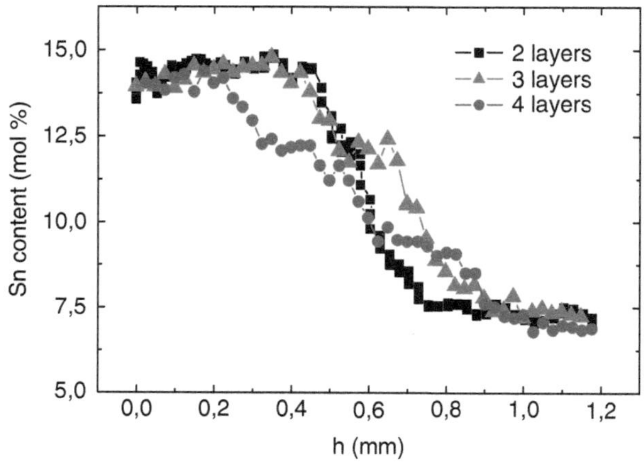

Fig. 14.1 Sn distribution along the thickness in BTS samples prepared with different number of powder layers

gradient plays the same role as the piezoelectric gradient in the final FGM – the better the gradient, the better the decrease in residual stresses because of the aforesaid mismatch. However, it should be noted that the strain mismatch is caused by thermal and not by electromechanical coupling. The thickness of a single powder layer is at least 0.25 mm. Using the foil pressing technique, the thickness of the single layer can be reduced. Foils prepared by the doctor blade technique (tape casting) have a thickness of 50–100 μm.

In Sect. 14.4.2, $Ba(Ti,Sn)O_3$ (BTS) ceramics with a gradient in Sn concentration will be discussed. The samples were prepared with up to four powder layers, wherein the Sn content was varied as 7.5, 10, 12.5, and 15 mol%, respectively. Here, the BTS*X* sample designation is used, where *X* is the amount of tin. The local content of Sn along the thickness was characterized by electron probe micro analysis (WDX-EPMA). The results of samples with 2, 3, and 4 powder layers are compared in Fig. 14.1. Because of the diffusion processes during sintering, the Sn concentration changes gradually at the interfaces between the different layers.

14.3 Modeling of the Poling Process

The poling process is the last and perhaps the most important step in preparing a functionally graded piezoelectric material. The chemical composition gradient has to be transformed into a gradient of piezoelectric coefficients. Since the piezoelectric coefficients are determined by the polarization state, a gradient in the local polarization has to be induced in the material.

The chemical composition of the basic materials used in the FGM governs not only the electromechanical but also the ferroelectric properties, and needs to be taken into account in designing an FGM. The dielectric permittivity as well as the spontaneous polarization, remnant polarization, and the coercive field strength strongly depend on the crystallographic symmetry, domain structure, and defects. All these parameters are influenced by the chemical composition of the ceramics, making an FGM formidably complex for theoretical analysis.

To understand the specifics of the poling processes in an FGM, it is helpful to develop a poling model. Or et al. introduced a simple model for a multilayer structure, which encapsulates all essential features of a graded piezoelectric (Or et al. 2003). This time-dependent model is based on an equivalent electrical circuit analysis of a system comprised of N layers. Each layer has homogeneous ferroelectric properties, and is considered as a resistive–capacitive (RC) element. Each RC element consists of a nonlinear (ferroelectric) capacitor and a resistance connected in parallel.

The dielectric displacement in each layer is:

$$D_k = \varepsilon_0 E_k + P_k, \quad k = 1, \ldots, N, \tag{14.1}$$

Where ε_0 is the permittivity of free space, E_k is the electric field strength and P_k is the dielectric polarization of the kth layer.

The current density j is composed of a conductive current and a displacement current, which are position independent in each layer. At the interface between neighboring layers, the current density is

$$j_k = j_{k+1} = \sigma_k E_k + \dot{D}_k = \sigma_{k+1} E_{k+1} + \dot{D}_{k+1}. \tag{14.2}$$

In a first approximation, the conductivity σ is supposed to be constant. In ferroelectrics the macroscopic polarization P varies nonlinearly with the applied electric field $E(t)$ and the time t. If the polarization responds to the electric field much faster than by any other factors, then the explicit time dependence can be neglected. Thus, the derivation with respect to time can be expressed as

$$\frac{\mathrm{d}P(E(t))}{\mathrm{d}t} = \frac{\partial P}{\partial E}\frac{\partial E}{\partial t} = P'(E)\dot{E}(t). \tag{14.3}$$

Equations (14.2) and (14.3) yield a system of $N-1$ differential equations, namely

$$\sigma_k E_k + \varepsilon_0 \dot{E}_k + P_k' \dot{E}_k = \sigma_{k+1} E_{k+1} + \varepsilon_0 \dot{E}_{k+1} + P_{k+1}' \dot{E}_{k+1}. \tag{14.4}$$

Since the RC elements are connected in series, the sum of the imposed voltages on each layer is equal to the applied voltage U_{app} in the multilayer structure. Differentiating the corresponding electric fields with respect to time results in

$$\dot{E}_{\mathrm{app}} = \frac{\dot{U}_{\mathrm{app}}}{d} = \sum_{k=1}^{N} \frac{d_k}{d}\dot{E}_k, \tag{14.5}$$

Where E_{app} is the applied electric field, d is the thickness of the multilayer structure, and d_k is the thickness of the kth layer. A special solution of this system of differential equations is the limiting case of saturation. The saturation is attained when the local electric field strength in all layers does not vary in time. Then the product of conductivity and electric field strength is constant for each layer and yields

$$\sigma_k E_k = \sigma_{k+1} E_{k+1}. \tag{14.6}$$

Equation (14.6) is approximately valid, if the values of polarization are within the saturated range and the conductive current densities are high in comparison with $\varepsilon_0 \dot{E}_k$ in both layers. Charges are collected at the interface between the layers to compensate the differences in the ferroelectric polarization, i.e., the differences in the dielectric displacement D_k. Consequently, the charge density q_k is calculated by

$$q_k = D_{k+1} - D_k. \tag{14.7}$$

Hence, the input parameters of the model are the conductivity and the ferroelectric hysteresis loop of each layer, and the applied electric field with respect to time. The output values are the electric field and the polarization within each layer, the interfacial charge density, and the current density through the multilayer structure.

In the literature, different models can be found those describe the dependence of the ferroelectric polarization on the applied electric field. Or et al. used in his work a mathematical model developed by Miller (Miller et al. 1990, 1991). Here, the slope of the polarization loop is calculated by

$$\frac{\partial P}{\partial E} = \left(1 - \tanh\sqrt{\frac{P - P_{\mathrm{sat}}}{\xi P_{\mathrm{s}} - P}}\right)\left(\frac{\partial P_{\mathrm{sat}}}{\partial E}\right), \tag{14.8}$$

where P_{s} is the magnitude of the spontaneous polarization and P_{sat} is the polarization of the saturated hysteresis loop, respectively. By using the values $+1$ or -1 for ξ, the behavior for increasing and decreasing electric field strength is simulated.

The disadvantage of this model has to be recognized: It makes an a priori assumption that saturation is reached in the hysteresis loop in each layer of the system to which it is applied. That, in turn, is an oversimplification and quite often one observes unsaturated loops in such multilayer structures because of excessive conduction losses.

A more flexible model is given by Pientschke (Pientschke et al. 2005) for describing multilayer structures using complete ferroelectric hysteretic behavior without considering first-principle calculations. They used the Preisach model (Preisach et al. 1935; Mayergoyz 1991), which provides an abstract approach for the calculation of hysteresis loops in general. Figure 14.2 shows saturated and nonsaturated polarization loops simulated by the Preisach model. The model is based on the weighted superposition of elementary hysteresis operators. The elementary hysteresis operator $\hat{\gamma}_{\alpha\beta}E$ depends on the input electric field E, which is rectangularly shaped, and characterized by an upper (α) and a lower (β) coercive field strength $(\alpha \geq \beta)$. Their output is an elementary polarization given by

$$\hat{\gamma}_{\alpha\beta}E = \begin{cases} +1, & \text{for}\, E > \alpha \\ -1, & \text{for}\, E < \beta \end{cases}. \tag{14.9}$$

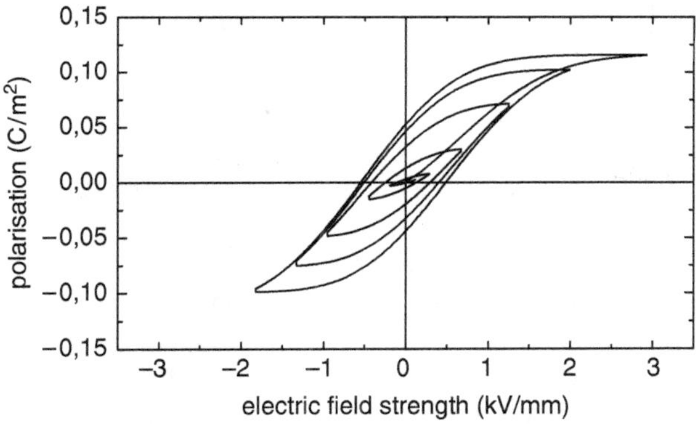

Fig. 14.2 Saturated and nonsaturated polarization loops simulated by the Preisach model

For electric fields between the limiting coercive fields, the last output of the elementary hysteresis operator is retained. The resulting hysteresis is constructed as a superposition of elementary hysteresis operators, according to

$$P = \iint_{\alpha \geq \beta} \mu(\alpha, \beta) \hat{\gamma}_{\alpha\beta} E \, d\alpha \, d\beta, \tag{14.10}$$

where $\mu(\alpha, \beta)$ is a weighting function that defines the shape of the resulting hysteresis loop. The output of the elementary hysteresis operator $\hat{\gamma}_{\alpha\beta}E$ depends on its previous output, which must be stored. Thus for numerical calculations, discretization of the electric field is necessary, which reduces the integrals in (14.10) to summations. Moreover, the derivative of $P'(E)$ is necessary in addition to polarization itself. Therefore, numerical methods are needed to determine the local slope of the $P(E)$ curve. For instance, a linear interpolation between the two adjacent grid points with higher or lower electric field can be used.

14.4 Different Concepts of Piezoelectric Graded Materials

As mentioned in the preceding sections, the purpose of utilizing graded piezoelectric materials is the reduction of mechanical stresses at the interface between materials with different electromechanical properties. A typical example of exploiting such an advantage offered by graded piezoelectrics is stress-reduced bending actuators (Hauke et al. 2000). In such actuators, the piezoelectric coefficients change from a high value to a low value or even to zero if a nonpiezoelectric layer is integrated. If the piezoelectric activity disappears, the material is usually still electrostrictive. In the case of bending actuators, it should be helpful to shift the sign of the piezoelectric coefficient to improve the bending deformation. Moreover, the performance of an actuator can be enhanced by matching the piezoelectric gradient with a dielectric one. Different concepts of monolithic graded materials will be discussed in what follows.

14.4.1 Combination of Piezoelectric and Electroconductive Ceramics

Bending actuators consisting of one piezoelectric active and one piezoelectric inactive layer are called unimorph or monomorph (Uchino 1997). A well-known representative of this group is Haertling's RAINBOW actuator as mentioned before (Haertling 1994). Electroconductive ceramics are not piezoelectric and have very small electrostrictive strains. Consequently, they can be used as an inactive layer for clamping the piezoelectric layer, thereby amplifying the piezoelectric deflection. The conductivity of $BaTiO_3$ ceramics can be substantially increased by doping with lanthanum (La^{3+}) – a donor dopant on the A site in the perovskite structure. At low concentrations ($\sim$0.2 mol % La^{3+}), the conductivity reaches about 10 $(\Omega m)^{-1}$. Thus, the La-doped layer can be used as a ceramic electrode since the voltage drop in this layer can be neglected in comparison to the $BaTiO_3$ layer. Hence, one may assume that the voltage across the piezoelectric layer is equal to the applied voltage.

The thickness of such a ceramic electrode influences the bending performance of the monolithic actuator. Therefore, samples were prepared with different thickness ratios of the piezoelectric layer (t_p) to the overall sample thickness t. The bending deflection of a one-side fixed bending actuator was measured with a capacitive displacement sensor (Hauke et al. 2000). If the applied electric field is constant, the bending deflection was found at a maximum if $r = t_p/t = 0.5$. This result can be confirmed for low electric field strengths by linear finite element modeling (FEM) (see inset in Fig. 14.3). The voltage applied to samples with a thinner piezoelectric layer is lower than that in samples with a higher ratio r to impose the same electric field strength. Figure 14.3 shows the bending deflection normalized to the applied voltage since the applied voltage is one of the significant parameters in applications. If the applied electric field strength increases, a nonlinear behavior is observed and the bending deflection increases.

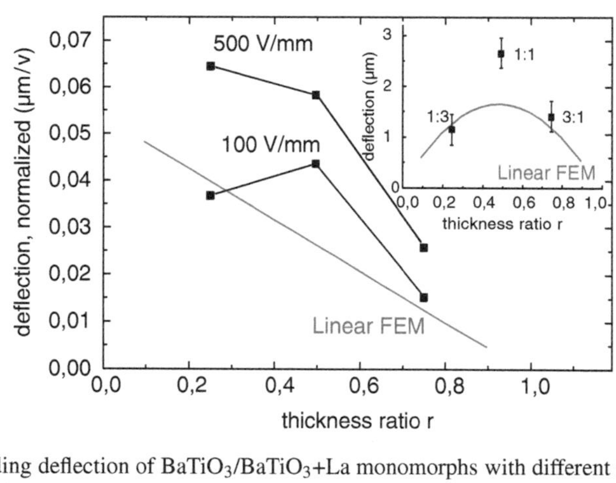

Fig. 14.3 Bending deflection of $BaTiO_3/BaTiO_3$+La monomorphs with different thickness ratios, normalized to the applied voltage (inserted: bending deflection vs. thickness ratio at a constant applied electric field)

14.4.2 Combination of Piezoelectric and Electrostrictive Ceramics

The electromechanical strain is the product of the piezoelectric coefficient and the electric field strength. Therefore, the performance of a graded piezoelectric actuator can be improved by an additional electric field gradient. This in turn requires that there is a gradient in permittivity. In the region of low permittivity, the higher voltage is applied. As a consequence, the induced electric field strength is lower in the layer with higher permittivity. Therefore, the gradient of the permittivity should be opposite to the piezoelectric gradient.

In the BTS system, the dielectric and piezoelectric coefficients strongly depend on the Sn content as shown in Fig. 14.4. In the region of interest (7.5–15 mol%), the piezoelectric coefficient d_{33} decreases and the dielectric permittivity increases with increasing Sn content.

The piezoelectric gradient in the chemically graded material should be induced by the poling process as discussed before. Hence, it is necessary to consider the polarization-electric field hysteresis loops of all ferroelectric compositions in their virgin state, also known as "electrical zero". The BTS7.5 ceramics have a rhombohedral symmetry, while BTS15 is a composition near the morphotropic phase boundary and has cubic phase symmetry (Yasuda et al. 1996). Correspondingly, the dielectric properties change from ferroelectric to paraelectric state as one crosses from one layer to another. The hysteresis loops of virgin BTS ceramics with different Sn content are shown in Fig. 14.5. The ferroelectric ceramics BTS7.5 exhibit a higher maximum polarization, and expectedly the $P(E)$ curve has a hysteretic shape. The field dependence of polarization of BTS15 is also nonlinear, but nearly nonhysteretic as it is not ferroelectric (no domains).

In the following, poling of a two-layer system will be discussed, which constituted a simple but yet instructive example. A DC electric field is applied to the system to pole it. The voltage is switched on and off by a ramp (see Fig. 14.6a). The local electric field strength in the layer with lower maximum polarization rises to a

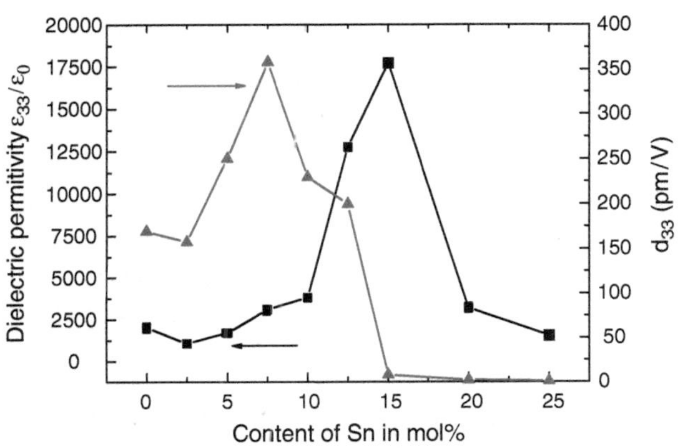

Fig. 14.4 Permittivity and piezoelectric coefficient d_{33} of $Ba(Ti,Sn)O_3$ ceramics

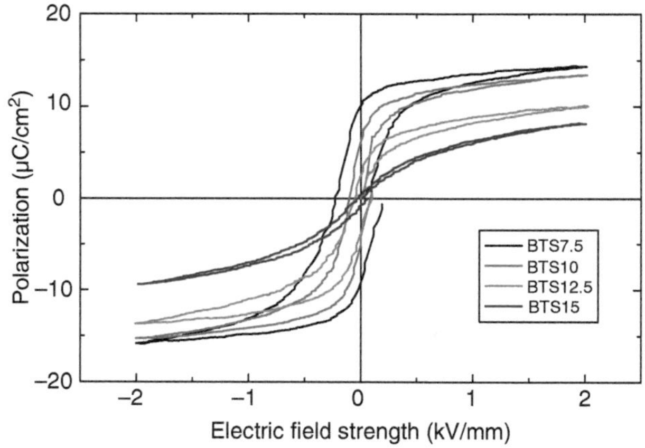

Fig. 14.5 Virgin loops of $Ba(Ti,Sn)O_3$ ceramics in dependence on the Sn content

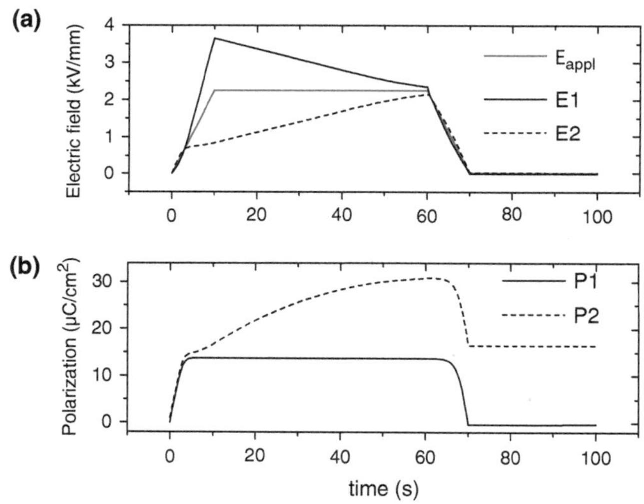

Fig. 14.6 Time dependence of the electric field strength, the poling field (**a**), and the local local polarization (**b**) in the two-layer system

maximum value very steeply, and then decreases slowly. In the other layer, the local electric field increases slowly to a saturation value. When saturation is reached, the electric field strength in each layer does not depend on time. The ratio of the saturated values of the electric field in both layers depends on the electrical conductivity of each material (see (14.6)), where it was assumed that the conductivity was approximately equal in each layer.

The monolithic BTS ceramics with a gradient of Sn content were poled with a electric field strength of 2 kV/mm for 4 min. In real ceramics, the conductivity slightly depends on the Sn content. The data shown in Fig. 14.7 (inserted) were

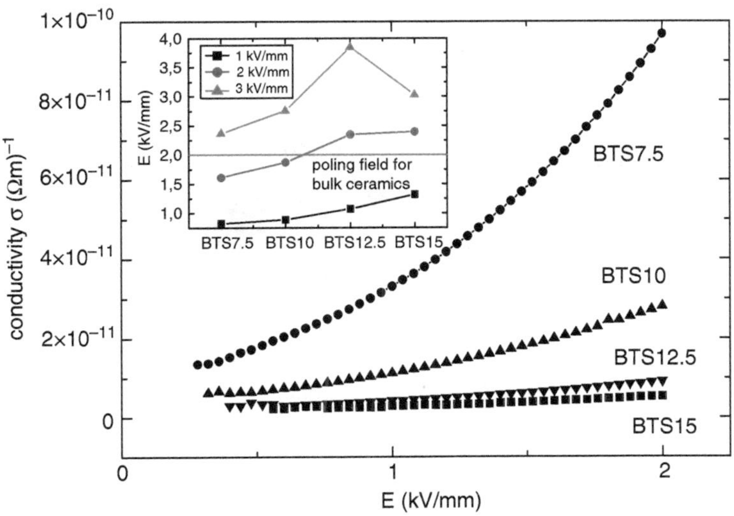

Fig. 14.7 Electrical conductivity of Ba(Ti,Sn)O_3 ceramics in dependence of the applied voltage. Inserted: Electric field distribution in a four-layer model structure consisting of BTS7.5-10-12.5-15

measured with a high resistance-meter. The conductivity increased with increasing voltage for all samples. Hence, one has to conclude that the constant-conductivities approach can provide qualitative results at best. The distribution of the electric field strength was measured at glued model structures with the same chemical composition and inner electrodes (Steinhausen et al. 2004). The electric field distribution of a four-layer sample (see Fig. 14.7) corresponds to field strengths up to 2 kV/mm, and relates to the measured data of the conductivity in consideration of (14.6).

After switching off the poling field, a relaxation of the local fields and of the polarization was observed. The polarization of the paraelectric material tends to zero as the field is switched off. On the other hand, in the ferroelectric layer a remnant polarization remains (see Fig. 14.6b). Consequently, the polarization varies from the remnant polarization to zero across a chemically graded material.

The polarization distribution in a monolithic graded ceramics across the thickness was characterized by the thermal wave method (Movchikova et al. 2008). The measured pyroelectric coefficient profile is shown in Fig. 14.8. The sample was prepared from four powder layers with BTS7.5, BTS10, BTS12.5, and BTS15, respectively. The induced polarization gradient is proportional to the pyroelectric signal. For comparison, the remnant polarization of the corresponding ceramics with the same chemical composition are plotted in arbitrary units (red lines). The obtained gradient is smoother than it was expected from the composition gradient (see Fig. 14.1). Furthermore, the degree of polarization of the bottom side with the low Sn content and the highest piezoelectric activity is not optimal.

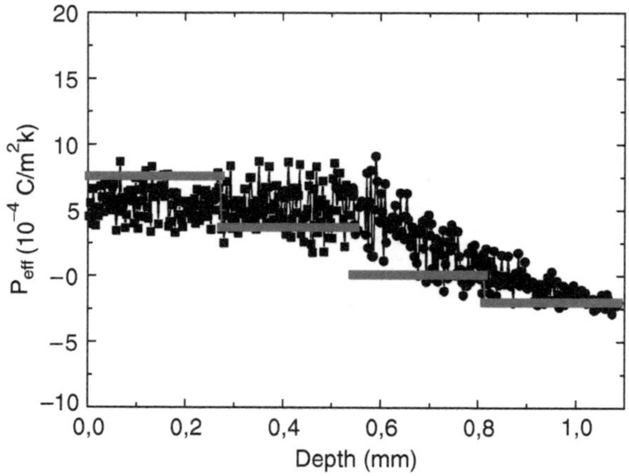

Fig. 14.8 Thickness profile of the pyroelectric coefficient of a monolithic BTS sample with a gradient of Sn (7.5 mol% at the *left*, 15 mol% at the *right* side)

14.4.3 Combination of Piezoelectric Hard and Soft Ceramics

In the first two concepts, the one side of the actuator consists of a piezoelectric non-active or a low-activity layer, with a poled piezoelectric layer on the opposite side. Usually, the driving field is applied in the poling direction of the bending actuator. In the opposite direction, the driving field is limited by the coercive field as the depolarization or switching of the polarization in the piezoelectric layer will partly deteriorate the performance of the actuator, and in the extreme case disintegrate it. Furthermore, the combination of ferroelectric and electrostrictive ceramics shows a nonlinear bending behavior when the driving field is applied in the opposite direction. The second-order electrostrictive effect causes a positive strain for negative fields. Consequently, the difference of the strain of bottom and top layer of the actuator slightly decreases with increasing negative applied voltage. For the negative voltages, the result is a nonlinear response of the bending loop (Fig. 14.9). The effect is reversible as long as the local field strength in the piezoelectric layer remains below the coercive field.

Higher bending deflections can be achieved if one side of the actuator contracts and the other side expands at the same time once the driving field is imposed. This can be accomplished if the layers can be poled in opposite directions. The idea of a polarization gradient with changing signs can be realized by using ferroelectric soft and hard materials. The soft material is characterized by a low coercive field strength and high spontaneous polarization, while the hard material exhibits a higher coercive field and lower polarization.

However, the poling process for such a system is more complicated. The applied voltage and the local electric field in the soft and hard material as well as the local polarization are shown in Fig. 14.10. A positive voltage is applied to the bilayer system. A ramp was used in the calculations to avoid numerical instabilities. It should

Fig. 14.9 Bending loops of a monolithic BTS actuator prepared from four powder layers

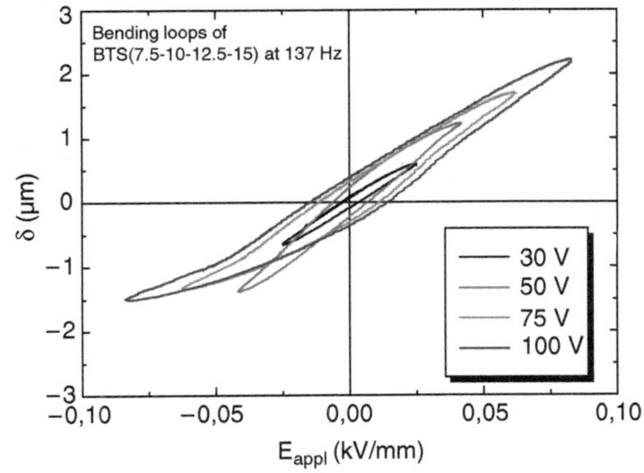

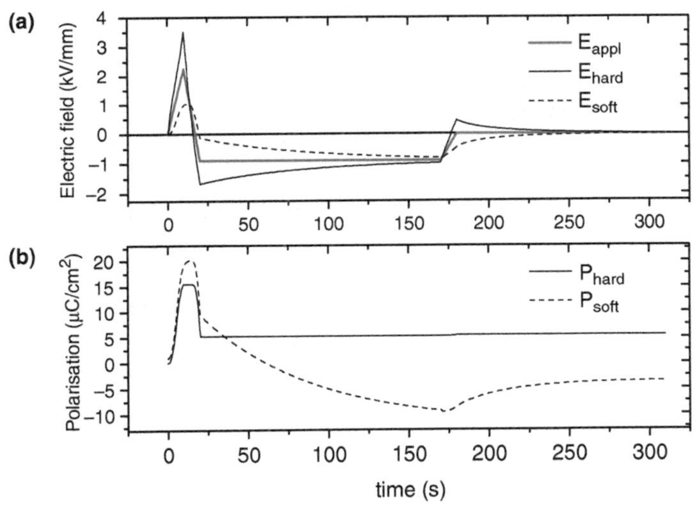

Fig. 14.10 Time-dependence of the local electric field (**a**) and of the polarization (**b**) in hard and soft layers during the poling process

also be noted that a DC voltage, applied for a short time, can be more useful. The poling time and the maximum value of the poling field should be high enough to completely pole a hard ferroelectric material.

Then the applied voltage is slowly reduced, the rate of which depends on the conductivity of the materials, i.e., the higher the conductivity, the faster the voltage decrease. If the conductivity is lower, more time is required to collect space charges at the interface, which compensate the difference in the local polarization. The electrical field strength is then reduced to a negative value. However, its absolute value should be higher than the coercive field strength of the soft material to switch its polarization, and yet lower than the coercive field strength of the hard material to protect the positive polarization in this layer. The negative DC voltage is applied

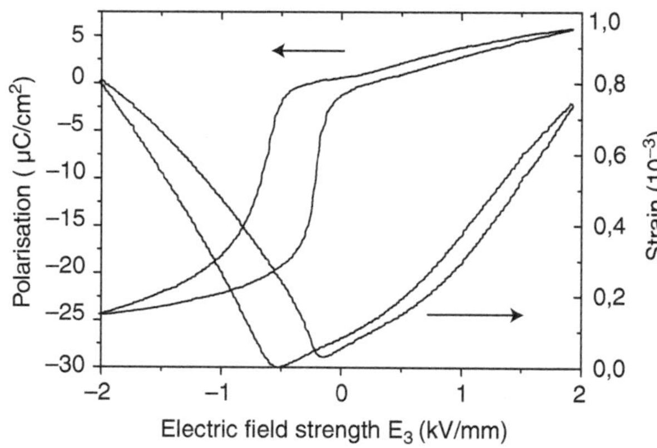

Fig. 14.11 Polarization and strain of poled $BaTiO_3$+Mn vs. electric field strength

for a longer time to permit the switching of polarization. The duration depends primarily on the conductivity of the materials. When the saturation is achieved, the voltage is switched off. The remnant polarization shows different signs after a short relaxation time.

Ferroelectric soft and hard materials based on $BaTiO_3$ are needed for preparing a gradient from positive to negative polarization. In a particular study, $BaTiO_3$ doped with 0.5 mol% Mn (BTMn) was used instead of a hard material. The sample was poled with 2 kV/mm, for 30 min, at 80°C. One day after poling, it was found that the strain loop has a small ferroelectric hysteresis with a shift to negative voltage (see Fig. 14.11). Thus, the piezoelectric coefficient ($d_{33} = \delta S_3/\delta E_3$) at small voltages remains positive after applying a high field strength of 2 kV/mm for a short time. The strain loop of the homogeneous ceramics was measured at 10 Hz by a capacitive displacement sensor for comparison.

A two-layer system can be used as a model structure to investigate the poling process of a monolithic graded ceramics and to characterize the local piezoelectric coefficients. Two single layers of $BaTiO_3$ and $BaTiO_3$–0.5 mol% Mn were connected by a wire and poled together. This allowed measuring the piezoelectric coefficients (d_{33}) of each single layer several times during the poling process. The poling field of 2 kV/mm was applied for 30 min. Negative electric field strength of 2 kV/mm was applied for a few seconds to switch the polarization in the soft ferroelectric layer. In Fig. 14.12, the strain of the $BaTiO_3$ and BTMn layers is shown depending on the applied voltage. The slope of the curve is equal to the piezoelectric coefficient d_{33}. After 20 s, the $BaTiO_3$ layer was nearly completely depolarized, implying that the absolute value of $d_{33,BT}$ was about zero. The piezoelectric coefficient of BTM decreased to half of the original value. After poling with −2 kV/mm for another 30 s, the polarization in the $BaTiO_3$ layer switched back, while the polarization in the BTM layer remained the same. This experiment has demonstrated that one could indeed prepare a piezoelectric gradient in which the sign of d_{33} switches from positive to negative across its thickness.

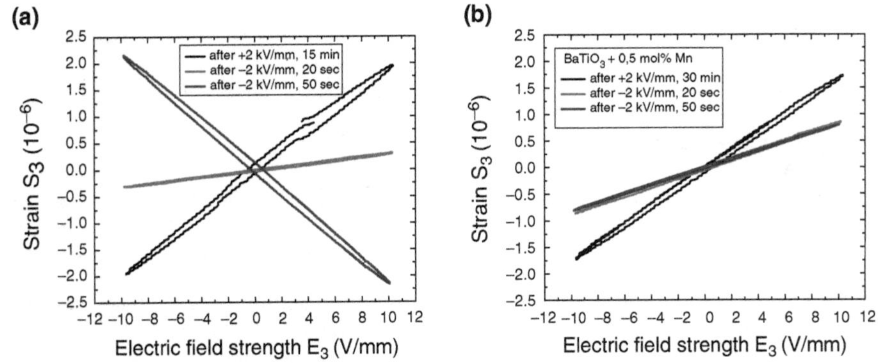

Fig. 14.12 Linear field dependence of the strain of the $BaTiO_3$ (**a**) and $BaTiO_3$ + Mn (**b**) layer for small voltage during the polarization switching

14.5 Summary and Conclusion

Monolithic ceramics with a gradient in piezoelectric properties along the thickness can be used for bending actuators with reduced internal mechanical stresses. In a first preparation step, a chemical gradient should be produced in the monolithic ceramic. A relatively simple method to obtain such gradients is the so-called powder pressing technique, where powder layers of different chemical composition are pressed together and sintered. The gradient in chemical composition, as obtained by sintering, results in the desired piezoelectric gradient once the specimen is poled. The poling process can be theoretically analyzed as a multilayer structure wherein each layer has a different electric conductivity. The bending deflection of such actuators depends on the difference of the induced strain in the top and bottom layers. Three types of material combinations based on pure and doped $BaTiO_3$ were presented, which have proven to be very effective. The combination of piezoelectric and electroconductive materials yield an actuator with good bending performance. However, it is difficult to prepare a continuous piezoelectric gradient, i.e., step-function profile. A smooth gradient in piezoelectric properties could be prepared if an actuator consisting of a piezoelectric and electrostrictive ceramics is formed.

Finally, it was shown that an actuator comprised of an electrically hard and soft piezoelectric material could also be fabricated. A special poling regime was discussed to induce a piezoelectric gradient, where the polarization, and consequently, the piezoelectric coefficient change their sign from the top to the bottom side of the actuator.

References

Arlt G and Pertsev NA (1991) Force constant and effective mass of 90° domain walls in ferroelectric ceramics. J. Appl. Phys. 70(4):2283–2289

Arlt G, Hennings D, de With G (1985) Dielectric properties of fine-grained barium titanate ceramics. J. Appl. Phys. 85(4):1619–1625

Haertling GH (1994) Rainbow ceramics – A new type of ultra-high-displacement actuator. Am. Ceram. Soc. Bull. 73(1):93

Hall A, Allahverdi M, Akdoğan EK, Safari A (2005) Piezoelectric/electrostrictive multimaterial PMN–PT monomorph actuators. J. Eur. Cer. Soc. 25:2991–2997.

Hauke T, Kouvatov AZ, Steinhausen R, Seifert W, Beige H, Langhammer HT, Abicht H (2000) Bending behavior of functionally gradient materials. Ferroelectrics 238:195–202

Jaffe B, Cook WR, Jaffe H (1971) Piezoelectric Ceramics. Academic Press, London

Jeon JH, Hahn YD, Kim HD (2001) Microstructure and dielectric properties of barium–strontium titanate with a functionally graded structure. J. Eur. Cer. Soc. 21:1653–1656

Mayergoyz ID (1991) Mathematical Models of Hysteresis. Springer, New York

Miller SL, Nasby RD, Schwank JR, Rodgers MS, Dressendorfer PV (1990) Device modeling of ferroelectric capacitors. J. Appl. Phys. 68:6463–6471

Movchikova AA, Malyshkina OV, Suchaneck G, Gerlach G, Steinhausen R, Langhammer HT, Pientschke C, Beige H (2008) Study of the pyroelectric behavior of $BaTi_{1-x}Sn_xO_3$ piezoceramics. J. Electroceramics 20:43–46

Or YT, Wong CK, Ploss B, Shin FG (2003) Polarization behavior of ferroelectric multilayered composite structures. J. Appl. Phys. 93(7):4112–4119

Pientschke C, Steinhausen R, Kouvatov A, Seifert W, Beige H (2005) Polarisation kinetic of electrically connected electroconductive ferroelectric multilayer structures. J. Eur. Cer. Soc. 25:2547–2551

Preisach F (1935) Ueber die magnetische Nachwirkung. Zeitschrift fuer Physik 94:277–302

Sakamura J, Yamada K, Nakamura K (2000) Equivalent network analysis of functionally graded piezoelectric transducers. Jpn. J. Appl. Phys. 39:3150–3151

Shvartsman VV, Emelyanov AY, Kholkin AL, Safari A (2002) Local hysteresis and grain size effect in $Pb(Mg_{1/3}Nb_{2/3})O_3$–$PbTiO_3$ thin films. Appl. Phys. Lett. 81(1):117–119

Steinhausen R, Kouvatov AZ, Pientschke C, Langhammer HT, Beige H (2004) Modelling of the poling process in functionally graded materials. In: Proc. IEEE Int. UFFC Joint 50th Anniversary Conf., Montreal, 118–121

Uchino K (1997) Piezoelectric actuators and ultrasonic motors. Kluwer, Boston

Wang SF, Kumar U, Huebner W, Marsh P, Kankul H, Oakley CG (1992) Grain size effect on the induced piezoelectric properties of 0.9PMN–0.1PT ceramic. In: Proc. of the Eighth IEEE ISAF, 148–151

Yasuda N, Ohwa H, Asano S (1996) Dielectric properties and phase transitions of $Ba(Ti_{1-x}Sn_x)O_3$ solid solution. Jpn. J. Appl. Phys. 35:5099–5103

Yasuoka M, Shirai T, Nishimura Y, Kinemuchi Y, Watari K (2006) Influence of microwave irradiation method on the sintering of barium titanate with liquid phase. J. Cer. Soc. Jpn. 114(5):377–379

Chapter 15
Robocasting of Three-Dimensional Piezoelectric Structures

James E. Smay,* Bruce Tuttle, and Joseph Cesarano III

Robocasting is a solid freeform fabrication (SFF) technique based on the direct writing of highly concentrated colloidal gels. Similar to other SFF techniques, robocasting offers facile assembly of complex three-dimensional geometries and a broad pallet of ceramic, metallic, and polymeric materials from which to design devices. Here, robocasting is used to assemble lead zirconate titanate (PZT) skeletons for direct use or to create epoxy-filled composites suitable for hydrostatic piezoelectric sensors. PZT composites labeled by connectivity (3–3, 3–2, 3–1) are demonstrated. For such composites, the piezoelectric hydrostatic figure of merit ($d_h\, g_h$) increases by up to 60-fold compared with bulk PZT. Ultimately, SFF allows for rapid iteration, prototyping of designs, and assembly of unique PZT skeletons. Sandia is a multiprogram laboratory operated by Sandia Corporation, a Lockheed Martin Company, for the U.S. Department of Energy's National Nuclear Security Administration under contract DE-ACO4 – 94AL85000.

15.1 Introduction

Traditional forming technologies for lead zirconate titanate (PZT)–polymer composites (Tressler et al. 1999) such as injection molding, dicing of monolithic blocks, and manual stacking of green and presintered ceramic elements are all viable routes to assemble complex structures; however, these processes may be labor intensive or suited for mass production and may be limited in the ability to explore theoretical designs for optimized figures of merit. Hence, solid freeform fabrication (SFF, also known as rapid prototyping) technologies offer a new alternative for device prototyping and may allow for the assembly of structures that are nonmanufacturable by traditional ceramic processes. Here, a SFF process known as robocasting

J.E. Smay
School of Chemical Engineering, Oklahoma State University, Stillwater, OK 74078, USA

A. Safari, E.K. Akdoğan (eds.) *Piezoelectric and Acoustic Materials for Transducer Applications.*

(Cesarano III et al. 1997; 1998) is described in the context of assembling complex electroceramic structures (Smay et al. 2002a,b; Tuttle et al. 2001).

This chapter first addresses the processing technology, including ink design, (Cesarano III et al. 1997; Smay et al. 2002a,b) machine operation, variants of the process, and the capabilities and limitations of robocasting (Cesarano III 1999; Cesarano III et al. 2000; Lewis et al. 2006). In the "Device Design" section, examples of PZT–polymer and porous-solid multilayer PZT composite structures are presented. Electrical and electromechanical characterization of these composite structures is then compared with those of similar devices formed by traditional processes. The chapter is then summarized and perspectives about the future for robocasting are given.

15.2 Processing Technology

In the robocasting process, a filament of ink (i.e., ceramic paste) is continuously extruded and deposited as a nozzle moves relative to a substrate. In Fig. 15.1, the ink reservoirs dispense ink at a rate proportional to the desired filament diameter and linear deposition speed. Discrete nozzle(s) may dispense one or more ink compositions or a mixing nozzle may dispense gradients of ink compositions.

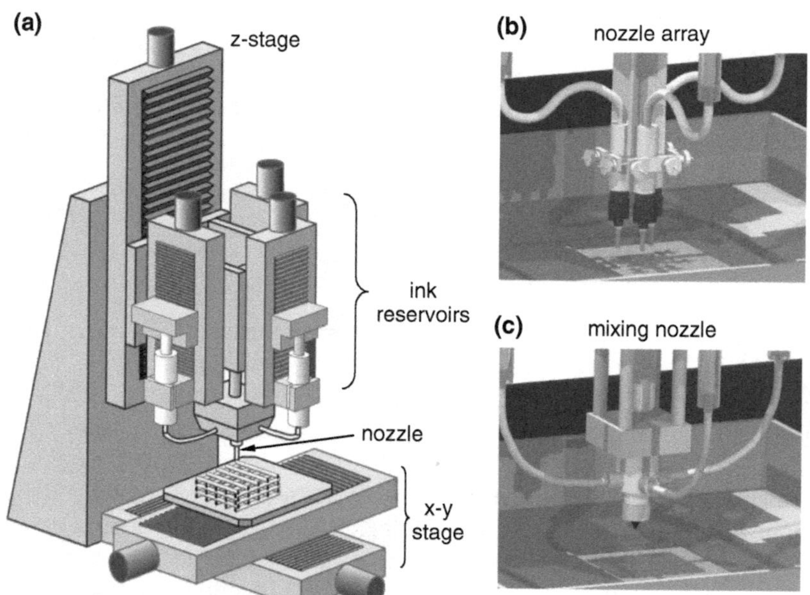

Fig. 15.1 (**a**) Schematic view of an x–y–z printing robot with three ink reservoirs (syringes) attached to the z axis. Each syringe is actuated by an independent syringe pump and the effluent maybe (**b**) deposited discretely through independent nozzles or (**c**) mixed and deposited through a common orifice

Table 15.1 Representative robocasting process parameters

Minimum tip diameter (D)	30 μm (Li and Lewis 2003)
Wall roughness	8 μm (Smay et al. 2004)
Linear write speed	4–20 mm/s
Layer thickness	$0.8D$
Print time for parts shown	<15 min
Drying time (convection oven)	≈ 20 min
Sintering ramp	$5–10^\circ\mathrm{C/min}$

The substrate may be a flat plate (plastic, ceramic, metal), previously deposited ink, or a curved object. The ink must flow through the nozzle with relative ease, but maintain a cylindrical cross section in the face of deforming stresses such as gravity-induced flow, surface tension, and capillary pressure during drying. Three-dimensional structures evolve as the ink filament(s) is patterned in two-dimensional layers that are printed one atop the other. The modus operandi is nearly identical to fused deposition modeling (FDM), with the exception of the ink composition and a staged solidification scheme. In robocasting, the ink may be deposited in air(Cesarano III 1999; Cesarano III et al. 1997, 1998, 2000) or in an oil bath to temporally separate the printing and drying steps (Smay et al. 2002a–c). Typical process properties are listed in Table 15.1.

The ink is a concentrated colloidal gel containing volume fractions of submicron solid particles in the range 40–60%, 1–5% polymeric processing aids, with the remaining fraction being solvent (typically water). Detailed formulations for a wide variety of technical ceramics have been described in the literature (Cesarano III 1999; Cesarano III et al. 1997, 1998, 2000, 2005; Dellinger 2005; Dellinger et al. 2006; He et al. 2000, 2001; Lewis 2002; Lewis and Gratson 2004; Nadkarni and Smay 2006; Smay et al. 2002a–c, 2006; Stuecker 1999; Stuecker et al. 2003; Tuttle et al. 2001; Xie et al. 2006). The ink formulation process consists of first creating a high concentration slurry by adsorbing a poly(electrolyte) onto the particle surfaces in the aqueous medium for dispersion. Next, rheology modifiers such as cellulosic thickeners are added in low concentration and finally the fluid slurry is gelled by weak flocculation of the particles using bridging flocculation, electrostatic screening, or solvent quality changes.

The ink rheology is best described as shear thinning with yield stress, or Herschel-Bulkley behavior (Smay et al. 2002a–c; Stuecker 1999; Stuecker et al. 2003). Yield stress, elastic shear modulus, and shear thinning exponent are tailored for the type of the structure desired. Figure 15.2 represents four primitive shapes in robocasting. In Fig. 15.2a, ink filaments should flow somewhat after deposition such that they fill the interstitial spaces. In Fig. 15.2b, a gap between adjacent filaments leads to the assembly of high aspect ratio walls. Higher yield stress prevents buckling, but some flowability helps minimize undulations in the wall (Cesarano III 1999; Cesarano III et al. 1997, 1998). In Fig. 15.2c, the layers are rotated by an angle $\theta = 90^\circ$ relative to the previous layer. The structure evolves as a three-dimensional periodic lattice. The ink must have a significant yield stress and rapid

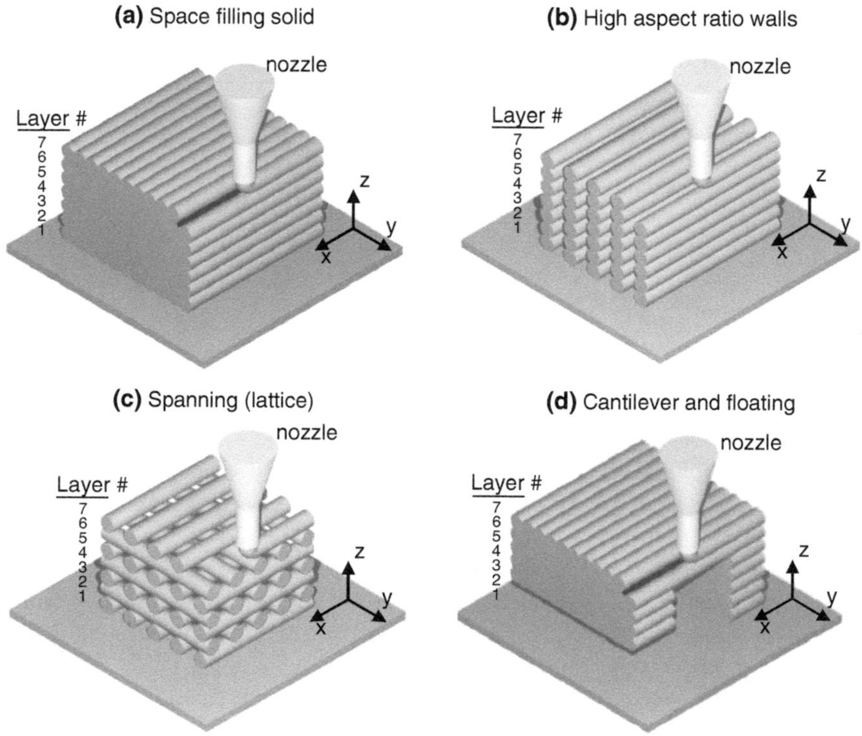

Fig. 15.2 Primitive layer patterns encountered during robocasting. (**a**) Space-filling solid where filaments touch to form continuous layers. (**b**) High aspect ratio walls formed by stacking filaments that do not touch in the layers. (**c**) Spanning structure formed by stacking layers of nontouching filaments rotated by 90°. (**d**) Cantilevered and floating elements are created when layers are unsupported by other ceramic layers

recovery of elastic properties after extrusion, yet be fluid enough to fuse at the intersections between filaments. In Fig. 15.2d, the structure contains cantilevered elements. Such structures require the use of fugitive support elements. This is an area of active development and will not be discussed further here. The detailed relation between shear rheological properties and as-printed span geometry was investigated thoroughly for $Pb(Zr_{0.53}Ti_{0.47})O_3$ colloidal gels (Smay et al. 2002a,b).

15.3 Device Design

The $Pb(Zr_xTi_{1-x})O_3$ examples highlighted here include a layered composite of $Pb(Zr_{0.95}Ti_{0.05})O_3$ (PZT 95/5) fabricated to probe effects of porosity on ferroelectric-to-antiferroelectric shock-induced transformations, linear and annular PZT arrays, and a variety of structures with periodic 3D lattices.

15.3.1 PZT 95/5

Monoliths and composites made of PZT 95/5 chemically prepared powder were prepared by single- and dual-nozzle robocasting, respectively (Tuttle et al. 2001). PZT 95/5 suddenly depoles at the ferroelectric to antiferroelectric transformation pressure and depends on temperature and microstructure of the ceramic. Previous experimentation revealed that regions with higher than expected density may cause higher depolarization pressures than those expected. Using robocasting, polyethylene spheres of diameter ranging from 5–100 μm was blended into a 45% by volume PZT ink and deposited along with pure inks to form a trilayer 2–2 composite, where the top and bottom layers contained porosity and the middle layer did not as illustrated in Fig. 15.3.

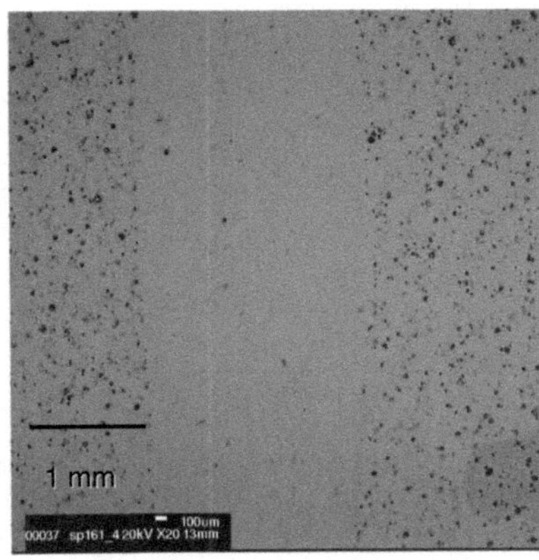

Fig. 15.3 Cross-sectional SEM image of a PZT 95/5 ceramic–ceramic composite with dense (96.1% of theoretical) and porous (93.9% of theoretical) layers prepared by robocasting of two discrete ink compositions

15.3.2 Linear and Annular Arrays

A soft piezoelectric ceramic, PZT-5H (Morgan Matroc, Electro Ceramics Division, Bedford, OH), was used to assemble high aspect ratio wall structures in the form of linear and annular arrays. For both the linear and annular arrays shown here, the deposited filament diameter was 200 μm and the ink contained 47% by volume PZT particles. The volume fraction of PZT in the structure as a whole was varied by controlling the spacing between linear or circular elements. Example structures are pictured in Fig. 15.4 and the corresponding parameters for the linear and annular arrays are given in Table 15.2. After sintering the arrays, the spaces between the high aspect ratio walls were filled with epoxy and cured to create a PZT–polymer composite. An example cross section of sample LA-2 from Fig. 15.4b is shown in Fig. 15.5 along with a measurement of the sidewall roughness due to the layered build sequence.

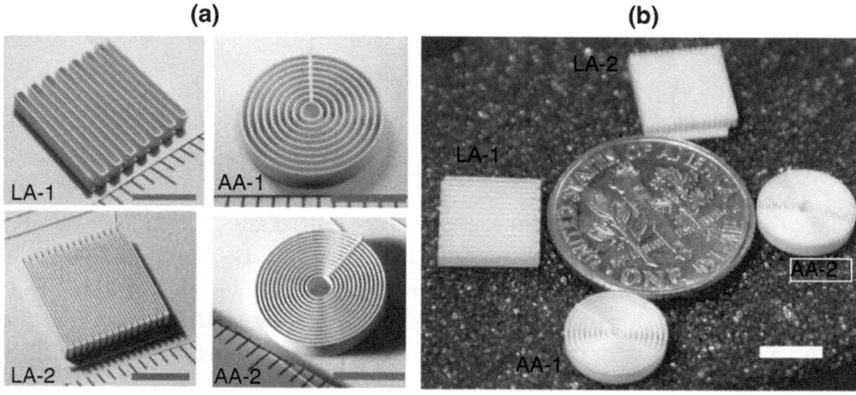

Fig. 15.4 Optical micrographs of (**a**) as-dried and (**b**) sintered linear and annular PZT arrays. All scale bars equal 5 mm

Table 15.2 Design parameters for PZT linear and annular arrays

Component	Height (mm)	Length or outer radius (mm)	Width or inner radius (mm)	Element spacing (mm)
LA-1	1.80	10	10	0.5
LA-2	1.80	10	10	0.3
AA-1	1.08	10	0.75	0.5
AA-2	2.08	10	1.0	0.3

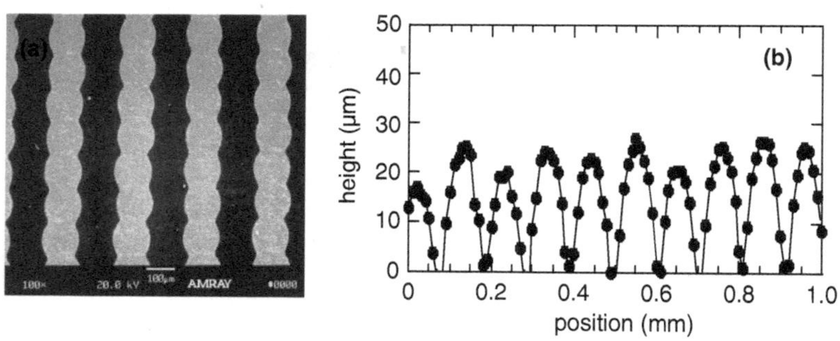

Fig. 15.5 (**a**) SEM micrograph of LA-2 composite cross section and (**b**) profilometry scan taken along one of the side (vertical) walls prior to epoxy infiltration

15.3.3 Lattice Composites

PZT5H ink was used to print three-dimensional, periodic lattices. An initial set of 3–3 composites were assembled by printing sets of orthogonal lines in adjacent layers, each contained within a circular perimeter as pictured in Fig. 15.6. After printing, drying, and sintering, the pore space of the lattices was infiltrated with epoxy and cured. The filament diameter was 160 μm after sintering. Figure. 15.6d reveals the continuous pillars of PZT formed at the overlap points of the lattice fila-

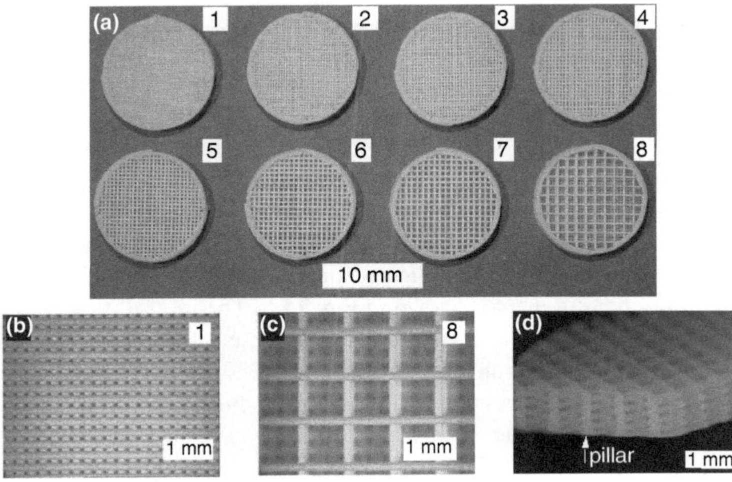

Fig. 15.6 Optical images of the (**a**) sintered PZT structures of varying lattice (rod) spacing, (**b**) and (**c**) top views of samples 1 and 8, respectively, and (**d**) cross-sectional view of sample 6

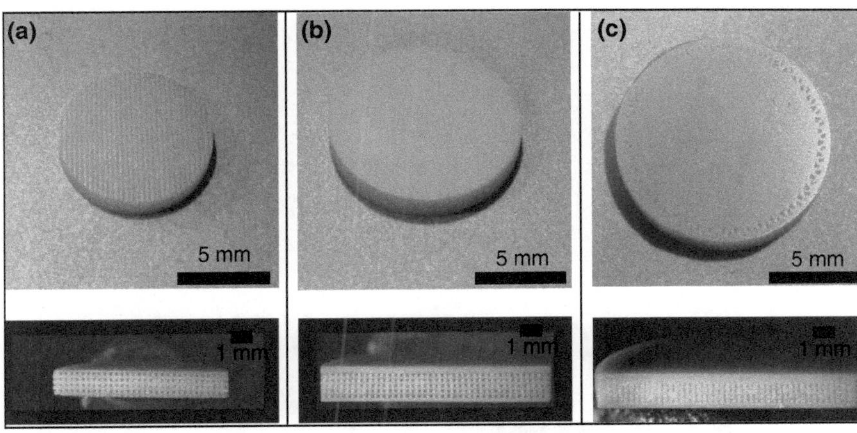

Fig. 15.7 Optical images of the top and cross-sectional views of (**a**) 3–3, (**b**) 3–2, and (**c**) 3–1 composites after polymer infiltration and machining. [Note: The ϕ_{PZT} in the lattice portion is ~0.6 for each sample]

ments. These pillars are the main load baring elements when stress is applied to the planar faces of the disk-shaped samples. Two modifications to the 3–3 lattice were also fabricated. First, solid PZT plates covering the planar faces were integrated into the print pattern, and second, solid PZT plates and a circumferential PZT ring were integrated into the design. The intent of the PZT plates was to promote uniform surface deflections of the underlying 3–3 lattice. These modifications limited the complete connectivity of the epoxy phase through the entire sample and the samples were dubbed 3–2 and 3–1, respectively. The isometric and cross sectional views of the 3–3, 3–2, and 3–1 samples are shown in Fig. 15.7.

15.4 Characterization

15.4.1 PZT 95/5

For the $Pb(Zr_{0.95}Ti_{0.05})O_3$ 2–2 layered composite, samples were encapsulated in urethane and subjected to an isostatic pressure cycle in pressure-transmission fluid from atmospheric pressure to ~500 MPa at a ramp rate of 10.3 MPa/s. The charge released upon depolarization was collected and measured by an 8-μF capacitor and a representative measurement is shown in Fig. 15.8. The inset illustrates the orientation of the dense and porous regions relative to the electrodes. Major depolarization took place between 250 and 350 MPa. Between 300 and 310 MPa, a change in depoling rate with applied pressure was attributed to the onset of ferroelectric to antiferroelectric transition in the dense slab of PZT 95/5. The total charge released by the sample and the fraction of that charge released by the dense region were in good agreement with predictions (Tuttle et al. 2001). The PZT 95/5 structures survived multiple poling–depoling test cycles without noticeable damage.

15.4.2 Linear and Annular Arrays

The epoxy-filled samples were electroded with the standing walls being orthogonal to the electrodes and poled at 25 kV/cm for 30 min, followed by short circuiting for 1 h. The dielectric constant (designated as K_{33} for the bulk material, and $\bar{K}_{33}$ for composites) was measured using an impedance analyzer (HP 4284A precision

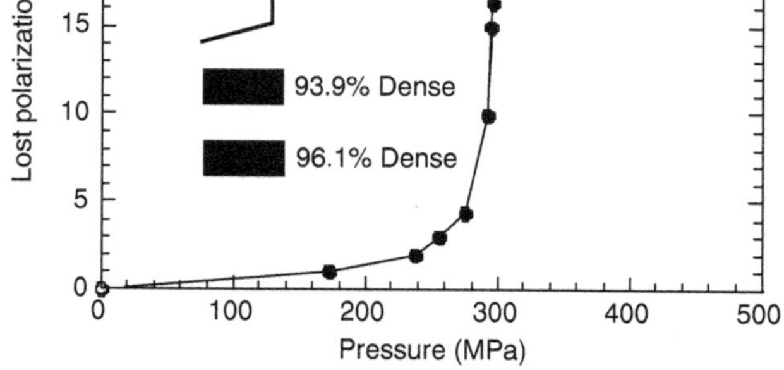

Fig. 15.8 Hydrostatic depoling characteristic of PZT 95/5 composite shown in Fig. 15.3. A schematic of the electrical test configuration is shown inset

Table 15.3 Electrical properties of PZT disks and 2–2 PZT–polymer composites

Component	ϕ_{PZT}	K_{33} or $\bar{K}_{33}$ at 1 kHz		P_r ($\mu C/cm^2$)	d_{33} or $\bar{d}_{33}$ (pC/N) at 100 Hz	
		Measured	Calculated		Measured	Calculated
CIP	1	2,991	–	31.5	627	–
RC	1	2,713	–	31.8	672	–
LA-1	0.31	831	841	7.1	388	633
LA-2	0.55	1,492	1,492	15.5	330	655
AA-1	0.39	1,081	1,058	10.0	496	643
AA-2	0.68	1,839	1,844	17.9	466	660

LCR meter, Yakogama-Hewlett-Packard, Tokyo, Japan) at a frequency of 1 kHz and field strength of 1 V/mm. Remanent polarization was measured using a ferroelectric tester (RT6000-HVS, Radiant Technologies, Albuquerque, NM) and piezoelectric coefficient of composites ($\bar{d}_{33}$) was measured with a Berlincourt meter (CPDT3300, Channel Industries, Santa Barbara, CA) with hemispherical contacts. For reference, robocast and isopressed monoliths were also tested. The measurement results are summarized in Table 15.3. Relative permittivity (ε_r), remanent polarization (P_r), and piezoelectric constant were in good agreement between the cold isostatically pressed (CIP) and the robocast (RC) samples.

The values in Table 15.3 indicate that ε_r scales in direct proportion to the volume fraction PZT (ϕ_{PZT}) and agrees well with calculation using a parallel dielectric model. Likewise, P_r scaled roughly in proportion to ϕ_{PZT}. The $\bar{d}_{33}$ coefficient deviated from expectations. The point contacts of the Berlincourt meter and the thin sample size ($\approx$1 mm) contributed to a less than predicted value of $\bar{d}_{33}$ due to nonuniform surface deflections. Nevertheless, a significant piezoelectric signal was measured from these samples, indicating they may be useful as ultrasonic arrays.

15.4.3 Lattice Composites

Characterization for 3–3, 3–2, and 3–1 samples included measurement of the composite dielectric constant ($\bar{K}_{33}$), the composite unidirectional piezoelectric constant ($\bar{d}_{33}$), and the composite hydrostatic piezoelectric constant ($\bar{d}_h$).

The $\bar{K}_{33}$ and $\bar{d}_{33}$ were measured with the same equipment and techniques as the 2–2 samples. The $\bar{d}_h$ was measured by immersing samples in pressure transfer fluid with a static head of 0.76 MPa. A pressure wave with 0.2 Hz was created by cyclically reducing the hydrostatic pressure to 0.62 MPa and returning to 0.76 MPa. The current output of the test sample was compared with a standard element and the $\bar{d}_h$ was calculated.

Composite dielectric constant decreases in proportion to ϕ_{PZT} as shown in Fig. 15.9. Dense, pure PZT-5H was expected to have a $K_{33} = 3{,}400$ value. The nonlinear decrease of $\bar{K}_{33}$ with ϕ_{PZT} is attributed to the increasing fraction of PZT

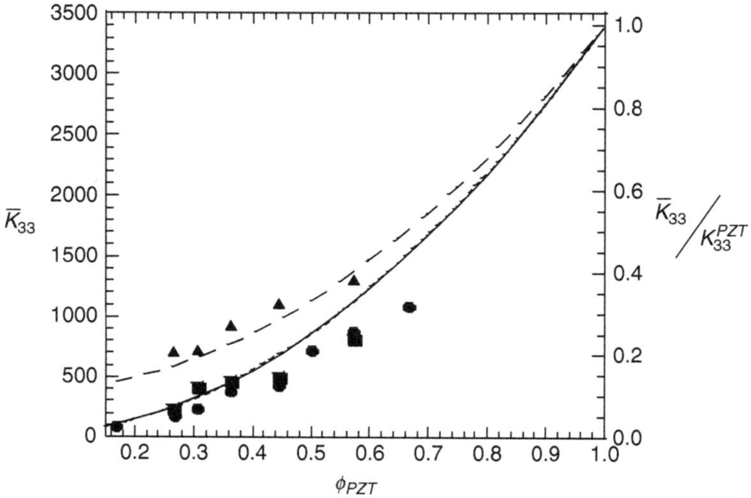

Fig. 15.9 Dielectric constant ($\bar{K}_{33}$) of 3–3 (*filled circle*), 3–2 (*filled square*), and 3–1 (*filled triangle*) composites as a function of PZT volume fraction (ϕ_{PZT})

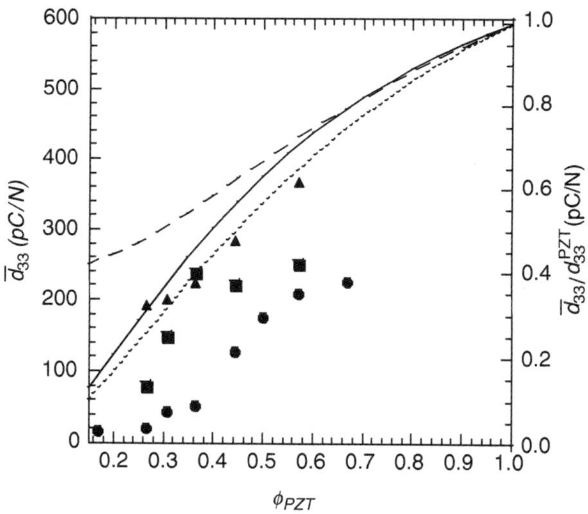

Fig. 15.10 Piezoelectric constant ($\bar{d}_{33}$) of 3–3 (*filled circle*), 3–2 (*filled square*), and 3–1 (*filled triangle*) composites as a function of PZT volume fraction (ϕ_{PZT})

in series with epoxy as the lattice spacing increases. Measured $\bar{d}_{33}$ values are plotted in Fig. 15.10 as a function of ϕ_{PZT}.

As with $\bar{K}_{33}$, $\bar{d}_{33}$ displayed a monotonic, nonlinear decrease with decreasing ϕ_{PZT}. The 3–1-type samples had higher values of $\bar{d}_{33}$ across the ϕ_{PZT} range explored while the 3–3 samples fared the poorest. 3–2 samples had intermediate values. These results are interpreted as an artifact of the nonuniform surface deformations such

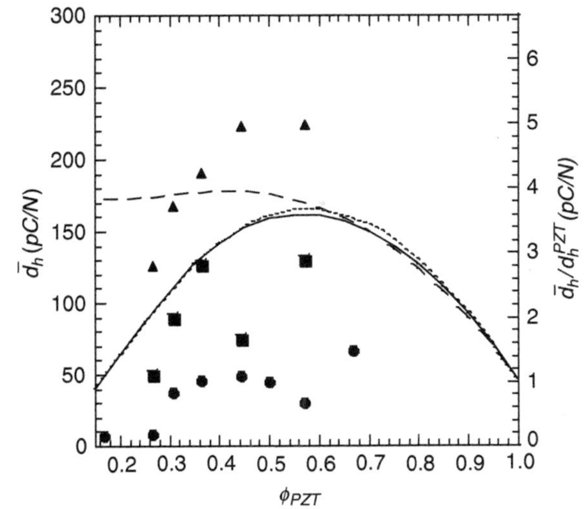

Fig. 15.11 Hydrostatic piezoelectric constant ($\bar{d}_h$) of 3–3 (*filled circle*), 3–2 (*filled square*), and 3–1 (*filled triangle*) composites as a function of PZT volume fraction (ϕ_{PZT})

thin (<1 mm) samples. The addition of the PZT faceplates distributed the point load of the Berlincourt meter more evenly in the 3–1 and 3–2 samples, whereas the open PZT–epoxy surface of the 3–3 samples may lead to nonisostrain conditions.

Measured $\bar{d}_h$ showed a maximum at intermediate values of ϕ_{PZT} for each of the sample types as illustrated in Fig. 15.11. As with the $\bar{d}_{33}$ measurements, $\bar{d}_h$ was highest for the 3–1 type samples and lowest for 3–3 with 3–2 having intermediate values. The significant improvement of 3–1 over 3–2 samples may be due to lateral stress shielding by the circumferential PZT ring. Notable is the comparison of $\bar{d}_h$ for the composite to that for bulk PZT on the secondary ordinate. For the 3–2 and 3–1 samples, the hydrostatic piezoelectric coefficient is enhanced over that for pure bulk PZT. The 3–1 samples showed up to 500% improvement between $\phi_{PZT} = 0.45$ and 0.6. Even the 3–3 composites, which had lower magnitude $\bar{d}_h$, were nearly equivalent to bulk PZT across a wide composition range.

A useful figure of merit (FOM) for hydrostatic applications is $\bar{d}_h\,\bar{g}_h$, where $\bar{g}_h = \bar{d}_h/\bar{K}_{33}\varepsilon_o$ and ε_o is the permittivity of free space. The FOM is plotted in Fig. 15.12 as a function of ϕ_{PZT}. For each type of sample, the FOM was at least an order of magnitude greater than that for bulk PZT. For 3–3 samples, the improvement was limited to samples with $\phi_{PZT} > 0.3$. Although $\bar{d}_h$ was not improved in these samples, the decreased $\bar{K}_{33}$ led to an improvement in the FOM. For the 3–2 and 3–1 designs, the enhanced $\bar{d}_h$ and reduced $\bar{K}_{33}$ lead to significant gains in the FOM. The maximum FOM was nearly a 70-fold improvement over bulk PZT and was observed for the 3–1 design at $\phi_{PZT} \approx 0.45$.

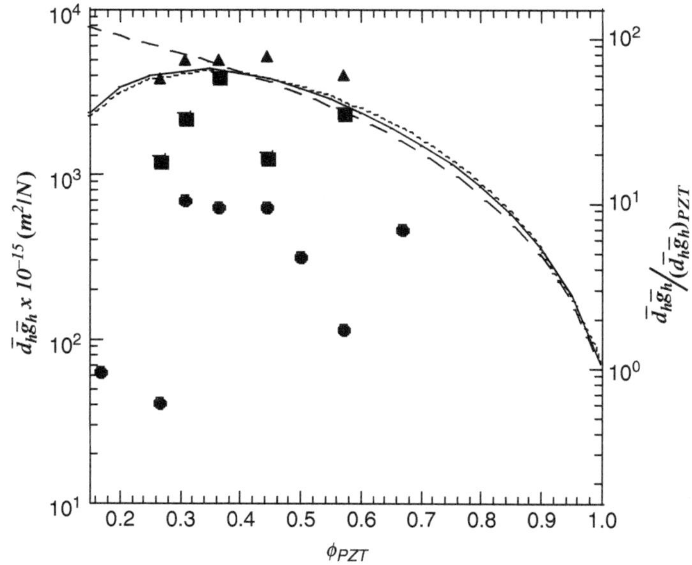

Fig. 15.12 Hydrostatic figure of merit ($\bar{d}_h\bar{g}_h$) for 3–3 (*filled circle*), 3–2 (*filled square*), and 3–1 (*filled triangle*) composites as a function of PZT volume fraction (ϕ_{PZT})

15.5 Summary

The processing technology of robocasting is a facile way to create complex geometries from PZT ceramics. The printed product is a green body that must be sintered, but contains little organic material so that the firing schedule may be accelerated.

The main advantage of the robocasting is flexibility in ink design and the potential for in situ blending of inks to create composition gradients, although this latter aspect has been little explored. For exploring PZT–polymer composite device designs, robocasting offers a quick way to establish ϕ_{PZT} dependencies that may not be obvious from simplistic models such as the one presented here. More sophisticated models may capture all the physical subtleties that belie the inaccuracies in our model. The production of idealized structures by robocasting allows experimental validation of such models and allows for rapid prototyping. The current resolution limits of robocasting are suitable for a wide variety of PZT–polymer composites, but as feature size is reduced, higher frequency composites may become accessible by robocasting.

References

Cesarano III J (1999) A Review of Robocasting Technology. Symposium on Solid Freeform and Additive Fabrication vol. 542: pp. 133–139.

Cesarano III J. and P. Calvert (2000) Freeforming Objects with Low-Binder Slurry, USA.

Cesarano III J., T.A. Baer, and P. Calvert (1997) Recent Developments in Freeform Fabrication of Dense Ceramics from Slurry Deposition. 8th Solid Freeform Fabrication (SFF) Symposium pp. 25–32.

Cesarano III J., R. Segalman, and P. Calvert (1998) Robocasting Provides Moldless Fabrication from Slurry Deposition. Ceramic Industry vol. 148: pp. 94–102.

Cesarano III J., J.G. Dellinger, M.P. Saavedra, D.D. Gill, R.D. Jamison, B.A. Grosser, J.M. Sinn-Hanlon, and M.S. Goldwasser (2005) Customization of Load-Bearing Hydroxyapatite Lattice Scaffolds. International Journal of Applied Ceramic Technology vol. 2: pp. 212–220.

Dellinger J.G (2005) "Development of model hydroxyapatite bone scaffolds with multiscale porosity for potential load bearing applications," in *Materials Science and Engineering*, vol. PhD. Urbana-Champaign: University of Illinois.

Dellinger J.G., J.A.C. Eurell, and R.D. Jamison (2006) Bone Response to 3D Periodic Hydroxyapatite Scaffolds with and without Tailored Microporosity to Deliver Morphogenetic Protein 2. Journal of Biomedical Materials Research Part A. vol. 76A: pp. 366–376.

He G., D.A. Hirschfeld, and J. Cesarano III (2000) Processing and Mechanical Properties of Si_3N_4 Formed by Robocasting Aqueous Slurries. 24th International Conference & Exposition on Engineering Ceramics and Structures.

He G., D.A. Hirschfeld, J. Cesarano III, and J.N. Stuecker (2000) "Processing of Silicon Nitride-Tungsten Prototypes," in *Ceramics Transactions*, vol. 114 Functionally Graded Materials, Trumble, K., K. Bowman, I. Remanis, and S. Sampath, Eds. Westerville, OH: American Ceramic Society, 2001, pp. 325–32.

Lewis J.A (2002) Direct-Write Assembly of Ceramics from Colloidal Inks. Current Opinion in Solid State & Materials Science vol. 6: pp. 245–250.

Lewis J.A. and G.M. Gratson (2004) Direct Writing in Three Dimensions. Materials Today vol. 7: pp. 32–39.

Lewis J.A., J.E. Smay, J. Cesarano III, and J.N. Stuecker (2006) Direct Ink Writing of Ceramic Structures. Journal of the America Ceramic Society vol. 16: pp. 1685–1693.

Li Q. and J.A. Lewis (2003) Nanoparticle Inks for Directed Assembly of Three-Dimensional Periodic Structures. Advanced Materials vol. 15: pp. 1639–1643.

Nadkarni S. and J.E. Smay (2006) Concentrated Barium Titanate Colloidal Gels Prepared by Bridging Flocculation for Use in Solid Freeform Fabrication. Journal of the America Ceramic Society vol. 89: pp. 96–103.

Smay J.E., J. Cesarano III, B.A. Tuttle, and J.A. Lewis (2002a) Piezoelectric Properties of 3-X Periodic Pb(ZrxTi1-x)O3-Polymer Composites. Journal of Applied Physics vol. 92: pp. 6119–6127.

Smay J.E., J. Cesarano III, and J.A. Lewis (2002b) Colloidal Inks for Directed Assembly of 3-D Periodic Structures. Langmuir vol. 18: pp. 5429–5437.

Smay J.E., G. Gratson, R.F. Shepard, J. Cesarano III, and J.A. Lewis (2002c) Directed Colloidal Assembly of 3D Periodic Structures. Advanced Materials vol. 14: pp. 1279–1283.

Smay J.E., J. Cesarano III, B.A. Tuttle, and J.A. Lewis (2004) Directed Colloidal Assembly of Linear and Annular Lead Zirconate Titanate Arrays. Journal of the America Ceramic Society vol. 87: pp. 293–295.

Smay J.E., S. Nadkarni, and J. Xu (2006) Direct Writing of Dielectric Ceramics and Base Metal Electrodes. International Journal of Applied Ceramic Technology vol. in review.

Stuecker J.N (1999) "Freeform Fabrication and Near Net Shape Processing of Composites from Mullite and Kaolin Ceramics," in *Materials Science and Engineering*, vol. Master of Science. Socorro, NM: New Mexico Institute of Mining and Technology.

Stuecker J.N., J. Cesarano III, and D.A. Hirschfeld (2003) Control of the Viscous Behavior of Highly Concetrated Mullite Suspensions for Robocasting. J. Mat. Proc. Technology vol. 142: pp. 318–325.

Tressler J.F., S. Alkpu, A. Doğan, and R.E. Newnham (1999) Functional composites for sensors, actuators and transducers. Composites Part A (Applied Science and Manufacturing) vol. 30A: pp. 477–482.

Tuttle B.A., J.E. Smay, J. Cesarano III, J.A. Voigt, T.W. Scofield, W.R. Olson, and J.A. Lewis (2001) Robocast $Pb(Zr_{0.95}Ti_{0.05})O_3$ Ceramic Monoliths and Composites. Journal of the America Ceramic Society vol. 84: pp. 872–874.

Xie B., R.L. Parkhill, W.L. Warren, and J.E. Smay (2006) Direct Writing of Three-Dimensional Polymer Scaffolds Using Colloidal Gels. Advanced Functional Materials vol. 16: pp. 1685–1693.

Chapter 16
Micropositioning

J. Juuti,* M. Leinonen, and H. Jantunen

16.1 Introduction

Micropositioning is required in many industrial segments as well as in everyday life. It is often utilized, without being recognized, in applications such as cars, cameras, or even when looking at internet pictures from space taken by telescopes. In such applications, small but accurate movements can significantly improve the performance and usability of the device. For scientific and engineering purposes micropositioning is widely used in research and production facilities such as AFM (atomic force microscope), SEM (scanning electron microscope), FIB (focused ion beam), micromanipulators (e.g., cell manipulators), active vibration dampers, and assembling and production devices, for example, in electronics and semiconductor manufacturing (Hubbard et al. 2006). Demand for wider and more efficient utilization of micropositioning is growing due to trends toward miniaturization in electronics as well as its wider exploitation in application fields.

There are several different materials and actuation schemes upon which micropositioning systems can be based. If especially small size, low forces, and high frequency are required for the system, electrostatic actuators are a good option. However, they are able to produce only a limited range of displacement with high voltage unless a comb structure instead of the basic capacitor plate configuration is employed. For low-voltage applications, thermal as well as magnetic actuators are widely used. Thermal actuators are usually comparable in size to electrostatic actuators but suffer also from a limited range of motion that usually has to be amplified mechanically. In contrast, magnetic and magnetostrictive actuators provide relatively large displacement but require the use of coils to generate the magnetic field and therefore can be bulky and expensive. Additionally, these approaches rely on actuation by current and therefore consume power while holding a static position.

J. Juuti
Microelectronics and Materials Physics Laboratories, EMPART Research Group of Infotech Oulu, University of Oulu, P.O. Box 4500, 90014 Oulu, Finland

A. Safari, E.K. Akdoğan (eds.) *Piezoelectric and Acoustic Materials for Transducer Applications.*

Shape memory alloys provide large deformations and forces, but their response time is usually very limited. Piezoelectric actuators, on the other hand, show high figure of merits in many presented characteristics but not the highest values, thus commonly outperforming the others by a combination of required characteristics rather than by one specific value (Hubbard et al. 2006; Pons and Rocon 2006).

Piezoactuators are known as one of the best established and the most mature technologies in the field of active materials having also one of the highest overall characteristics and widest range of existing applications. Applicability of different actuators greatly depends on manufacturing technologies and feasible actuator structures, which enable tailoring of properties according to application. Therefore, it is natural that piezoactuators are widely exploited, especially in micropositioning, as they can be utilized by an extensive range of manufacturing methods and technologies.

In this chapter, the general properties and requirements of piezoelectric micropositioners, their control and sensor techniques, and some commercial applications are discussed.

16.2 General Properties and Requirements

Piezoelectric actuators are widely used in applications utilizing micropositioning because they provide the desired features such as high resolution, operation frequency, forces and durability, and fast response time. Additionally, the actuators are very rugged, compact, and exhibit high stability and are practically immune to electromagnetic interference. Compositions of actuator materials have a significant effect on actuator properties, and by tailoring composition and doping, application-specific requirements such as Curie temperature, coercive electric field, or different piezoelectric coefficients can be attained (Davis et al. 2005, 2006; Morozov et al. 2005; Park and Shrout 1997; Randall et al. 2005). To utilize material properties efficiently, careful design of the actuator structures is required. Additionally the actuators should commonly be rugged, provide integration possibilities (e.g., sensors), and enable shielding from external interference (humidity, electromagnetic interference, chemicals, stresses, etc.). These features are readily available in many commercial piezoactuators that even meet the requirements of extreme conditions where many other schemes for micropositioning are not feasible. At the same time piezoelectric actuators have the widest manufacturing and shape giving process possibilities in the field, thus giving clear advantage over other materials in terms of cost and flexibility (Allahverdi et al. 2001; Niezrecki et al. 2001; Pearce et al. 2002; Wolny 2000). In spite of the very good properties of the piezoelectric ceramics, they also have drawbacks, which are typically related to their large hysteresis and creep characteristics. However, these problems can be diminished by the proper selection of materials and structures in addition to error-reducing controlling schemes.

16.3 Different Types of Piezoelectric Actuators

Numerous actuators and micropositioning systems based on piezoelectric phenomena have been developed. However, some general categorization can be made by operation principles and characteristics as shown in Fig. 16.1, where the piezoelectric material in actuators is marked as light gray.

In certain piezoactuator structures some moving parts do exist, which, however, are utilized in a frictionless manner via a hinge or by the bending of passive materials (in the centre of Fig. 16.1) that makes them suitable for producing displacements of nanometers resolution and thus feasible for micropositioning. However, accuracy and repeatability of the actuators are other issues that strongly depend on the behavior of piezoelectric material, actuator structure and its material selections, bonding methods, environment, operation frequency, control system, and so on. Here the most typical actuator structures are generally introduced first, followed by the common principles for their displacement, force, and frequency behavior.

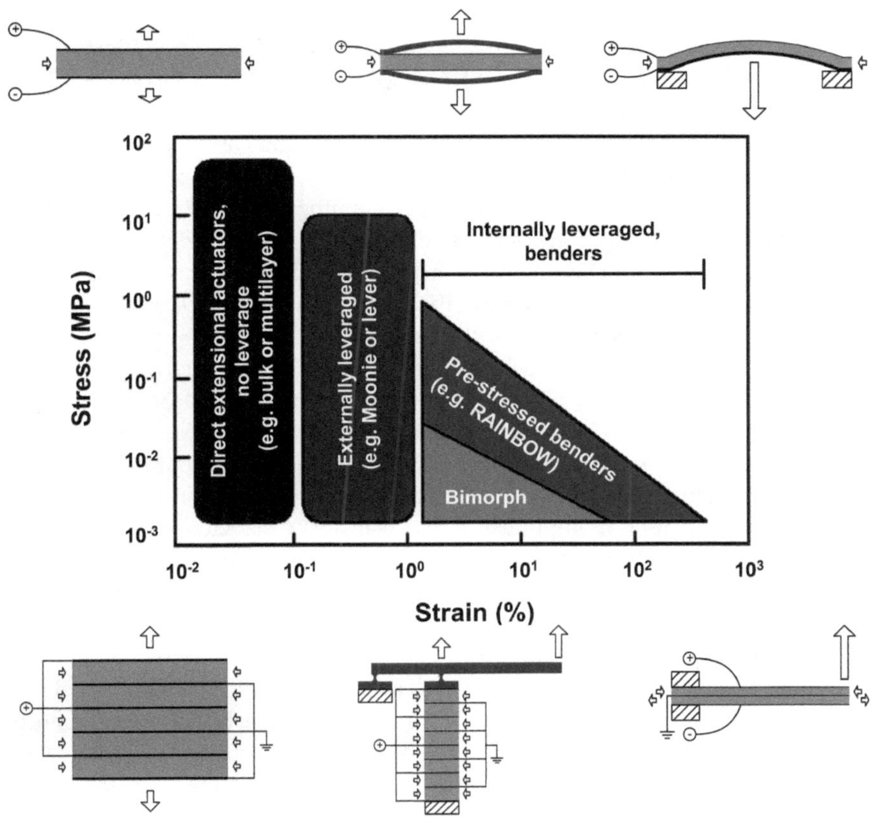

Fig. 16.1 Relationship between stress and strain for different types of piezoelectric actuators and their schematic examples (Haertling 1994, 1997)

Direct extensional actuators utilize expansion and contraction of piezoelectric materials directly without any leverage. They are basically structures of bulk or multilayer solid ceramic materials or stacks of these (left side of Fig. 16.1) depending upon the categorization. Direct extensional actuators can sustain and produce the highest stresses (up to 100 MPa) and forces (up to 80 kN with commercial actuators) among the piezoactuators while their typical sizes are from dozens of cubic millimeter to hundreds of cubic centimeter (i.e., from a sugar cube to about six D-size batteries) (Dynamic Structures & Materials 2007; PI GmbH & Co., 2001; Piezo Systems Inc. 2006; Piezomechanik, 2006; Piezosystem jena GmbH 2007). On the other hand, such actuators produce low strains (the ratio of displacement to thickness in Fig. 16.1) in the range of 0.1–0.2% for polycrystalline PZT and up to 1.7% for single crystal PZN-PT (Park and Shrout 1997). The actuators are typically designed to be subjected to uniform stresses, which make them very stiff and resilient for high loads. Direct extensional actuators have been utilized in, e.g., camber control in missile fins, vibration and suspension control, drum brake actuation, diesel fuel injection, and micropositioning in general (Niezrecki et al. 2001).

Actuators using direct extension of the material have excellent characteristics but in many practical applications, their usage is severely limited due to the small displacement. Hence, different actuator schemes for displacement amplification have been widely researched and developed. One of them is called the externally leveraged actuator where direct extension of the material is amplified by an external mechanism. The mechanism amplifying the displacement can be hydraulic or mechanical, i.e., levers and Moonie-type structures as shown in the center of Fig. 16.1 (Kallio et al. 1996; Niezrecki et al. 2001). The amplification increases displacement of the actuator at the cost of the obtainable force. One can consider that piezoelectric materials have a certain capacity for producing strain and stress or displacement and force. Such capacity is often described as free displacement and as blocking force, thus giving two extremities of the actuator. The free displacement describes displacement under an electric field without load and is constrained only by internal stresses of the actuator. The blocking force determines the level of external force that will nullify the actuator movement against the load. In direct extensional actuators, the capacity of the material is directly utilized as an output of the actuator while in amplified actuators some of the capacity is consumed in the deformation of the passive structure of amplification mechanisms. Characteristics of the externally leveraged actuators are described in more detail later. This actuator scheme has been utilized, for example, in the control of the trailing edge flat of helicopter rotor blades, cell manipulators, proof-mass actuation, active/passive vibration control, micropositioning stages, and adjustable optics (Niezrecki et al. 2001).

To obtain even higher displacements with smaller size, internally leveraged actuators have been developed including, for example, unimorph, bimorph, monomorph, C-block, and recurve actuators. Instead of an external mechanism, high displacements of such actuators are obtained by an internal mechanism that creates bending of the actuator structure itself. In these actuators, active material creates a nonuniform stress distribution by its expansion and contraction thus forming an internal bending moment that finds equilibrium via bending of structure. Bending-type actuators are typically produced as a bimorph with two piezoelectric layers (lower right corner of Fig. 16.1) or a unimorph with one piezoelectric and accompanying passive layer (Germano 1971; Niezrecki et al. 2001; Sawyer 1931). In addition, a branch of the unimorph type of actuators called monomorphs has been developed utilizing a structure without a separate bonding layer between the active and passive materials as is the case with conventional unimorph and bimorph (Nakamura et al. 1987; Uchino 1997; Uchino et al. 1987). Furthermore, mechanically prestressed internally leveraged bending actuators utilizing unimorph or monomorph configuration have been developed. In such actuators, the basic structure is the same, but mechanical prestress is applied during fabrication, which alters the shape of the actuator, as shown in Fig. 16.1, upper right corner. Sufficient tensile prestress in the piezoelectric material enhances its properties, which was first implemented for the actuators known as RAINBOW (reduced and internally biased oxide wafer) in 1992 (Haertling 1999; Li et al. 1997; Schwartz 2001; Schwartz and Moon 2001). The benders have been utilized in robotic applications, spoilers of missile fins, actuation for a quick-focusing lens, control of vibration of a helicopter blade, and the vertical tail of the F/A-18 airframe. They have also been used in various sport goods such as bicycle shock absorbers, snow skis, baseball bats, and water skis, and in synthetic jets and sound sources for noise cancellation (Niezrecki et al. 2001).

In addition, various piezoelectric motors have been developed utilizing the previously mentioned actuator schemes (bimorph, multilayer, bulk, etc.) to obtain practically unlimited displacement capabilities (Choi et al. 2005; Uchino 1997, 1998). Basically, operation of the motors relies on many small steps taken at high frequency, thus they are also considered as frequency-leveraged actuators. In contrast to the aforementioned actuators, the motors are friction-driven devices, transferring the mechanical energy of the actuators to rotational or linear movement via friction of mechanical contacts. Advantages and disadvantages of the motors derived from the operating principle and piezoelectric material are listed later (Niezrecki et al. 2001; Spanner 2006; Tenzer and Mrad 2004; Uchino 1998).

Advantages are as follows:

- Quick response
- Silent operation
- Wide range of velocity
- High-power off torque
- Compact size, simple structure, and lightweight
- High power/weight ratio and efficiency
- Direct drive without gears enabled by low speed and high torque
- Negligible influence or generation of magnetic fields
- Good controllability and high resolution due to hard brake

Disadvantages are as follows:

- Frictional drive decreases durability
- Drooping torque–speed characteristics
- Require a high-frequency power supply

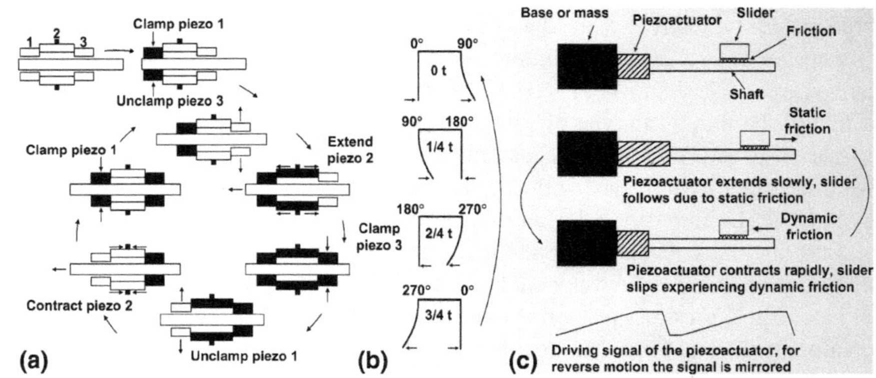

Fig. 16.2 Operation principles of (**a**) inchworm, (**b**) stepping, and (**c**) inertial motor (http://konicaminolta.com; May, 1975; Uchino, 1998)

Extensive research and development have been made to utilize vibration of piezoactuators in the motors by different methods. Here only a brief overview is given while more detailed information can be found in other publications. The motors can be categorized in many different ways, such as into two groups as defined by their operating frequency. Ultrasonic motors are driven in resonance at ultrasonic frequencies (20 kHz – 10 MHz) while quasistatic motors are typically driven at lower frequencies. The quasistatic motors are mainly used to produce linear movement while the ultrasonic motors have mainly been utilized for rotational movement, which, however, can be transformed into linear displacement if required (Niezrecki et al. 2001; Spanner 2006; Uchino 1998).

Inchworm® or clamping principle motors utilize a sequenced operation of the piezoactuators in the same manner as an inchworm moves. The full operation sequence of the inchworm motor is illustrated in Fig. 16.2a. Similarly, a motor utilizing the stepping principle is taking its exemplar from nature and imitating walking, as shown in Fig. 16.2b. In the inertial motor shown in Fig. 16.2c, a moving part is first displaced in the required direction by slow motion of a piezoactuator so that the part follows the movement due to static friction between the drive shaft and the part. Then the actuator is rapidly retracted by the driving voltage, causing the moving part to slip along the shaft and experience dynamic friction when the force enabling the movement exceeds the friction force. To obtain movement in the opposite direction the driving waveform is mirrored (http://konicaminolta.com; Ko et al. 2006; May 1975; Spanner 2006; Tenzer and Mrad 2004; Uchino 1998).

The ultrasonic motors are driven at resonance in different vibration modes whose frequencies depend on the dimensions and shape of the actuator. In a standing wave motor, a flat elliptical movement is generated when the tip of the resonator (piezoactuator and mechanical structure shown in Fig. 16.3a) impacts against the rotor. Because of the mechanical restriction, the tip slides along the rotor and creates a thrust via friction. The impulses of the resonator are directed to the rotor with a slight cant angle to transform the bending into a torsional vibration. The coupled transverse

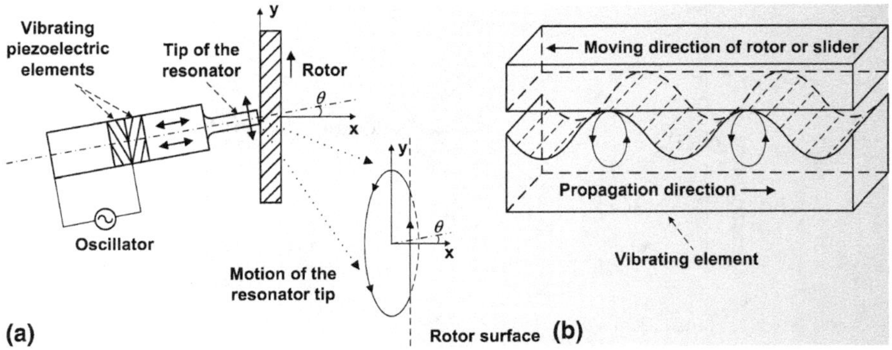

Fig. 16.3 Operation principle of (**a**) standing wave and (**b**) traveling wave ultrasonic motors (Uchino, 1998)

moment and up-and-down bending motion introduce elliptical rotation on the resonator tip that produces rotation of the rotor in one direction.

In contrast to the standing wave motor, in traveling wave motors rotation on both directions can be created. In the motor, the superposition of multiple standing waves is used to produce traveling acoustic waves along an actuator structure. Traveling waves can be created in continuous structures such as a ring or cylinder, by combining two propagating waves having a 90°-phase difference in time and space. In such cases, a bending wave travels through the actuator creating elliptical movement for the particles on the surface or circumference, thus giving a rotational thrust for the rotor as shown in Fig. 16.3b (Spanner 2006; Uchino 1998; Uchino et al. 2004).

Although resolution of the piezoelectric motors is usually not as high as with rigid actuators, they still offer resolution in the range of dozens of nanometers with very high displacement and stable positioning after actuation, which are wanted characteristics for various microspositioning applications (Spanner 2006; Uchino 1998).

16.4 Mechanics of Micropositioning Piezoactuators

The piezoelectric actuators exhibit very different characteristics and operation principles although the actuator material used could be the same in each case. A common factor relating the strain and stress or displacement and force characteristics of the different actuators is the stiffness. In other words, decreased stress and increased strain (Fig. 16.1) can be described by decrease in stiffness across the different actuator schemes. For example, direct extensional actuators have the highest load-bearing capabilities and lowest strains because they have the highest stiffness, while the benders produce the highest displacements but have the lowest load-bearing capabilities and stiffness. In reality, regions of strain and stress are not as precise as presented in

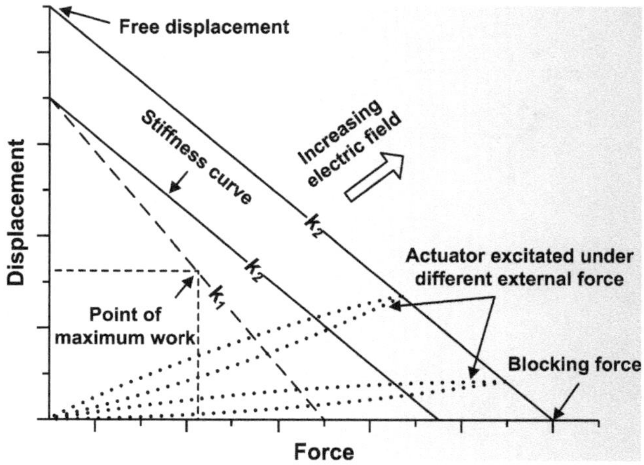

Fig. 16.4 Schematic displacement as a function of force of two actuators with different stiffness

Fig. 16.1, but instead they overlap due to different actuator materials and structure designs. The same rule of stiffness also applies within specific actuator schemes or structures. Hence, by decreasing actuator rigidity (e.g., making a bending actuator thinner) the achievable force is decreased but the displacement is increased. The same behavior is present also in the piezoelectric motors where the effect can be seen, for example, in rotation speed, velocity, and thrust or torque, which are the sum of each steps of an actuator (Bexell and Johansson 1999; Park et al. 2002).

Furthermore, the displacement of the actuators as a function of force is defined by their stiffness, which defines the slope of the displacement vs. force characteristic, as shown in Fig. 16.4.

In the figure, the displacement of two actuators with lower (k_1) and higher (k_2) stiffness and previously mentioned extremes, i.e., free displacement and blocking force is shown. The maximum work is obtained at the point of the stiffness curve, which defines the largest area under the curve by force and displacement axis. Moreover, the stiffness curve shifts to the right and left with higher or lower electric fields, respectively (PI GmbH & Co. 2001). At the same time, the slope can slightly change its angle because the stiffness and displacement of the piezoelectric ceramic are nonlinear functions of electric field, especially at high electric fields. Similarly, it should be noted that the applied prestress changes the actuator stiffness and piezoelectric coefficients depending on the material and level and type of the prestress (Li et al. 1999; Wang and Cross 1998; Wolff et al. 1994; Zhang and Zhao 1999). Therefore, force–displacement curves can also exhibit nonlinear behavior as a consequence of anisotropic and nonlinear material characteristics, structure, and its clamping.

To clarify the behavior, the source of actuator stiffness and some formulations to calculate it are described later. The stiffness of the actuator is a consequence of the elastic properties of the material and the geometric moment of inertia of the structure. The modulus of elasticity or Young's modulus E can be described according

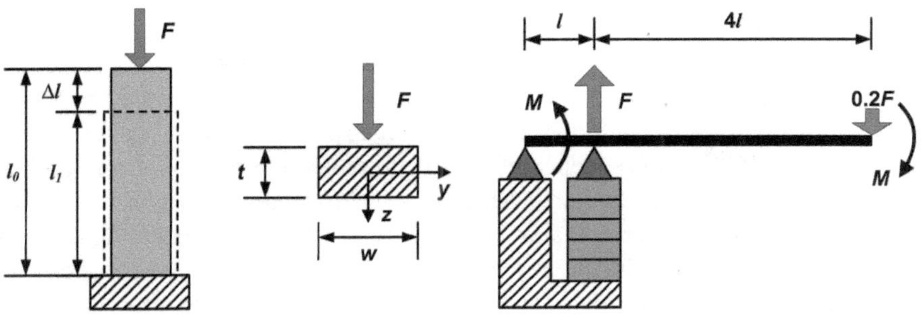

Fig. 16.5 (**a**) Axially loaded actuator, (**b**) rectangular beam, and (**c**) external leverage with equal moments canceling each other

to Hooke's law $E = \sigma/\varepsilon$, where σ is stress and ε is strain. Furthermore, $\sigma = F/A$ and $\varepsilon = (l_1 - l_0)/l_0$ where F is force, A is area, l_1 is the changed length under force, and l_0 is the original length. If the force is applied axially so that bending of the actuator will not occur (as shown in Fig. 16.5a) its stiffness or spring constant k can be directly calculated with (16.1) derived from the Hooke's law, where $\Delta l = l_1 - l_0$

$$k = \frac{F}{\Delta l} = \frac{EA}{l}. \tag{16.1}$$

If bending occurs, the geometric moment of inertia needs to be taken into account. This depends upon geometry and the direction of the force or moment applied to the structure. For example, the geometric moment of inertia I_y of a rectangular beam (Fig. 16.5b) for a force F, creating a moment on y axis, is $I_y = wt^3/12$, where w is the width and t is the thickness of the beam. Moreover, the tip deflection Δz of a beam supported from one end (structure shown in Fig. 16.1. lower right corner) and under a point load is described as follows $\Delta z = Fl^3/3EI_y$. Correspondingly, the stiffness of the beam or internally leveraged actuator can be calculated in the same way as in (16.1) by knowing the applied force F and deflection Δz. It should be noted additionally that the obtained results and formulations strongly depend on boundary conditions, i.e., type of clamping, applied force or momentum, and actuator structure. Also anisotropic material parameters (e.g., Young's modulus) should be taken into account, especially in the case of the internally leveraged bending actuators where the stiffness is a sum of the combined passive and active material parameters and their bonding.

In the externally leveraged actuators, the load is applied to the piezoactuator via the passive structure. By this means, the stiffness of the piezoactuator and the external structure is coupled and the overall stiffness is the result of both of these. Furthermore, displacement amplification will produce amplified loads for the piezoactuator. To clarify this, one can consider a simple rule of leverage as presented in Fig. 16.5c. In the ideal case, force F generated by the actuator produces a moment $M = Fl$, while an equal, but opposite moment created by the load would be $M = 0.2F{,}\cdot 5l = Fl$; hence, the tip of the lever would not move. Thus, the lever has an initial amplification factor of 5 for displacement (without load) and for the external load recognized by the actuator (Chopra 2002). In practice, however, the obtainable displacement and force are reduced by the force required for deformation of the passive structures, e.g., hinge or curved bars shown in the centre of Fig. 16.1.

Additionally, in dynamic actuation, the mechanism will have resonances at certain frequencies as a consequence of its stiffness, moving mass, and mechanical losses. In the resonance, amplitude of the mechanism is significantly increased thus increasing the loads for the actuator by following the Newton's second law of motion $F = ma$, where m is the moving mass and a is acceleration. It should be noted also that amplification mechanisms can produce nonuniform stress distribution in the piezoactuator, especially in the region where they are connected. The mechanisms may induce shear, tensile, and compressive stresses that can have a significant influence on the lifetime of the actuator and its reliability. Such occurrence can be avoided or minimized by proper design of the external mechanism and the actuator through stress analysis of the structure.

The analogy for displacement amplification in internally leveraged actuators is somewhat different because the lever amplifying the displacement can be considered to be inside the structure. In contrast to the externally leveraged structure, in this case the active material is in intimate contact with another material, either active or passive. As a consequence, actuation of the piezoelectric material will result in an internal bending moment instead of a simple force, as in the case shown in Fig. 16.5c. Depending upon the structure, material thicknesses, and Young's modulus of the materials, a neutral axis is formed in the structure representing a plane with zero longitudinal strain and stress. Furthermore, the axis divides the regions of tensile and compressive stresses as can be seen in Fig. 16.6a. The highest displacements of internally leveraged structures have been typically obtained by selecting materials and thicknesses in such a way that the neutral axis is located outside the piezoelectric element (Kim et al. 2005; Li et al. 1999; Wang and Cross 1999).

From Fig. 16.6b the leverage can be imagined along the thickness of the material where the pivot point of a lever would be located in the plane of the neutral axis. If thickness of the actuator is decreased the deflection and angle around the pivot point is increased if the same strains (expansion and contraction) are maintained. In such a system, however, the strain and moment thus created by piezoelectric material are resisted by passive material (in unimorph) or by another piezoelectric material (in bimorph), producing opposite effects and resulting in the equilibrium of moments after bending, as in the case of Fig. 16.5c. Now, instead of having one

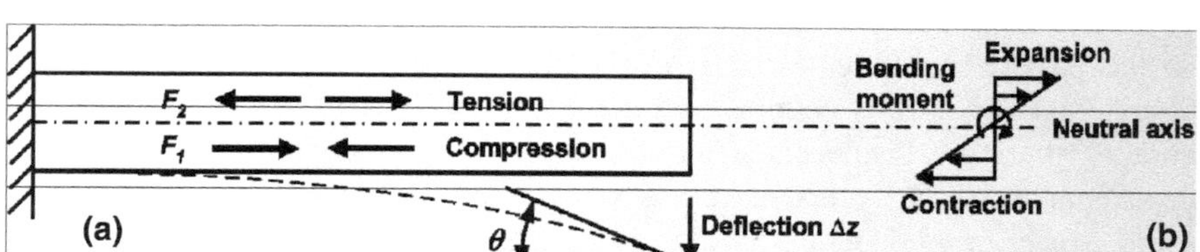

Fig. 16.6 (**a**) Deflection, compressive, and tensile stresses and (**b**) bending moment around the neutral axis in the internally leveraged actuator

actuator that produces leverage or angular change, amplification obtained in benders can be thought of as the sum of many infinitely small angle producing actuators. The total angle θ and maximum deflection Δz obtained at the tip of the bender can be calculated by (16.2).

$$\theta = \arctan\left(-\frac{Ml}{EI_y}\right), \quad \Delta z = \frac{Ml^2}{2EI_y}. \tag{16.2}$$

As mentioned earlier, the stiffness of the total actuator structure has a significant impact on the micropositioning attributes. In general, higher stiffness results in higher acceleration, velocity, force, pressure, operation frequency, and faster response time and lower displacement. The significance of stiffness will be emphasized in dynamic actuation in applications where response time and operation frequency are crucial parameters. To obtain high resolution and accuracy, the increased load and inertia of the moving mass due to high acceleration and velocity should be addressed during design of the system. In many cases, the objects to be moved are very light, which, however, can still produce significant loads in operation. For example, in static conditions 1-g mass will produce only ~10-mN load for an actuator due to gravity, but under sinusoidal movement with 5-μm amplitude at 10-kHz frequency the mass will produce ~20-N force at maximum acceleration, as can be calculated using Newton's second law of motion and $a = -A(2\pi f)^2 \sin(2\pi ft)$, where t is time, A is amplitude, and f is frequency. Depending on actuator size, structure, and operating principle, significant stresses can be produced for actuators by the dynamic loading, e.g., 20 MPa in previously mentioned case for an actuator with 1-mm^2 cross section. Additionally, the mass of the actuator itself should also be taken into account. Furthermore, if continuous excitation is suddenly ended in the earlier example, an actuator with stiffness of 20 N/μm can have a maximum backlash or overshoot of ~2.2 μm depending on the phase of the driving signal. The result can be calculated by equilibrium of kinetic energy (E_k) and energy stored into the spring (E_s)$E_k = E_s = (1/2)mv^2 = (1/2)kx^2$, where v is the velocity and x is displacement, and $v = A(2\pi f)\cos(2\pi ft)$. Depending on the damping coefficient, i.e., the mechanical losses of the system, the stored spring energy is dissipated as heat and the displacement response will decay as a function of time, as shown in Fig. 16.7, where a higher damping coefficient results in faster attenuation of the vibration and vice versa.

To avoid ringing and to obtain a fast response time or short period for the determined steady state, designing for optimal damping is a necessity. The damping ratio of a spring mass dashpot system can be calculated using (16.3).

$$\zeta = \frac{c}{2\sqrt{k(m_s + m)}}; \quad c = \frac{F}{v}, \tag{16.3}$$

where ζ is the damping ratio, c is the damping coefficient, m_s is the effective mass of the spring (dependent on the shape of the spring or actuator), and F is the force resisting the movement with the velocity v in the dashpot. A dashpot is usually typified as a shock absorber of a car, but it can also embrace different energy dissipating

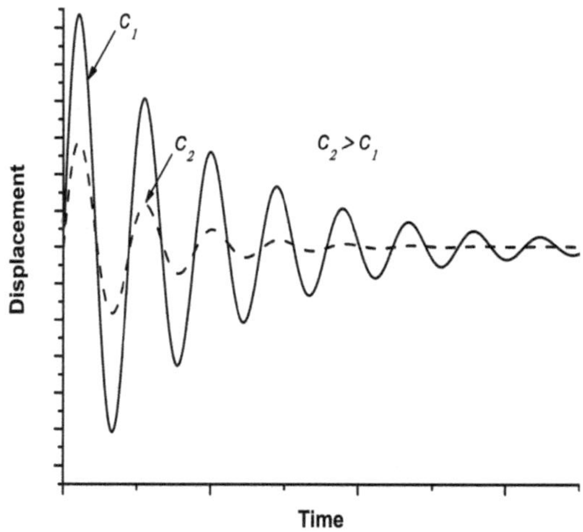

Fig. 16.7 Attenuation of displacement of spring mass dashpot system as a function of time with different damping coefficients after excitation

mechanisms such as friction or deformation. A system is overdamped when $\zeta > 1$, underdamped when $\zeta < 1$, and critically damped when $\zeta = 1$, this being the point where the fastest response is typically obtained and under which the system will oscillate, instead of attenuation. Damping of the system can be adjusted mechanically or electrically, thus the dynamic properties of the system can be significantly improved by proper actuator designing or electrical driving of the actuator via a control system, discussed later on.

Stiffness and damping coefficient also have an impact on the resonance frequency, f_r, of the actuator as can be seen from (16.4).

$$f_\mathrm{r} = \frac{1}{2\pi}\sqrt{\frac{k}{(m_\mathrm{s}+m)} - \frac{c^2}{2(m_\mathrm{s}+m)^2}} \tag{16.4}$$

The equation determines the first resonance frequency of a spring mass dashpot system, which is an important parameter in actuator and control design because some configurations try to avoid it while others exploit it, depending upon their operation principles and application. To illustrate this behavior of the system, displacement and phase difference as a function of frequency with different stiffness are shown in Fig. 16.8a, b, respectively. The example is a simplified mechanical model where the output of the piezoactuator is assumed to be a constant force. As can be seen from Fig. 16.8a, a higher stiffness (k_2) results in a higher resonance frequency but decreased displacement. Simultaneously, displacement around the resonance frequency flattens and a wider linear range is obtained below the resonance.

Additionally, the displacement is strongly suppressed above the resonance frequency, especially in the case of k_1. Therefore, the operational frequency of the actuator is often limited by the mechanical resonance frequency of the structure. The reason for such behavior can be found from Fig. 16.8b where the phase differ-

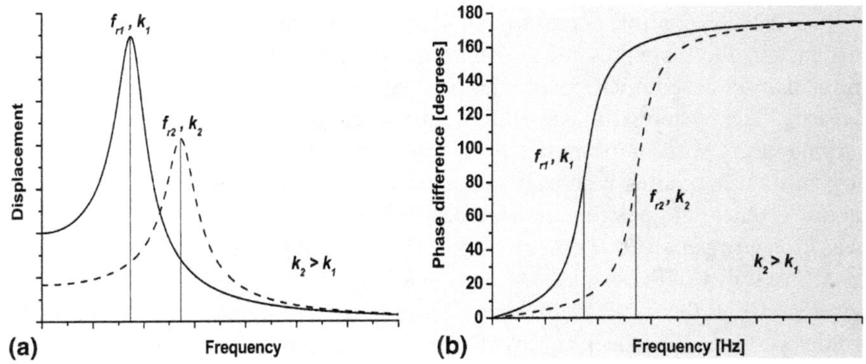

Fig. 16.8 General behavior of (**a**) displacement and (**b**) phase compared with driving signal as a function of frequency of piezoactuator

ence of the input force and output displacement is shown. As a consequence of the kinetic energy of the mass and energy stored in the spring, the vibrating mass cannot follow the excitation signal as the frequency increases, hence a phase difference appears. A large phase difference between the moving mass and the actuator causes nonsimultaneous movement, thus reducing the output displacement level. Furthermore, the large phase difference will cause difficulties in certain aspects of control, which are to be discussed later. By increased stiffness (k_2) a wider linear range of phase difference can be accumulated before the resonance frequency (Fig. 16.8b). It is noteworthy also that decreased mechanical losses would make the resonance peak narrower and higher and shift it to the right while increased losses would affect the response in the opposite way.

The fastest response is usually obtained within the resonance period and therefore the stiffness also limits the response time of the actuator. The ringing can be minimized by setting the rise time of the driving signal and the resonance frequency to correspond with each other. Alternatively, integer multiples of the resonance period can be used (Uchino 1997). Moreover, it should be noted that spring mass dashpot system (Kelvin-Voigt model) used here for illustrative purposes utilizes constant stiffness and Newtonian dashpot, i.e., linear viscoelasticity whereas in reality, these parameters are linear only at limited area. More realistic models taking into account, e.g., nonlinearity are presented by Schaller et al. (2001).

16.5 Control and Sensor Techniques

16.5.1 Control Overview

Because of the nature of the piezoelectric phenomena, any actuator based on a piezoelectric material needs a control system to achieve an accurate positioning performance. Challenges are usually posed by hysteresis, saturation, and creep. In this chapter, charge control, open-loop and feedforward method, and closed-loop and feedback methods are presented as solutions to improve the accuracy of a piezoelectric actuation-based micropositioning system. Hysteresis is the most serious nonlinearity. The hysteresis makes it difficult to control the displacement in a case of varying target position because the voltage needed for the desired displacement is not constant but varies with each actuation and also depends on the history of the motion. Hence, the positioning accuracy is typically only 15% without any compensation (Ang et al. 2003). Another problem relating to the hysteresis is that it can cause instability of the system. Methods to reduce the effects of hysteresis have been developed, and these can be divided into three categories: charge control, feedforward control, and feedback control. Saturation is generally not as serious a problem as hysteresis because the piezoelectric actuators are rarely used in this area; however, such occasions can rise if the actuators are driven by a high electric field. This is usually caused by the controller in the micropositioning system, and can thus be avoided by a proper control design. Drift of a piezoelectric actuator, or *creep* can be described as a slow variation of displacement under a constant DC voltage. This can create a problem for feedforward and open-loop methods, where no measured feedback from the actual displacement is present, whereas closed-loop methods usually provide this information and therefore can compensate for any drift present. The severity of this problem, however, is application specific.

16.5.2 Charge Control

When driving a piezoelectrical actuator with charge instead of voltage, hysteresis is greatly reduced. The general idea in charge control is to measure the charge or current flowing to the piezoactuator and feed back this information to the controller (Comstock 1981; Ronkanen et al. 2002). The simplest method to measure the charge of the actuator is to include a series capacitor with the piezoactuator and measure the voltage across the known capacitor (Comstock 1981). This method, however, does not take into account the change of capacitance of the piezoelectric element under deformation, and can therefore result in erroneous reading. A more reliable method is to measure the current flowing to the piezoactuator and then integrate over time to determine the charge. This method gives a hysteresis reduction from 27% of the initial hysteresis to 2% of current controlled hysteresis (Ronkanen et al. 2002). Additionally, the charge control can reduce the hysteresis of the piezoactuator, but other disturbances remain uncorrected because no information about the actual displacement is used in this control method. Furthermore, the hardware associated with charge control is more cumbersome than the usual voltage drive.

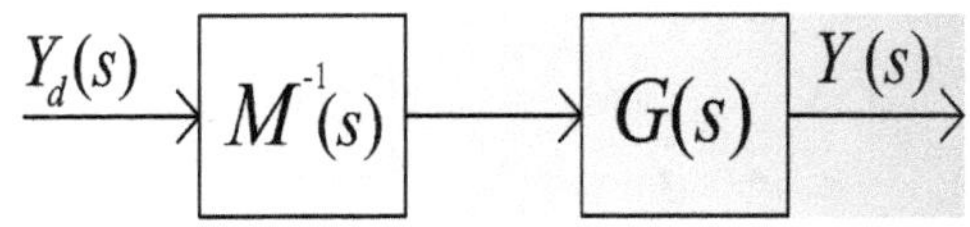

Fig. 16.9 Inverse model control

16.5.3 Open-Loop and Feedforward Methods

Various open-loop and feedforward methods have been studied and applied to micropositioning applications. The open-loop method, as the name suggests, utilizes, instead of the output signal of the system (usually the displacement), various models of the system. Accurate position sensors, not needed in the open-loop method, can be expensive and relatively bulky for some applications, making open-loop control the more attractive option. Another advantage is that open-loop methods are immune to instability problems present in closed-loop systems. One open-loop method is the inverse model configuration. In this, the system $G(s)$ is modeled as $M(s)$ and the inversion of the system is used as the controller $M^{-1}(s)$. The system in Fig. 16.9 realizes the transfer function of identity in the case when $M(s) = G(s)$.

In some cases only the nonlinear part of the system is modeled, thus resulting in a linear response but still retaining the dynamics of the system intact. It has to be noted, however, that this method is only applicable when the model can be inverted, i.e., when $M^{-1}(s)$ exists. In this type of control where the hysteresis was modeled with an adaptive LMS method, the nonlinearity error was reduced from 12 to 2% (Ru et al. 2005). In the case of hysteresis cancellation, various models of the nonlinearity have been proposed where the most popular is the Preisach model of the hysteresis. Prandtl-Ishlinskii model has also been used in micropositioning applications where the error caused by the hysteresis was reduced from 17.5 to 2% (Sun et al. 2004). This clearly demonstrates the potential of the feedforward methods. Prandtl-Ishlinskii modeling can also be modified to reduce the rate dependence of the hysteresis (Ang et al. 2003).

16.5.4 Closed-Loop Methods

Closed-loop methods have been used widely in micropositioning applications to compensate drift and hysteresis. However, a disadvantage of this method is the requirement for accurate displacement metering, and the possible introduction of instability into the system. With a good control system these problem can be solved.

The piezoelectric actuator can usually be described as a second order system with a relatively high Q-value. This is apparent from Fig. 16.8a, which shows the displacement vs. frequency plot of an equivalent mechanical spring mass dashpot system. Therefore, the poles of the system have only a small negative real part and if the closing of the feedback loop results in a positive real part of the poles, the system becomes unstable. In other words, the poles of the system reside in the negative

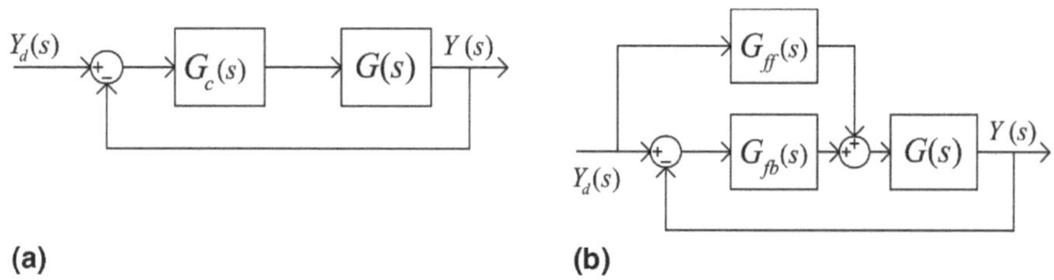

Fig. 16.10 (**a**) The closed-loop control system diagram and (**b**) feedback controller with a feedforward block

part of the complex s-plane near the imaginary axis, and in the case of an unsuccessful controller design these poles travel to the positive side of the complex s-plane. The accuracy of these methods is naturally highly dependent upon the sensors used in the feedback process. The system cannot be more accurate than the displacement measured. Additionally, the noise of the measurement is fed back to the system and therefore can cause instability in some cases (Aloisi et al. 2006). The motion of the actuator can be sensed either directly with a CCD-camera or indirectly with positioning sensors. A variety of sensors have been used such as inductive LVDT sensors, LASER, and capacitive and strain gauge sensors (Dinulovic and Gatzen 2006; Furuta et al. 2002; Harb et al. 1992; Sacconi et al. 1999).

The basic feedback control system is shown in Fig. 16.10a. It consists of the controller G_c and the actuator G. The output signal Y is subtracted from the input signal Y_d, thus creating a negative feedback loop. The most used control block is the PID-controller (PID), which consists of proportional gain (P), integrator (I), and a derivator (D). This type of controller is usually tuned to match the performance criteria (overshoot, rise-time, etc.) of the application. Many commercial piezo-controllers feature a tunable PID-controller (PI GmbH & Co. 2001; Piezosystem jena GmbH 2007). Other controllers that have been used in micropositioning applications include, among others, various learning methods, neural network-based methods, and robust controllers (Adda et al. 2005; Ku et al. 2000; Lv and Wei 2004; Tan et al. 2004).

The performance of the feedback system, especially its speed, can be improved by incorporating a feedforward controller in the system as shown in Fig. 16.10b (Shen and Wei 2006). The feedforward controller can, for example, be a model-based feedforward controller designed to improve the hysteresis performance.

Another closed-loop method used in micropositioning is visual servoing (Kallio et al. 1998). This method relies on the use of a CCD camera as the sensor. In this case, the camera with its accompanying optics produces a picture of the micropositioner and the positioning target. From this picture, with the aid of machine vision methods, the motion and the position of the target are measured. The output coordinates are usually two dimensional, but with the use of out-of-focus information in the picture, the z-coordinates can also be determined (Feddema and Simon 1998). The downside of this method is the need for a bulky camera, and it is also computationally intensive and can therefore be slow. The benefits include the actual position of the positioning target and the possibility of teleoperation where an operator controls the positioner using the image information via, for example, a computer screen (Kallio et al. 1998).

16.6 Piezoelectric Micropositioning Applications

In large earthbound telescope systems, piezoelectric actuators are used to compensate atmospheric turbulence and vibration by the micropositioning of mirrors. The effective resolution of such systems is up to ten times higher by a real-time correction of image shifts than without. In this case, the actuators are relatively large in size but can typically obtain a resolution in the nanoradian range with a tip/tilt range of 150–600 μrad and resonance frequencies in the range of 280–900 Hz while moving the mirror with a size of 100–314 mm in diameter.

The correction of image fluctuations requires high accelerations that produce significant inertia and forces toward the telescope structure via moving mass of the large mirrors. To avoid resulting in unwanted vibrations, integrated momentum compensation systems in the tip/tilt platforms are used. These piezoelectric systems can compensate vibrations caused by movement of the mirrors up 95% efficiency. In addition to telescopes, piezoelectric hexapods (Fig. 16.11a) have been utilized in, e.g., nondestructive testing, biomechanical dental applications, satellite antenna control, surgical robot, and optical fiber alignment (http://www.hexapods.net; PI GmbH & Co. 2001).

In diesel engines of cars, injection units based on piezoelectric actuators have recently been elaborated (Fig. 16.11b). For example, Bosch, Siemens, and Delphi

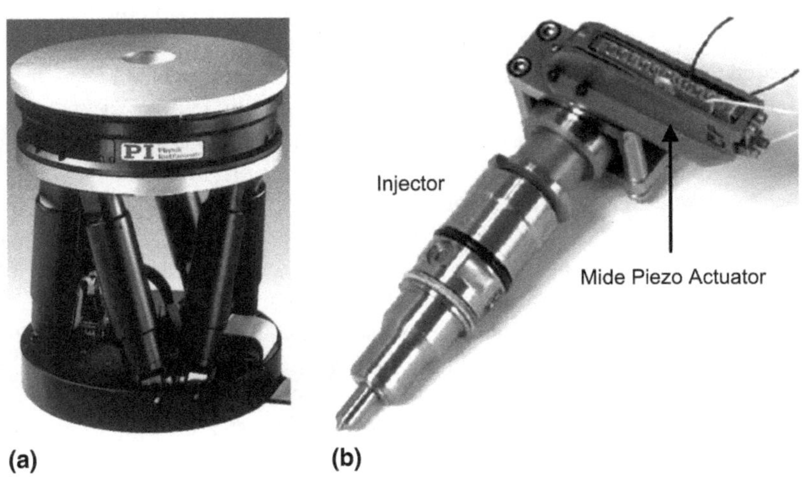

Fig. 16.11 (**a**) Piezoelectric hexapod used for active secondary mirror in NASA Infrared Telescope Facility (IRTF) on Mauna Kea, Hawaii (© PI GmbH & Co. All rights reserved), (**b**) piezoelectric injection unit (© 2004 Society of Photo-Optical Instrumentation Engineers) (MacLachlan et al. 2004)

have introduced piezoactuator-based diesel fuel injection systems. The piezoelectric actuators offer short switching times with a fast response, high actuation force delivery, and minimum waste of electrical energy (MacLachlan et al. 2004).

Faster actuation enables more accurate injection of fuel, which can also be carried out in different stages of combustion to optimize the process. By this means, the process becomes environmentally cleaner as lower amounts of CO_X and NO_X are produced by the combustion. Additionally, optimized combustion increases engine power and decreases consumption of fuel together with decreased engine weight (MacLachlan et al. 2004; Randall et al. 2005; Yanagihara et al. 1997). Typically, these applications require advanced actuator structures, which provide moderate force and displacement characteristics. Therefore the development work is steered toward direct extensional and externally leveraged piezoactuators obtaining stiffness above 80 N/mm and displacements of several hundred micrometers. To maximize actuator bandwidth and work efficiency, the stiffness of the actuator is maximized, and it is set to work against a spring with equal stiffness (MacLachlan et al. 2004; Randall et al. 2005).

In the AFM or scanning probe microscope (SPM), piezoelectric micropositioners are used to scan features down to atomic scale and adjust the distance of the probe toward the measured surface via signals from a photodetector (Giessibl 2003). Stable operation and high resolution of the scanner is a necessity for the system, otherwise, drift and low resolution can produce poor targeting of the sample, bad image, and false measurement results. Additionally, since a large number of measurement points are needed for picture scanning, the scanning frequency should be as high as possible, i.e., not limited by actuator but the measurement method itself. Conventionally the control of the piezoactuator tube in an AFM is based on open-loop scanning with all X, Y, and Z directions. For example, the range of the horizontal and vertical movements in Veeco dimension 3100 AFM scanners is $\sim$90 and 6 μm, respectively. With this scanning level and a 16-bit A/D converter, the resolution of horizontal and vertical displacement is commonly better than $\sim$1.4 and $\sim$0.1 nm, respectively. However, with smaller scanning areas, the resolution can be even down to $\sim$0.015 nm (Veeco Instruments Inc. 2003, 2004). Good control systems are also crucial since the movement of the tip is not linear, with the control voltage being most sensitive at its lower values. These devices thus form a good example where two different control systems are required at the same time.

In the camera industry, piezoelectric actuators have been used in several applications. The lenses on single lens reflex (SLR) cameras can utilize integrated piezoelectric ultrasonic motors instead of the traditional in-body motor drive. This improves the focusing speed and makes the focusing very silent. The sound is reduced by means of the high inherent torque and low rotational speed making reduction gears unnecessary. Also the operational frequencies are in the ultrasonic range making the focusing inaudible (Canon Inc. 2006; Uchino 1998). Other applications for piezoelectric actuators are the antishake mechanisms employed in some cameras. The imaging sensor on these cameras is moved by a piezoelectric actuator to reduce the effect of hand shake when making an exposure without a tripod. The required displacement range in this application is usually too high for a simple piezoactuator, thus different displacement amplification schemes are used.

16.7 Future Opportunities

There is a continuous demand for improvement of micropositioning performance through novel piezoelectric structures and materials. The basic issues such as lower hysteresis and creep together with larger bandwidth, degrees of freedom, forces and displacements form everlasting challenges in the future. In recent year, however, nanoscale positioning has become more and more important in a variety of applications, thus also accelerating the development of piezoelectric materials and structures as well as their associated control and sensor systems. Multiple stage positioning, where all X, Y, and Z-directions positioning and both nanoscale and microscale positioning are already performed, is able to offer novel, accurate, or at least more convenient use of commercial application and research equipment. A great potential, however, will be in the area of automation and miniaturization, where compact and inexpensive micropositioning devices with well-developed integrated, efficient control and sensor systems are required. An important fact will surely be the price at which the systems can be produced in large quantities. In this sense, the number of parts to be assembled, the cost of materials and structures, the overall reliability and multifunctionality together with the opportunities for customization and a wide choice of displacement amplification mechanisms will enable a *quantum jump* in system performance and reliability at all levels.

References

Adda C, Laurent GJ, Le Fort-Piat N (2005) Learning to control a real micropositioning system in the STM-Q framework. In: Proc 2005 IEEE Int Conf Robotics and Automation, IEEE, Piscataway, NJ, 4569–4574.

Allahverdi M, Danforth SC, Jafari M, Safari A (2001) Processing of advanced electroceramic components by fused deposition technique. J Eur Ceram Soc 21: 1485–1490.

Aloisi G, Santucci A, Carla M, Dolci D, Lanzi L (2006) Electronic linearization of piezoelectric actuators and noise budget in scanning probe microscopy. Rev Sci Instrum 77: 073701.

Ang WT, Garmón A, Khosla PK, Riviere CN (2003) Modeling rate-dependent hysteresis in piezoelectric actuators. In: Proc 2003 IEEE/RSJ Int Conf Intelligent Robots and Systems, IEEE, Piscataway, NJ, 1975–1980.

Bexell M, Johansson S (1999) Characteristics of a piezoelectric miniature motor. Sens Actuator A Phys 75: 118–130.

Canon Inc. (2006) EF lens work III. http://www.canon-europe.com/support/documents/digital_slr_educational_tools/en/ef_lens_work_iii_en.asp (accessed 14 March 2007).

Choi HD, Kim JH, Kim S, Kwak YK (2005) Development of the piezoelectric motor using momentum generated by bimorphs. Rev Sci Instrum 76: 105109.

Chopra I (2002) Review of state of art of smart structures and integrated systems. AIAA J 40(11): 2145–2187.

Comstock RH (1981) Charge control of piezoelectric actuators to reduce hysteresis effects. U.S. Patent 4,263,527.

Davis M, Damjanovic D, Hayem D, Setter N (2005) Domain engineering of the transverse piezoelectric coefficient in perovskite ferroelectrics. J Appl Phys 98: 014102.

Davis M, Damjanovic D, Setter N (2006) Temperature dependence of the direct piezoelectric effect in relaxor-ferroelectric single crystals: Intrinsic and extrinsic contributions. J Appl Phys 100: 084103.

Dinulovic D, Gatzen HH (2006) Microfabricated inductive micropositioning sensor for measurement of a linear movement. IEEE Sens J 6(6): 1482–1487.

Dynamic Structures & Materials LLC, Catalog. http://www.dynamic-structures.com/piezo_actuators.html (accessed 13 March 2007).

Feddema JT, Simon RW (1998) Visual servoing and CAD-driven microassembly. IEEE Robot Autom Mag 5(4): 18–24.

Furuta A, Munekata M, Higuchi T (2002) Precise positioning stage driven by multilayer piezo actuator using strain gauge. Jpn J Appl Phys 41: 6283–6286.

Germano C (1971) Flexure mode piezoelectric transducers. IEEE Trans Audio Electroacoust 19(1): 6–12.

Giessibl FJ (2003) Advances in atomic force microscopy. Rev Mod Phys 75(3): 949–982.

Haertling GH (1994) Rainbow ceramics – A new type of ultra-high-displacement actuator. Am Ceram Soc Bull 73(1): 93–96.

Haertling GH (1997) Rainbow actuators and sensors: A new smart technology. In: Proc SPIE Int Soc for Optical Eng, SPIE, Bellingham, WA, 3040: 81–92.

Haertling GH (1999) Ferroelectric ceramics: History and technology. J Am Ceram Soc 82(4): 797–818.

Harb S, Smith ST, Chetwynd DG (1992) Subnanometer behavior of a capacitive feedback, piezoelectric displacement actuator. Rev Sci Instrum 63(2): 1680–1689. http://konicaminolta.com/about/research/core_technology/picture/antiblur.html (accessed 8 February 2007). http://www.hexapods.net (accessed 18 April 2007).

Hubbard NB, Culpepper ML, Howell LL (2006) Actuators for micropositioners and nanopositioners. Appl Mech Rev 59(6): 324–334.

Kallio P, Lind M, Kojola H, Zhou Q, Koivo HN (1996) An actuation system for parallel link micromanipulators. In: Proc 1996 IEEE/RJS Int Conf Intelligent Robots and System, IEEE, New York, 2: 856–862.

Kallio P, Zhou Q, Koivo HN (1998) Control issues in micromanipulation. In: Proc 1998 Int Symp Micromechatronics and Human Science, IEEE, Piscataway, NJ, 135–141.

Kim KY, Park KH, Park HC, Goo NS, Yoon KJ (2005) Performance evaluation of lightweight piezo-composite actuators. Sens Actuator A Phys 120: 123–129.

Ko H-P, Kang C-Y, Kim J-S, Borodin SN, Kim S, Yoon S-J (2006) Constructions and characteristics of a tiny piezoelectric linear motor using radial mode vibrations. J Electroceram 17: 603–608.

Ku S-S, Pinsopon U, Cetinkunt S, Nakajima S-I (2000) Design, fabrication, and real-time neural network control of a three-degrees-of-freedom nanopositioner. IEEE-ASME Trans Mechatron 5(3): 273–280.

Li G, Furman E, Haertling GH (1997) Stress-enhanced displacements in PLZT Rainbow actuators. J Am Ceram Soc 80(6): 1382–1388.

Li X, Shih WY, Aksay IA (1999) Electromechanical behaviour of PZT-brass unimorphs. J Am Ceram Soc 82(7): 1733–1740.

Lv Y, Wei Y (2004) Study on open-loop precision positioning control of a micropositioning platform using a piezoelectrical actuator. In: Proc Fifth World Congress on Intelligent Control and Automation, IEEE, Piscataway, 1255–1259.

MacLachlan BJ, Elvin N, Blaurock C, Keegan NJ (2004) Piezoelectric valve actuator for flexible diesel operation. In: Proc SPIE Int Soc Optical Eng, SPIE, Bellingham WA, 5388: 167–178.

May WG (1975) Piezoelectric electromechanical translation apparatus. U.S. Patent 3,902,084.

Morozov M, Damjanovic D, Setter N (2005) The nonlinearity and subswitching hysteresis in hard and soft PZT. J Eur Ceram Soc 25: 2483–2486.

Nakamura K, Ando H, Shimizu H (1987) Ferroelectric domain inversion caused in LiNbO3 plates by heat treatment. Appl Phys Lett 50(20): 1413–1414.

Niezrecki C, Brei D, Balakrishnan S, Moskalik A (2001) Piezoelectric actuation: State of the art. Shock Vib Dig 33(4): 269–280.
Park SE, Shrout TR (1997) Ultrahigh strain and piezoelectric behaviour in relaxor based ferroelectric single crystals. J Appl Phys 82: 1804–1811.
Park T, Kim B, Kim M-H, Uchino K (2002) Characteristics of the first longitudinal-fourth bending mode linear ultrasonic motors. Jpn J Appl Phys 41: 7139–7143.
Pearce DH, Hooley A, Button TW (2002) On piezoelectric super-helix actuators. Sens Actuator A Phys 100: 281–286.
PI GmbH & Co (2001) Micropositioning, nanopositioning, nanoautomation solutions for cutting-edge technologies. PI catalog.
Piezo Systems Inc. (2006) Catalog no. 7. http://www.piezo.com (accessed 13 March 2007).
Piezomechanik GmbH (2006) Catalogs. http://www.piezomechanik.com (accessed 13 March 2007).
Piezosystem jena GmbH Datasheets. http://www.piezojena.com (accessed 13 March 2007).
Pons JL, Rocon E (2006) Scaling of piezoelectric actuators: A comparision with traditional and other technologies. Bol Soc Esp Ceram Vidr 45(3): 132–138.
Randall CA, Kelnberger A, Yang GY, Eitel RE, Shrout TR (2005) High strain piezoelectric multilayer actuators – A material science and engineering challenge. J Electroceram 14: 177–191.
Ronkanen P, Kallio P, Koivo H (2002) Current control of piezoelectric actuators with power loss compensation. In: Proc 2002 IEEE/RSJ Int Conf Intelligent Robots and System, IEEE, Piscataway, NJ, 2: 1948–1953.
Ru C-H, Sun L, Kong M-X (2005) Adaptive inverse control for piezoelectric actuator based on hysteresis model. In: Proc Fourth Int Conf Machine Learning and Cybernetics, IEEE, Piscataway, NJ, 3189–3193.
Sacconi A, Picotto GB, Pasin W (1999) The IMGC calibration setup for microdisplacement actuators. IEEE Trans Instrum Meas 48(2): 483–487.
Sawyer CB (1931) Piezoelectric device. U.S. Patent 1,802,782.
Schaller R, Fantozzi G, Gremaud G (2001) Mechanical spectroscopy Q^{-1} 2001: With applications to materials science. In: Materials Science Forum, vol. 366–368, Trans Tech, Switzerland.
Schwartz RW, Moon Y-W (2001) Domain configuration and switching contributions to the enhanced performance of Rainbow actuators. In: Proc Int Soc Optical Eng, SPIE, Bellingham WA, 4333: 408–417.
Schwartz RW, Cross LE, Wang QM (2001) Estimation of the effective d31 coefficients of the piezoelectric layer in rainbow actuators. J Am Ceram Soc 84(11): 2563–2569.
Shen G, Wei Y (2006) Study on nonlinear model of piezoelectric actuator and accurate positioning control strategy. In: Proc Sixth World Congress on Intelligent Control and Automation, IEEE, Piscataway, NJ, 8356–8360.
Spanner K (2006) Survey of the various operating principles of ultrasonic piezomotors. White paper available at http://www.piusa.us/technotes/actuator2006_surveyofthevariousoperatingprinciplesofultrasonicpiezomotors_c.pdf.
Sun L, Ru C, Rong W (2004) Hysteresis compensation for piezoelectric actuator based on adaptive inverse control. In: Proc Fifth World Congress on Intelligent Control and Automation, IEEE, Piscataway, NJ, 5036–5039.
Tan KK, Ng SC, Huang SN (2004) Assisted reproduction system using piezo actuator. In: Proc Int Conf Communications, Circuits and Systems IEEE, Piscataway, NJ, 2: 1200–1203.
Tenzer PE, Mrad RB (2004) On amplification in inchworm precision positioners. Mechatronics 14: 515–531.
Uchino K (1997) Piezoelectric actuators and ultrasonic motors. Kluwer Academic, Boston.
Uchino K (1998) Piezoelectric ultrasonic motors: Overview. Smart Mater Struct 7: 273–285.
Uchino K, Yoshizaki M, Kasai K, Yamamura H, Sakai N, Asakura H (1987) "Monomorph actuators" using semiconductive ferroelectrics. Jpn J Appl Phys 26(7): 1046–1049.
Uchino K, Cagatay S, Koc B, Dong S, Bouchilloux P, Strauss M (2004) Micro piezoelectric ultrasonic motors. J Electroceram 13: 393–401.
Veeco Instruments Inc. (2003) SPM Training Notebook 004–130–00 Revision E.

Veeco Instruments Inc. (2004) Dimension 3100 Manual Revision D.
Wang Q-M, Cross LE (1998) Determination of Young's modulus of the reduced layer of a piezoelectric RAINBOW actuator. J Appl Phys 83 (10): 5358–5363.
Wang Q-M, Cross LE (1999) Tip deflection and blocking force of soft PZT-based cantilever RAINBOW actuators. J Am Ceram Soc 82(1): 103–110.
Wolff A, Cramer D, Hellebrand H, Probst I, Lubitz K (1994) Optical two channel elongation measurement of PZT piezoelectric multilayer stack actuators. In Proc Ninth IEEE Int Symp Appl Ferroelectrics, IEEE, New York, pp 755–757.
Wolny WW (2000) Piezoceramic thick films – Technology and applications state of the art in Europe. In: Proc IEEE 12th Int Symp Appl Ferroelectrics, IEEE, Piscataway, NJ, 1: 257–262.
Yanagihara H, Sato Y, Mizuta J (1997) A study of DI diesel combustion under uniform higher-dispersed mixture formation. JSAE Rev 18(3): 247–254.
Zhang QM, Zhao J (1999) Electromechanical properties of lead zirconate titanate piezoceramics under the influence of mechanical stresses. IEEE Trans Ultrason Ferroelectr Freq Control 46(6): 1518–1526.

Chapter 17
Piezoelectric Actuator Designs

Aydin Doğan* and Erman Uzgur

Piezoelectric actuators are getting immiscible parts of many important electromechanical and smart systems. Smart systems consist mainly of sensors, actuators, and signal processing units. Actuators are the responding units of many smart systems including those for active vibration and noise control, valve, shutter, focal lens, and many others. Increased demand for actuators with high displacement, high generative force, and quick response time has led to a search for new actuator materials and new designs. In this chapter, first, piezoelectric actuators were compared with magnetically active and thermally active actuators. Second, piezoelectric actuators and their design, especially traditional piezoelectric transducers with newly designed flextensional transducers were compared. Finally, application-related issues have been reviewed.

17.1 Introduction

One of the qualities that distinguish living systems from inanimate matter is the ability to adapt to changes in the environment. Smart materials have the ability to perform both sensing and actuating functions and are, therefore, capable of imitating this rudimentary aspect of life. In this review, we focus on actuator materials and solid-state actuator design. The need for high-accuracy displacement elements in such fields as optics, precision machinery, and small motors has led to renewed emphasis on displacement devices. Since the last decade, there is a tremendous increase in the demand for precise motion in optical systems and fluid flow control, which can only be provided by piezoelectric actuators. It would be nice to see and compare all the actuator materials in general perspective and focus on the piezoelectric ones.

A. Doğan
Department of Materials Science & Engineering, Anadolu University, Eskisehir, Turkey

A. Safari, E.K. Akdoğan (eds.) *Piezoelectric and Acoustic Materials for Transducer Applications.*

17.2 Displacement Transducers: Actuators

Displacement transducers, in other words actuators, can be defined as the devices that exhibit deformation or elongation under applied energy input. In other words, they convert various energy inputs to the mechanical displacement or strain. This energy can be electric field, magnetic field, and pressure or temperature change. Actuators can be classified into two main groups: conventional (large displacement) and solid-state (small but precise displacement) actuators (Fig. 17.1).

There are three types of conventional displacement transducers: oil pressure cylinders, servo or step motors, and voice coils. The oil pressure cylinder operates on hydraulic principles, where the displacement piston is driven by oil pressure from another piston. The principal disadvantages are large space requirements and long response times. The servo or step motor converts the revolutions of an electromagnetic motor into linear movement by using a gear mechanism. Mechanical backlash is a difficult problem for servo/step motor displacement transducers.

The voice coil (speaker) operates through a combination of electromagnetic coils and springs. The generative force of the voice coil is small compared with its input power, and its response time is slow. Each of the conventional displacement devices is capable of generating large displacements in the millimeter to centimeter range, but they are inadequate for precise positioning in the micrometer range.

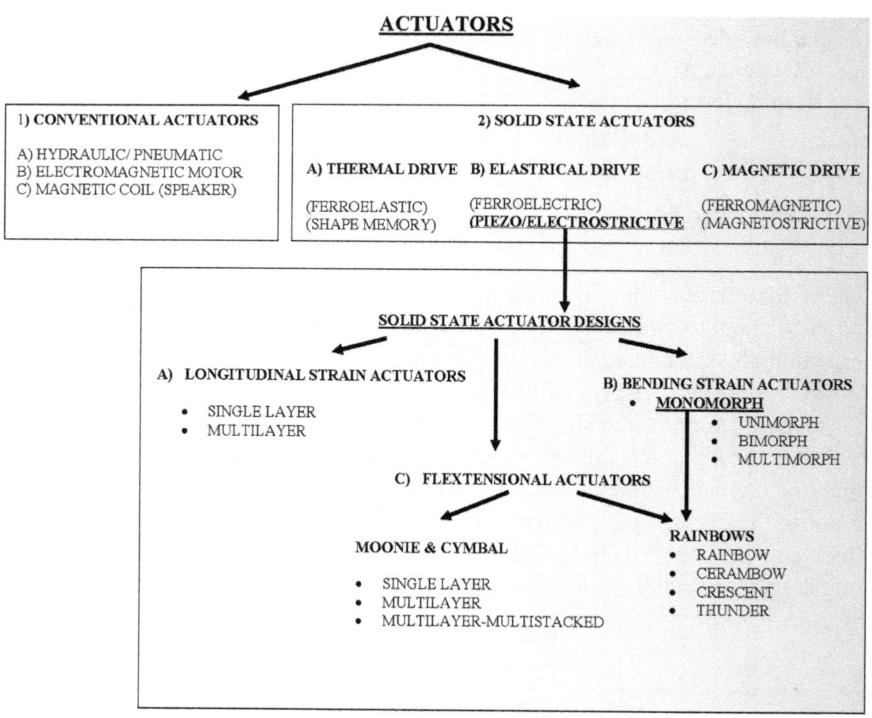

Fig. 17.1 Classification of actuators

Table 17.1 Comparison of solid-state actuators

Characteristic	Piezoelectric	Electrostrictive	Magnetostrictive	Shape memory
Ferroic class	Ferroelectric	Ferroelectric	Ferromagnetic	Ferroelastic
Switching force	Electric field	Electric field	Magnetic field	Thermal stress
Maximum strain (%) (Ref.)	0.1 (Jaffe et al. 1971; furuka and Uchino 1986)	0.1–0.2 (Cross et al. 1980; Uchino et al. 1980)	0.1–0.2 (Hathaway and Clark 1993)	7–10 (Wayman and Shimizu 1972)
Fastest response	μs	μs	μs	s
Generative force (N/cm^2)	3,500	5,000	3,500	20,000
Efficiency (%)	50	50	80–90	3
Example	$Pb(Zr, Ti)O_3$	$Pb(Mg_{1/3}, Nb_{2/3})O_3$	$(Tb, Dy)Fe_2$	$Ni_{1-x}Ti_x$

A second category of displacement transducers consists of a number of different solid-state actuators. The characteristics of the solid-state actuators are summarized in Table 17.1. This group can be further classified into three categories based on the type of driving mechanism:

1. Shape memory effect (thermal drive)
2. Magnetostriction (magnetic drive)
3. Piezostriction/electrostriction (electrical drive)

17.2.1 Shape Memory Effect

Materials that exhibit shape memory effects are the most important thermally driven actuators. The shape memory effect is a consequence of a crystallographically reversible martensitic phase transformation occurring in the solid state (Wayman and Shimizu 1972; Wayman 1993). After deformation at low temperature, a shape memory alloy will regain its original shape when heated above the phase transition temperature. The shape memory effect can be classified as either a *one-way* memory or a *two-way* memory. For the one-way memory effect, the deformed sample recovers its undeformed shape upon heating in a one-time only operation. This undeformed shape then remains constant when the sample is subjected to thermal cycling. In the two-way memory, the shape memory effect is tied to the occurrence of a martensitic phase transformation during cooling and its subsequent reversal during heating. Thermal cycling can be repeated indefinitely. A schematic comparing one-way memory with two-way memory is shown in Fig. 17.2.

Shape memory alloys can exhibit large recoverable strains as large as 7%. Because of these large recoverable strains, shape memory alloys have been proposed as alternatives for solenoids, motors, and bimorph-type actuators, as well as for use in flexible robot structures (Duering et al. 1990). However, the large thermal energy

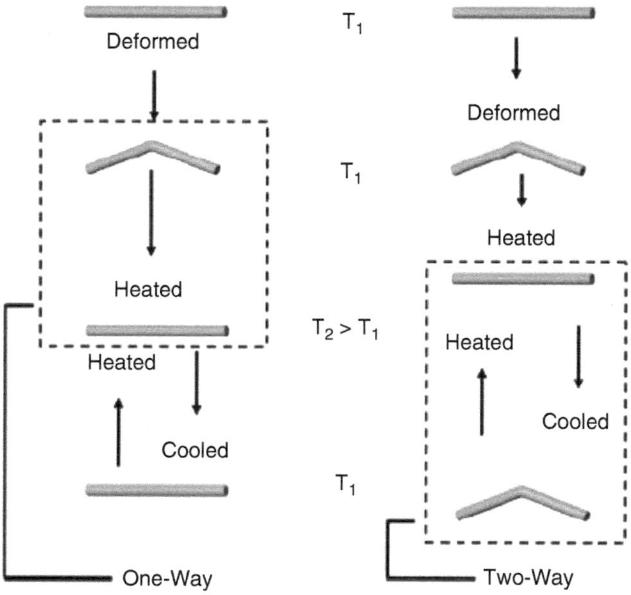

Fig. 17.2 Comparison of one-way and two-way shape memory effects (Wayman and Shimizu 1972)

requirements for martensitic phase transformation together with large hysteresis and long response times are important drawbacks for shape memory alloys. Some of the important shape memory materials are Nitinol, a Ni–Ti alloy, as well as alloys of Cu–Zn–Al, and Cu–Al–Ni.

17.2.2 Magnetostriction

Magnetostriction is defined as the change in dimension of a magnetic material caused by a change in its magnetic state (Clark 1980). Magnetostrictive materials can be used as both sensors and actuators. The magnetostrictive strain arises from a reorientation of the atomic magnetic moments. High-power magnetostrictive actuators can produce forces exceeding 50 MPa with strains in the order of 0.6% (Hathaway and Clark 1993). High reliability, ruggedness, and imperviousness to adverse environmental conditions are the advantages of magnetostrictive materials. However, magnetostrictive materials have the problems of requiring a coil to create the necessary magnetic field. This, in turn, induces noise into adjacent electronic circuits and devices. The most important commercial room temperature magnetostrictive material is Terfenol-D, a $Tb_xDy_{1-x}Fe_2$ alloy. The strain vs. magnetic field curves for Terfenol-D rods under various conditions of prestress are shown in Fig. 17.3 (Clark et al. 1988).

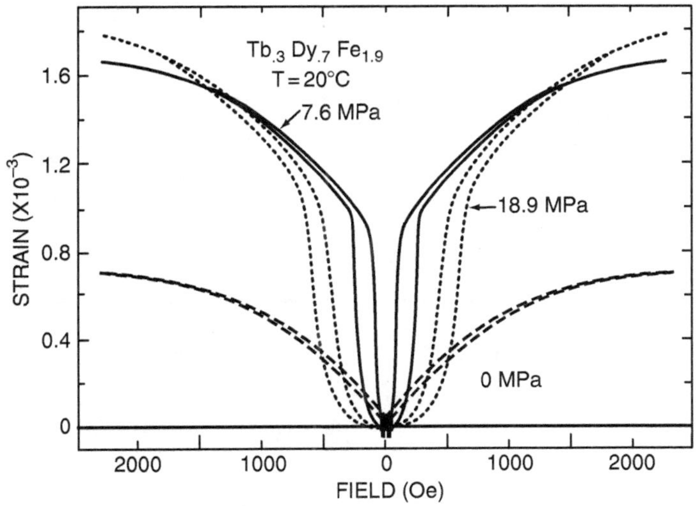

Fig. 17.3 Strain hysteresis of a magnetostrictor (Terfonal D) under varying compressive stress (Clark et al. 1988)

17.2.3 Piezoelectricity

Piezoelectricity is the phenomenon whereby electric polarization is generated in certain acentric crystals when they are subjected to mechanical stress (i.e., the direct effect). Materials showing this phenomenon must also show a geometrical strain (x), which is proportional to the applied electric field E, which is known as the converse effect (i.e., $x = \mathrm{dE}$).

Piezoelectric materials have got some certain advantages over magnetostrictives and shape memory alloys. Since they do not require any coil for driving, magnetic shielding is not necessary for piezoelectric actuators to avoid noise. Besides, they have high efficiency and large stored energy. Thus, they are most convenient to be miniaturized among the solid-state actuators. They have also short response time that makes them suitable for high-precision applications.

Natural crystals such as quartz, tourmaline, and zinc blende are the classical piezoelectric materials. For many years, these materials have served as transducers for converting mechanical energy into electrical energy and vice versa (Cady 1964). In general, natural crystals have rather low piezoelectric coefficients. Ceramic piezoelectric materials were developed in the second half of the twentieth century and have been constantly improved since then. Modified lead zirconate titanate $[\mathrm{Pb(Zr,Ti)O_3}]$ ceramics (commonly known as PZT) are the leading materials for piezoelectric applications (Jaffe et al. 1971).

The strain curve of a lanthanum-modified lead zirconate titanate, PLZT $[(\mathrm{Pb,La})(\mathrm{Zr,Ti})\mathrm{O_3}]$ (7/62/38), is shown in Fig. 17.4a (Haertling and Land, 1971; Furuka and Uchino 1986). The induced strain is nearly proportional to the applied

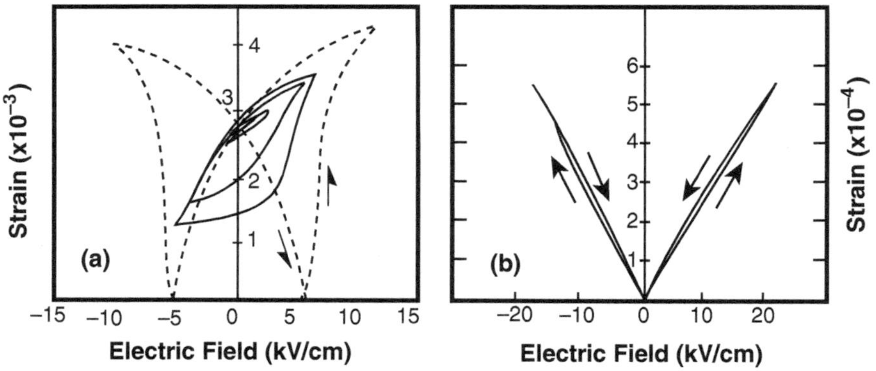

Fig. 17.4 Electric field-induced strain hysteresis of piezostrictors (**a**) (Pb, La)(Zr, Ti)O_3 and (**b**) Ba(Sn, Ti)O_3 (Haertling and Land 1971; Furuka and Uchino 1986; Cieminski and Beige 1991)

field for low field levels. However, the strain curve deviates from this linear trend, and significant hysteresis is exhibited because of domain reorientation as the field becomes larger.

The magnitude of the piezoelectric coefficients depends markedly on dopants and defect structure because of their influence on domain wall motion. This in turn controls the nature of the hysteresis loop in the piezoelectric ceramic. The interaction of the domains and defects leads to so-called soft and hard piezoelectric compositions. Piezoelectrically soft materials are characterized by high piezoelectric constants and high hysteresis as a result of relatively mobile domain walls. In hard piezoelectric ceramics the domain wall motion is inhibited, resulting in lower piezoelectric constants and reduced hysteresis. Soft piezoelectric materials are preferred for most multilayer and bimorph actuator applications because of their high strain. For some actuator applications, which require nonhysteretic response, hard piezoelectric ceramic can be preferable.

Solid solutions of barium tin titanate [Ba(Sn, Ti)O_3] are an interesting new family for actuator applications (Cieminski and Beige 1991). Because of its small coercive field, the composition [Ba($Sn_{0.15}$ $Ti_{0.85}$)O_3] exhibits an unusual strain curve, in which domain reorientation occurs only at low fields. This is followed by a long linear range at higher fields (Fig. 17.4b).

17.2.4 Electrostriction

When an electric field is applied to a centrosymmetric dielectric material, it produces a strain proportional to the square of the field. This phenomenon is called the electrostrictive effect. The strain, x, is approximately equal to ME^2 where M is the field electrostrictive coefficient and E is the electric field. For high-permittivity solids, however, electrostrictive strain is proportional to the square of the polarization, i.e., $x = QP^2$ where Q is the electrostriction coefficient and P is the polarization.

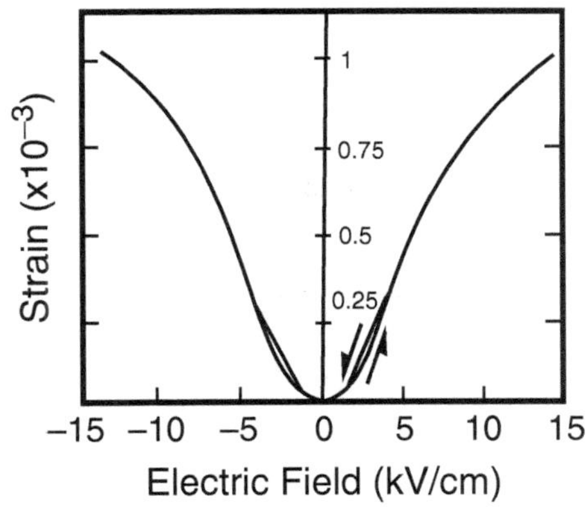

Fig. 17.5 Electric field-induced strain hysteresis of electrostrictor $Pb(Mg_{1/3}Nb_{2/3}Ti)O_3$ (Cross et al. 1980)

Because electrostriction is a result of the polarization induced by the applied field, electrostriction may occur in all crystals whether or not the crystals have polarity.

Electrostriction in PMN [Pb $(Mg_{1/3}Nb_{2/3}O_3]$ ceramics is extraordinarily large, with strains exceeding 0.1% (Cross et al. 1980). An interesting feature of these materials is the near absence of hysteresis. Figure 17.5 shows the strain hysteresis of a PMN-PT [(Pb $(Mg_{1/3}Nb_{2/3}O_3) - (PbTiO_3)$] ceramic. Displacement of the piezoelectric ceramics can be positive or negative depending on applied electric field. Electrostrictive oxide ceramics generally elongate parallel to the applied field and contract perpendicular to the field but a polymer such as PVDF contracts in the field direction.

17.2.5 Electric-Field-Induced Phase-Transition-Related Strain

The phase transition from an antiferroelectric to a ferroelectric phase can induce larger strains than those found in either piezoelectric or electrostrictive materials (Kittel 1951; Cross 1967). A unique characteristic of the phase-transition-induced strain in the antiferroelectric phase is its isotropic volume expansion, similar to thermal expansion (Uchino et al. 1980; Uchino 1989). With appropriate compositions, piezoelectric/electrostrictive materials can exhibit two kinds of phase transitions related to strain: a shape memory effect and digital displacement.

Figure 17.6a shows the strain hysteresis of $Pb_{0.99}Nb_{0.02}[(Zr_{0.6}Sn_{0.4})_{1-y}Ti_y]_{0.98}O_3$ (also known as PNZST), which exhibits a shape memory effect (Oh et al. 1990). Once the ferroelectric phase has been induced, it is retained even at zero electric field. It can, however, be erased with the application of small reverse bias fields. This shape memory ceramic is used in energy-saving actuators (Uchino 1993).

On the other hand, for the phase-transition-related digital displacement, two well-defined (on/off) strain states can be achieved. A rectangular-shaped hysteresis for a digital displacement transducer is shown in Fig. 17.6b. The longitudinally

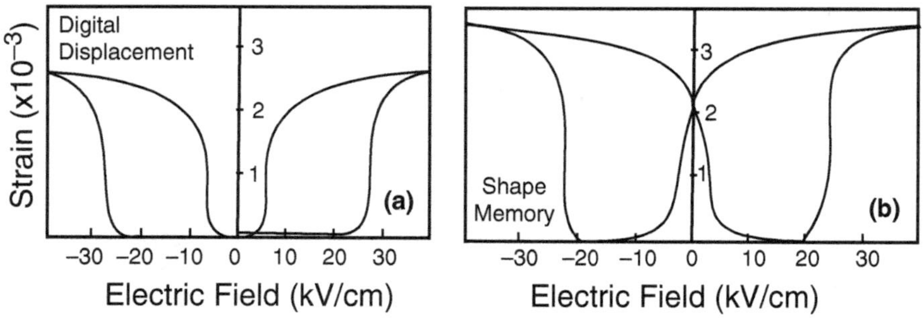

Fig. 17.6 Phase transition-related strain in Nb-doped Pb(Zr, Sn, Ti)O_3 – *Y*-axis same for both graphs (Cross 1967; Uchino et al. 1980; Uchino 1989; Oh et al. 1990)

induced strain approaches 0.1%, which is rather large when compared with the normal piezostrictors and electrostrictors.

The fundamental characteristics of solid-state ceramic actuators can be summarized as follows:

- Displacements of up to several tens of microns that can be controlled with a precision of 0.01 μm
- Response speeds in the order of 1–5 μs
- Generative forces as large as 3,500 N
- Driving power an order of magnitude smaller than electromagnetic motors

17.3 Piezoelectric Actuator Designs

There are certain rules that we have to keep in our mind for the piezoactuator designs. Piezoactuator can show linear and bending actuations, at certain speed with certain generative force, depending on the design. To achieve desired output from the piezoactuator it is necessary to apply certain voltage with certain waveform. Single-plate piezoactuator, which is a linear actuator with longitudinal displacement, can only give small displacement at high voltages. Hence, single-block piezoactuators have limited applications. For that reason multilayer linear actuators have been developed (Takahashi 1989). Multilayer linear piezoactuators exhibit longitudinal displacement with lateral contraction because of the poison effect. On the other hand, a bending-type piezoactuator exhibits bending displacement and reaches the highest value at the tip of the free end of the actuator. Monomorph, unimorph, bimorph, and multimorph-type bending actuators have been developed to simplify the applied electrical field and enhance the displacement.

Figure 17.7 shows the structure of multilayer actuators with various electrode patterning. The multilayer actuator is composed of a number of ceramic layers alternating with internal electrodes. Individual internal electrodes are electrically connected in parallel with two external electrodes attached to the sides. When an electric

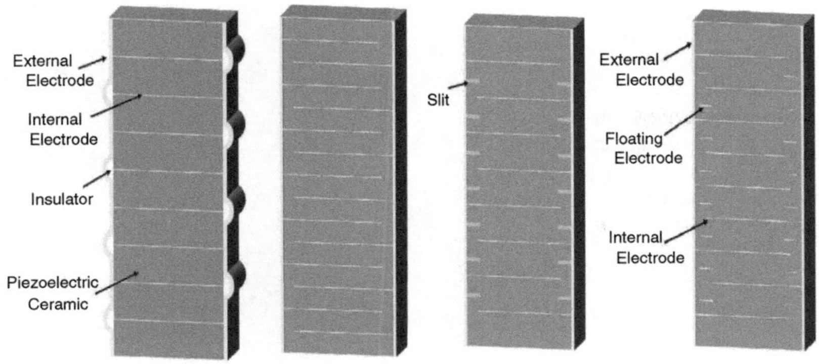

Fig. 17.7 Multilayer actuators with various electrode patterning

field is applied to the element, the element expands along the longitudinal direction in accordance with the converse piezoelectric effect. On the linear multilayer actuator, d_{33} is the dominant piezoelectric coefficient, which affects the displacement performance of the actuator. Depending on the application area, both soft and hard piezoelectric materials can be used. As it is well known ceramic materials are not strong under tensile stresses. Applied electric field creates strain, meanwhile local stresses have only been created. Under cycling loading these stress-concentrated regions have potential to cause failure. For that reason, various electrode patterns have been applied on piezoactuators, and extensive studies have been conducted for reliable actuators. Important features of multilayer ceramic actuators are low driving voltage (100 V), quick response (10 μs), high generative force (3.0 kN), and high electromechanical coupling (Uchino 1993, 1997; Takahashi 1989; Aburatani and Uchino 1995). However, they only exhibit displacements in the range of 10 μm, which may not be sufficient for some applications.

Bending displacement can be created in two ways on piezoceramics. One of them is the nonuniform field distribution generated in the piezoelectric that causes an overall bending of the entire plate. Their operating principle is based on the combination of semiconductor contact phenomenon and the piezoelectric or electrostrictive effect (Fig. 17.8). The electric field is concentrated on one side that leads to Schottky barrier formation under certain applied voltage. This kind of actuator is called monomorph (Uchino and Giniewicz 2002).

In the other way, bending actuators basically consist of a piezoelectric plate or disc, which is glued together with nonelectroactive elements such as metal shims or disc. The simple bending actuator is unimorph. Under applied electrical field, piezoelectric element will extend in the longitudinal direction and shrink in the transverse direction, and this motion will cause bending action on passive metal or composite shim. Strain created on the active layer (piezoelectric or electrostrictive material) leads to bending of the whole device. Under expansion in the poling direction, the strain in the plane perpendicular to the poling direction undergoes a contraction (Fig. 17.9).

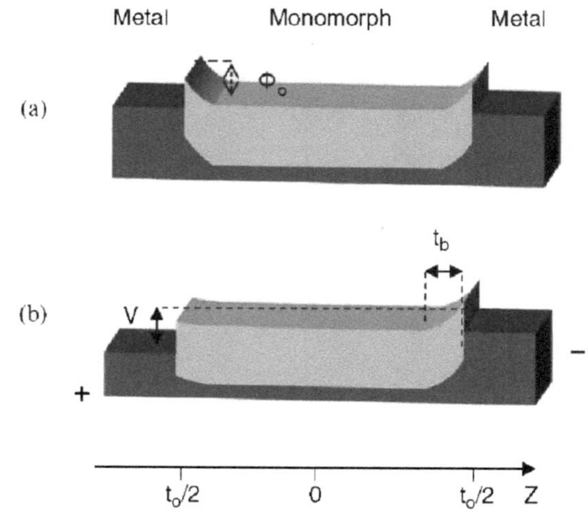

Fig. 17.8 Schottky barrier generated at the interface between a semiconductive (n-type) piezoceramic and metal electrodes. Energy barrier structure in monomorph (**a**) before and (**b**) after the electric field (After Uchino and Giniewicz 2002)

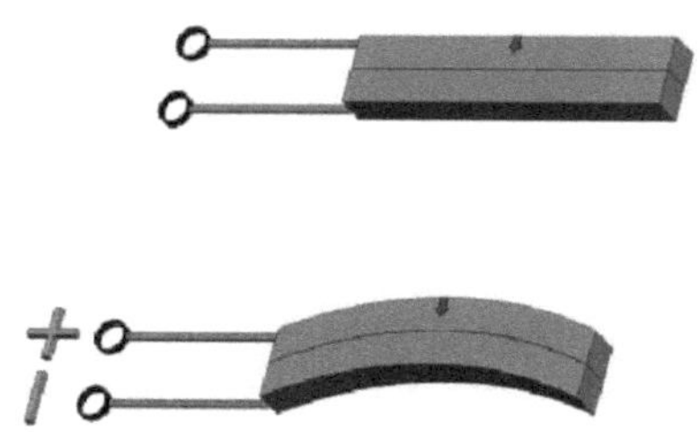

Fig. 17.9 Unimorph actuator

Displacement amplitude increases with the lateral dimensions. In the bending actuators, dominant piezoelectric coefficient is the d_{31}. Electric field applied to the piezoelectric elements parallel to the polarization direction, but contraction under applied electric filed will create the bending action. Best examples for this kind of actuator are the piezobuzzers. In this structure, 100-μm-thick piezoceramic is glued on the brass metal shim. Under applied electrical field, it created bending motion. This kind of actuator has lot of applications such as air acoustic device, micropumps, etc.

Bimorphs usually consist of two thin ceramic sheets bonded together with their poling directions opposed and normal to the interface. When an electric field is applied to a bimorph, one of the plates expands while the other contracts. This mechanism creates a bending displacement. Bimorph actuators can be constructed as parallel and series bimorphs (Fig. 17.10). In parallel bimorph, the polarization direction of the piezoplates is in the same direction, so one of the actuators is driven under reverse bias. On the series bimorph structure, polarization direction of the piezoplates is opposite to each other.

In bimorph transducers, for a total thickness of t and a transverse piezoelectric coefficient of d_{31}, the tip displacement is proportional to the voltage V and is expressed as follows:

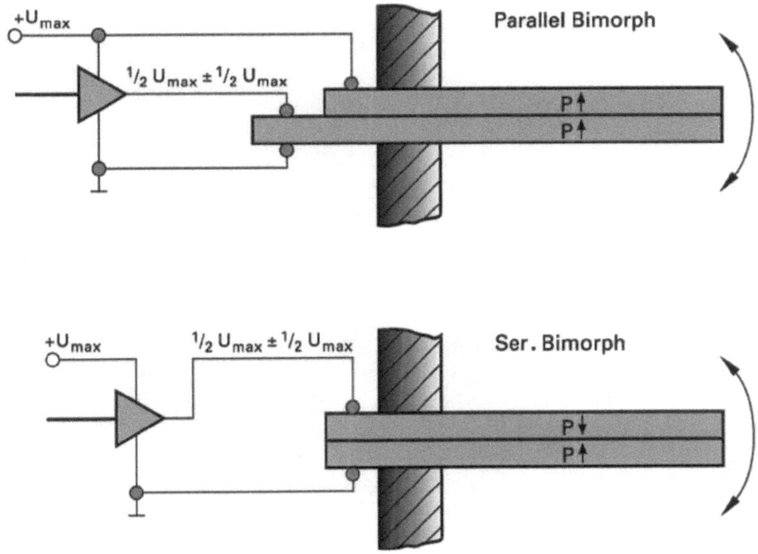

Fig. 17.10 Parallel and series bimorph actuators

$$\Delta u = \frac{3 d_{31} V L^2}{t^2}. \tag{17.1}$$

In contrast to the multilayer actuator, bimorphs are capable of generating large bending displacements of several hundred micrometers, but the response time (1 ms) and the generative force (1.0 N) are low.

When we compare multilayer and bimorph actuators based on three important characteristics (displacement, generative force, and response time), there is a sizeable gap between them with respect to performance. Neither provides both a relatively large displacement and an intermediate level of generative force, but recently there has been extensive effort to design new actuators to fill this gap.

The moonie, cymbal, RAINBOW, thunder, and cerambow are new designs developed within the past 15 years. These new designs provide moderate displacements in conjunction with moderate generative force (Fig. 17.11). Basically, all the new designs exhibit bending or flextensional displacements.

17.4 Flextensional Transducers

Flextensional transducers are mechanical amplifiers that couple the longitudinal strains in a ceramic bar or disk to the radial flexure of a metal or ceramic shell. The concept of the flextensional transducer originated from the electroacoustic foghorn used for ship navigation in the early 1920s (Rolt 1990). The theory and design of flextensional transducers was explained by Hayes in 1936 (Hayes 1936). Hayes used

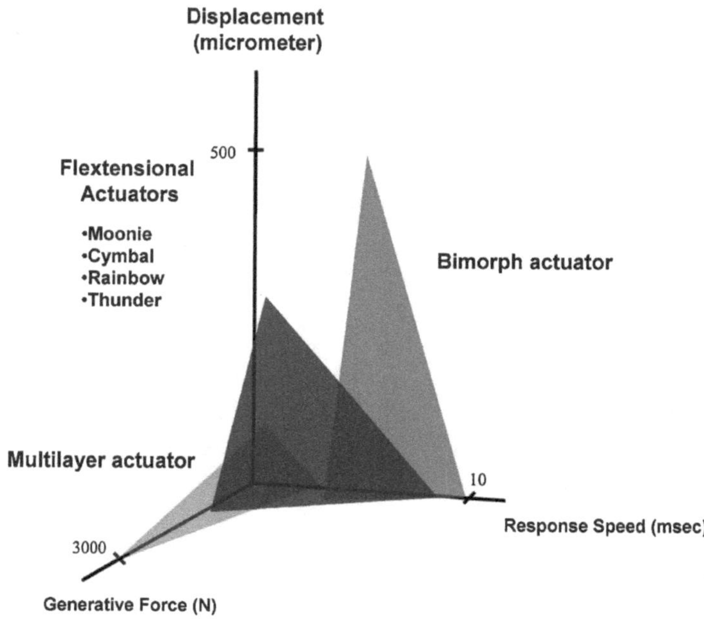

Fig. 17.11 Comparison of the features of several solid-state actuator designs (approximate values)

magnetostrictive materials as the drive element in the foghorn design. Toulis applied this concept to the underwater electroacoustic transducer (Toulis 1966). Flextensional transducers were classified into five different classes based on their shapes and mode of operation (Rolt 1990; Hayes 1936; Toulis 1966; Royster 1968). Class IV transducers are the most intensively studied, best understood, and most widely used underwater transducers. Class V extensional transducers contain two spherical cap shells joined to a radially vibrating ring or dish.

A simplified version of the flextensional or bending transducer emerged in early 1990s. The basic idea behind the flexural transducer is to convert some of the lateral dilation in the planar direction to the longitudinal direction and thereby increasing the longitudinal displacement. Newnham devised a simple, compact type of flextensional transducer, the moonie (Newnham et al. 1991).With its unique properties the moonie has been investigated for many potential applications. The cymbal is the improved version of the moonie transducer, with higher efficiency, displacement, and generative force (Doğan et al. 1997). RAINBOW is another new type of transducer made of reduced and internally biased dome-shaped oxide wafers (Haertling 1994). Flextensional designs referred to as cerambow, thunder, and crescent are the latest generation transducers derived from RAINBOW structure (Kugel et al. 1997).

17.4.1 Flextensional Moonie Transducer

A moonie consists of a piezoelectric or electrostrictive ceramic disk sandwiched between two metal endcaps, each having a crescent-moon-shaped cavity on its inner surface (hence the name *moonie*); see Fig. 17.12. These metal endcaps serve as mechanical transformers for converting and amplifying the lateral displacement of the ceramic into an axial motion normal to the endcaps. Both the d_{33} and d_{31} piezoelectric coefficients contribute to the axial displacement of the composite (Xu et al. 1994; Sugawara et al. 1992). The design is analogous to a continuous ring of bimorphs whose free ends merge at the center of a circle. Furthermore, since the structure is three dimensional, generative forces larger than those produced by a bimorph alone are generated. Thus, the moonie actuator shows higher generative forces than a bimorph and higher displacements than a multilayer actuator.

The geometry of the endcap affects the displacement characteristics of the moonie actuator. Cavity diameter, cavity depth, and endcap thickness are the main geometrical parameters that play important roles in moonie actuator performance. The displacement of a moonie actuator increases rapidly with increasing cavity diameter (Doğan et al. 1994). Displacement of the moonie actuator is inversely proportional to the endcap thickness. A maximum displacement of 22 μm can be obtained at the center of the moonie actuator 12.7 mm in diameter and 1.7 mm in total thickness. Incorporating a multilayer ceramic in the moonie design reduces the drive voltage to a more practical 100 V (Fig. 17.13a). By stacking the multilayer moonie actuators together, higher displacement values can be obtained (e.g., for five multilayer moonie stacks, it was 105 μm). Displacement also increased with increasing compliance of the endcap material (Doğan et al. 1994).

Generative force and displacement values of the moonie actuator are position dependent. Generative force increases when moving from the center to the edge, where it approaches that of the PZT itself. Conversely, the displacement value decreases when moving from center to the edge of the endcap.

The fastest response time of the moonie actuator is in the range of 5–50 μs depending on the cavity size and endcap thickness. The response time of the moonie

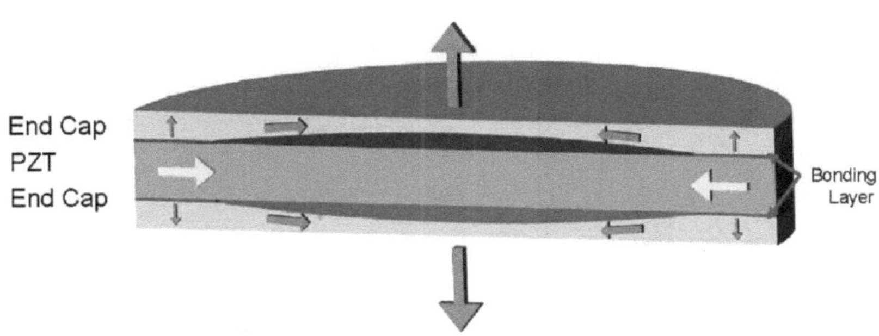

Fig. 17.12 Moonie actuator (arrows shows displacement directions)

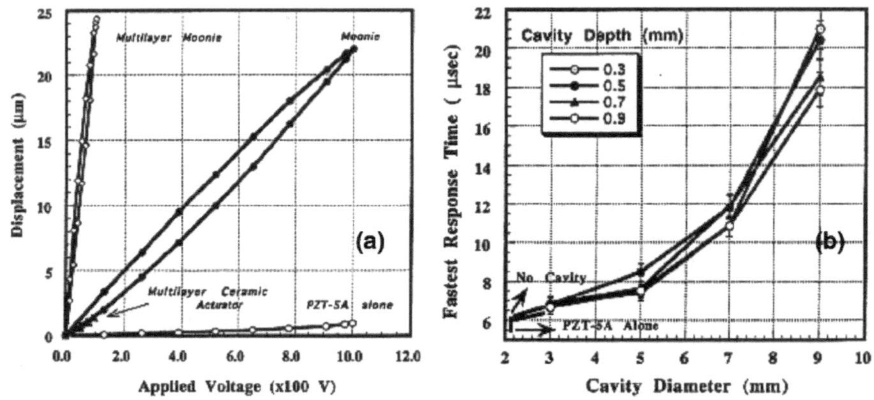

Fig. 17.13 (**a**) Displacement characteristic of moonie actuator with multilayer and single-layer ceramic driving elements, and (**b**) effect of cavity diameter on the fastest response time of the moonie transducers at various cavity depths

actuator increases with increasing endcap compliance and increasing cavity diameter, but changes only slightly with increasing cavity depth (Fig. 17.13b). The fastest response time is inversely proportional to the endcap thickness.

The bonding layer between the metal endcaps and ceramic driving element is a critical region in moonie structure, since it is under severe shear stresses. Despite this concern, reliability test results show that there is no degradation of the displacement and resonance spectrum of the moonie transducers (Doğan et al. 1994). Temperature dependence and cyclic fatigue tests were performed to investigate the reliability of moonie transducer for long-term usage. Less than $\pm 0.1\%$ deviations were observed on displacement during the fatigue test up to 10^7 cycles under applied field of 1 kV/mm at 100-Hz sinusoidal wave form (Fig. 17.14). Destructive (peel tests) and nondestructive (scanning acoustic microscopy and resonance spectroscopy) techniques were used to investigate the reliability and strength of the moonie actuator. Eccobond and Indium alloy gave the best performance. For the moonie actuators fabricated with brass endcaps and Eccobond epoxy, rather good temperature characteristics were obtained between -20 and $+70°$C. However, because of the difference in coefficient of thermal expansion between brass endcaps and ceramic element, thermally induced displacement was observed. This problem has been solved by using caps with similar thermal expansion coefficients to PZT, such as kovar or tungsten.

17.4.2 Roto-Flextensional Cymbal Transducer

From finite element analysis, it was found that a large stress concentration is developed in the thick moonie endcaps in the region above the bonding layer (Onitsuka et al. 1995). A new endcap design *cymbal* has subsequently been developed to

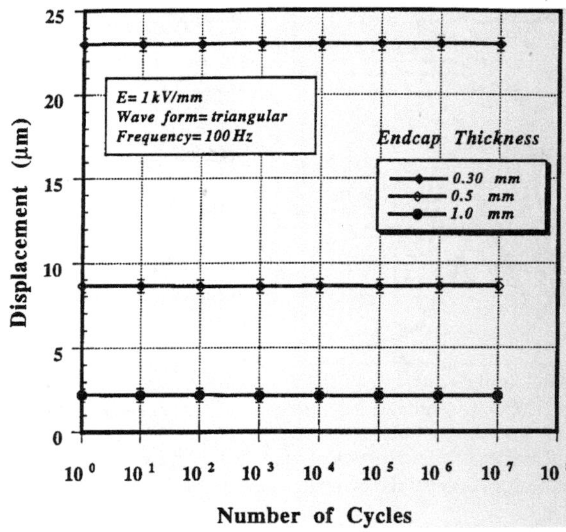

Fig. 17.14 Fatigue characteristics of moonie actuators

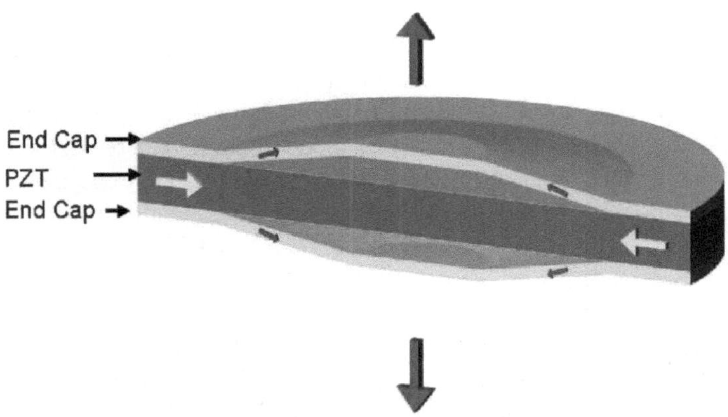

Fig. 17.15 Cymbal actuator (arrows shows displacement directions)

eliminate this stress concentration and the position-dependent displacement behavior of the moonie (Doğan et al. 1997). This new endcap is referred to as the cymbal due to its similarity in shape to the musical instrument of the same name. The cymbal cap is thinner than the moonie cap and has the added advantage of being able to be mass-produced since it is simultaneously cut, pressed, and shaped from a punch/die assembly. The cymbal transducer operates on the same principle as the moonie with one exception. The displacement of the moonie is generated solely from a flexural motion of the caps, whereas the cymbal has an added contribution from rotational motion (Fig. 17.15). The cymbal also exhibits less position-dependent behavior with a more homogeneous displacement and generative force

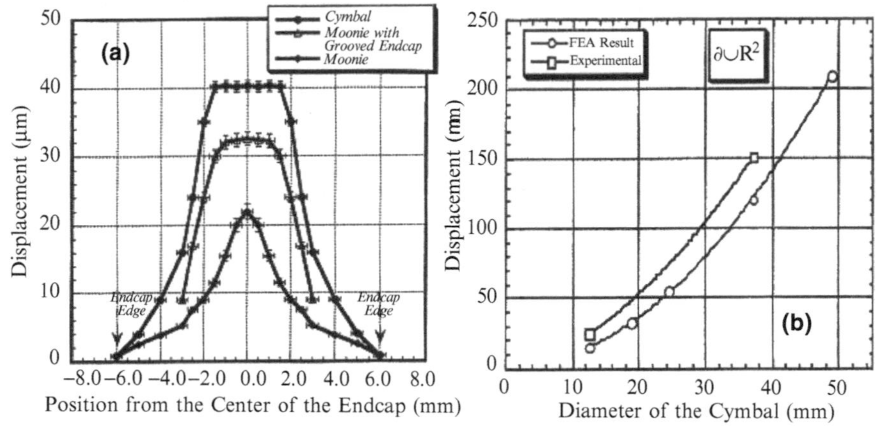

Fig. 17.16 (**a**) Displacement characteristics of moonie and cymbal actuators, (**b**) size dependence of the displacement performance of the cymbal actuator

characteristics across the top of the cap (Fig. 17.16a). A 12.7-mm diameter cymbal transducer with brass caps 0.20-mm thick and a cavity depth of 200 μm exhibits a displacement of 40 μm (under a drive field of 1 kV/mm) and a maximum generative force of 15 N (Doğan et al. 1997). The amplification factor is approximately equal to the radius of the cavity divided by the height of the cavity. One of the main advantages of the cymbal transducer is that transducer properties can be easily tailored by altering ceramic driving element, metal endcap material, and endcap size parameters. In the cymbal transducer design, the flexibility and durability of the metals and the driving power of the ceramic element are combined. Each of the piezoelectric coefficients, d_{31}, d_{33}, and d_{15}, can be used as the driving element of a composite cymbal transducer. In addition to material and size effects, polarization and electric field directions of the samples can be altered systematically to achieve desired properties for different applications (Doğan and Uzgur 2006; Uzgur et al. 2002, 2004).

The shape and mechanical behavior of the cymbal resembles the Bellville washer (spring). The modulus of elasticity of the endcap material plays an important role in the performance of the cymbal transducer. Displacement, generative force, and response time can be tailored by altering the endcap metal (Doğan and Uzgur 2006; Uzgur et al. 2002, 2004). Cymbal actuator can easily be built in 3–50 mm (even higher) diameters with displacement values of one micrometer to several hundred micrometers, respectively (Fig. 17.16b). Cymbal actuators also have higher generative force than moonie actuators because of the enlarged active surface on endcap. By choosing stiffer material and thicker endcaps, very high generative forces can be achieved while sacrificing some of the displacement.

In the moonie and cymbal designs ceramic materials are used as driving elements and metals or polymer as endcaps. The ceramic elements in both designs are mainly under compressive stress. As is well known the tensile strength of ceramics is much lower than their compressive strength due to their brittle nature. They are not reliable under tensile stress. Metal endcaps can withstand both tensile and compressive forces. For this reason the ceramic driving element in the moonie and cymbal designs is kept flat and under compressive stress as much as possible.

17.4.3 RAINBOW

The RAINBOW is actually an acronym for reduced and internally biased oxide wafer. It consists of an electromechanically active layer, such as PZT, PMN, PLZT (lead lanthanum zirconate titanate), PBZT (lead barium zirconate titanate), or PSZT (lead stannate zirconate titanate), in direct contact with a lead-rich constraining layer (Haertling and Land 1971). This constraining layer is formed by exposing one side of a lead-containing ceramic to a reducing atmosphere at high temperature produced by placing a ceramic in contact with a carbon block. The reduction of the active layer occurs as a result of oxidation of the solid carbon block, first to CO(g) then to CO_2(g). This reduced layer is no longer piezoelectric, and is in fact, a good electrical conductor. Because of the thermal expansion mismatch between the reduced and oxide layer, a curvature develops in the structure, giving it a dome shape, with the oxide layer in compression throughout its volume (Fig. 17.17). Rainbow actuators consist of an electromechanical active layer and a constraining layer, similar to the conventional unimorphs. Unlike the unimorph, however, the rainbow is a monolithic structure. The rainbow disks are typically 0.5-mm thick or less and can range in diameter from 1 to 10 cm. Displacement performance of

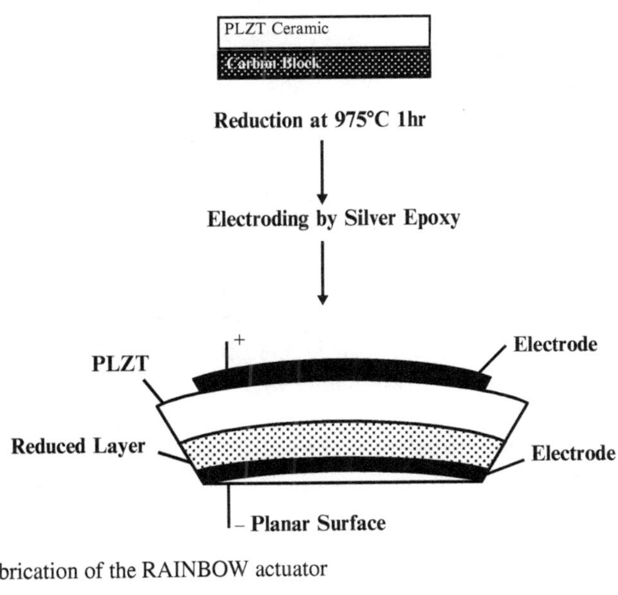

Fig. 17.17 Fabrication of the RAINBOW actuator

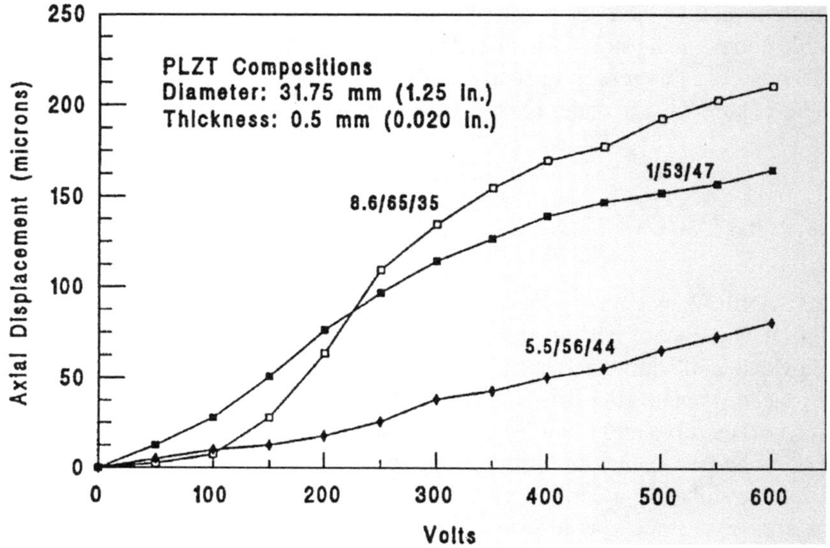

Fig. 17.18 Displacement characteristics of the RAINBOW actuator in various sizes

the rainbow highly depends on the composition of ceramic material (Fig. 17.18; Dausch 1997; Haertling 1994). Even though it shows flexural motion, the rainbow can also be categorized as a monomorph type of actuator (Fig. 17.1).

Rainbows are stress-biased structures. One section of the structure is under high compressive stress, and the other under high tensile forces. Hence, they may have long-term stability problems. Rainbows of PZT-5H, PZT-5A, and PLZT 9/65/35 compositions exhibit polarization fatigue under cycling loading. A 53% decrease in remnant polarization has been observed after 10^8 cycles under 1 kV/mm in the PZT-5A rainbow repoling (Dausch and Wise 1998). The resonance spectrum revealed a significant decrease in the mechanical coupling factor as well. This behavior may be caused by degradation of the interface between the oxide and reduced layer. Depolarization will lead to a decrease in electromechanical coupling and create displacements in the lateral and longitudinal directions.

Long-term tests of displacement show that the under cyclic load Rainbow actuators lose performance significantly (Dausch and Wise 1998). Rainbows made of soft piezoelectric ceramics lose between 10 and 20% of their original displacement value after 10^7 cycles under an applied field of 0.5 kV/mm at 20 Hz and zero force conditions. Fatigue tests under applied force of 3 N show even more catastrophic results. After 10^4 cycles most of the displacement disappears. Repoling the samples recovers around 90% of the displacement value of virgin samples. However, Rainbow actuator made of electrostrictive materials could not be restored by repoling (Dausch and Wise 1998).

17.4.4 THUNDER

THUNDER, which was developed at the Langley Research Center, is an improved version of rainbow actuator and based on coefficient of thermal expansion mismatch between materials (Dausch 1997). Thunder stands for thin layer composite unimorph ferroelectric driver and sensor. The mechanical advantage of the Thunder or Rainbow design is due to the flexibility of the device and the radial expansion created by the pairing of preselected thermally mismatched materials. Thunder transducers consist of an integrated sandwich of three layers. The first and third layers are the prestressed metal layers that sandwich the piezoelectric layer with epoxy bonding between layers. A schematic of the thunder is shown in Fig. 17.19.

Thunder actuators can be fabricated in virtually any size and thickness (Dausch 1997; Hallbaum et al. 1997). Initially, electrical contacts are deposited or plated on the active areas of conventional piezoelectric wafers. Next, alternating layers of adhesive and aluminum foil are placed on one side of the piezoelectric wafer. The package is vacuum-bagged on a flat fixture and placed in an oven for processing at around 325°C with a pressure of 100 psi. After removal from the oven, the wafers are domed or curved during the cool-down process utilizing the difference in thermal expansion rates of the various materials (layers). Even though it has been claimed that it can go beyond a billion cycles without failure, no experimental data has been published for the dynamic performance and reliability. On the contrary, it has been reported that, after 2-weeks testing period, the overall displacement performance of the thunder wafer began to noticeably degrade. A 33% drop in the capacitance has been observed (Pinkerton and Moses 1997).

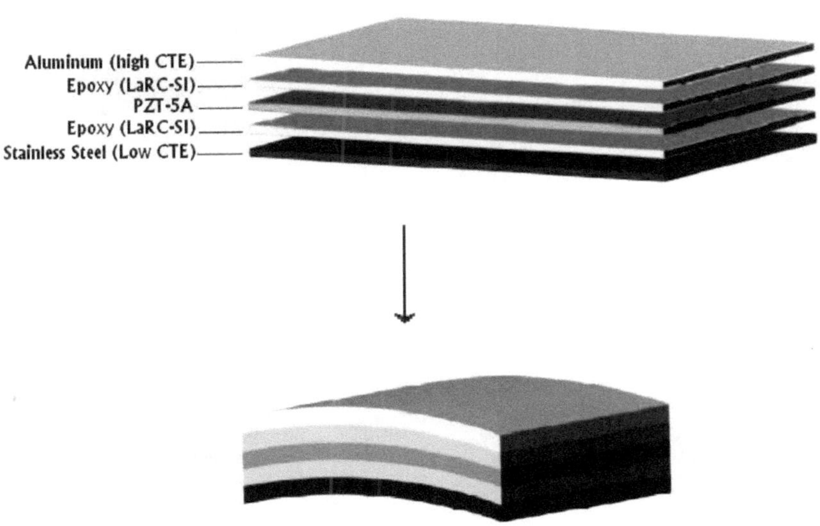

Fig. 17.19 Schematic of THUNDER actuator

A prototype thunder actuator measuring 25.4 × 25.4 mm^2, which consists of 170 – μm thick PZT layer (0.325 total thickness), exhibits 60 – μm displacement in its center position under an applied electrical field of 1 kV/mm (Taleghani and Camphell 1999). Both rainbow and thunder are highly position-dependent actuators. They exhibit highest displacement at the center of the dome, and displacement decreases drastically toward the age of the actuators.

17.5 Comparison of the Solid-State Actuator Designs

Several features of the various solid-state actuator designs are listed in Table 17.2. Because of insufficient data, Thunder does not appear in this table. It is rather difficult to compare different actuators because of different geometry and various operating conditions for specific applications. To make a fair comparison, similar

Table 17.2 Comparison of the solid-state actuator designs

Features	Multilayer (Uchino and Takahashi 1996)	Bimorph (Uchino and Takahashi 1996)	RAINBOW (Shih et al. 1997)	(Cymbal Doğan et al. 1997)	Moonie (Newnham et al. 1991)
Dimensions (mm)	5 × 5 × 12.7 ($L \times W \times T$)	12.7 × 10 × 1 ($L \times W \times T$)	$\emptyset$ = 12.7; T = 0.5 mm	$\emptyset$ = 12.7; T = 1.7	$\emptyset$ = 12.7; T = 1.7
Driving voltage (V)	100	100	450	100	100
Displacement (μm)	10	35	20	40	20
Displacement direction	Positive	Positive	Negative	Positive	Positive
Contact surface (mm^2)	25	1	1	3	1
Generative force (N)	900	0.5–1	1–3	15–100	3
Position dependent of displacement	No	Maximum at the tip	Maximum at the center	Maximum at the center but more diffuse	Maximum at the center
Stability under loading	Very high	Very low	Low	High	Low
Fastest response (μs)	1–5	100	100	5–50	5–50
Fabrication method	Type casting and cofiring at 1,200°C	Bonding ceramic element with metal shim	Reducing ceramic element at 950°C	Bonding ceramic element with metal endcaps	Bonding ceramic element with metal endcaps
Cost	High	Medium	Medium	Low	Low

Table 17.3 Displacement performance of various flextensional composite actuators

Feature	Moonie	Cymbal	RAINBOW	THUNDER
Dimensions (mm)	25.4 – ∅ Disk	25.4 – ∅ Disk	25.4 – ∅ Disk	25.4-mm^2 Square
PZT	PZT-5A	PZT-5A	PZT-5A	PZT-5A
Applied field (kV/mm)	Unipolar 1.0	Unipolar 1.0	Bipolar ± 0.65	Unipolar 1.0
Thickness of PZT (mm)	0.500	0.500	380	0.325
Displacement (μm)	50	80	88	60

dimensions for each actuator were selected, and the measurement conditions are those specified in the table.

Displacement mechanisms of cymbal, moonie, and rainbow class transducers are based on the same physical principle: converting the radial displacement to the axial displacement by flexing or bending the structure. In principle, moonie, cymbal, and the rainbows can exhibit comparatively high displacement if an appropriate diameter (R) to thickness (t) ratio is chosen. So far, the rainbows have achieved higher axial displacement than the moonie or cymbal because of differences in the R/t ratio. Scaling analysis shows that, being fair in comparison and choosing the same size, moonie and rainbow will exhibit similar displacements (Shih et al. 1997).

To exhibit a positive displacement, the applied field is in the opposite direction to the polarization in the rainbows but in the same direction as the polarization for the moonie and cymbal designs. For flexural actuators the displacement is given by a simple formula.

$$\delta \approx \pm \frac{d_{31} E l^2}{2t}. \tag{17.2}$$

In this expression, l stands for length or diameter, and t stands for thickness for the rainbow or thunder designs, and for cavity height for the cymbal and moonie. The sign is (+) for rainbows, and (−) for the moonie and cymbal. Displacement performances of the flexural transducers are compared in Table 17.3. For a fair comparison, similar actuator dimensions (25.4 mm diameter or side length) and identical electric field were chosen (1 kV/mm).

17.6 Drive Techniques for Piezoelectric Actuators

Based on the drive techniques piezoelectric/electrostrictive actuators have been classified into two main categories: unidirectional strain induced parallel to the applied DC field, and AC-field-excited devices, which are resonated at their mechanical resonance frequency (Fig. 17.20). Unidirectional strain induced can be divided into two

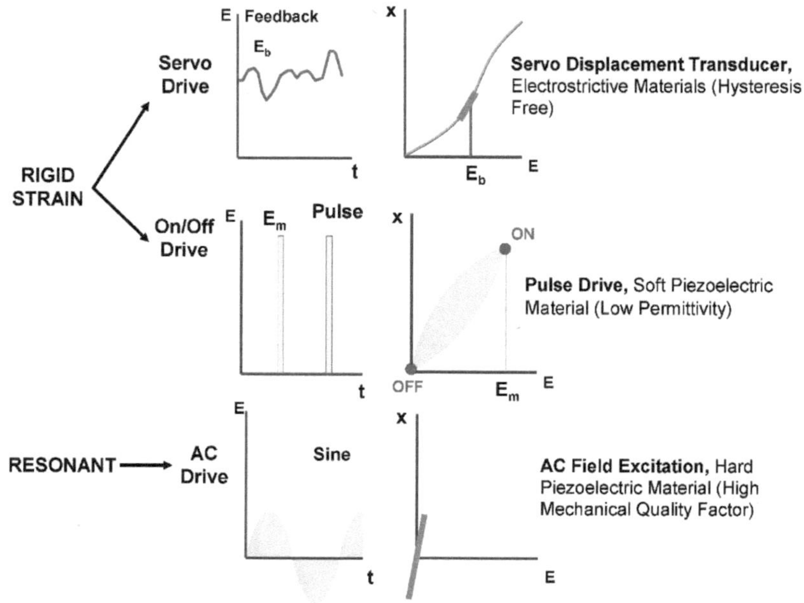

Fig. 17.20 Classification of piezoelectric/electrostrictive actuators based on driving techniques (Uchino and Giniewicz 2002)

subclasses as servo displacement transducers such as positioners and pulse-drive systems such as dot-matrix printers. Servo displacement transducers are controlled by a closed system through feedback unit. On the other hand, pulse-echo actuators are driven in on/off switching mode (Uchino and Giniewicz 2002).

Such driving techniques have been improved not just based on different applications but also on material requirements and limitations. Although having high losses and hysteresis compared with the hysteresis-free electrostrictive PMN, soft PZT materials are preferred for pulse-drive applications because of their relatively low permittivity and better response time rather than PMN. On the other hand, electrostrictive PMN ($Pb(Mg_{1/3}Nb_{2/3})O_3 - PbTiO_3$) materials are most convenient for servo-drive applications because of their hysteresis-free characteristic. The ultrasonic motor applications are required for low heat generation and high AC strain during the process that can be obtained by using hard piezoelectric materials with very high mechanical quality factor Q.

Because servo-drive transducers pose high accuracy, they are used for optical and precise machinery. They are usually controlled by feedback voltage superimposed to DC bias to increase the accuracy. On the other hand, pulse-drive actuators are used as impact elements such as in inkjet printers.

Drive components of the resonant strain actuators are both applied voltage and applied frequency. They require high AC displacement and low loss. Their AC resonant displacement is characterized by

$$d \times E \times l \times Q_m, \tag{17.3}$$

where d is the piezoelectric strain coefficient, E is the applied electric field, l is the sample length, and Q_m is the mechanical quality factor. Although hard PZT materials have low piezoelectric strain coefficient, they are characterized by high mechanical quality factor that increases the AC displacement and low losses. Resonant strain actuators are used for high-frequency requirement applications such as ultrasonic motors (Uchino and Giniewicz 2002).

17.7 Applications of the Solid-State Ceramic Actuators

Multilayer actuators and bimorphs are used extensively in applications such as precision machining, dot matrix printers, active vibration control, deformable mirrors, optical choppers, electroacoustic transducers, and ultrasonic motors (Uchino and Takahashi 1996; Tokin Co 1992)

Moonie and cymbal actuators have great potential in the automotive industry. They can be utilized as sensing and vibration suppression elements and as switching elements in valve designs (Tressler et al. 1994). Rainbows can be used for air acoustic cancellation, pumps, and as switches (Haertling 1994).

The moonie and cymbal actuators can also be used as a micropositioner for applications requiring small size and quick response time, but only for modest loads. The Omron Corporation has already succeeded in using the multilayer moonie actuator for an optical scanner (Goto and Imanaka 1991; Imanaka 1992). The high-density memory storage driver such as CD-ROM driver and magneto-optic memory storage driver are some other possible applications for moonie or cymbal actuators capable of delivering precise positioning (Newnham and Doğan 1998). In this part, some applications based on design principles have been reviewed.

17.7.1 Positioners

One of the servo-drive applications of piezoelectric actuators is positioners. Particularly, because of fast development in semiconductor technology; demand for high-precision systems in the range of micron to nanosensitivity have been increased. Because of their high rigidity, piezoelectric actuators are ideal for fast, precise adjustment of tools or even workpieces (Fig. 17.21). By applying fixed voltage patterns, in phase with the rotation of a lath, high precision out-of-round workpieces become possible (Mamiya 2006).

The piezoelectric actuators are also used for the optical axis alignment of optical fiber. Optical axis of the fiber is aligned by piezoelectric actuator is placed to center of alignment system. This system can make the alignment in nanometric sensitivity, and it is one of the systems in which piezoelectric actuators are utilized most effectively (Mamiya 2006).

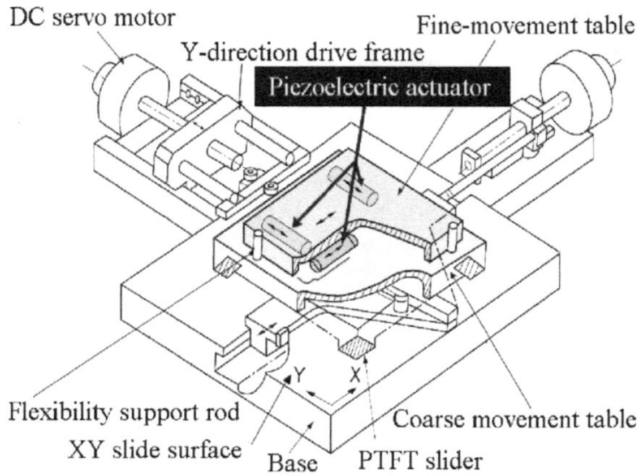

Fig. 17.21 Precision stage controlled by piezoelectric actuator (Mamiya 2006)

17.7.2 Deformable Mirrors

Adaptive optics, a recently emerging technology, are used to correct higher order aberrations in the optical systems. They are commonly used in telescope systems. They consist of deformable mirrors to measure and rapidly compensate optical distortions, which are caused by several effects such as atmospheric distortion. The deformable mirror has a surface shape that can be altered to control wavefont phase. Figure 17.22 shows the typical adaptive optic system consisting of a deformable mirror and wavefront sensor for telescope systems (Welch et al. 2003). Incoming light is picked off and distortions in the wavefront are measured by the wavefront sensor, typically with reference to a guide star. The processor system can then calculate the correction that is needed in the optical path and the necessary shape at the mirror. Many types of the deformable mirrors are wavefront modulators in use, such as membrane mirrors that depend on electrostatic distortion of a thin metallic membrane. On the other hand, most two common deformable mirror systems consist of piezoelectric actuators: the thin glass face sheets deformed by axial piezoelectric actuators and the bimorph mirrors. These mirrors range from a few centimeters to a few tens of centimeters across.

Thin glass face sheet mirrors can be operated by using bulk piezoelectric actuator or cofired multilayer stacks of piezoelectrics (Fig. 17.23). The shape of the plate is controlled by individual piezoelectric actuators bonded to back side of the mirror. Actuator stroke, interactuator displacement, and actuator pitch are the major parameters for mirror performance. Because they produce high stroke and they allow smooth wavefront control with large number of degrees of freedom, they are considered as best among deformable mirrors. Besides, cofired stacks are preferred because of their low operating voltage.

Bimorph deformable mirrors are formed by two or more layers. They were operated based on lateral motion of active layers, which is fabricated by piezoelectric or electrostrictive materials (Fig. 17.23) (Dalimier and Dainty 2005).

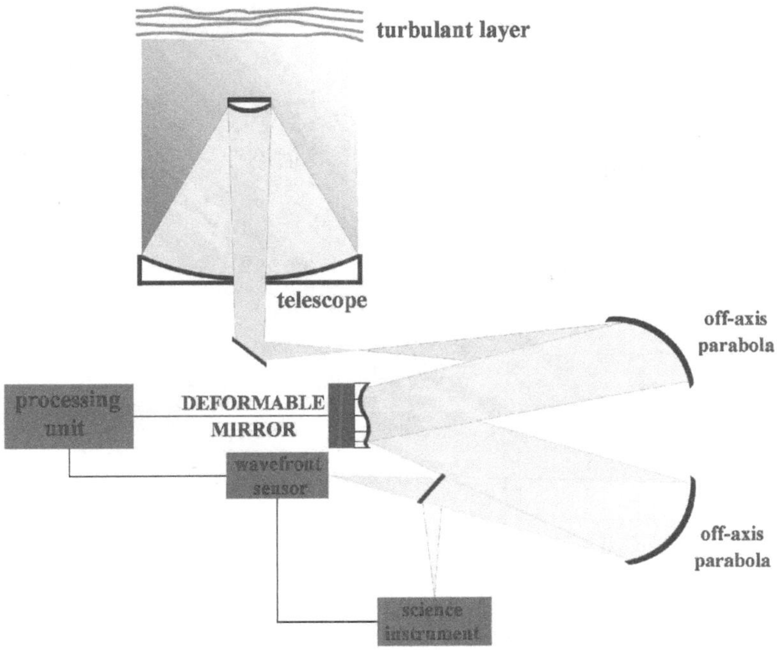

Fig. 17.22 Classical adaptive optics system used to correct atmospheric distortion (Welch et al. 2003)

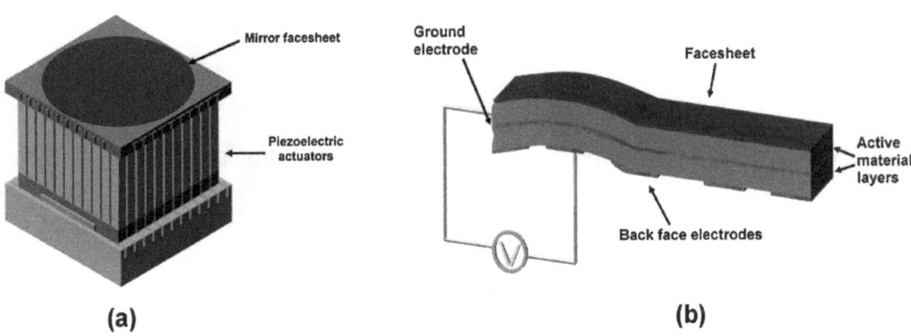

Fig. 17.23 (**a**) Typical view of piezoelectric deformable mirror and working principle and (**b**) bimorph deformable mirror working principle (Developed by AOptix Inc.) (Dalimier and Dainty 2005)

17.7.3 Diesel Injection Valves

Diesel engine efficiency is mainly increased by using injection valves consisting of piezoelectric multilayer actuators (Uchino and Giniewicz 2002). Figure 17.24 shows the common injection valve produced by Bosh. They are driven simply by pulse-drive systems.

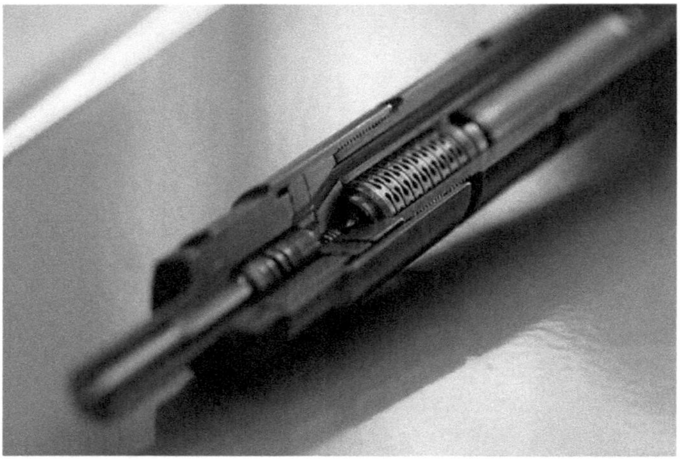

Fig. 17.24 Common rail-type diesel injection valve with a piezoelectric multilayer actuator (courtesy by Bosh)

Such application requirements that are around 50-μm displacement for 50-mm length, %5 temperature stability for -40 to $150°C$, $>10^9$ cycle life time, and low fabrication cost have been achieved by using multilayer piezoelectric actuators. Diesel injection valves have been widely commercialized by Siemens, Bosh, and Denso Corp. (Uchino and Giniewicz 2002).

17.7.4 Dot Matrix and Inkjet Printers

Although their usage decreases, dot matrix printers are still popular in ATM machines and office applications. Dot matrix printers are the example of one of the most successful piezoelectric multilayer actuator applications. Bimorph type has been developed but only multilayer type has been commercialized particularly because of its high impact forces (Uchino and Giniewicz 2002). Schematic display of the dot matrix impact printer is shown in Fig. 17.25. In dot matrix printers, 10-μm displacements of multilayer actuators can be amplified to the about 500-μm stroke by using hinge lever mechanisms.

Another pulse-drive system application is inkjet printers. All Epson and most commercial inkjet printers use piezoelectric actuators. Following the dot matrix printer technology they satisfy the need for better image quality and faster printing. They are also very cost effective.

Inkjet printer applications of piezoelectric actuators are operated by pulse drive. When the piezoactuator has been driven, ink stream starts to shake leading to break off in very small fine droplets as the ink leaves the nozzle (Fig. 17.26). Depending on the application, droplet can be charged or not charged.

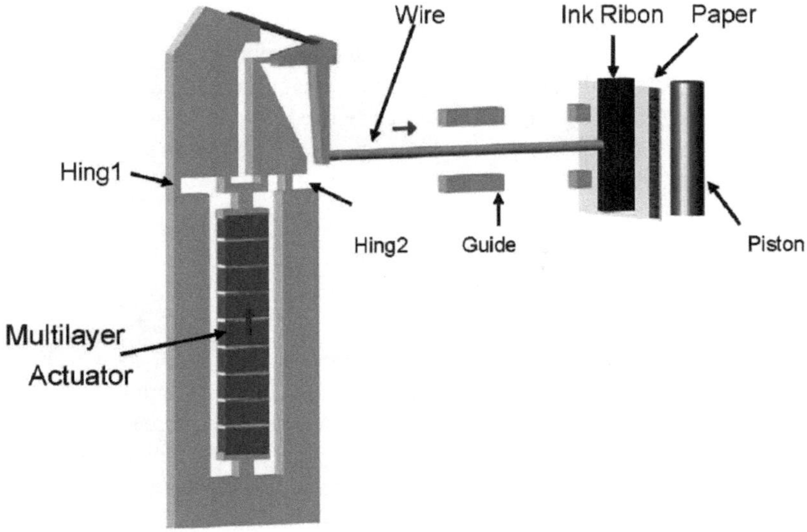

Fig. 17.25 Working principle of dot matrix printer

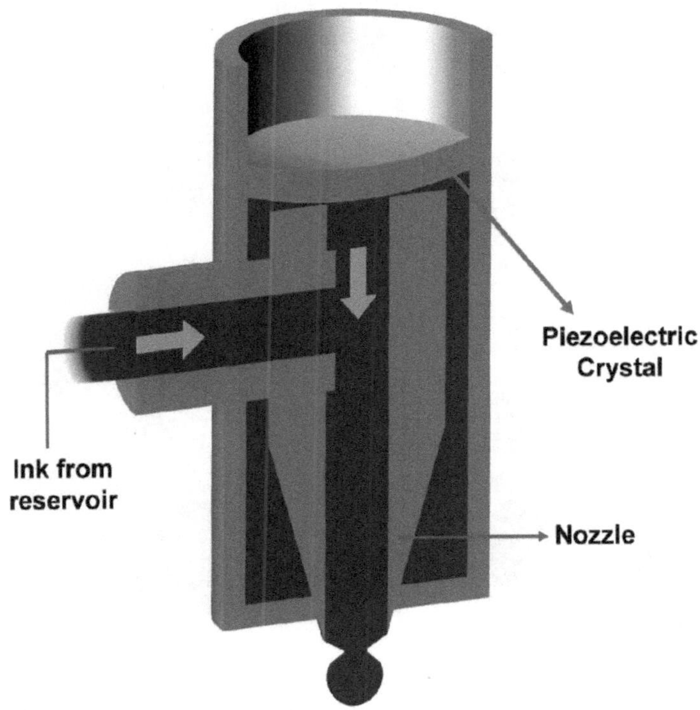

Fig. 17.26 Piezoelectric inkjet printer

Epson commercialized inkjet printer consisting of integrated piezo segment printer head. Inkjet movement has been created by bending mode of piezoelectric unimorph. The so-called multilayer ceramic with hyperintegrated piezoelectric segment (MLChips) units in 10 mm × 10 mm size is composed of two arrays of 48 ink chambers. This laminated structure allows mass production at low costs (Uchino and Giniewicz 2002).

17.7.5 Imaging System Applications

One of the typical applications for piezoelectric actuators is for the CCD (charge coupled device). CCD drive technique is a pixel deviation technique, which shifts the CCD by a half pixel in the horizontal direction in order to increase the pixels and resolution. CCD shifts can be arranged by using piezoelectric actuator. Pixel deviation technique is used in camcorders and LCD projection systems (Mamiya 2006).

Recently, reduction in the size and driving voltage increase in the displacement in multilayer actuator systems lead to the usage of piezoelectric actuators in hand blurring correction systems in digital cameras. Thus, compact electronic systems have been improving strictly. Figure 17.27 explains the working principle of the hand blurring correction systems. CCD is moved by piezoelectric actuator based on the signal detected by sensor (Mamiya 2006).

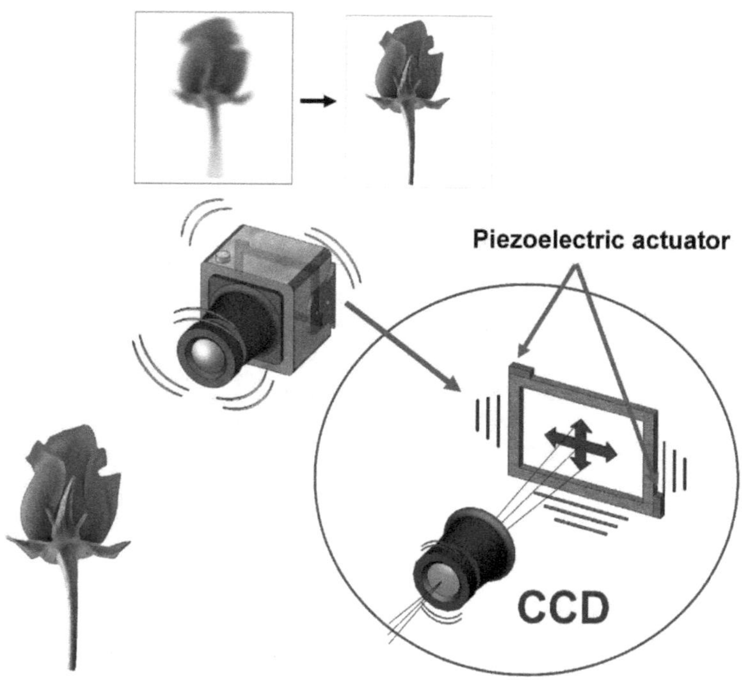

Fig. 17.27 Hand blurring correction with piezoelectric actuator (Mamiya 2006)

17.8 Summary

Ceramic materials are not strong under tensile forces due to their structure, but rather good under compressive forces. In the moonie and cymbal structure, electroactive ceramic elements are kept flat, under compressive forces over the bonding layer and small tensile forces under the cavity beneath the endcaps. On the other hand, rainbows and thunder are stress-biased structures. Part of the structure is under high compressive force and others are under high tensile forces. Rainbow and thunder structures are based on coefficient of thermal expansion differences between different parts of the actuator, and the displacement is highly temperature dependent. By double-layering and using different ceramic compositions, temperature-dependent behavior can be eliminated to a certain extent. Flextensional moonie, cymbal, rainbow, and thunder actuators all give moderate generative forces and displacement values that fill the gap between multilayer and bimorph actuators. Each solid-state actuator design has attractive features that can be exploited for certain applications. Advantages of the moonie and cymbal actuators are the easy tailoring of the desired actuator properties by altering the cavity size and endcap dimensions. Easy fabrication is another advantage. The rainbow actuator also partially covers this gap. For that type of actuator, a reduction step during the processing of the ceramic element at high temperature results in a semiconducting layer and stress bias, which is a potential problem for long-term usage. The fatigue behavior of the actuators depends markedly on the driving conditions. Dynamical studies under uniaxial prestress under cycling electric field should be performed on all flexural transducers for aerospace and automobile applications. Thermal cycling tests under various humidity levels will also be very useful, particularly for aerospace applications. The real concern now is the reliability for long-term usage of such actuators.

References

Aburatani H, Uchino K (1995) Destruction Mechanism and Destruction Detection Technique for Multilayer Ceramic Actuators. IEEE, Piscataway, NJ, 750–752.

Cady WG (1964) Piezoelectricity. Mc Graw-Hill, New York.

Cieminski VJ, Beige HJ (1991) High-Signal Electrostriction in Ferroelectric Materials. Phys. D 24:1182–1186.

Clark AE (1980) Magnetostrictive Materials. Ferromagn. Mater. 1:531.

Clark AE, Teter JP, McMasters OD (1988) Magnetostriction 'Jumps' in Twinned $Tb_{0.3}Dy_{0.7}Fe_{1.9}$. J. Appl. Phys. 63:3910–3912.

Cross LE (1967) Antiferroelectric-Ferroelectric Switching in a Simple 'Kittel' Antiferroelectric. J. Phys. Soc. Jpn. 23:77–82.

Cross LE, Young SJ, Newnham RE, Nomura S, Uchino K (1980) Large Electrostrictive Effects in Relaxor Ferroelectrics. Ferroelectrics. 23:187.

Dalimier E, Dainty C (2005) Comparative Analysis of Deformable Mirrors for Ocular Adaptive Optics. Opt. Exp. 13:4275–4285.

Dausch DE (1997) Ferroelectric Polarization Fatigue in PZT-Based RAINBOWs and Bulk Ceramics. J. Am. Ceram. Soc. 80:2355–2360.

Dausch DE, Wise SA (1998) Composition Effects on Electromechanical Degradation of RAINBOW Actuators. NASA, Hampton, VA, NASA/TM-1998-206282, 2–3.
Doğan A, Uchino K, Newnham RE (1997) Composite Piezoelectric Transducer with Truncated Conical Endcaps, Cymbal. IEEE UFFC 44:597–605.
Doğan A, Uzgur E (2006) Size and Material Effects on Cymbal Transducer for Actuator Applications. Ferroelectrics 331:53–63.
Doğan A, Xu QC, Yoshikawa S, Uchino K, Newnham RE (1994) High Displacement Ceramic Metal Composite Actuators Moonies. Ferroelectrics 156:1–6.
Doğan A, Yoshikawa S, Uchino K, Newnham RE (1994) The Effect of Geometry on the Characteristics of the Moonies Transducer and Reliability Issue. ISAF-IEEE Proc. 2:935–939.
Duering T, Melton KN, Stockel D, Wayman CM (1990) Engineering Aspects of Shape Memory Alloys. Butterworth-Heinemann, Boston, MA.
Furuka K, Uchino K (1986) Electric-Field-Induced Strain in (Pb,La)(Zr,Ti)O_3 Ceramics. Adv. Ceram. Mater. 1:61–63.
Goto H, Imanaka K (1991) Super Compact Dual Axis optical Scanning Unit Applying a Torsional Spring Resonator Driven by a Piezoelectric Actuator. SPIE Proc.1544:272–281.
Haertling GH (1994) A New Type of Ultra-High-Displacement Actuator. Am. Ceram. Soc. Bull. 73:93–96.
Haertling GH (1994) Chemically Reduced PLZT Ceramics for Ultra-High Displacement Actuators. Ferroelectrics 154:101–106.
Haertling GH, Land C (1971) Hot-Pressed (Pb, La) (Zr, Ti) O_3 Ferroelectric Ceramics for Electrooptic Applications. J. Am. Ceram. Soc. 54:1–11.
Hallbaum FR, Bryant RG, Fox RL (1997) Thin Layer Composite Unimorph Ferroelectric Driver and Sensor. US Patent 5,632,841.
Hathaway KB, Clark AE (1993) Magnetostrictive Materials. MRS Bull. 18:34–41
Hayes HC (1936) Sound Generating and Directing Apparatus. US Patent 2,064,911.
Imanaka K (1992) Microhybrid Integrated Devices and Components, Micro Photonic Devices. SPIE Proc. 1751:343–353.
Jaffe B, Cook WR Jr, Jaffe H (1971) Piezoelectric Ceramics. Academic Press, New York.
Kittel C (1951) Theory of Antiferroelectrics. Phys. Rev. 82:729.
Kugel VD, Chandren S, Cross LE (1997) A Comparative Analysis of Piezoelectric Bending-Mode Actuators. SPIE Proc. Smart Struct. Mater. Smart Mater. Technol. 3040:70–80.
Mamiya Y (2006) Application of Piezoelectric Actuator. NEC Technol. J. 1:82–86.
Newnham RE, Doğan A (1998) Metal-Electroactive Ceramic Composite Transducer. US Patent 5,729,07.
Newnham RE, Xu QC, Yoshikawa S (1991) US Patent 999,819.
Oh KY, Furuta A, Uchino K (1990) Shape Memory Unimorph Actuators Using Lead Zirconate-Based Antiferroelectrics. J. Jpn. Ceram. Soc. 98:905–908.
Onitsuka K, Doğan A, Tressler JF, Xu QC, Yoshikawa S, Newnham RE (1995) Metal-Ceramic Composite Transducer, The Moonie. J. Int. Mater. Syst. Struct. 6:447–455.
Pinkerton JL, Moses RW (1997) A feasibility Study to Control Airfoil Shape Using THUNDER. NASA, Hampton, VA, NASA TM 4767:6–7.
Rolt KD (1990) History of Flextensional Electro-Acoustic Transducers. J. Acoust. Soc. Am. 87:1340–1345.
Royster LH (1968) Flextensional Underwater Acoustic Transducers. J. Acoust. Soc. Am. 45: 671–683.
Shih WY, Shih WH, Aksay I (1997) Scaling Analysis for the Axial Displacement and Pressure of Flextensional Transducers. J. Am. Ceram. Soc. 80:1073–1078.
Sugawara Y, Onitsuka K, Yoshikawa S, Xu QC, Newnham RE, Uchino K (1992) Metal-Ceramic Composite Actuator. J. Am. Ceram. Soc. 75:996–998.
Takahashi S (1989) Recent Developments in Multilayer Piezoelectric Ceramic Actuators and Their Applications. Ferroelectrics 91:293–302.
Taleghani BK, Camphell JF (1999) Non-Linear FEM of Thunder Piezoelectric Actuators. NASA, Hampton, VA, NASA/TM-1999-209322:8–9.

Tokin Co (1992) Multilayer Piezoelectric Actuators. Short Form Catalog, Tokyo.
Toulis WJ (1966) Flexual-Extensional Electromechanical Transducer. US Patent 3,277,433.
Tressler JF, Xu QC, Yoshikawa S, Uchino K, Newnham RE (1994) Composite Flextensional Transducers for Sensing and Actuating. Ferroelectrics 156:67–72.
Uchino K (1993) Ceramic Actuators Principles and Applications. MRS Bull. 18:42–48.
Uchino K (1997) Piezoelectric Actuators and Ultrasonic Motors. Kluwer Academic, Boston.
Uchino K (1989) Recent Topics of Ceramic Actuators. Ferroelectrics 91:281–292.
Uchino K, Giniewicz JR (2002) Micromechatronics. Marcel Dekker, New York/Basel.
Uchino K, Nomura S, Cross LE, Newnham RE (1980) Electrostriction in Perovskite Crystal and Its Applications for Transducers. J. Phys. Soc. Jpn. 49:45–48.
Uchino K, Takahashi S (1996) Multilayer Ceramic Actuators. Solid State Mater. Sci. 1:698–705.
Uzgur E, Doğan A, Markley D, Meyer RJ, Hladky-Henion AC, Newnham RE (2004) Materials for High Performance Cymbal Transducer. J. Electroceram. 13:403–407.
Uzgur E, Doğan A, Newnham RE (2002) Design Optimization of Piezoelectric Composite Transducers using Finite Element Analysis. Key Eng. Mater. 206–213:129–1300.
Xu QC, Doğan A, Tressler JF, Yoshikawa S. Newnham RE (1994) Ceramic Metal Composite Actuator. Ferroelectrics. 160:337–346.
Wayman CM (1993) Shape Memory Alloys. MRS Bull. 18:49–56.
Wayman CM, Shimizu K (1972) Shape Memory effects in Alloys. Met Sci. J. 6:175.
Welch S, Doel P, Greenaway A, Love G (2003) Smart Optics in Astronomy and Space. Smart Opt. 44:26–29.

Chapter 18
Piezoelectric Energy Harvesting using Bulk Transducers

Shashank Priya,* Rachit Taneja, Robert Myers, and Rashed Islam

Self-powered nodes can be developed by harvesting wasted mechanical vibration energy available in the environment. Piezoelectric converters have been found to be most suitable for transforming mechanical energy into electric energy where the charge generation is directly related to the extent to which the element is deformed. This chapter provides the strategy for the selection of the piezoelectric transducer depending upon the frequency and amplitude of the mechanical stress. The figure of merit for the material selection was found to be directly proportional to product $(d \cdot g)$, where d is the piezoelectric strain constant and g is the piezoelectric voltage constant. The criterion for maximization of the product $(d \cdot g)$ was found to be given as $|d| = \varepsilon^n$, where n is the material constant which is fixed by the magnitude of the piezoelectric and dielectric constants. Results are reported on various devices utilizing piezoelectric bimorph transducers.

18.1 Introduction

In recent times, several applications have emerged where batteries are inferior option to power the microdevices and sensors. The examples for these applications include structural health monitoring of aircraft and industrial machines which requires powering corrosion, strain, temperature, and pressure sensors; perimeter security surveillance of sensitive sites and buildings which requires powering devices such as microcamera, acoustic sensor, touch sensor, and gas sensor; and weather monitoring in remote locations which requires powering wind speed and humidity sensors. All of these applications deploy sensors in multiple locations (nodes) which communicate wirelessly with each other and base station (network). The powering of the densely populated sensor nodes in a network is a critical problem owing to the high cost of wiring or replacing batteries. In cases mentioned above, these

S. Priya
Materials Science and Engineering, Virginia Tech, Blacksburg, VA 24061, USA

A. Safari, E.K. Akdoğan (eds.) *Piezoelectric and Acoustic Materials for Transducer Applications.*

operations are restricted by the infrastructure. In order for the sensor nodes to be conveniently placed and used, they should be as small as possible which puts an upper limit on their life time. Out of various possible solutions the most efficient and practical method is to develop self-powered nodes by scavenging energy from the wasted mechanical vibration energy.

The power densities for various environmental sources are available in literature (Roundy et al. 2003, 2004; Calaway 2004; Shenck and Paradiso 2001; Meninger et al. 2001; Paradiso and Starner 2005). The outside solar energy has the capability of providing power density of $15,000\,\mu\text{Wcm}^{-3}$ which is about two orders of magnitudes higher than other sources. Definitely, solar energy is a very attractive source for powering the sensor array and the solar technology has matured over the years. One of the major challenges in implementation of solar technology on the "energy on demand" platform has been the requirement of bulky electronics. Further, the variation in light intensity (cloudy vs. sunny day) can drop the efficiency significantly. However, as the power requirements for the sensor nodes have dropped down to a low level of $100\,\mu\text{W}$ (and are decreasing continuously) an elegant solution to powering would be finding simpler and cheaper technology. Air flow and mechanical vibrations are the other attractive alternatives in locations where these are available. Recent research on smart shoes and bag pack has shown that combined generated electrical power could be in the range of $\sim$1 W (Starner and Paradiso 2004; Rome 2006).

Mechanical energy can be converted into electric energy using piezoelectric, electromagnetic, or electrostatic mechanism (other transduction mechanisms are also possible utilizing conducting polymer-based composites and dielectric elastomers but these are not discussed here). It can be easily shown that piezoelectric transducers are more suitable mechanical to electrical energy converters. In addition to the advantage of being smaller and lighter, piezoelectrics have at least three times higher energy density as compared to their counterparts electrostatics and electromagnetics. The transformation from mechanical to electrical energy is obtained through direct piezoelectric effect and using a storage circuit the generated electrical energy is stored. The mechanical output can be in the form of a burst or continuous signal depending on the cyclic mechanical amplifier assembly. In this transformation, there is mechanical loss because of mismatch in mechanical impedance represented by damping factor and there is electromechanical loss depending upon the magnitude of coupling factor.

18.2 Piezoelectric Material

The piezoelectric material is the first prime factor affecting the performance of generator. Commonly used piezoelectric materials are based on lead zirconate titanate (PZT) ceramics, relaxor ferroelectric single crystal, and piezoelectric polymer polyvinylidene fluoride (PVDF). The primary factor for the selection of a piezoelectric material for energy harvesting device or sensor application is the transduction rate. The magnitude of transduction is governed by the effective piezoelectric stress constant, d, and the effective piezoelectric voltage constant, g. Using the linear constitutive piezoelectric equations, a relation between the energy density of the piezoelectric material and the transduction coefficient $(d \cdot g)$ under an applied stress X can be derived. Under an applied force ($F = XA$, where A is the area), the open circuit output voltage (V) of the ceramic is given as

$$V = Et = -gXt = -\frac{gFt}{A}, \quad (18.1)$$

where t is the thickness of the ceramic, E is the electric field, and g is the piezoelectric voltage coefficient given as

$$g = \frac{d}{\varepsilon_0 \varepsilon^X}, \quad (18.2)$$

where ε is the dielectric constant. At low frequencies the electric energy available under AC stress excitation is given as

$$U = \frac{1}{2}CV^2 \quad \text{or} \quad u = \frac{1}{2}(d \cdot g)\left(\frac{F}{A}\right)^2, \quad (18.3)$$

where the latter expression in (18.3) above is for unit volume. Equations (18.1)–(18.3) show that under certain experimental conditions and for a given material of fixed area and thickness, a high $(d \cdot g)$ product and a high g constant will generate high voltage and power when the piezoelectric ceramic is directly employed for energy harvesting and sensing. The criterion for maximization of the product $(d \cdot g)$ with respect to the microstructure variable x (such as grain size) can be determined as following (Islam and Priya 2006a, b):

$$|d| = \varepsilon^n. \quad (18.4)$$

Using (18.1)–(18.4) the maximum magnitude of the product $(d \cdot g)$ is given as

$$\max[d \cdot g] = \varepsilon^{2n-1}, \quad (18.5)$$

where n is the material constant which is fixed by the magnitude of piezoelectric and dielectric constants. It has been shown that for most of the commercial polycrystalline ceramic materials the magnitude of n lies in range of 1.1–1.3. Equations (18.4) and (18.5) represent a deterministic rule for realizing the high energy density piezoelectric materials. According to this equation a giant enhancement in the magnitude of the g constant will be obtained by reducing the magnitude of n. As the magnitude of n decreases and approaches the theoretically minimum possible value of 0.5 the magnitude of product $(d \cdot g)$ reaches the maximum possible magnitude of unity. In order to obtain large magnitude of the product $(d \cdot g)$, the piezoelectric compositions can be tailored by modifying with the dopant or processing technique such that the change in piezoelectric and dielectric constant leads to a decrement in the magnitude of constant n (Nishida and Ouchi 1976, 1977; Ando and Hisaki 1998; Kawano and Sube 1997; Yoo et al. 2001).

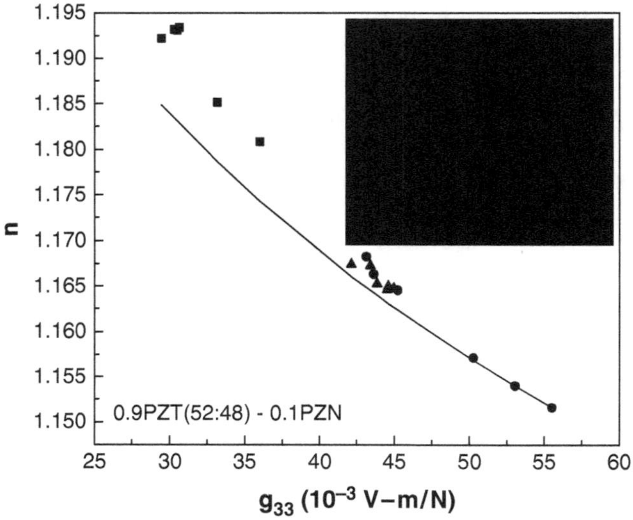

Fig. 18.1 Variation of material constant, n, as a function of piezoelectric voltage coefficient, g_{33} for the composition 0.9PZT $(52:48) - 0.1\text{PZN} + y\,\text{wt}\%\text{MnCO}_3$ (© Wiley-Blackwell Publishing Ltd., Islam and Priya (2006b))

We have been successful in developing the synthesis route that allows us to fabricate polycrystalline ceramic composition with desired grain size and shape for the $\text{Pb}(\text{Zr}_{1-x}\text{Ti}_x)\text{O}_3 - \text{Pb}[(\text{Zn}_{1-y}\text{Ni}_y)_{1/3}\text{Nb}_{2/3}]\text{O}_3$ (PZT–PZNN)-based compositions. The $d_{33} \cdot g_{33}$ values of the samples having composition 0.9PZT $(\text{Zr}/\text{Ti} = 56/44) - 0.1\ \text{PZNN} + 2\,\text{mol}\%\,\text{MnO}_2$ (sintered in two steps at $1{,}100 - 1{,}000°\text{C}$) was found to be as $18{,}456.2 \times 10^{-15}\text{m}^2\,\text{N}^{-1}$, which is one of the highest values reported for polycrystalline ceramics (Islam and Priya 2006a, b). This composition was also found to exhibit a high magnitude of g_{33} as $83.1\,(\text{V m})\text{N}^{-1}$, corresponding to the magnitude of n as 1.126 (Islam and Priya 2006a, b). The developed material has the potential to provide an energy density of $80.84\,\text{mJ}\,\text{cm}^{-3}$ at the applied stress equal to one-sixth of the compressive strength for the piezoelectric ceramic ($\sim$100 MPa). Figure 18.1 shows the relationship between the magnitude of n and piezoelectric voltage coefficient for 0.9PZT $(52:48) - 0.1\text{PZN} + y\,\text{wt}\%\,\text{Mn}$. From this figure it can be clearly seen that as the magnitude of the n approaches toward unity a giant enhancement in the magnitude of g constant is obtained.

The efficiency of ceramic element in energy harvesting applications is dependent on the electromechanical coupling coefficient (k) and mechanical quality factor (Q_m). Results from a study conducted by Umeda et al. has shown that the efficiency increased with the increase in mechanical quality factor (Q_m), coupling factor (k), and decrease in dielectric loss factor (tan δ_d). Recently, a quantitative model derives the efficiency (η) of the mechanical to electric conversion process as (Richards et al. 2004)

$$\eta = \left(\frac{1}{2}\right)\frac{k^2}{1-k^2} \Bigg/ \left(\frac{1}{Q_\text{m}} + \frac{1}{2}\frac{k^2}{1-k^2}\right). \qquad (18.6)$$

18.3 Selection of Transducer for Energy Harvesting

Piezoelectric transducer selection plays a significant role in the efficient energy harvesting from a given vibration spectrum. The transducer design should be tailored for a specific frequency range (ideally the working frequency coincides with the resonance frequency of transducer) and mounting mechanism should be designed such that an effective transfer of stress between the vibration source and transducer structure can be obtained. Transducers such as bimorph, rainbows, unimorph, and diskbenders can be strained to maximum at very low magnitudes of stress and thus they are not suitable to harvest energy from high amplitude vibrations (Colla and Morita 2002). These transducers have lower operating frequency where the resonance can be easily changed by varying the mounting mechanism and adding extra mass to the transducer. On the other hand, transducers such as stack, mechanically amplified composite structures, and raw ceramic require large stresses to provide significant deflection or to generate electric charge and thus they are not suitable to harvest energy from low amplitude vibrations. Cymbal and Moonie structures fall in the middle of the stress – strain curve and as such present an interesting possibility to harvest energy over wide band (Doğan et al. 1997; Newnham et al. 2002). Mechanical impedance of the transducer plays an important role in effectively transferring the stress from the source to the generator.

The recent progress in the field of piezoelectric fiber composites and thin profile flexible transducers allows the implementation of energy harvesters in variety of scenarios including complex surfaces and noisy environments. Some representative examples of the fiber composites and thin profile transducers are shown in Fig. 18.2. The transducers shown in this figure are attractive for energy harvesting applications for several reasons including light weight, flexibility, easy mounting, large response, and low frequency operation. Prestressed transducers, such as Thunder, consist of a sandwich structure where the bottom layer is stainless steel, the middle layer is PZT ceramic, and the top layer is aluminum. The prestress in the structure results in higher charge generation as compared to the ceramics, however, they are equally brittle. Fiber composites [Active Fiber Composite (AFC) and Macro Fiber Composite (MFC)] overcome the drawback of rigidity and brittleness by using high aspect ratio structure and interdigital electrode pattern. The fibers in MFCs are machined from low-cost piezoelectric ceramic wafers using a computer-controlled dicing saw. MFCs can be operated in either d_{33}-mode or d_{31}-mode by designing the two different electrode patterns. It has been shown that MFC operating in d_{33}-mode has higher energy conversion rate but lower electrical current as compared to that d_{31}-mode (Daue 2006). Further, MFC transducers have been reported to have the reliability of over 10^9 cycles operating at maximum strain.

NASA LaRC's Radial Field Diaphragm (RFD) comprises of a thin circular piezoelectric ceramic disk sandwiched between two polyimide dielectric films with copper-etched spiral electrodes (Bryant et al. 2002). This novel electrode pattern induces an electric field into the piezoelectric ceramic that extends out radially from the center of the wafer. At low frequencies of DC – 10 Hz, RFDs have been shown to provide 2–3 times higher displacement as compared to conventional unimorphs

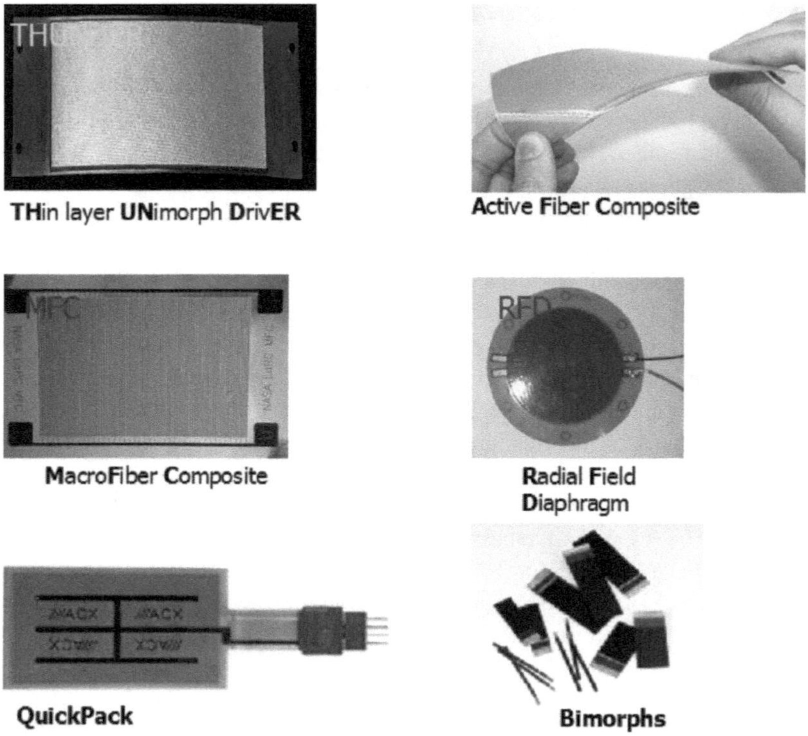

Fig. 18.2 Various flexible thin profile piezoelectric transducers. Thunder – Face International Corp., Norfolk, VA; AFC – Advanced Cerametrics Inc., Lambertville, NJ; MFC – Smart Materials Corp., Sarasota, FL; RFD – NASA Langley Research Center, Hapton, VA; QuickPack – Mide Technology, Medford, MA; Bimorphs – APC International, Mackeyville, PA

(Bryant et al. 2002). Experimental demonstration of the RFDs in energy harvesting application has not yet been reported. QuickPack (rigid) and PowerAct (flexible) transducer developed by Mide Technology Corporation are packaged bimorphs in a protective skin with preattached electrical leads, producing a highly reliable component with no soldered wires (Website: http://www.mide.com). QuickPack and PowerAct transducer has been widely utilized in the energy harvesting applications in both on-resonance and off-resonance vibration conditions.

If the vibration spectrum is known beforehand (normally characterized by using accelerometers) than the energy harvesting transducer can be tuned to operate at the resonance frequency of the system. The maximum generated power from the energy harvester can be approximated by using the model based on damped mass-spring system and it has been shown to be given as (Williams et al. 2001)

$$P_{\max} = \frac{mY_o^2 \omega_n^3}{4\varsigma}, \tag{18.7}$$

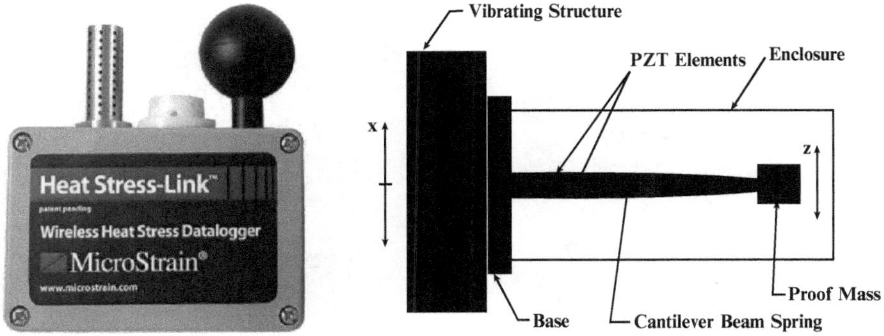

Fig. 18.3 "Wireless heat stress sensor node." The node includes a dry bulb temperature sensor, black body globe temperature sensor, and relative humidity sensor (copyright: Steve Arms, MicroStrain). The schematic shows the design of vibration energy harvester. (Presented at UTA Energy Harvesting Workshop, January 29, 2007)

where $\omega_n^2 = k/m$ is the system resonant frequency, $\varsigma = d/2\sqrt{mk}$ is the damping ratio, m is the seismic mass (electrical analog inductance, L), d is the damping constant (electrical analog resistance, R), k is the spring constant (electrical analog inverse capacitance, $1/C$), and Y_o is the maximum amplitude of the vibration. Equation (18.7) implies that the power is inversely proportional to the damping ratio which should be minimized through proper selection of materials and design and directly proportional to the system resonance frequency.

There are various ways of tuning the resonance frequency of the energy harvester to the available vibrations. Figure 18.3a,b shows an example of the energy harvesting device developed by MicroStrain and the schematic of layout of piezoelectric elements onto a cantilever beam with a tip mass (Website: http://www.microstrain.com). The piezoelectric elements are laid out as elements in order to maintain the flexibility of the beam. In this design, the length of the cantilever beam and tip mass can be changed to obtain the desired working frequency.

The concept of multimode transducers is very much applicable to the design of wide-band energy harvesters. Figure 18.4a shows the schematic of the piezoelectric "Tunnel" transducer consisting of two curved piezoelectric plates and a flat piezoelectric plate, all of them bonded together. The asymmetry in the dimensions, shape, and poling of the three plates allows to overlap the impedance curve closer to each other as shown in Fig. 18.4b. This extends the range of the response of the piezoelectric. This concept can be easily applied to design the structures operating in the low frequency ranges to obtain wide-band spectrums. Figure 18.4d shows the schematic of the output power as a function of the frequency for a multimode energy harvester depicting that the energy harvester can operate over wide vibration range.

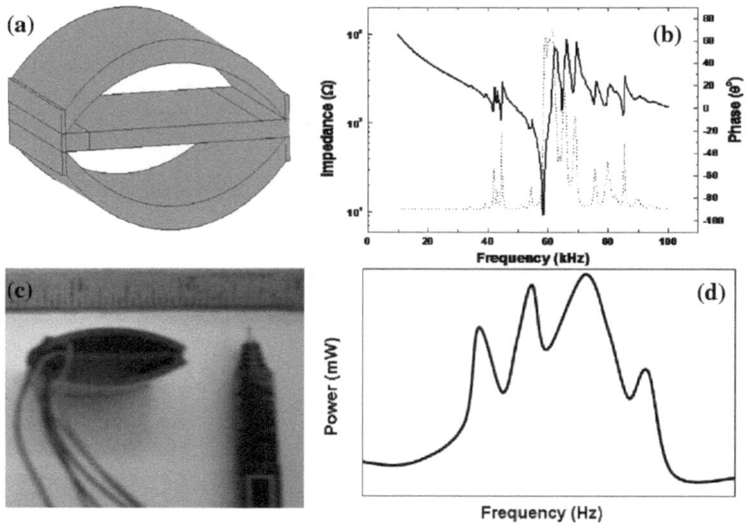

Fig. 18.4 (**a**) Schematic of the multimode "Tunnel" transducer, (**b**) typical impedance spectrum of the "Tunnel" transducer, (**c**) picture of the fabricated prototype, and (**d**) schematic representation of the power spectrum for the multimode energy harvester

18.4 Energy Harvesting from Low Amplitude: Low Frequency Vibrations

In common scenarios, the energy harvesting is done from low amplitude to low frequency vibrations. Piezoelectric bimorph transducer structure is one of the prominent choices for the low vibration conditions. We have developed several devices using the bimorph of the size: $35 \times 2.5 \times 0.60\,mm^3$, which has the maximum deflection of 0.7 mm. The free length of the bimorph was 28 mm, blocking force >0.06 N, first resonance frequency ~245 Hz, and capacitance at 1 kHz was 12 nF. The bimorph transducer can be loaded in several ways depending upon the available vibrations including cantilever with a tip mass, three-point bending, and vertical shaft with a tip plate.

The vibration testing results on the bimorph loaded in the cantilever form are shown in the Fig. 18.5. Figure 18.5a shows the picture of the bimorph transducer loaded with a fixed magnitude of tip mass. Figure 18.5b shows the natural resonance of the test prototype. The resonance frequency was found to be 20 Hz with a tip mass of 2.5 g. The testing was conducted by moving the bimorph transducer along the vertical axis or perpendicular to the beam length. As the bimorph is moved along the vertical or Z axes, flexing occurs causing charge generation through direct piezoelectric effect. The mass was increased in the steps of 0.5 g to determine the mechanical strength of the bimorph and voltage response. As the mass was increased to 2.5 g the output voltage obtained from the bimorph increased ($2 - 10\,V^{PP}$), however,

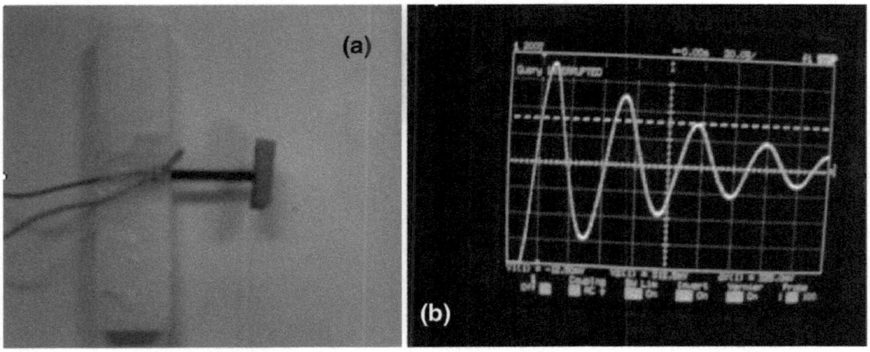

Fig. 18.5 (**a**) Picture of the fabricated prototype and (**b**) natural response

the mechanical strength of the bimorph decreased. For given size of bimorph which is $35 \times 2.5 \times 0.60\,\text{mm}^3$, a weight of 2.5 g was found to be the upper bound. When the bimorph was moved along the vertical axis over a height of 1 ft. at a frequency of 3 Hz a high voltage of $10\,\text{V}^{\text{PP}}$ was obtained with the maximum deflection of 0.5 mm in the bimorph. This is a significant response considering the dimension of the device and the range of the motion which was fixed to be 1 ft. Similar type of response was obtained from the testing samples in the X and Y direction motion. In this section, we present examples of the energy harvesting devices using the bimorph transducers loaded in cantilever form to illustrate the practical applications.

18.4.1 Piezoelectric Vibration Energy Harvester for Broadband Applications

This design utilizes the piezoelectric cantilever structure and has several features that allow exploiting the asymmetry in the vibrations (wide-band vibrations) with high efficiency. Figure 18.6 explains the architecture of this design (Fig. 18.6a), which shows the piezoelectric cantilevers of varying length and cross section with a tip mass to capture energy from wide-band vibrations. This type of device can be easily tuned to operate in low to high frequency range (broad band operation) by modulating the length of the cantilever arms, number of the cantilever arms, and tip mass. The model starts with a 3D stereolithography printed base plate and two types of piezoelectric plates as shown in Fig. 18.6b, c.

The piezoelectric plates are laser micromachined and assembled in a cubical box as shown in Fig. 18.7a. The electrical connections are made using six strips of silver electrode paint as shown in Fig. 18.7b and the top plate is attached to protect the sensitive "fingers." The piezoelectric fingers are machined using the laser micromachining (details of the laser machining of PZT can be found in Uppal et al. 2005). Femtosecond micromachining has emerged as the viable alternative for rapid prototyping of microcomponents. The main advantages of femtosecond

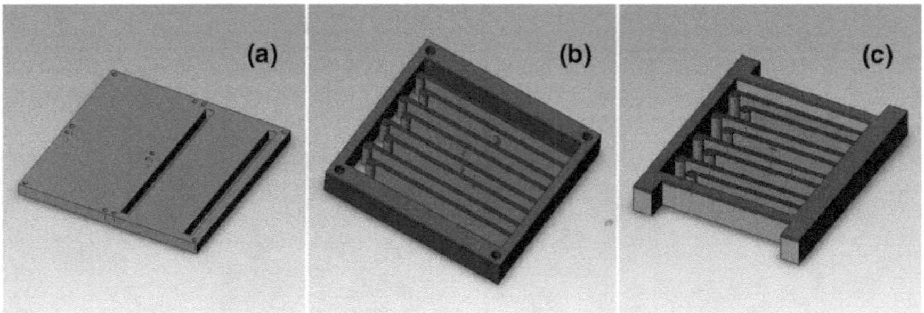

Fig. 18.6 (**a**) Base and the top plate, (**b**) piezoelectric plate designed to capture the motion along the $X-$ and Y-axis, and (**c**) piezoelectric plate designed to capture the motion along the Z-axis. The plates are fabricated from one piece of poled ceramic using the laser micromachining

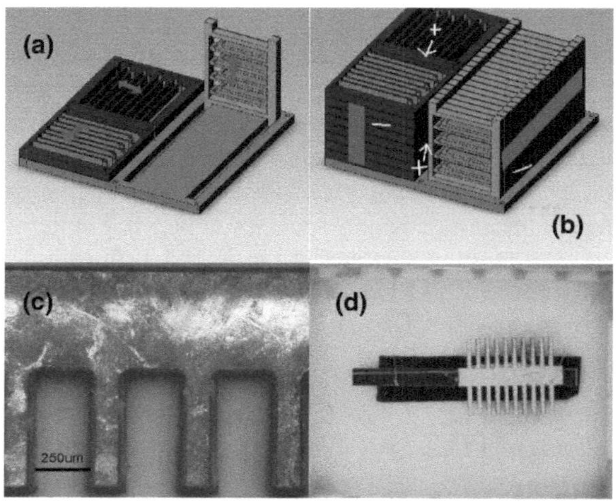

Fig. 18.7 (**a**) Direction in which the motion will be captured, (**b**) plates assembled and wired, (**c**) laser machined part showing the one directional fingers, and (**d**) laser machined bidirectional fingers

laser as compared to other lasers is the high specific power ($\sim 10^{13}$ W cm^{-2}) delivered in one pulse ($\sim$106 fs) that ablates a material upon contact. The instantaneous ablation restricts the heat affected zone (HAZ) into an extremely small region, thus preventing damage such as microcracks to the material surrounding the machining area. It has been shown that the variation in the dielectric constant and longitudinal piezoelectric constant (d_{33}) of the PZT before and after micromachining is negligible. Figure 18.7c, d shows the example of the machined cantilever structure. These plates can be stacked in the configuration shown in Fig. 18.7a, b.

Figure 18.8a shows another example of the energy harvesting device utilizing the "LEGO" concept. This model mounts the three bimorph transducers into a rectangular cavity designed to harvest the electrical energy from the motion along X, Y, and Z axes. The design easily accommodates three bimorphs for each X, Y, and Z

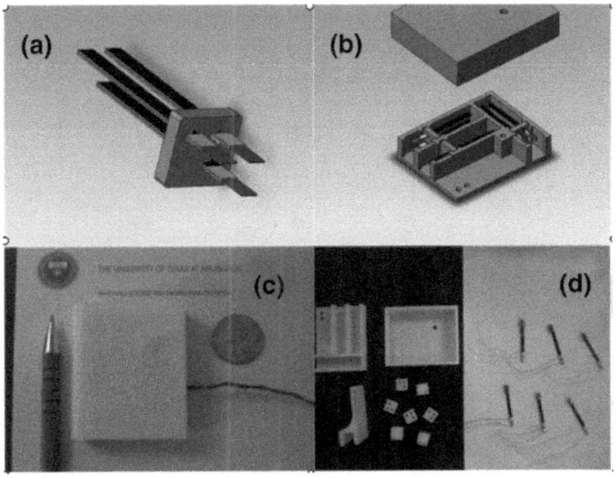

Fig. 18.8 (**a**) Assembly of the bimorphs into the holding fixture, (**b**) schematic of the bimorphs aligned along the X, Y and Z axes. The empty space is for accommodating the circuit. Three bimorphs are used along each of the axes. (**c**) Complete package of the energy harvester. The packaged assembly consists of all the wiring and (**d**) Picture of all the components inside the packaged assembly

axes motion in a small volume as shown in the Fig. 18.8b. The flat area in Fig. 18.8b with two holes represents the space for energy harvesting circuit which consists of rectifier, filter capacitor, and matching load. The total power harvested from the generator along each direction of motion was of the order of 300 μW with a tip mass of 2.5 g. The rated power for the complete device is 1 mW with operating frequency of 27 Hz.

18.4.2 Piezoelectric Wind Energy Harvester

Recently, we have demonstrated that practical levels of electric power can be generated from wind energy using piezoelectric bimorph transducers at small scale (Priya 2005; Priya et al. 2005; Chen and Priya 2006). Generally windmills are designed with rated speed of 15–18 mph operating at an efficiency of 20%. At smaller scale the rated speed will be further reduced implying reduction in the efficiency.

The dimensions of each individual bimorph used in the design of the piezoelectric windmills were $60 \times 20 \times 0.6\,\text{mm}^3$ with a free length of 53 mm. The resonance frequency and capacitance for this size of bimorph was measured to be 65 Hz and 170 nF, respectively. The power generation from the piezoelectric bimorphs loaded in the cantilever form can be modeled using the bending beam theory developed by Timoshenko. The voltage generated across two oppositely poled plates connected in parallel is given as (Priya 2005)

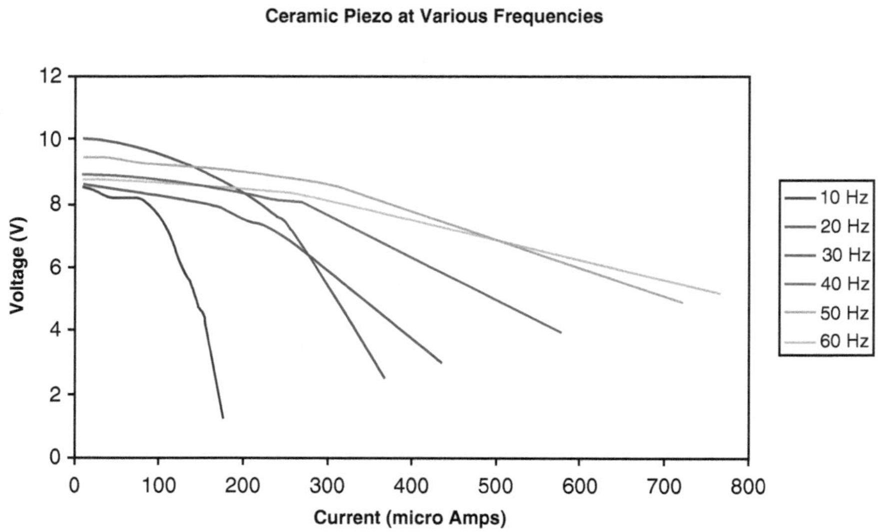

Fig. 18.9 Load line curve for the piezoelectric bimorph of dimension $60 \times 20 \times 0.6\,\text{mm}^3$ (free length = 53 mm)

$$V_{\text{OC}} = \frac{3}{8}\left(\frac{H}{L}\right)^2 h_{31}\delta z, \tag{18.8}$$

where h_{31} is the piezoelectric constant, H is the total thickness, and L is the length of the bimorph, and δz is the deflection under given AC stress. Using (18.8) the voltage generated for the bimorph with parameters $H = 0.6\,\text{mm}$, $L = 53\,\text{mm}$, $h_{31} = -1.67 \times 10^8\,\text{V m}^{-1}$, and $\delta z = 1.5\,\text{mm}$ can be determined to be of the order of 12.04 V. This value has been found to be in agreement with the measurement. Figure 18.9 shows the voltage–current relationship for the bimorph of the dimension $60 \times 20 \times 0.6\,mm^3$ (free length = 53 mm) at various frequencies. The matching load for the transducer can be found from the load line curve shown in this figure and corresponds to the centre point. Further, it has been found that by enhancing the pretension a higher voltage and correspondingly higher power ($V_{\text{rms}}^2/R_{\text{load}}$) can be obtained. Prestress also enhances the fatigue characteristics of the bimorphs.

A detailed description of the windmill fabrication technique is described elsewhere (Priya et al. 2005; Myers et al. 2007). Figure 18.10 shows the picture of the optimized windmill design which provides significant power density. The windmill was operated at an average speed of 10 mph. The output voltage was measured from one row of nine piezoelectric bimorphs (dimension: $60 \times 20 \times 0.6\,\text{mm}^3$) connected in parallel. The magnitude of the output voltage was $30\,\text{V}_{\text{pp}}$ at the oscillating frequency of 2 Hz. A rectification circuit with full bridge diodes and a capacitor with a magnitude of $200\,\mu\text{F}$ were used to characterize the maximum power from the windmill as a function of the resistive load. The maximum power from one row of the bimorphs was around 3 mW at the optimum load of $20\,\text{k}\Omega$ (Myers et al. 2007). One of the projected applications of piezoelectric windmill is for powering the wireless

Fig. 18.10 Picture of the three-blade piezoelectric windmill with rated power of 5 mW

sensor networks for weather monitoring in remote locations. This application requires an approximately 80 mW of power to receive and transmit typical set of data. In order to accomplish this requirement, a simplified circuit was designed and incorporated in the windmill to increase the efficiency of power transfer to the supercapacitor. The circuit simply consists of full bridge rectifier, storage capacitor, and matching load. It was found that supercapacitors are the suitable storage media in the windmill applications. Supercapacitors employ plates with extremely high surface areas which provide a higher storage capacity. The charging/discharging is not limited by ionic condition into the bulk electrode so supercapacitors can provide very high specific powers in a short period.

Figure 18.11 shows the charging and discharging cycle from the windmill shown in Fig. 18.10 operating at 10 mph. The size of the capacitor was 0.1 F with saturation at 5.5 V. It was found that the capacitor was almost charged and saturated for continuous operating time of 30 min. The discharging time of the capacitor was measured as a function of the resistive load and the output voltage during discharging was set by the impedance. Figure 18.11c shows the generated output power from the capacitor measured during discharging depending on resistive load. The maximum output current was 35 mA with 100-Ω resistive load while the generated output power was around 130 mW. The capacitor was found to discharge extremely fast for the small output loads. For the higher resistive load in the range of 1,000 Ω, the output power from the capacitor was stable for much longer period. At this time the small current was spent, so the output power was constant at 30 mW for more than for 2 min. To generate stable power for the sensor network, stored energy in the capacitor can be discharged continuously to a rechargeable battery during sleep mode. This device clearly demonstrates all the steps followed in the design and implementation of the energy harvesting system.

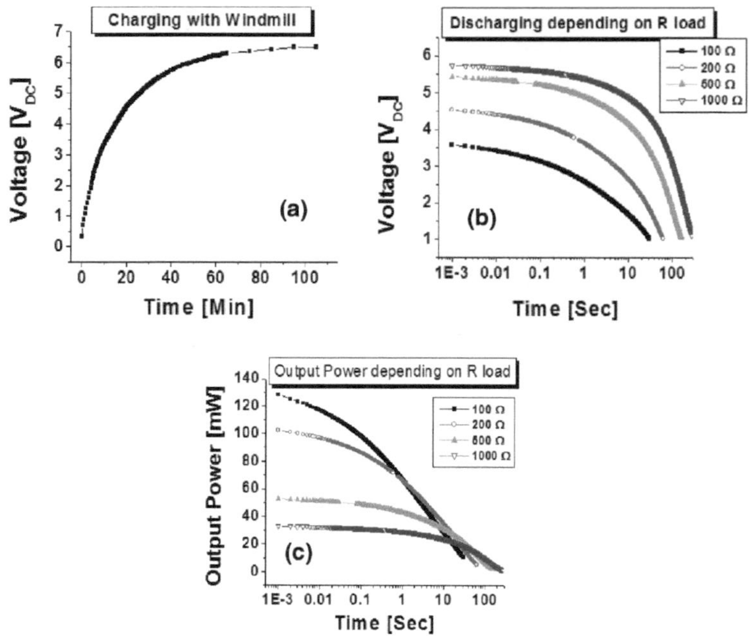

Fig. 18.11 Energy storage from piezoelectric windmill: (**a**) Charging of a supercapacitor, (**b**) discharging depending on the resistive load, and (**c**) output power of the capacitor depending on the resistive load

18.5 Piezoelectric Energy Harvesting Circuits

The energy harvesting circuit has two functions (1) to effectively transfer the electrical charge from piezoelectric transducer into an effective storage mechanism such as supercapacitor and (2) to match the impedance of the generator to the external load. In the bulk harvester cases under high vibration conditions, a DC-to-DC converter circuit is commonly used to harvest the generated electric energy and transfer it across the output load (Kim et al. 2004, 2005, 2006; Kim 2006; Ottoman et al. 2002, 2003; Guyomar et al. 2005).The commonly used converter circuit is called a buck converter and it operates in the discontinuous current mode (DCM). Several other techniques are used in conjunction with the circuit for reducing the matching impedance such as multilayering the piezoelectric transducer structure and increasing the area aspect ratio (area/thickness).

Recently, Guyomar et al. have presented a nonlinear processing technique "Synchronized Switch Harvesting on Inductor" (SSHI) for harvesting energy, which consists of a switching device in parallel with piezoelectric element (Guyomar et al. 2005). This nonlinear technique has been shown to significantly enhance the performance of the energy harvesting circuit and will be well suited for the resonating structures (Guyomar et al. 2005). In case of the off-resonance condition the energy harvesting circuit can be simply designed using the full wave rectifier, matching load, and storage capacitor to reduce the electrical components.

18.6 Conclusion

In this chapter, we presented the analysis and results on piezoelectric energy harvesting under low and high vibration conditions to illustrate the basic steps required in the design of the devices. Material, transducer structure, and energy storage circuit all play an important role in effectively transforming the mechanical energy into electrical energy. Results on the piezoelectric material with high energy density and high piezoelectric voltage coefficient in the PZT–PZNN family was presented. Parameters controlling the response of the piezoelectric transducer were evaluated in a given vibration scenario and the effective stiffness of the system. The electrical impedance matching methods under high vibration conditions are important to maximize the efficiency of the power transfer to the small load.

Acknowledgment The authors are thankful for the support of the various graduate students and postdoc working on the energy harvesting projects. The authors also acknowledge the financial support from Texas ARP and Nokia.

References

Ando A, Hisaki T (1998) Piezoelectric ceramic and method for producing piezoelectric ceramic element. US Patent No: 5,798,052

Bryant RG, Effinger IV RT, Aranda Jr I, Copeland Jr BM, Covington III EW (2002) Active Piezoelectric Diaphragms. Proc. SPIE, Smart Struc. and Materials – Active Materials: Behavior and Mechanics, Paper 4699-40, San Diego, CA

Calaway Jr EH (2004) Wireless Sensor Networks. Auerbach, New York, NY

Chen C, Priya S (2006) Electric energy generator. IEEE Trans. Ultrason. Ferroelectr. Freq. Control 53(3): 656–661

Colla EL, Morita T (2002) Piezoelectric technology for active vibration control. In: Piezoelectric Materials in Devices, N. Setter, ed., Ceramics Laboratory, EPFL

Daue T (2006) Overview Using the MFC for Energy Harvesting and other Commercial Applications. UTA Workshop on Energy Harvesting, Jan 27, 2006

Doğan A, Uchino K, Newnham RE (1997) Composite piezoelectric transducer with truncated conical endcaps "Cymbal". IEEE Trans. Ultrason. Ferroelectr. Freq. Control 44, 597–605

Guyomar D, Badel A, Lefeuvre E, Richard C (2005) Toward energy harvesting using active materials and conversion improvement by nonlinear processing. IEEE Trans. Ultrason. Ferroelectr. Freq. Control 52: 584–595

Islam RA, Priya S (2006a) Realization of high-energy density polycrystalline piezoelectric ceramics. Appl. Phys. Lett. 88: 032903

Islam RA, Priya S (2006b) High energy density composition in the System PZT – PZNN. J. Am. Ceram. Soc. 89: 3147–3156

Kawano K, Sube M (1997) Piezoelectric ceramic compositions. US Patent No: 5,607,614

Kim H (2006) Piezoelectric Energy Harvesting. PhD dissertation, The Pennsylvania State University, University Park, PA

Kim H, Batra A, Priya S, Uchino K, Markley D, Newnham RE, Hofmann HF (2004) Energy harvesting a using piezoelectric "Cymbal" transducer in dynamic environment. Jpn. J. Appl. Phys. 43: 6178–6183

Kim H, Priya S, Uchino K, Newnham RE (2005) Piezoelectric energy harvesting under high pre-stressed cyclic vibrations. J. Electroceram. 15: 27–34

Kim H, Priya S, Uchino K (2006) Modeling of piezoelectric energy harvesting using cymbal transducers. Jpn. J. Appl. Phys. Part 1-Reg. Papers 45(7): 5836–5840

Meninger S, Mur-Miranda JO, Amirtharajah R, Chandrakasan AP, Lang JH (2001) Vibration-to-electric energy conversion. IEEE Trans. VLSI Syst. 9: 64–76

Myers R, Vickers M, Kim H, Priya S (2007) Small scale windmill. Appl. Phys. Lett. 90: 054106

Newnham RE, Doğan A, Markley DC, Tressler JF, Zhang J, Uzgur E, Meyer Jr RJ, Hladky-Hennion AC, Hughes WJ (2002) Size Effects in Capped Ceramic Underwater Sound Projectors. Oceans '02 MTS/IEEE 4: 2315–2321

Nishida M, Ouchi H (1976) Piezoelectric ceramic compositions. US Patent No: 3,998,748

Nishida M, Ouchi H (1977) Piezoelectric ceramic compositions. US Patent No: 4,062,790

Ottoman GK, Hoffmann HF, Bhatt AC, Lesieture GA (2002) Adaptive piezoelectric energy harvesting circuit using wireless remote power supply. IEEE Trans. Power Electron. 17: 669–676

Ottoman GK, Hoffmann HF, Lesieture GA (2003) Optimized piezoelectric energy harvesting circuit using step-down converter in discontinuous conduction mode. IEEE Trans. Power Electron. 18: 696–703

Paradiso J, Starner T (2005) Energy scavenging for mobile and wireless electronics. Pervasive Comput. 4(1): 18–27

Priya S (2005) Modeling of electric energy harvesting using piezoelectric windmill. Appl. Phys. Lett. 87: 184101

Priya S, Chen CT, Fye D, Zhand J (2005) Piezoelectric windmill – A novel solution to remote sensing. Jpn. J. Appl. Phys. 44: L104–L107

Richards CD, Anderson MJ, Bahr DF, Richards RF (2004) Efficiency of energy conversion for devices containing a piezoelectric component. J. Micromech. Microeng. 14: 717–721

Rome LC (2006) Backpack for harvesting electrical energy during walking and for minimizing shoulder strain. US Patent No: 20060192386

Roundy S, Wright PK, Rabaey J (2003) A study of low level vibrations as a power source for wireless sensor nodes. Comput. Commun. 26: 1131–1144

Roundy S, Wright PK, Rabaey JM (2004) Energy Scavenging for Wireless Sensor Networks. Kluwer, Boston, MA

Shenck N, Paradiso JA (2001) Energy scavenging with shoe-mounted piezoelectrics. IEEE Micro 21: 30–42

Starner T, Paradiso JA (2004) Human-generated power for mobile electronics. In: Low-Power Electronics Design, C. Piguet, ed., CRC, Boca Raton, FL, Chapter 45, pp. 1–35

Uppal N, Shiakolas P, Priya S (2005) Femtosecond laser micromachining of PZT ceramics. Ferroelect. Lett. Sect. 32(3–4): 67–77

Williams CB, Sherwood C, Harradine MA, Mellor PH, Birch TS, Yates RB (2001) Development of an electromagnetic micro-generator. IEE Proc. Circuits Devices Syst. 148(6): 337–342

Yoo J, Lee Y, Yoon K, Hwang S, Suh S, Kim J, Yoo C (2001) Microstructural, electrical properties and temperature stability of resonant frequency in $Pb(Ni_{1/2}W_{1/2})O_3 - Pb(Mn_{1/3}Nb_{2/3})O_3 - Pb(Zr,Ti)O_3$ ceramics for high-power piezoelectric transformer. Jpn. J. Appl. Phys. 40: 3256–3259

Chapter 19
Piezocomposite Ultrasonic Transducers for High-Frequency Wire Bonding of Semiconductor Packages

Siu Wing Or* and Helen Lai Wa Chan

Piezocomposite ultrasonic transducers for high-frequency wire bonding of semiconductor packages are developed to alleviate the strong mode coupling and high mechanical quality factor intrinsic in piezoceramic transducers by using ring-shaped lead zirconate titanate (PZT)/epoxy 1–3 piezocomposites as the driving elements. In this chapter, the background and challenges involved in the state-of-the-art high-frequency wire-bonding process are described. The fabrication, resonance characteristics, and material properties of the piezocomposite rings having high PZT volume fractions in excess of 0.8 and with a small epoxy width of 77 μm are reported. The structure, electrical characteristics, vibrational characteristics, and wire-bonding performance of a 136-kHz piezocomposite transducer are presented, together with those of a PZT piezoceramic transducer of similar structure. With the guide of a finite-element modal analysis, the nature of most experimental resonance modes in the two transducers is identified. The low lateral coupling of the piezocomposite rings effectively suppresses the nonaxial and many other spurious resonances in the piezocomposite transducer, retaining only the axial-mode resonances. Because of the effect of epoxy damping, the piezocomposite transducer exhibits a 2.4-times reduction in mechanical quality factor to a desired low value of 296. The process study confirms the value of the piezocomposite transducer in commercial wire bonders for enabling high-frequency wire-bonding technology.

19.1 High-Frequency Ultrasonic Wire Bonding

High-frequency ($\geq$100 kHz) ultrasonic wire bonding is an urgently needed technology to produce smaller wire bonds of higher quality in a shorter time, as integrated-circuit (IC) chips and their associated packages in the semiconductor industry are

S.W. Or
Department of Applied Physics, The Hong Kong Polytechnic University, Hung Hom, Kowloon, Hong Kong

A. Safari, E.K. Akdoğan (eds.) *Piezoelectric and Acoustic Materials for Transducer Applications.*

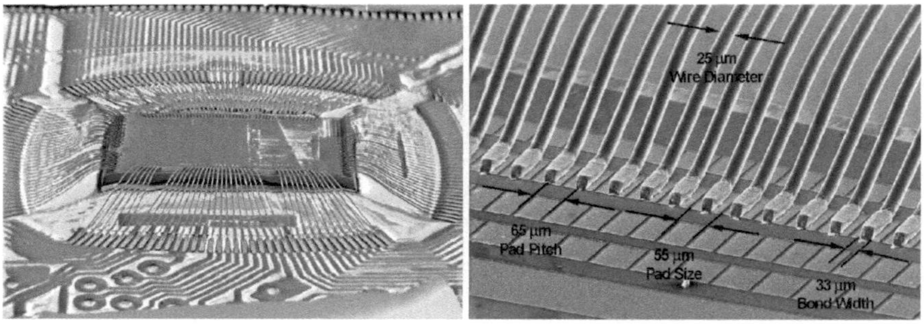

Fig. 19.1 Ultrasonic bonding of aluminum (Al) wires on an ASM standard chip-on-board (COB) package (Or et al. 2007)

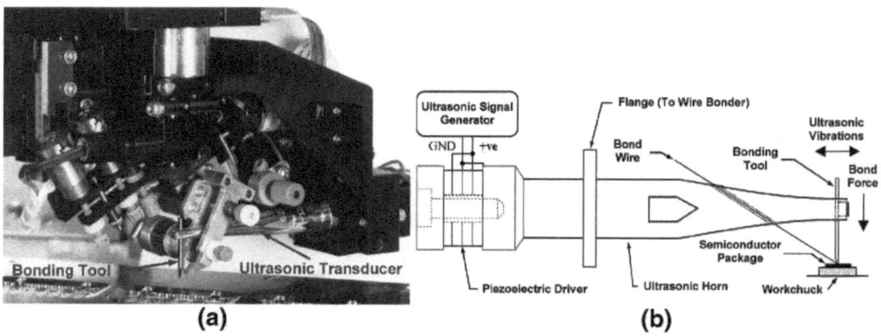

Fig. 19.2 A 137-kHz PZT piezoceramic transducer and its wire bonder. (**a**) Photograph, (**b**) Schematic (Or 2001)

reducing in size and increasing in volume production at a very rapid rate (Fig. 19.1) (Ramsey and Alfaro 1991; Cheung et al. 1999; Or 2001; ITRS 2007). The wire-bonding process is essentially based on the proper transmittal of ultrasonic vibration energy under pressure to plasticize the metallic bond wires (typically Al or Au wires of $\leq$25-µm diameter) for forming metallurgical bonds (Fig. 19.1) by means of an ultrasonic transducer mounted on an automatic wire bonder (Fig. 19.2a). As a key component of the wire bonder, the transducer consists of four main parts (Fig. 19.2b) as follows:

1. A Langevin-type piezoelectric driver (in the form of a stack of piezoelectric rings sandwiched between two metal slabs under compression via a central prestress screw) converts electrical driving signals supplied from an ultrasonic signal generator into ultrasonic axial vibrations.
2. An ultrasonic horn couples and amplifies the said vibrations with the maximum amplitude at its tip.
3. A bonding tool holds the bond wire at its tip and thereby couples the amplified vibrations to the bond wire.

4. A flange allows the mounting of the whole transducer onto the wire bonder so that a predetermined bond force can be applied, together with the ultrasonic vibrations, to both the bond wire and the bonding site (on the IC chip or on the chip substrate) of the semiconductor package via the tip of the bonding tool.

To fully enable fine-pitch ($<50\,\mu$m) and high-speed (>10 wires/s) processes, axial-mode transducers operating in the 100-kHz range with moderately high mechanical quality factors Q_m ($\sim$300) are of prime need (Cheung et al. 1999; Or 2001). While the hard PZT piezoceramic rings used in state-of-the-art transducers have large axial coupling coefficients k_t ($\sim$0.38 for PZT8), they also have large lateral coupling coefficients k_{31} ($\sim$0.38) and high thickness Q_m ($\sim$1500) (Fig. 19.3). The resulting transducers thus exhibit strong lateral coupling with many nonaxial and spurious modes distributed over the 100-kHz range (Fig. 19.4). These cause the supplied energy to be drained from the desired axial modes into these undesirable modes, thereby reducing the purity of axial motion of transducers. Moreover, high-Q_m transducers have sharp resonances with limited bandwidths; they are sensitive to load changes and impose difficulties in frequency tuning and mode locking (Or 2001).

Therefore, we have developed novel ultrasonic transducers possessing low lateral coupling and broadband characteristics to alleviate the drawbacks intrinsic in PZT piezoceramic transducers based on the PZT/epoxy 1–3 piezocomposite technology (Or 2001; Or and Chan 2001; Chan et al. 2001; Or et al. 2007). In the following sections, we will describe the fabrication, resonance characteristics, and material properties of PZT/epoxy 1–3 piezocomposite rings with high PZT volume fractions of 0.82–0.94, and present the structure, electrical characteristics, vibrational characteristics, and wire-bonding performance of a 136-kHz piezocomposite transducer. A PZT transducer of similar structure is also fabricated and characterized for comparison.

19.2 1–3 Piezocomposite Rings

19.2.1 Fabrication

The 1–3 piezocomposite rings were fabricated using the dice-and-fill technique (Chan 1987; Or 2001; Or and Chan 2001). PKI804 PZT piezoceramic rings (12.7-mm outer diameter, 5.0-mm inner diameter, and 2.5-mm thickness) and Araldite LY5210/HY2954 epoxy were used as the active and passive phases, respectively. Two sets of equally spaced, parallel grooves were cut in the PZT rings in orthogonal directions using an automatic dicing saw (Disco DAD 341) equipped with a diamond saw blade of 70-µm thickness. The resulting grooves in the ring were about 77-µm wide. Approximately 0.19 mm of PZT was allowed to remain on the base so that the whole ring structure could be retained after cutting. By applying a different number of cuts (5, 7, 9, 11, 13, and 15 cuts/direction) to the PZT rings,

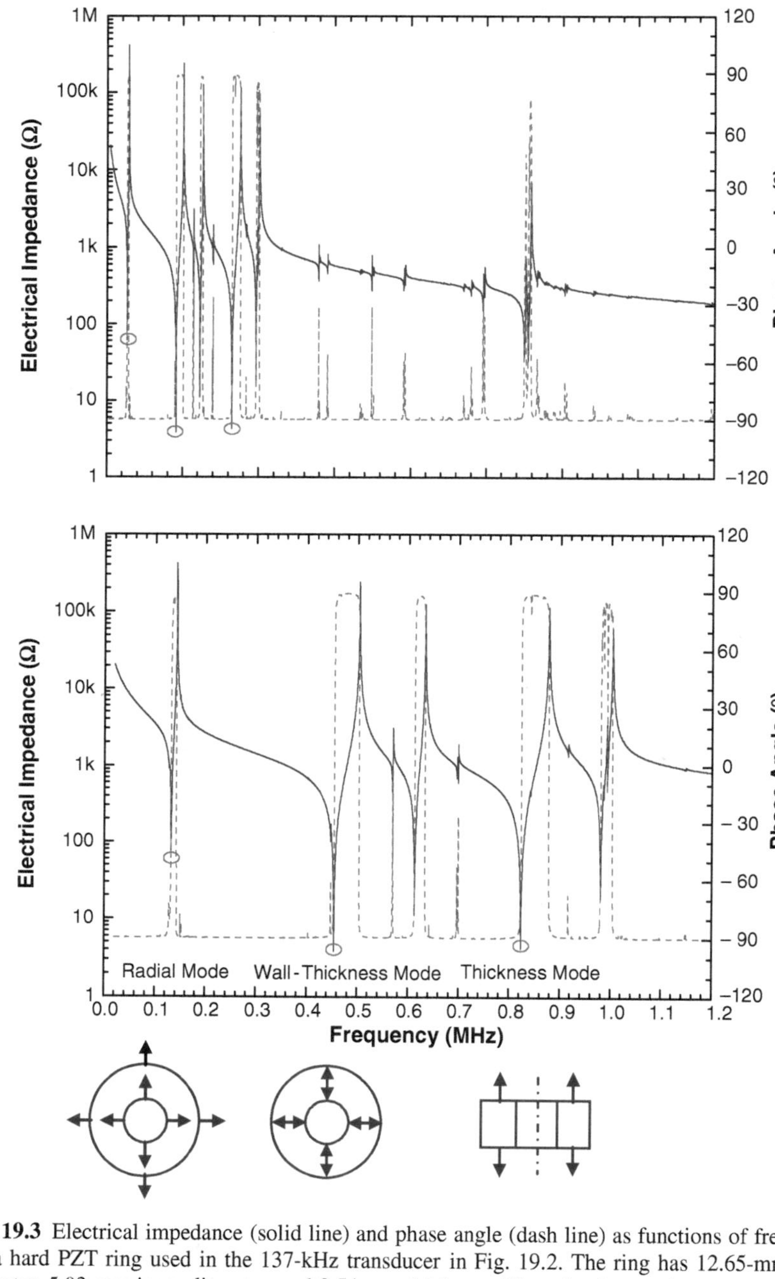

Fig. 19.3 Electrical impedance (solid line) and phase angle (dash line) as functions of frequency for a hard PZT ring used in the 137-kHz transducer in Fig. 19.2. The ring has 12.65-mm outer diameter, 5.03-mm inner diameter, and 2.54-mm thickness. Three fundamental resonance modes, including the radial (132 kHz), wall-thickness (454 kHz), and thickness (823 kHz) modes, are observed, besides many other complex modes (Or 2001)

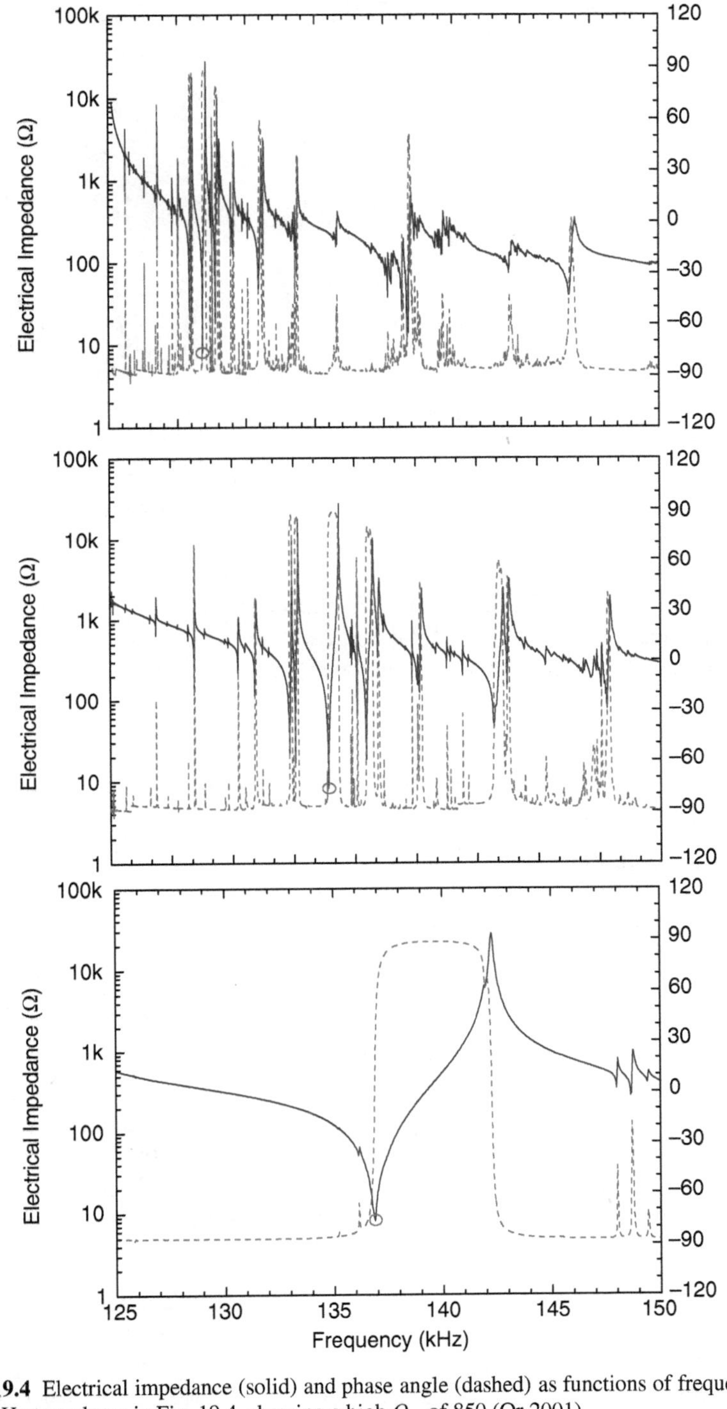

Fig. 19.4 Electrical impedance (solid) and phase angle (dashed) as functions of frequency for the 137-kHz transducer in Fig. 19.4, showing a high Q_m of 850 (Or 2001)

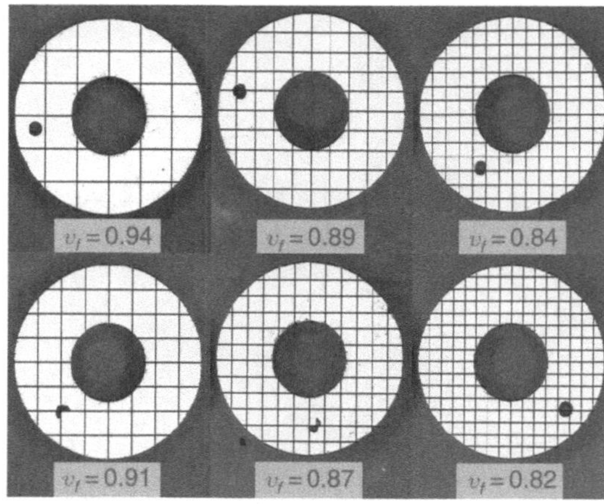

Fig. 19.5 Fabricated piezocomposite rings with six different υ_f. The black dot indicates the anode surface (connected to positive voltage during poling) (Or and Chan 2001)

piezocomposite rings with different PZT element widths L of 2.03, 1.50, 1.19, 0.98, 0.83, and 0.71 mm and with different PZT volume fractions υ_f of 0.94, 0.91, 0.89, 0.87, 0.84, and 0.82 were obtained, respectively (Fig. 19.5). Filling of epoxy under a vacuum was carried out. The cured piezocomposite rings were lapped to grind off the PZT base and to obtain a pair of flat and parallel surfaces. It was found that a thickness of 2.30 ± 0.02 mm could generally be obtained after lapping.

19.2.2 Resonance Characteristics

19.2.2.1 Allowable Resonance Modes

Resonance modes of a 1–3 piezocomposite ring can be classified into two main categories as follows (Or and Chan 2001):

1. Characteristic resonances of individual PZT elements inside the ring comprise the first category.
   (a) The coupled longitudinal-thickness mode resonance f_H, which is essentially determined by the height H of the PZT elements, is the mode of practical interest in ultrasonic transducers as it corresponds to the thickness-mode resonance f_t of the whole ring with an axial moment of the ring's thickness.
   (b) The coupled lateral-mode resonances f_{L1} and f_{L2} are mainly determined by the width L of the PZT elements with square cross section.

When the piezocomposite ring is thick, the PZT elements are tall thin bars, and f_H is well separated from f_{L1} and f_{L2}. When the elements approach the shape of a cube, coupling of these resonance modes occurs. This mode coupling effect can be predicted using the mode coupling theory (Chan 1987; Or 2001; Or and Chan 2001).

2. Cooperative resonances of the whole ring comprise the second category.
   (a) The radial f_R and wall-thickness f_W mode resonances are the lateral resonances of the ring as a whole and are usually the two low-frequency resonance modes in a thin ring. They are essentially determined by the mean diameter and wall thickness of the ring and are characterized by a symmetric- and an asymmetric-radial movement of the inner and outer diameters of the ring, respectively.
   (b) The stopband resonances f_{S1} and f_{S2} result from the periodic structure of a piezocomposite and occur at frequencies near where scattering of laterally propagating shear waves in the epoxy matrix from the planes of the regularly spaced PZT elements occur. They are related to the lateral periodicity d (one epoxy width d_p+ one element width L) if the piezocomposite has low values of υ_f (<0.25) with large d_p (Chan 1987; Hayward and Bennett 1996). In contrast, they are mainly determined by d_p for piezocomposites having relatively high values of υ_f with sufficiently small d_p (Hayward and Bennett 1996; Geng and Zhang 1999). Our rings belong to the latter case (Or and Chan 2001).

19.2.2.2 Thinning Test

Thinning test was carried out to study the resonance characteristics of the piezocomposite rings. It involves lapping each ring sample to different thicknesses in steps and at each thickness applying a thin silver layer of 5–10 μm on the two major surfaces to form electrodes and then measuring the electrical impedance spectrum using an impedance analyzer (Agilent 4294A). f_H of the PZT elements is a piezoelectric stiffened mode, while the other resonance modes are regarded as piezoelectric unstiffened modes (Or and Chan 2001).

Characteristic Resonances of Individual PZT Elements

Figure 19.6 shows the thinning test results of a piezocomposite ring with $\upsilon_f = 0.82$ and $L = 0.71$ mm. The electrical impedance spectrum of the ring with $H = 2.30$ mm (configuration ratio $G = L/H = 0.31$) is plotted in Fig. 19.6a. Guided by the mode coupling theory (Chan 1987; Or 2001; Or and Chan 2001), the strongest resonance mode at 0.93 MHz is identified as f_H. The mode at about 2.70 MHz is its third harmonic and is labeled as $3f_H$ since it appears at about three times f_H. Even harmonics are not expected to be observed because the applied electric field has opposite polarity on the two faces of the ring. The two lateral modes f_{L1} and f_{L2} are weak resonances found at 3.73 and 2.48 MHz, respectively. They are often mixed with other high-frequency resonances. For example, f_{L2} has merged with $3f_H$ in Fig. 19.6a. Besides, there are several weak resonances above f_{L1} and f_{L2}. These may be other lateral modes, which arise from the PZT elements with smaller lateral sizes L such

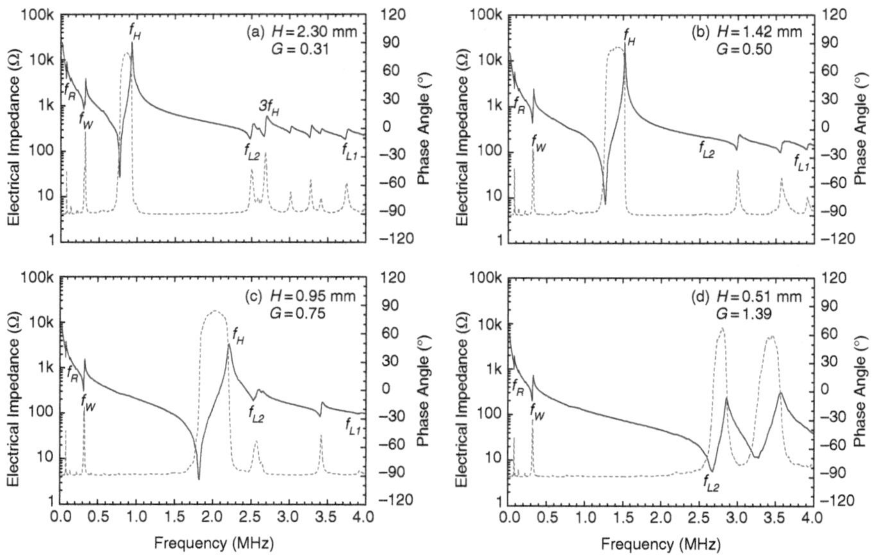

Fig. 19.6 Electrical impedance (solid line) and phase angle (dot line) vs. frequency plots for piezocomposite ring with $\upsilon_f = 0.82$ and $L = 0.71$ mm at different thicknesses H or configuration ratios G (Or and Chan 2001)

as those elements located near the inner and outer circumferences of the ring as shown in Fig. 19.5. Moreover, the two small resonances that appeared in the lower frequency band of 75 and 306 kHz are f_R and f_W of the whole ring, respectively. Detailed behavior of these modes will be discussed in a later paragraph. When the ring is subsequently thinned down (Fig. 19.6b–d), f_H increases accordingly while f_{L1} and f_{L2} stay quite constant. Also, $3f_H$ increases at approximately three times the rate of f_H. If there is no other resonance close to f_{L1} and f_{L2}, the resonance peaks associated with them will be very weak and sometimes even disappear (Fig. 19.6b). For $H < 0.95$ mm $(G > 0.75)$, f_H starts to couple with f_{L2} (Fig. 19.6c) and later continues to merge with the subsequent lateral modes (Fig. 19.6d), thereby forming a strong but complex resonance mode covering a broad range of frequency. Consequently, f_{L2} serves to impose an upper frequency bound where mode coupling begins.

The general behaviors of f_H, as well as f_{L1} and f_{L2}, at different ring thicknesses for the remaining five batches of rings with υ_f values of 0.84, 0.87, 0.89, 0.91, and 0.94 are quite similar to each other and also to the one with $\upsilon_f = 0.82$. The corresponding spectra for the electrical impedance are shown in Figs. 19.7–19.11. The observed results are summarized as follows:

1. For the ring with $\upsilon_f = 0.84$ and $L = 0.83$ mm (Fig. 19.7), its initial thickness is $H = 2.31$ mm $(G = 0.36)$ (Fig. 19.7a). A clean f_H is observed at 0.93 MHz, and then no other resonance is detected until f_{L2} appears at 2.22 MHz. Resonance modes following f_{L2} include other high-frequency lateral modes, $3f_H$ (that has mixed with these lateral modes at about 2.80 MHz), and f_{L1} (3.35 MHz). All

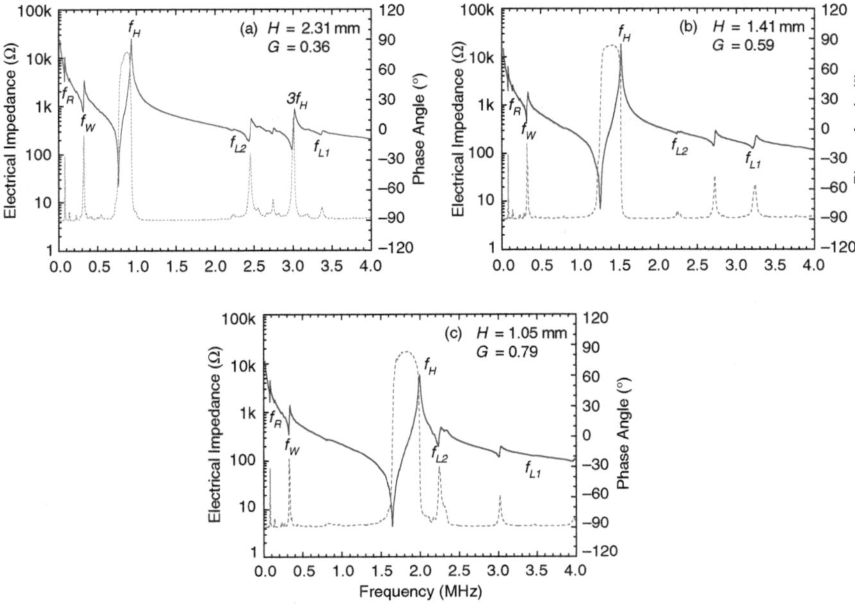

Fig. 19.7 Electrical impedance (solid line) and phase angle (dot line) vs. frequency plots for piezocomposite ring with $\upsilon_f = 0.84$ and $L = 0.83$ mm at different thicknesses H or configuration ratios G (Or and Chan 2001)

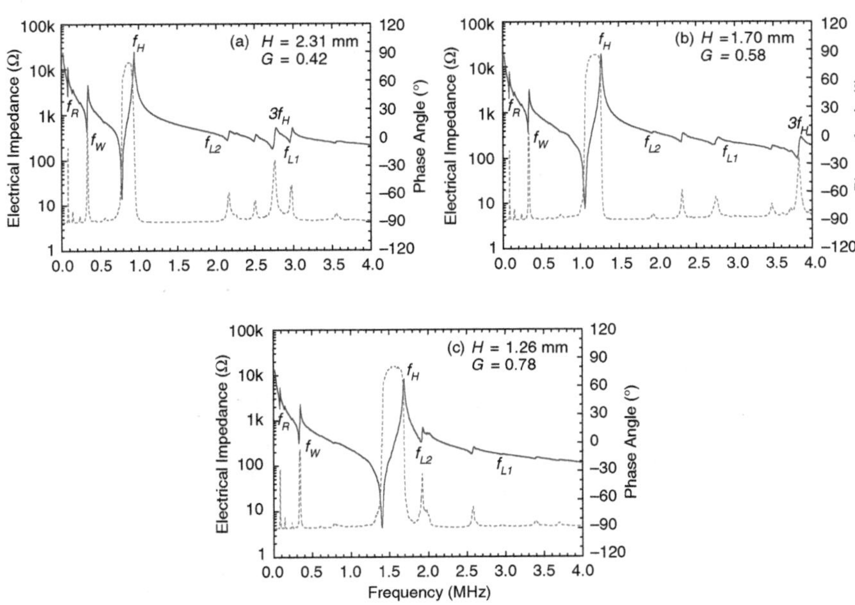

Fig. 19.8 Electrical impedance (solid line) and phase angle (dot line) vs. frequency plots for piezocomposite ring with $\upsilon_f = 0.87$ and $L = 0.98$ mm at different thicknesses H or configuration ratios G (Or and Chan 2001)

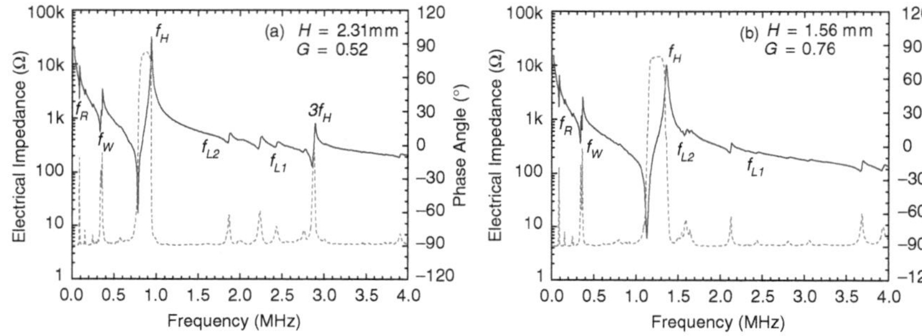

Fig. 19.9 Electrical impedance (solid line) and phase angle (dot line) vs. frequency plots for piezocomposite ring with $\upsilon_f = 0.89$ and $L = 1.19\,\text{mm}$ at different thicknesses H or configuration ratios G (Or and Chan 2001)

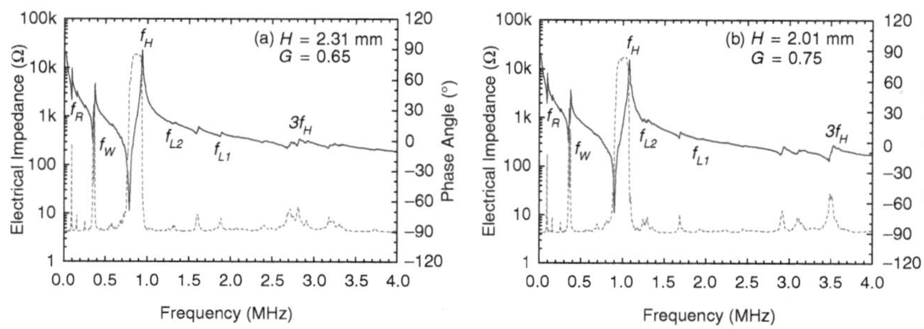

Fig. 19.10 Electrical impedance (solid line) and phase angle (dot line) vs. frequency plots for piezocomposite ring with $\upsilon_f = 0.91$ and $L = 1.50\,\text{mm}$ at different thicknesses H or configuration ratios G (Or and Chan 2001)

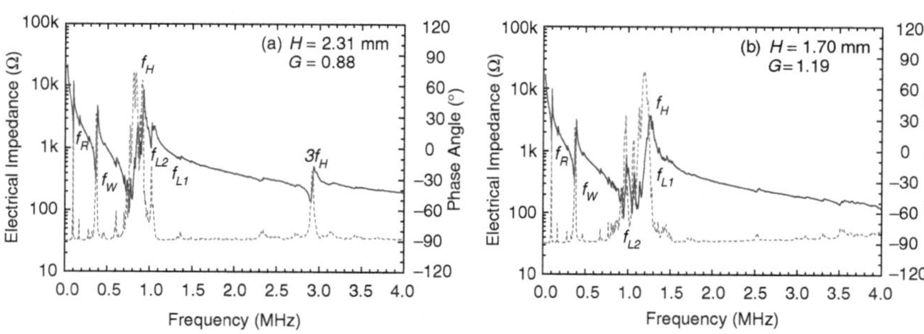

Fig. 19.11 Electrical impedance (solid line) and phase angle (dot line) vs. frequency plots for piezocomposite ring with $\upsilon_f = 0.94$ and $L = 2.03\,\text{mm}$ at different thicknesses H or configuration ratios G (Or and Chan 2001)

these lateral resonances are found at frequencies lower than those of the ring with $\upsilon_f = 0.82$ due to an increment of L from 0.71 to 0.83 mm. When H is reduced, f_H and $3f_H$ rise almost linearly while f_{L1} and f_{L2} remain nearly unchanged (Figs. 19.7b–c). Coupling of f_H with f_{L2}, as well as the subsequent lateral modes, occurs at $H < 1.05\,\text{mm}$ $(G > 0.79)$ (Fig. 19.7c).

2. For the ring with $\upsilon_f = 0.87$ and $L = 0.98$ mm (Fig. 19.8), f_H, $3f_H$, and f_{L1} are found at 0.94, 2.78, and 2.96 MHz, respectively for $H = 2.31$ mm $(G = 0.42)$ (Fig. 19.8a). f_{L2} is not observed because of no nearby resonances but reappears at 1.92 MHz when the ring is thinned down to 1.70 mm $(G = 0.58)$ (Fig. 19.8b). Clearly, the lateral mode resonances, including f_{L1} and f_{L2}, have lower frequencies compared with the previous batches, but f_H is similar. It increases continuously in response to the thinning process and tends to couple with f_{L2} at $H = 1.26$ mm $(G = 0.78)$ (Fig. 19.8c).
3. For the ring with $\upsilon_f = 0.89$ and $L = 1.19$ mm (Fig. 19.9), f_H and $3f_H$ are clearly observed at 0.94 and 2.89 MHz when $H = 2.31$ mm $(G = 0.52)$, and they increase with decreasing ring thickness (Fig. 19.9a). Both f_{L1} and f_{L2} remain quite stationary at lower frequencies of 2.42 and 1.61 MHz, respectively. Merging of f_H with f_{L2} results at a comparatively larger thickness of $H = 1.56$ mm $(G = 0.76)$ (Fig. 19.9b).
4. For the ring with $\upsilon_f = 0.91$ and $L = 1.50$ mm (Fig. 19.10), there are no major changes in f_H (0.94 MHz) and $3f_H$ (2.83 MHz) at $H = 2.31$ mm $(G = 0.65)$ (Fig. 19.10a). However, the operating bandwidth of f_H is even smaller since f_{L2} is detected at a frequency as low as 1.30 MHz and f_{L1} at 1.87 MHz. The corresponding sample thickness of f_H that starts to couple with f_{L2} is $H < 2.01$ mm $(G > 0.75)$ (Fig. 19.10b).
5. For the ring with $\upsilon_f = 0.94$ and $L = 2.03$ mm (Fig. 19.11), mode coupling is apparent, and complex peaks are found even for ring as thick as 2.31 mm $(G = 0.88)$ (Fig. 19.11a, b). Nevertheless, the mode coupling theory can provide a good guide for locating the position of f_H, f_{L1}, and f_{L2}.

Cooperative Resonances of Piezocomposite Rings

The cooperative resonances of piezocomposite rings are reported as follows:

1. For the radial and wall-thickness mode resonances, f_R and f_W are the two low-frequency resonance modes in a thin ring resonator as shown in Figs. 19.6–19.11. The several weak resonances following f_R and f_W are their harmonics. They have frequencies approximately equal to an integral multiple of them. Because of the mechanical absorption of the epoxy matrix, both f_R and f_W become weaker in comparison with the pure PZT ring (Fig. 19.12). Therefore, it is an advantage to use 1–3 piezocomposites instead of PZT to fabricate transducers with low lateral-mode coupling characteristics. Figure 19.13 shows the variation of f_R and f_W (taken at the series resonance frequencies) with ring thickness for the piezocomposite rings at different υ_f. Both f_R and f_W decrease slightly as the ring thickness increases but increase as υ_f increases. The dependence of f_R and f_W on υ_f elucidates the existence of different degrees of lateral clamping of the PZT elements by the epoxy matrix at different υ_f. If the thickness of a piezocomposite ring becomes comparable with its wall thickness (3.85 mm), f_H will couple with f_W. Hence, f_W acts as a lower frequency bound, which limits the selection of the thickness of the ring.

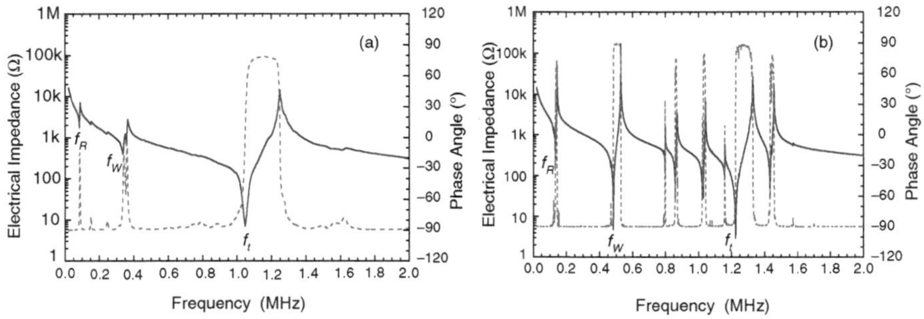

Fig. 19.12 Comparison of the electrical resonance characteristics between (**a**) a piezocomposite ring ($\upsilon_f = 0.89$) and (**b**) its PZT ring, both with the same outer diameter of 12.7 mm, inner diameter of 5.0 mm, and thickness of 1.7 mm. f_R, f_W, and f_t denote the fundamental radial, wall-thickness, and thickness-mode resonances of the rings, respectively (Or and Chan 2001)

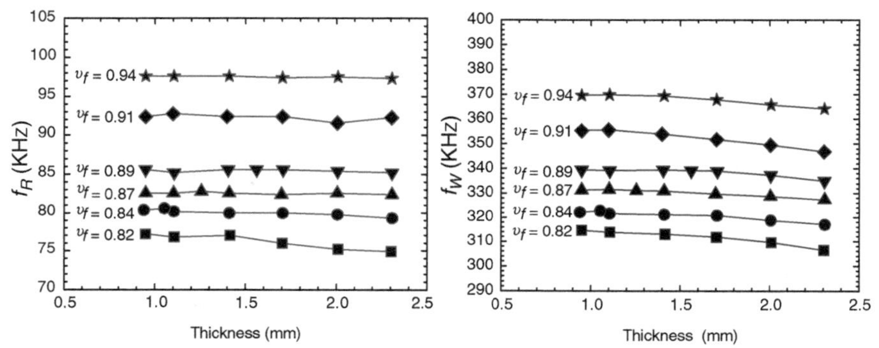

Fig. 19.13 Variation of radial f_R and wall-thickness f_W mode resonance frequencies with ring thickness for the piezocomposite rings at different υ_f (Or and Chan 2001)

2. For the stopband resonances, the empirical resonance frequency equations proposed by Geng and Zhang (1999) were employed to estimate the two lateral stopband resonances f_{S1} and f_{S2}. They are given as follows:

$$f_{S1} = \frac{f_{S2}}{\sqrt{2}}, \tag{19.1}$$

$$f_{S2} = \frac{\upsilon_S}{2d_P}, \tag{19.2}$$

where f_{S2} can be viewed as a half-wave resonance in the epoxy gap width d_P; f_{S1} can be considered as another half-wave resonance across the diagonal direction $\sqrt{2}d_P$, and υ_S denotes the shear wave velocity in the epoxy. Using $\upsilon_S = 1207$ m/s for Araldite LY5210/HY2954 epoxy (Or 2001), f_{S1} and f_{S2} were calculated to be 5.54 and 7.84 MHz, respectively. However, an investigation into the electrical impedance spectra for frequencies up to 10 MHz found no observable resonance peaks for all the rings studied. The reasons are probably that the PZT elements are much larger than the epoxy gap width and are spaced so closely that the possible reflection of shear waves is weakened and the resulting standing waves cannot be greatly enhanced in this limited epoxy gap width.

19.2.2.3 Practical Implication for Transducer Design

In designing a piezocomposite transducer for high-frequency wire bonding, the use of low-υ_f, large-d_P piezocomposite rings is not recommended as the embedded PZT elements tend to oscillate at the frequency associated with the oscillations of the elements, giving rise to uneven surface displacements and inhomogeneous material properties. In addition to optimizing the homogeneity, material properties, and manufacturing cost of the rings, it should avoid other resonance modes being coupled to f_H. This in turn requires a ring thickness not too thin in order to keep f_H smaller than all lateral modes of the PZT elements ($G < 0.8$) but not too thick to avoid coupling of f_H with f_W and f_R of the ring.

For the batches with υ_f equal to 0.85, 0.87, and 0.89, a strong and clean thickness-mode resonance can generally be obtained even for the reduced ring thicknesses of 1.0, 1.2, and 1.5 mm, respectively. This undisturbed thickness-mode resonance makes these rings suitable for use in most high-frequency transducers where ring thicknesses varying from 1 to 2 mm are typically employed. Among them, the one with $\upsilon_f = 0.89$ would provide an optimal performance for tailoring a 136-kHz transducer for use in a commercial ultrasonic wire bonder (ASM AB559) (Or et al. 2007). Having been shown in Fig. 19.12a, this batch of rings has sufficiently pure resonance characteristics with only three fundamental resonance modes, namely, the radial mode f_R (a symmetric-radial movement of the inner and outer diameters of the ring), wall-thickness mode f_W (an asymmetric-radial movement of the inner and outer diameters of the ring), and thickness mode f_t (an axial moment of the thickness of the ring) for frequencies up to 2 MHz. Moreover, it has a larger k_t (0.476 vs. 0.378 for PKI804), a smaller k_{31} (0.221 vs. 0.380 for PKI804), and a lower thickness Q_m (212 vs. 1560 for PKI804) without deteriorating the original thickness-mode activities of the PKI804 (Table 19.1 in Sect. 19.2.3). In addition, the number of cuts required is reasonably less (nine cuts/direction) so that the manufacturing is quite feasible. These distinct properties/features suggest that broadband transducers with low lateral coupling can be fabricated by using the piezocomposite ring instead of the PZT piezoceramic ring. Details of using this piezocomposite ring in the development of a 136-kHz transducer will be disclosed in Sect. 19.3.

19.2.3 Material Properties

In the previous section, PKI804/Araldite LY5210/HY2954 1–3 piezocomposite rings with υ_f varying from 0.82 to 0.94 have been fabricated, and their resonance characteristics have been studied theoretically and experimentally to elucidate the

Table 19.1 Material parameters of PKI804/Araldite LY5210/HY2954 1–3 piezocomposite rings and their PZT piezoceramic ring ($\upsilon_f = 1$) (Or 2001)

0	PZT volume fraction (υ_f)						
	0.82	0.84	0.87	0.89	0.91	0.94	1
Dielectric properties							
κ_{33}^{T}, 1 kHz	853	873	899	919	947	973	1,042
κ_{33}^{S}, 5 kHz	506	517	529	539	552	563	577
tan δ_e, 1 kHz	0.005	0.005	0.005	0.005	0.005	0.005	0.005
Electromechanical properties							
k_t	0.490	0.487	0.481	0.476	0.466	0.453	0.378
k_{33}	0.610	0.610	0.611	0.611	0.611	0.611	0.612
k_{31}	−0.193	−0.201	−0.211	−0.221	−0.239	−0.261	−0.380
d_{33} (pm/V)	252	253	253	253	254	254	255
d_{31} (pm/V)	−109	−110	−111	−111	−112	−112	−114
e_{33} (C/m^2)	11.1	11.3	11.5	11.6	11.7	11.7	10.6
e_{31} (C/m^2)	−1.5	−1.7	−1.9	−2.1	−2.6	−3.1	−6.2
g_{33} (mV m/N)	33.4	32.7	31.8	31.1	30.3	29.5	27.6
g_{31} (mV m/N)	−14.5	−14.2	−13.9	−13.7	−13.3	−13.0	−12.4
Elastic properties							
ρ(kg/m^3)	6,538	6,666	6,823	6,945	7,117	7,280	7,700
υ_{3t}^{D} (m/s)	4,190	4,203	4,220	4,235	4,261	4,293	4,457
υ_{3t}^{E} (m/s)	3,651	3,671	3,699	3,725	3,770	3,827	4,126
υ_{33}^{D} (m/s)	4,267	4,273	4,281	4,286	4,294	4,301	4,317

υ_{33}^{E} (m/s)	3,381	3,385	3,390	3,394	3,399	3,404	3,415
Z (kg/m^2/s)	27.4×10^6	28.0×10^6	28.8×10^6	29.4×10^6	30.3×10^6	31.3×10^6	34.3×10^6
c_{11}^{D} (GPa)	35.0	38.2	42.9	47.5	55.9	67.2	142.7
c_{33}^{D} (GPa)	114.8	117.8	121.5	124.6	129.2	134.1	153.0
c_{12}^{D} (GPa)	19.2	20.8	23.1	25.3	29.1	34.0	55.6
c_{13}^{D} (GPa)	5.9	6.2	6.7	7.2	8.2	9.4	16.6
c_{11}^{E} (GPa)	34.7	37.7	42.3	46.7	54.7	65.5	135.3
c_{33}^{E} (GPa)	87.2	89.8	93.4	96.4	101.2	106.6	131.1
c_{12}^{E} (GPa)	18.8	20.3	22.4	24.5	28.0	32.3	48.2
c_{13}^{E} (GPa)	18.2	19.8	22.0	24.1	28.0	33.0	61.5
s_{11}^{D} (pm^2/N)	40.85	37.25	32.87	29.45	24.61	20.06	8.32
s_{33}^{D} (pm^2/N)	8.40	8.21	8.00	7.84	7.62	7.43	6.97
s_{12}^{D} (pm^2/N)	−22.21	−20.10	−17.55	−15.55	−12.72	−10.05	−3.17
s_{13}^{D} (pm^2/N)	−0.91	−0.87	−0.82	−0.78	−0.73	−0.69	−0.58
s_{11}^{E} (pm^2/N)	42.44	38.81	34.41	30.97	26.10	21.52	9.72
s_{13}^{E} (pm^2/N)	13.38	13.09	12.75	12.50	12.16	11.85	11.13
s_{12}^{E} (pm^2/N)	−20.62	−18.54	−16.01	−14.03	−11.23	−8.59	−1.77
s_{13}^{E} (pm^2/N)	−4.56	−4.46	−4.33	−4.24	−4.11	−4.00	−3.73
Q_m	147	162	187	212	262	338	1,560

mode coupling behavior for high-frequency ultrasonic wire-bonding transducer applications. It is known that these rings have sufficiently small epoxy width of about 77 μm and can be treated as effective homogeneous media for frequencies up to 10 MHz.

On the basis of physical parallel models for calculating effective material properties of homogeneous 1–3 piezocomposites (Smith et al. 1985; Chan 1987; Or 2001), the effective material parameters of the piezocomposite rings as functions of υ_f were calculated (Or 2001) and are tabulated in Table 19.1, in conjunction with the PZT piezoceramic ring parameters (the column with $\upsilon_f = 1$). To obtain a valid set of material parameters, the piezocomposite rings cannot be too thin or too thick to avoid coupling of the thickness mode to the other resonance modes. With the guide provided in Sect. 19.2.2, piezocomposite rings with G in the range of 0.3–0.8 can generally fulfill such requirement. This is different from the conventional disk- or plate-shaped piezocomposites in which samples need to be thick enough so that the resulting G has to be sufficiently small (Smith et al. 1985; Chan 1987; Hayward and Hossack 1990; Smith and Auld 1991).

19.3 Piezocomposite Ultrasonic Transducers

19.3.1 Structure

Figure 19.14a shows the solid model of the 136-kHz piezocomposite transducer (denoted as *composite transducer*). The composite transducer has a length of 47.38 mm and consists of a Langevin-type piezoelectric driver (12.46-mm long), an ultrasonic horn (34.92-mm long), and a bonding tool. The piezoelectric driver has a

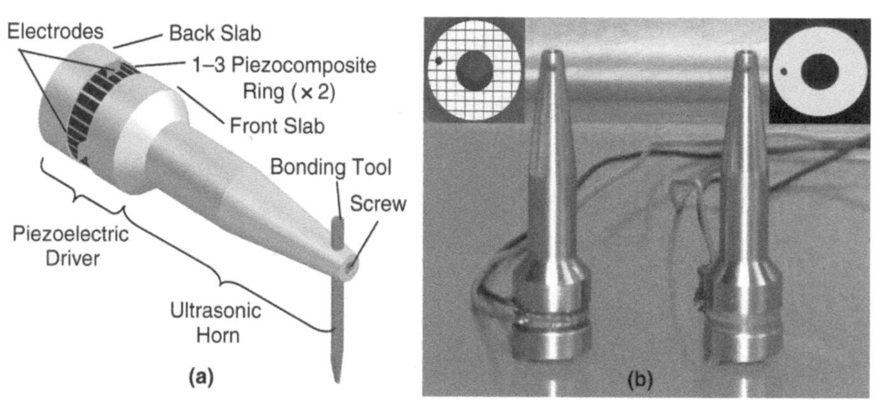

Fig. 19.14 (**a**) Solid model of the 136-kHz composite transducer, and (**b**) photograph of the fabricated composite (*left*) and PZT (*right*) transducers. The two insets in (**b**) are their composite and PZT rings (Or et al. 2007)

pair of piezocomposite rings, each of 12.65-mm outer diameter, 5.03-mm inner diameter, and 1.70-mm thickness, connected electrically in parallel and mechanically in series, and preloaded by a prestress mechanism formed by two 4.47-mm-thick stainless-steel-304 (SS304) slabs and a central prestress screw (not shown). Three pieces of 0.04-mm-thick, ring-shaped copper electrodes are inserted between the components of piezoelectric driver for electrical connection purposes. The piezocomposite rings are the self-made PKI804/Araldite LY5210/HY2954 1–3 piezocomposites with 0.89 PZT volume fraction (the inset at the upper-left corner of Fig. 19.14b) as have been presented in Sect. 19.2. The ultrasonic horn of the composite transducer is formed by two SS304 half-wave sectional horns. The section next to the piezoelectric driver is a first half-wave horn of conical (3-mm long, 13.2-mm larger diameter, 8.2-mm smaller diameter) and cylindrical (13.06-mm long, 8.2-mm diameter) shape. The second section is another half-wave horn of conical shape (18.86-mm long, 8.2-mm larger diameter, 3.6-mm smaller diameter). The entire ultrasonic horn is capable of amplifying 2.6-folds the axial displacement output from the piezoelectric driver to its tip (3.6-mm diameter) affixed with a bonding tool by a screw.

The fabricated composite transducer is displayed in Fig. 19.14b, along with a PKI804 PZT piezoceramic transducer (denoted as *PZT transducer*) having a similar structure. To fairly compare the performance of both transducers, the PZT piezoceramic rings used have the same dimensions as the piezocomposite rings. However, as the properties of both materials are different (Table 19.1), the lengths of the resulting front and back slabs in the PZT driver are 0.34-mm longer, thereby increasing the length of PZT transducer to 48.06 mm (Or et al. 2007).

19.3.2 Electrical Characteristics

The electrical impedance spectra of the composite and PZT transducers were measured using an impedance analyzer (Agilent 4294A). By associating the series and parallel resonance frequencies depicted in the measured impedance spectra with the resonance and antiresonance frequencies computed using an ANSYS finite-element modal analysis, the nature of most experimental resonance modes was identified and grouped into three categories according to the excitations governed by the piezoelectric ring stack as illustrated in Fig. 19.15. These categories are as follows:

1. Axial excitation, which is due to the longitudinal-axial expansion and contraction of the length of the vibrating stack (Fig. 19.15a)
2. Radial excitation, which is due to the symmetric-radial expansion and contraction of the inner and outer diameters of the vibrating stack (Fig. 19.15b)
3. Wall-thickness excitation, which is due to the asymmetric-radial expansion and contraction of the inner and outer diameters of the vibrating stack (Fig. 19.15c)

From a physical standpoint, a pure axial excitation of the ring stack produces a piston-like motion of the transducer in the axial direction that, in turn, causes the

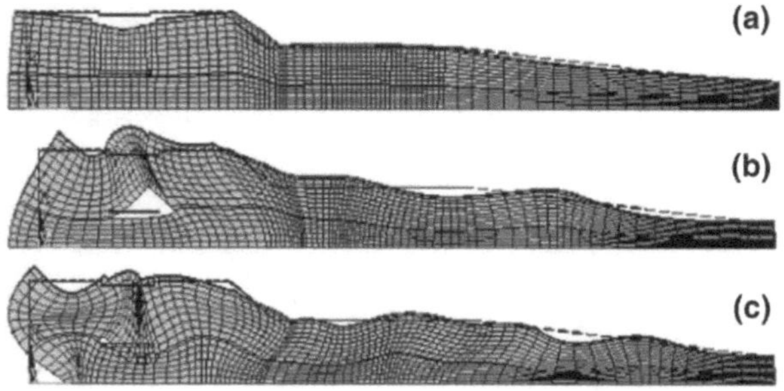

Fig. 19.15 Deformed shapes of three major natural modes governed by (**a**) axial, (**b**) radial, and (**c**) wall-thickness excitations of the piezoelectric ring stack (Or et al. 2007)

tip of the ultrasonic horn (and also that of the bonding tool) to move back and forth axially for facilitating the largest axial displacement. As this motion is essentially in line with the wire to be bonded and hence the long-and-narrow shape of the resulting wire bonds (Fig. 19.1), axial modes are the most desirable resonance modes in ultrasonic wire bonding. Any nonaxial excitation (e.g., radial and wall-thickness excitations) tends to deteriorate the bond quality as a result of diminishing the purity of axial excitation through the loss of ultrasonic energy in other vibration directions. To achieve a high-quality bonding, it is imperative to minimize the excitation of nonaxial modes (Or 2001).

As depicted in Fig. 19.16a, the composite transducer has the strongest resonance at 135.1 kHz, which not only is the designated working mode of the transducer but also is the third axial mode (M3) with 1.5 longitudinal wavelengths (Fig. 19.15a). The two lower modes at 54.5 (M1) and 103.5 kHz (M2) are the first and second axial modes with 0.5 and 1 longitudinal wavelength, respectively. The two higher modes at 181.8 (M4) and 193.7 kHz (M5) are a pair of the fourth quasi-axial modes. The one observed at 225.9 kHz (M6) is the fifth quasi-axial mode. After that, no prominent resonances are found. Consequently, the impedance spectrum is very pure in that the axial modes are promoted by impeding nonaxial and spurious modes in the piezocomposite ring stack.

For the PZT transducer shown in Fig. 19.16b, the working axial mode is found at 135.7 kHz (M3), while the two lower axial modes, M1 and M2, and the three higher quasi-axial modes, M4, M5, and M6, are located at 54.4, 105.1, 184.5, 195.1, and 227.6 kHz, respectively. Besides, very complicated resonance characteristics covering two broad frequency ranges of 250–350 kHz and 400–600 kHz originate from the radial (Fig. 19.15b) and wall-thickness (Fig. 19.15c) activities of the PZT ring stack. Thus, when this highly laterally coupled transducer is used in a bonding operation, the nonaxial and spurious resonance modes, especially for those with frequencies very close to the working axial mode or equal to its higher frequency harmonics, are likely to be excited simultaneously, leading to a reduction in the purity of axial motion and hence in the bonding stability and yield.

Owing to the effect of epoxy damping, the measured Q_m of the composite transducer at M3 is 296, corresponding to a 3-dB bandwidth of 456 Hz (the inset of

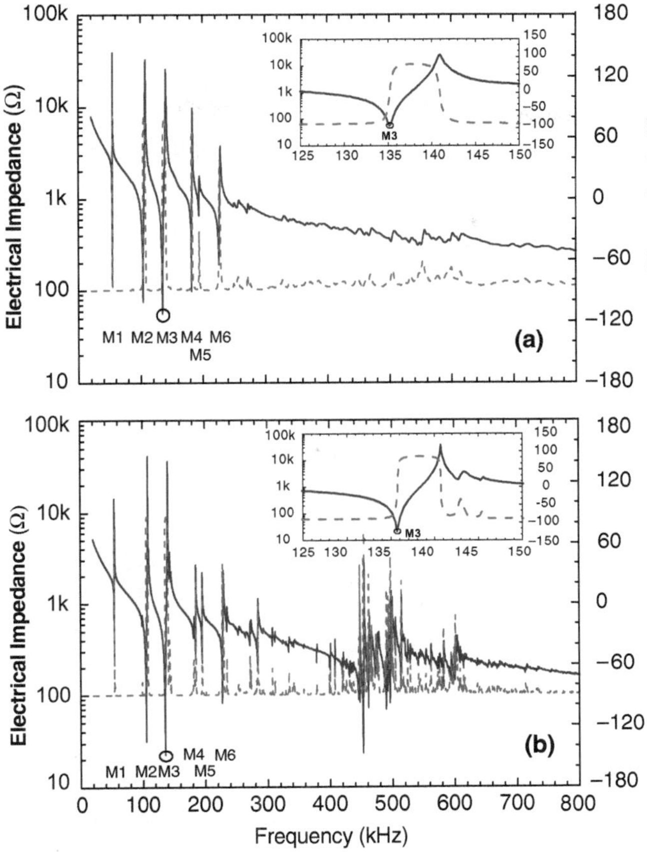

Fig. 19.16 Electrical impedance (solid line) and phase angle (dash line) vs. frequency for (**a**) the composite transducer, and (**b**) the PZT transducer (Or et al. 2007)

Fig. 19.16a). The PZT transducer has about 2.4 times increased Q_m (=710), thus giving a smaller 3-dB bandwidth of 191 Hz (the inset of Fig. 19.16b). Under such a low Q_m operation, the composite transducer will be more adaptable to various types of bonding surfaces so that a reliable working platform can be created for high-frequency bonding.

19.3.3 Vibrational Characteristics

The vibrational characteristics of both transducers were evaluated with a laser Doppler vibrometer (Polytec OFV-303/OFV-3001). An ultrasonic generator and controller board (ASM PC-BQM) operating in a constant voltage, digital phase-locked-loop mode were used to drive the transducers. To initiate an optimal bonding condition, a 0.15-W burst in 15 ms with a duty cycle of 13% was supplied to each transducer at resonance (∼136 kHz). The temporal profile of axial displacement at the tip of the ultrasonic horn was examined by measuring the temporal profile of axial velocity. By applying fast Fourier transform (FFT) to the steady-state portion of the displacement waveform, the corresponding amplitude spectrum (in dBm re 50 Ω, 1 mW) was obtained.

From the waveforms in Fig. 19.17, the composite transducer has a rise/fall time of about 50% shorter than the PZT transducer due to its heavily damped nature as has been seen in its lower Q_m (Fig. 19.16). Importantly, shortening the rise time can directly shorten the time required to form a bond. This is definitely an advantage for promoting high-speed bonding.

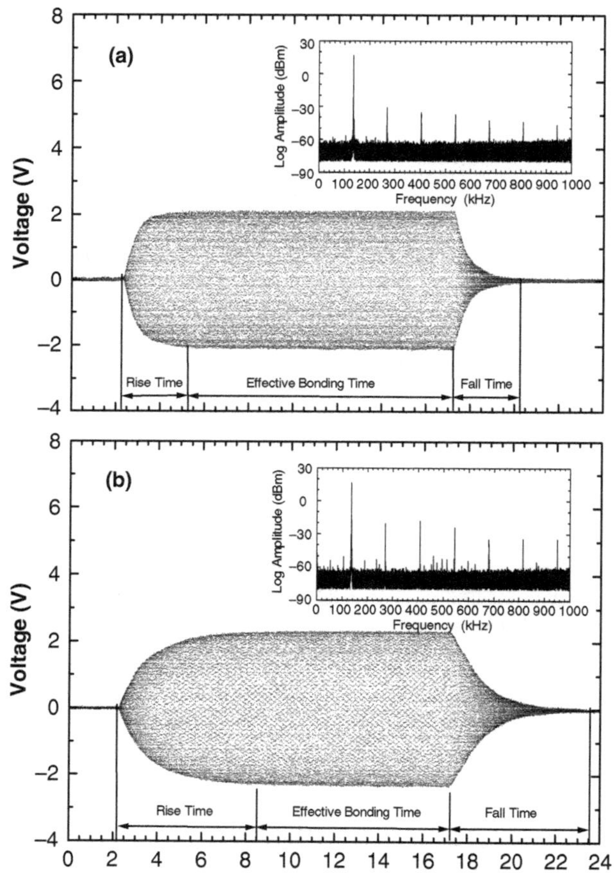

Fig. 19.17 Axial vibration responses at the horn tip for (**a**) the composite transducer, and (**b**) the PZT transducer. Waveform (*left*) and amplitude spectrum (*right*) of the detected signal, where 1 V = 10 dBm = 147 nm (Or et al. 2007)

The amplitude spectra (the insets of Fig. 19.17) reveal that excitations of other frequency components (i.e., higher-order harmonics and subharmonics to the fundamental) are about 10 dBm lower in the composite transducer, except for the fundamental amplitude (16 dBm or 295 nm), which is almost equal to that of the PZT transducer (16.1 dBm or 297 nm). If we refer to their waveforms (Fig. 19.17), it is obvious that the composite transducer has steady-state amplitude (2.1 V or 311 nm) slightly lower than that of the PZT transducer (2.3 V or 344 nm). Hence, the purity of axial excitation, which is defined as the amplitude ratio of the fundamental component to the summation of all components, is determined to be about 95% for the composite transducer, but only about 86% is obtained for the PZT transducer. The result reflects again the low lateral coupling nature of the composite-based transducers.

19.3.4 Wire-Bonding Performance

Each of the two prototype transducers was mounted on an automatic wire bonder (ASM AB559) for evaluating its high-frequency wire-bonding performance (Fig. 19.18). A total of 120 aluminum wires, each with the same diameter of 25 μm, the same loop height of 360 μm, and the same wire length of 2 mm, were bonded on the gold-plated surface of a gold–nickel–copper–FR-4 layered printed circuit board (PCB) of 50-mm length and 20-mm width. This PCB fulfilled the requirements of the ANSI/IPC-SM-784 standard (1990) for chip-on-board (COB) applications. For each bond point, the transducers were energized using the same power of 0.15 W at their resonance frequencies (~136 kHz) for the same duration of 15 ms under the

Fig. 19.18 Photograph showing the composite transducer mounted on the bondhead assembly of an automatic wire bonder (ASM AB559) for evaluating high-frequency wire-bonding performance (Or 2001)

same bond force of 0.3 N. The quality of each bonded wire was examined by an optical projector (Nikon V-12A) with a 200-times magnification for a visual inspection and then by a pull tester (Royce Instruments System 550) for a standard pull test. The visual inspection involves visual checking of the appearance and deformation ratio of each bond point, while the pull test is a destructive test aiming to measure the pull force of each bonded wire by pulling it at its loop top vertically until rupture occurs (Or 2001). Hence, the term *good bond* is defined as a wire bond possessing both higher pull force and smaller deformation ratio with smaller standard deviations. The most desirable case is breaking at the wire span followed by breaking at the bond neck. For a poor bond, it may be completely peeled off from the bonding surface in the pull test because of the weak intermetallic adhesion between the wire and the bonding surface. For the nonstick bond, the wire cannot stick on the bonding surface on the completion of the whole bonding period.

Table 19.2 summarizes the process-test results for the composite and PZT transducers. It is clear that wire bonds produced by the composite transducer can withstand higher pull forces and have smaller deformation ratios, both with smaller standard deviations. From the observed modes of failure after the pull test, the failure is mostly breaking at the neck of the first bond (~84%); the next majority of failure is breaking at the neck of the second bond (~13%), and the remaining failure mode is breaking at the mid-span (~3%). All these breakages indicate the existence of good adhesion in the interfaces of the bond wire and the bonding surface even though the mean deformation ratio is as small as 1.39. Moreover, the bonds are developed quite consistently as reflected by the smaller standard deviations associated with the higher pull forces and smaller deformation ratios. For the PZT transducer, it is obvious that the occurrence of fatal failure modes, including weld-off of the first and second bonds, is as high as 12.5%. This indicates that the use of this transducer may not guarantee secure bonds on this challenging PCB.

Table 19.2 Comparison of the ultrasonic wire-bonding performance between the composite and PZT transducers

		Composite transducer	PZT transducer
Pull force (mN):	Mean	75.6	68.2
	±1 Standard deviation	9.8	12.2
Number of failures for various failure modes:	First weld-off	0	3
	First neck break	101	95
	Span break	3	0
	Second weld-off	0	12
	Second neck break	16	16
Deformation Ration (D):	Mean	1.39	1.47
	±1 Standard deviation	0.04	0.05

$D = 25\,\mu\text{m}$ is the diameter of aluminum wire (Or 2001)

19.4 Conclusion

PKI804/Araldite LY5210/HY2954 1–3 piezocomposite rings having sufficiently small epoxy width of about 77 μm and with υ_f varying from 0.82 to 0.94 have been fabricated, and their resonance characteristics and material properties have been studied theoretically and experimentally. Piezocomposite rings with $\upsilon_f = 0.89$ have shown the optimal resonance characteristics and material properties for fabricating into a 136-kHz wire-bonding transducer. The evaluation results have confirmed that the superior behavior of low mode coupling and reduced Q_m in the piezocomposite transducer compared with its piezoceramic transducer has made it a promising transducer for use in high-frequency wire bonding. An ultrasonic wire bonder when equipped with the piezocomposite transducer possesses a wider operating window and a shorter bonding time, thereby promoting the fine-pitch and high-speed capabilities of the equipment (Chan et al. 2001).

Acknowledgment The authors would like to acknowledge the support of the Centre for Smart Materials of The Hong Kong Polytechnic University and ASM Assembly Automation Ltd. Special thanks are due to Dr. Peter Chou Kee Liu, the Chief Technical Officer of ASM, for providing collaborative efforts to the development of the current innovative technology and products.

References

ANSI/IPC-SM-784 (1990) Guidelines for chip-on-board technology implementation. The Institute for Interconnecting and Packaging Electronic Circuits, Northbrook, IL

Chan HLW (1987) Piezoelectric ceramic/polymer 1–3 composites for ultrasonic transducer applications. PhD Thesis, Macquarie University, Sydney, Australia

Chan HLW, Or SW, Cheng KC, Choy CL (2001) Ultrasonic transducer. US Patent: US 6,190, 497 B1

Cheung YM, Or SW, Ching S (1999) Low temperature gold wire bonding. Proceedings IEEE/CPMT International Electronics Manufacturing Technology Symposium at SEMICON® Southwest 1999, Austin, TX, pp 196–202

Geng X, Zhang QM (1999) Resonance modes and losses in 1–3 piezocomposites for ultrasonic transducer applications. Journal of Applied Physics 85(3):1342–1350

Hayward G, Bennett J (1996) Assessing the influence of pillar aspect ratio on the behavior of 1–3 connectivity composite transducers. IEEE Transactions on Ultrasonics, Ferroelectrics, and Frequency Control 43(1):98–108

Hayward G, Hossack JA (1990) Unidirectional modeling of 1–3 composite transducers. Journal of Acoustical Society of America 88(2):599–608

International Technology Roadmap for Semiconductors (ITRS) (2007). http://www.itrs.net

Or SW (2001) High frequency transducer for ultrasonic bonding. PhD Thesis, The Hong Kong Polytechnic University, Hong Kong

Or SW, Chan HLW (2001) Mode coupling in lead zirconate titanate/epoxy 1–3 piezocomposite rings. Journal of Applied Physics 90(8):4122–4129

Or SW, Chan HLW, Liu PCK (2007) Piezocomposite ultrasonic transducer for high-frequency wire-bonding of microelectronics devices. Sensors and Actuators A 133:195–199

Ramsey TH, Alfaro C (1991) The effect of ultrasonic frequency on intermetallic reactivity of Au–Al bonds. Solid State Technology 12:37–38

Smith WA, Auld BA (1991) Modeling 1–3 composite piezoelectrics: thickness-mode oscillations. IEEE Transactions on Ultrasonics, Ferroelectrics, and Frequency Control 38(1):40–47
Smith WA, Shaulov A, Auld BA (1985) Tailoring the properties of composite piezoelectric materials for medical ultrasonic transducers. Proceedings of IEEE Ultrasonics Symposium 1985, San Francisco, CA, pp 642–647

Chapter 20
Piezoelectric MEMS: Materials and Devices

Nazanin Bassiri-Gharb

20.1 Introduction

Microelectromechanical systems (MEMS) are miniaturized devices that when coupled with IC (integrated circuit) components have the ability to interact with the environment or other on-chip components, allowing creation of sensor, actuator, and/or transducer systems thereby. Although originally the term referred to only miniaturized *mechanical* devices, MEMS is now a generic expression for any miniaturized device that has interactive capabilities. Therefore, optoelectronic and microfluidic miniaturized systems, including lab-on-chip, are also currently described as MEMS.

MEMS technology emerged in early 1990s by leveraging many of the materials and processes from the IC industry, and thus it was mainly based on the silicon technology. As the field has grown over the years, concepts of more complex devices have been proposed and demonstrated, many based on new, functional materials with new fabrication processes, which allow creation of systems consuming less power, with faster and more reliable response, and capable of incorporating more complex functions.

An important family of functional materials are piezoelectric, and in general, polar materials.[1] Piezoelectric thin films are especially useful in those applications that require low power and low noise, high frequency and large output signals. Among the many proposed chemistries, the most used are AlN, ZnO, $Pb(Zr_{1-x}Ti_x)O_3$ (PZT). This chapter reviews the literature on piezoelectric MEMS, with specific attention to the materials used and the reported applications.

N. Bassiri-Gharb
George W. Woodruff School of Mechanical Engineering, Georgia Institute of Technology, Atlanta GA 30332-0405, USA

[1] A unique exception is quartz, which is used only in single crystalline form for piezoelectric applications.

A. Safari, E.K. Akdoğan (eds.) *Piezoelectric and Acoustic Materials for Transducer Applications.*

20.2 Piezoelectric Thin Films

Piezoelectric materials develop charge on the sample surfaces when exposed to applied stresses (direct effect) or equally, under applied electric field, they undergo a change in their dimensions (converse effect). The electric displacement and stress/strain are related by a linear dependence at low fields, expressed by the following equations:

$$D_i = \sum_k d_{ik}\sigma_k \quad \text{and } D_i = \sum_k e_{ik}x_k (\text{direct effect}), \tag{20.1}$$

$$x_\text{i} = \sum_k d_{ki}E_\text{k} \text{ and } \sigma_\text{i} = -\sum_k e_{ki}E_\text{k} (\text{converse effect}). \tag{20.2}$$

D_i is the electric displacement and E_k is the electric field tensor; σ_k and x_k are the stress and strain tensors, respectively, and d_{ik} and e_{ik} are the piezoelectric coefficient tensors. All tensors are shown in Newtonian indices. It is important to note that the piezoelectric response is anisotropic as it is defined by a third rank tensor. The indices used for definition of the piezoelectric coefficient tensors usually refer to the crystallographic axes of the crystal. The index "3" is typically associated with the polar axis direction (IEEE 2003).

In general piezoelectric thin films are used in a composite structure, with silicon, silicon nitride or oxide membranes. Furthermore, the clamping to the substrate itself during the processing steps creates an anisotropic response. As a consequence, the actual piezoelectric response of thin films is quite different from that of the bulk piezoelectric materials. This is usually reflected in the use of *effective* piezoelectric coefficients for the films:

$$d_{33,\text{f}} = d_{33} - \frac{2s_{13}^\text{E}}{s_{11}^\text{E} + s_{12}^\text{E}} d_{31} \quad \text{and} \quad e_{31,\text{f}} = \frac{d_{31}}{s_{11}^\text{E} + s_{12}^\text{E}} = e_{31} - \frac{c_{13}^\text{E}}{c_{33}^\text{E}} e_{33}, \tag{20.3}$$

where s_{ij} and c_{ij} represent components of the elastic and stiffness tensor, respectively. The practical effect is that the $d_{33,\text{f}}$ and $e_{31,\text{f}}$ coefficients are smaller than the bulk values (note that the absolute value of $e_{31,\text{f}}$ is actually bigger than e_{31}) (Trolier-Mckinstry and Muralt 2004).

Although 20 different crystal systems can exhibit piezoelectricity, the most used piezoelectric thin film materials in MEMS applications have either the wurtzite (AlN and ZnO) or the perovskite (PZT) crystal structure. Properties of ZnO, AlN, and PZT films are reported in Table 20.1. The major difference between the previous two crystal structures is the possibility of reorienting the polarization direction between multiple crystallographic directions (ferroelectricity) or lack thereof. In the polycrystalline samples, as is the case of most of the piezoelectric thin films in MEMS applications, the piezoelectric response is averaged over all the grains. Therefore, in nonferroelectric films, such as AlN and ZnO, the growth of the films has to be controlled to provide for the alignment for the polar direction across the film.

In ferroelectric materials, the polarization vector can be reoriented along different directions. Consequently, an increased alignment of the polarization can be obtained

Table 20.1 Piezoelectric thin film properties

Parameter	ZnO	AlN	PZT (1–3 μm)
$d_{33,f}$ (pm/V)	5.9	3.9–5.5	60–130[a]
$e_{31,f}$ (C/m^2)	−1.0	−1.05	−8.0 to −17.7
$e_{31,f}/\varepsilon_0\varepsilon_{33}$ (GV/m)	−10.3	−11.3	−0.7 to −1.8
ε_{33}	10.9	8.5–10.5	300–1,620
tan δ (1–10 kHz, 10^5 V/m)	0.01–0.1	0.003	0.01–0.03
$c^E{}_{33}$ (GPa)	208	395	98

[a] $d_{33,f}$ of the PZT films from Calame and Muralt (2007) are unknown. *References*: Carlotti et al., 1987; Gualtieri et al., 1994; Chen et al., 2007; Tonisch et al., 2006; Calame and Muralt, 2007; Ledermann et al., 2003

by application of high enough electric fields along the required direction, i.e., poling, after the deposition of the film. PZT ceramics are ferroelectric and therefore, do not require specific processing to obtain alignment of the polarization direction, but usually require poling after deposition, to exhibit a net piezoelectric effect.

20.2.1 AlN and ZnO

Aluminum nitride and zinc oxide are both wurtzite-structured materials, with the polar direction, and therefore the piezoelectric response, lying along the $[0001]_{\text{hexagonal}}$ crystallographic direction. As mentioned earlier, these materials are not ferroelectric and consequently, to achieve good piezoelectric coefficients, particular deposition processes are required to obtain homogeneous nucleation of grains with same $(0001)_{\text{hexagonal}}$ out of plane texture and identical polar direction. There is compelling evidence that a more perfect $\{0001\}_{\text{hexagonal}}$ growth is directly correlated to a higher selectivity of the polar orientation, indicating that the same mechanism is responsible for the crystalline and polar orientation (Hickernell et al. 1997; Loebl et al. 2004).

The transverse piezoelectric coefficient $e_{31,f}$ of AlN and ZnO are very similar ($\sim -1\,C/m^2$), while the longitudinal coefficient is slightly smaller in AlN than in ZnO. Nevertheless, ZnO has two major limitations with respect to AlN for use in piezoelectric MEMS. First, ZnO is essentially a semiconductor, and therefore, any off-stoichiometry in the deposited films will potentially lead to increased conductivity (high dielectric loss at low frequencies) due to Zn interstitials. Secondly, Zn is a fast diffusing species and therefore with major integration issues with silicon IC processing. The deposition temperatures for ZnO are therefore kept usually close to room temperature. High-quality AlN films can as well be obtained by sputtering at relatively low temperatures (below 200 °C), offering complete integration capability with CMOS technology. The low processing temperature for both these materials offers also the possibility of use of standard Al for metallization layers (electrodes).

AlN is suitable for use is sensor applications, where a better signal to noise ratio and higher output voltage can be obtained from its low permittivity and low dielectric loss. For actuator and transducer applications, where high output force or sensitivity (piezoelectric coefficients) is required, ferroelectric thin films are more appropriate.

20.2.2 Perovskite Ferroelectric Thin Films

Among the perovskite structure thin films, PZT has been the prominent chemistry for use in MEMS applications. Relaxor ferroelectric films have also been proposed for various applications because of their higher piezoelectric coefficients, but most of the suggested compositions (PMN-PT) suffer from low transition temperatures, leading to early depoling and instability of the piezoelectric response at increasing temperatures. Furthermore, while processing of PZT has been strongly leveraged on the know-how of the ferroelectric memory industry, fabrication and full evaluation of new compositions have yet to be assessed.

Piezoelectric response of ferroelectric thin films depends on many factors, among which are composition, crystallographic orientation, crystalline quality of the films, and grain size. In bulk ferroelectrics, maxima in the piezoelectric properties are usually reported in proximity of the morphotropic phase boundary (Campbell and Mutharasan 2005). MPB is a nearly temperature-independent phase transition and is observed in many of $PbTiO_3$-based solid solutions. The increase in the piezoelectric coefficients in proximity of the MPB, often reported also in ferroelectric thin films, is usually attributed to the increased ease in poling, because of the increased available polarization directions.

Rhombohedral PZT films of composition near the MPB show the highest response in {100} textured films (Ledermann et al. 2003). Indeed, the same piezoelectric enhancement has been observed in many rhombohedrally distorted bulk perovskites when poled and driven along the [001] crystallographic direction (Park and Shrout 1997). Piezoelectric response can, therefore, be enhanced by the appropriate choice of composition and crystalline orientation. Texture in thin films can be controlled by use of specific seed layers (Gharb and Trolier-Mckinstry 2005b; Maeder et al. 1996) and substrates, deposition conditions (Eun Sun et al. 2005; Hongxia et al. 2000), and heating rates, including excess lead content (Wang et al. 2006), ambient humidity (Schwartz et al. 1995), pyrolysis temperature, and heating ramp speed to the crystallization temperature (Jiang et al. 2006; Wang et al. 2006).

It has to be noted that although phenomenological calculations predict orientation dependence of the piezoelectric coefficients for epitaxial PZT thin films (Ouyang et al. 2005), these are limited to only the intrinsic linear response in monodomain single crystalline film. In practice, monodomain ferroelectric films are extremely unlikely to occur in films of thickness superior to 50 nm, which is seldom if ever the case for MEMS applications. Additionally, nonlinear components, due to *intrinsic* and *extrinsic* contributions, should be added to obtain the overall response of a ceramic thin film. Specifically, the extrinsic response is mostly due to the contributions from vibration and motion of domain walls and phase boundaries and reorientation of defects. These contributions can amount to a significant component of the overall piezoelectric response, especially in MEMS devices, where the fields (electric, stress/strain) are particularly strong (Bassiri-Gharb et al. 2007; Hall 2001). Extrinsic contributions lead also to creation of nonlinear and hysteretic behaviors at field values well below the coercive field (Taylor and Damjanovic 2000), above which intrinsic nonlinearities are predominant.

The piezoelectric response of many ferroelectric thin films is much smaller than that of their equivalent bulk compositions. This has been partially attributed to the reduced ferroelastic domain wall motion in thin films. Factors influencing the domain state and therefore extrinsic contributions in ferroelectric thin films include stress at cooling through the transition temperature, the grain size, and film thickness (Tuttle et al. 1992; Xu et al. 2001; Bassiri-Gharb et al. 2007). Tensile stress at cooling in the ferroelectric film tends to pull the polarization direction in the plane parallel to the substrate; conversely, a compressive stress experienced during the cooling translates into a polarization direction more normal to the substrate. Once at room temperature, the mobility of the ferroelastic domain walls is limited, and the poling of the polarization vector out of the plane of the substrate is more arduous (Kholkin et al. 1996; Xu et al., 2001). The remanent polarization, P_r, in ferroelectric thin films is therefore a good measure of the effective piezoelectric response: films with square hysteresis and high remanent polarization tend to exhibit also higher piezoelectric response, while films with rotated hysteresis loops and lower magnitude of P_r display also lower piezoelectric response, $d_{33,f}$ (Zhang et al. 2000). Films with larger grains, where creation of complicated domain patterns is possible, could allow higher mobility of non-180° domain walls and therefore be less susceptible to the reduction of the remanent polarization and piezoelectric response. It has also to be noted that films of submicron thickness show much more reduced extrinsic contributions to the piezoelectric response than thicker films. At increasing film thickness, the domain walls' motion is *unclamped* from the substrate, increasing the extrinsic contributions and thus piezoelectric response, which then saturates at a thickness of few microns (Bassiri-Gharb et al. 2007).

20.2.2.1 Fabrication and Integration Issues

Many of the basic Si microfabrication technologies such as photolithography, deposition, and etching of different thin films chemistries such as SiO_2, Si_3N_4, poly and amorphous Si, Al, Cr, and Au are well known and documented in literature. However, introduction of ferroelectric thin films to the MEMS production lines requires specific micromachining processes that are not part of the IC technologies. New needs include deposition of appropriate seed layers ($PbTiO_3$), diffusion barrier layers (TiO_2, ZrO_2), electrodes (Pt, $LaNiO_3$, Ir, IrO_2), and micromachining techniques (including etching methods) for these new materials and the ferroelectric materials themselves.

Barrier layers between PZT and the substrate are specifically needed to reduce interdiffusion and oxidation. Seed layers are required to obtain the appropriate

Table 20.2 Typical materials used in PZT-based MEMS devices

Material	Function	Deposition method	Patterning method	Thickness (μm)	Stress (MPa)
SiO_2	Etch stop	Thermal growth	Wet etch (BOE)	<2	−350 to 0
	Passive layer	CVD	ICP etch		
ZrO_2	Etch stop barrier layer	Sol–gel	IBE	0.3	0–270 (w/o anneal)
					−40 to −250 (w/anneal)
TiO_2	Barrier layer	Sputtering	ICP wet etch	–	About +320
Si_3N_4	Passive layer	LPCVD	ICP etch	<0.5	–
Pt	Bottom electrode	Sputtering	ICP etch	0.1–0.3	+400–800
$LaNiO_3$	Bottom electrode Seed layer	Sputtering	ICP etch	0.05–0.15	–
		Sol–gel			
PZT	Active material		IBE/RIBE	0.5–4	+100–200 (unpoled)
		Sol–gel	DRIE		
		Sputtering	Wet etching (HF/HCl)		
			ICP		+150–250 (poled)
Cr/Au	Top electrode	Evaporation	Liftoff	0.1/0.015	+45–280
			RIE		

References for stress and thickness: Baborowski 2004; Chang-Il et al. 2006; Gross 2004; Gross et al. 2003; Hong 2004; Hong et al. 2006; Kennedy et al. 2004; Lee et al. 1999; Miller et al. 2007; Zhou et al. 2005

crystalline phase and orientation. Noble metal or conductive metal oxide electrodes are needed for ferroelectric thin films in MEMS and memory applications. Some of the deposition and patterning methods for thin film used in ferroelectric thin film-based MEMS are reported in Table 20.2.

One of the most challenging aspects for the flow process is the etching of the PZT and Pt electrodes. PZT thin films can be patterned by wet etch, using hydrofluoric acid. The fluorinated by-products are removed by use of HCl (mixed with the HF solution or in a separate step) (Trolier 1987). Although wet etch is a very versatile patterning method for thick PZT films, the etching solution must be limited in HF concentration, as high acid concentrations are not selective to SiO_2 or TiO_2 and can create delamination. Additionally, undercuts of 2:1 up to 7:1 have been reported (Baborowski et al. 2001; Wang et al. 2001), limiting the design rules for the MEMS devices. Ion beam etches (IBE) allow better step definitions and reduced undercuts; however, exposure of PZT films to ion bombardment, even when covered by photoresist masks, causes degradation of the ferroelectric properties (Gross 2004). Although appropriate post-etch anneals allow good recovery of these properties, major damage to the photoresist edge is observed in most cases, leading to poorly defined structure boundaries (Gross 2004). Suggested use of Cr hard mask circumvents the earlier problems, but requires an additional deposition process. Furthermore, electrical isolation between the Cr and the top electrode (Pt) is required[2] for selective removal of Cr by wet etching without undercutting caused by galvanic effects (Gross 2004). To overcome the low selectivity to the Pt bottom electrode and material re-deposition along the walls by ion milling, reactive ion etching (Coudevylle et al. 2001) with halogenous gases has been suggested, but in many cases the by-products of the process have low volatility, or boiling points in a wide range of temperatures, limiting the operation conditions (Pan et al. 1998). Additionally, to avoid damage to the photoresist (hardening by bombardment) the RF power has to be dramatically reduced for thicker films ($>250\,\text{nm}$), which translates into much lower etch rates. More recently inductively coupled plasma (ICP) (Subasinghe et al. 2006; Zeto et al. 1998) and reactive ion beam etching (RIBE) in a dual-frequency ECR/RF reactor (Baborowski et al. 2000) have been reported for successful etching of the PZT, with high etch rate and steep etch profiles.

20.3 Piezoelectric MEMS Devices

The vast majority of the piezoelectric MEMS sensor and actuator devices are based on structures that amplify the displacement. Although the force produced by the piezoelectric material can be considerable, the actual strain is usually limited to 0.1%. The most common examples of these *amplifiers* are as follows:

- *Simple cantilevers* (clamped beams): Atomic force microscopy (AFM) tips, RF switches, and biosensors (lab-on-chip)
- *Beams and bridges* (clamped-clamped beams): RF switches, acoustic and temperature sensors
- *Suspended membranes*: Micromotors, micropumps and liquid injectors, micromirrors, pressure (acoustic and tactile) sensors, ultrasonic transducers, and accelerometers

Some examples of these devices and applications are given in what follows.

20.3.1 Piezoelectric Microactuators

The piezoelectric microactuators can be classified into two general groups: linear and bimorph structures. In a linear actuator, the forces generated by the application of the electric field are used directly to expand or contract the piezoelectric element linearly. In a bimorph structure, the piezoelectric material is used in a composite configuration and the generated displacement is mostly by way of the lateral or bending motion.

---

[2] This is usually achieved by oxygen doping during the first few minutes of Cr deposition and therefore creation of a thin ($<100\,\text{nm}$) CrO_x film.

As mentioned earlier, the linear displacement is usually extremely small, even for microdevice applications, but the actuation speed is considerably high, enabling stepper motors accumulating small displacements very rapidly (Judy et al. 1990). Nevertheless and although the examples of linear actuators are present in literature (mostly cutting tools), the bimorph structures are preferred for creating larger displacements.

A bimorph is a composite structure made of two or more elements where at least one of the elements undergoes a linear expansion or contraction. Bimorph actuators can be made of two active elements, or one active and one passive element. The devices based on the latter, heterogeneous structure are also known as unimorphs or monomorphs. The constraint to a passive component translates the piezoelectric strain of the active element into a large bending displacement in the perpendicular direction. The devices can work in either d_{33} or d_{31} mode as shown in Fig. 20.1.

In both d_{33} and d_{31} actuation modes, upon application of electric field, the active element creates stress and strain. The passive element resists the created strain with the final outcome of creation of a bending moment, which causes displacement in the direction perpendicular to the plane of the constraint. Upon removal of the field, the elastic restoring forces bring the structure to the original curvature. The deflection at the tip of a unimorph cantilever or at the center of a diaphragm is a function of the structural dimensions, piezoelectric coefficient, applied electric field, and the elastic properties of the passive element.

If both layers in the bimorph structure are made by piezoelectrically active material, the electrical configuration allows both series and parallel connection of the active layers, as shown in Fig. 20.2. In a series configuration of active bimorphs, the two layers have to be polarized in opposite direction to create expansion in

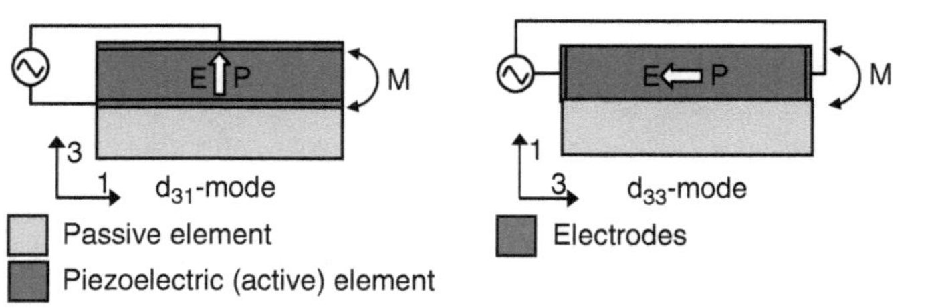

Fig. 20.1 Basic bending principles of d_{31} and d_{33} unimorph piezoelectric actuators. The directions of the electric field, E, dielectric polarization, P, and created bending moment, M are also indicted

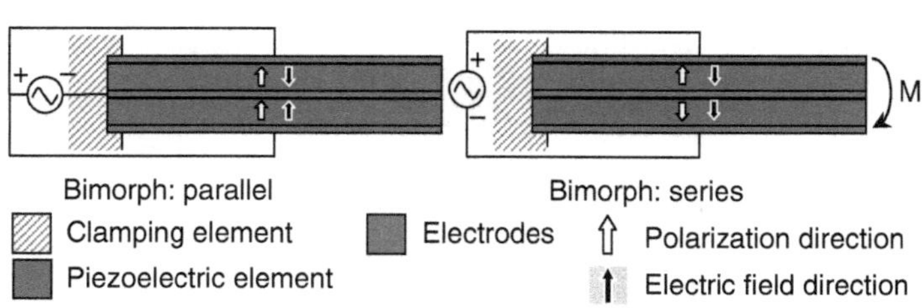

Fig. 20.2 d_{31}-Mode cantilever bimorphs with electrical connection in parallel and in series

one layer and contraction in the other, under applied electric field. In the parallel configuration, both layers are polarized in the same direction, while the applied electric field has opposite direction in the two layers. At a given voltage, the applied field for each piezoelectric layer is twice as big in the parallel configuration than in the series configuration.

For a simple cantilever configuration as illustrated in Fig. 20.2, this translates in twice-bigger tip deflection and generated force. The parallel configuration is therefore usually favored in active bimorph structures operating in d_{31} mode.

20.3.1.1 AFM Probe Tips and Hard Disk Drive Scanners

AFM is a powerful surface imaging tool based on the measurement of the deflection of a probe tip (ceramic or semiconductor) scanning the surface. In both contact and tapping mode, the surface forces deflect the cantilever. The deflection is registered by the movement of a laser deflecting at oblique angle on the other extreme of the probe. An integrated piezoelectric actuator allows high resonant frequencies and therefore increased scanning speeds.

Integration of a piezoelectric element within the tip, as schematically shown in Fig. 20.3, can also simplify the force-sensing system, as the laser alignment procedure can be complicated. Both PZT and ZnO integrated tips have been reported in literature, combining displacement sensing and actuation (Miyahara et al. 1999; Rogers et al. 2004). The former is usually favored due to the higher piezoelectric constant and higher actuation range.

Another application of piezoelectric cantilever actuator is in servo systems for data access in hard drive disks (Tsaur et al. 2002). A single stage servo system is inadequate for the precise positioning, high operating speed, and frequencies required for the high track densities. To overcome the problem, two stage systems integrating a voice coil motor (VCM) with a microactuator have been developed. The VCM performs low-frequency coarse track seeking, while the microactuator allows high-frequency tracking. The actuator requires high displacement strokes and high resonant frequencies for operation. Piezoelectric actuators allow higher stiffness than the interdigitated (Fabula et al. 1994) electrodes (IDT) necessary for electrostatic actuation. Furthermore, piezoelectric actuation allows creation of large forces with low power consumption. Conversely, piezoelectric actuators lack the precise positioning and accuracy of electrostatic actuators, and the two actuation systems are equally used in industry.

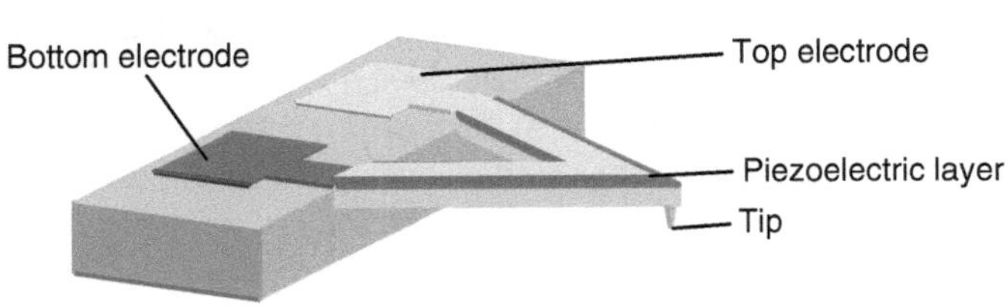

Fig. 20.3 Piezoelectrically active atomic force microscopy (AFM) tip

20.3.1.2 RF Switches

Radio frequency (RF) mechanical microswitches offer several advantages over semiconductor switches including reduced insertion loss (in the on state) and increased isolation (in the off state) (Yao et al. 1999). Although to date electrostatic actuation for MEMS RF switches has prevailed, this can be mainly accredited to the ease of fabrication of such devices.

RF MEMS switches are based on the motion of a free-standing element to create either capacitive coupling (shunt) or metal–metal (DC) contact with the transmission line. Various designs have been proposed, some based on simple cantilevers, others on clamped-clamped beams (simple and modified beams), and yet others on structures with increased complexity (Hee-Chul et al. 2007; Kugeler et al. 2005). Unimorph cantilever structures are very susceptible to residual (thermal) stresses in the various layers, which result into unwanted curvature of the beams. Bimorph clamped-clamped beams are less prone to show initial curvature, but the deflection of the beams is also restricted by the increased stiffness of the structure.

Piezoelectric MEMS RF switches offer advantages over electrostatically actuated ones, mainly in the low power consumption and low actuation voltages fundamental for applications in mobile devices. The switching speed of piezoelectric switches is also higher than that of the electrostatic, thermal, or electromagnetic MEMS devices over the gaps (1–3 μm) between the insertion and transmission lines necessary for good isolation values. In fact, the resonance frequency is inversely related to the dimensions of the device. High piezoelectric coefficients allow reduction of the lateral dimensions while offering large strokes at relatively modest driving voltages.

20.3.1.3 Micromirrors

Micromirrors have received increased attention over the last decade with the continuous development of the optical information system. Applications include barcode readers, laser positioning components, image scanning projectors, and optical switching arrays. With respect to the electrostatic actuators, the piezoelectric ones allow relatively lower operation voltage with drastically lower power consumption. Furthermore, the electrostatic actuation shows inherent nonlinearities of the deflection angle as a function of the pull-in voltage that are larger than the ones observed for the piezoelectric actuators.

Most of the piezoelectric MEMS designs for micromirrors are based on one or more cantilevers that allow rotation (1D) and twisting (2D) of a connected micromirror. Other approaches include deformable mirrors using subsurface piezoelectric actuation (Kanno et al. 2007a, b; Kotera et al. 2005); see Fig. 20.4. The deformable mirrors are based on unimorph structures (the mirror is a passive element clamped to the piezoelectric actuator), while the cantilever actuator designs have been reported in both unimorph and bimorph modes. These structures are highly susceptible to the residual stress of the films after deposition. As reported in Table 20.2, such stresses can reach considerable amounts, and accurate design and processing steps

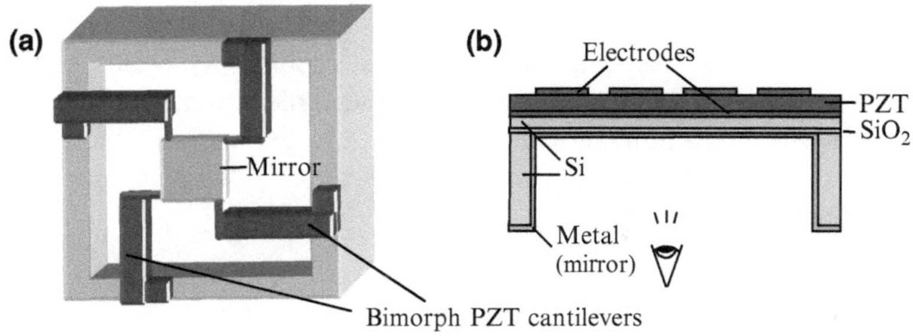

Fig. 20.4 (**a**) Cross section of a deformable micromirror, (**b**) top view of a scanning micromirror actuated by four piezoelectric bimorph cantilever actuators

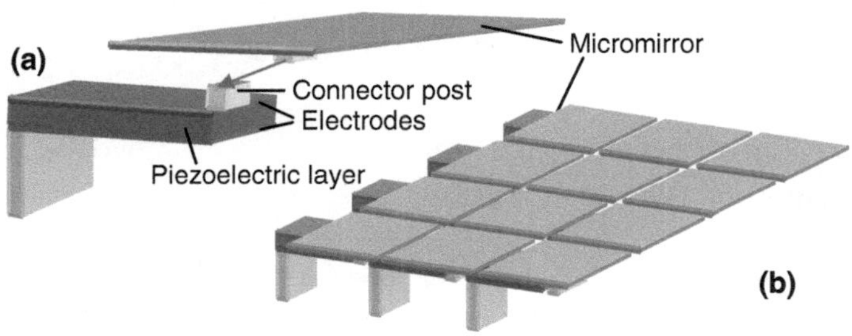

Fig. 20.5 (**a**) Thin film micromirror array (TMA) micromirror and actuator, (**b**) an array of micromirrors. For clarity the micromirror is shown separately from the actuator: the arrow shows the location of the connection of one to the other

are required to reduce the final residual stress of the composite structures and thus lessen any undesired curvature in the final device. One solution in decoupling mirror and piezoelectric actuator for large reflective area has been offered by the TMA (thin film micromirror array) from Daewoo (Sang-Gook and Kyu-Ho 2000). Each micromirror is centrally connected through a support post to the tip of a unimorph piezoelectric cantilever; see Fig. 20.5. Upon application of electric field, the d_{31}-actuated cantilever tip moves upward, tilting the micromirror. Light modulation is achieved by rotation angle of the micromirror: an unaddressed mirror (flat) does not reflect any light (dark state), while a fully tilted mirror reflects the light fully (bright state).A precise control of the tilt angle of the micromirror allows gray scale achievement.

20.3.1.4 Micropumps and Microvalves

Microactuators used for micropumps and microvalves are widely used in control of microfluidic, lab-on-chip systems, and microreactors (Varadan and Varadan 2000). Historically many of the proposed piezoelectrically actuated systems worked with bulk piezoelectric materials bonded to the Si micromachined devices by use of epoxy glue. Different groups have also suggested micropumps and microvalves that are based on thick piezoelectric films, created by stereolithography and screen printing and therefore with increased possibility for mass production. More recently, micropumps based on thin-film fabrication methods (sol–gel for PZT films) that allow complete integration with micromachining processes have also been reported (Hong et al. 2005; Polla et al. 2000).

20.3.2 *Piezoelectric Microsensors*

Piezoelectric microsensors can be generally divided into two major groups: devices that use the piezoelectric response, and the measured quantity is transformed in variation of voltage or current, and devices that use the variations in resonant frequencies to detect a variable.

20.3.2.1 Motion and Vibration Sensors

Use of motion microsensors has rapidly increased over the last decade with increasing need for low-cost miniaturized units for applications in automotive industry, robotics, and navigation systems. Specific applications include, but are not limited to, accelerometers, tilt and shock sensing (airbag deployment, safety hard drive head parking in laptops, gaming industry, image stabilizer for mobile recording), gyroscopes (global positioning systems (GPS), rollover detection, 3D mouse, medical products such as body movement monitoring), and vibration and strain/pressure sensors (for health monitoring of structural components, intrusion sensors).

The advantage of piezoelectric sensing for MEMS accelerometers is in high-quality factors (Q), low damping, and high output impedance, while offering large bandwidth and high sensitivity. Many piezoaccelerometer MEMS are based on simple elastic beam structures loaded with a proof (seismic) mass at the free end (cantilevers) or center of beam (bridges), as shown in Fig. 20.6. Other groups have proposed more sophisticated designs, including trampolines with suspended proof mass at junction of cross beams and circular diaphragm with seismic mass at the center of the membrane. The sensing limit of these devices is limited by the

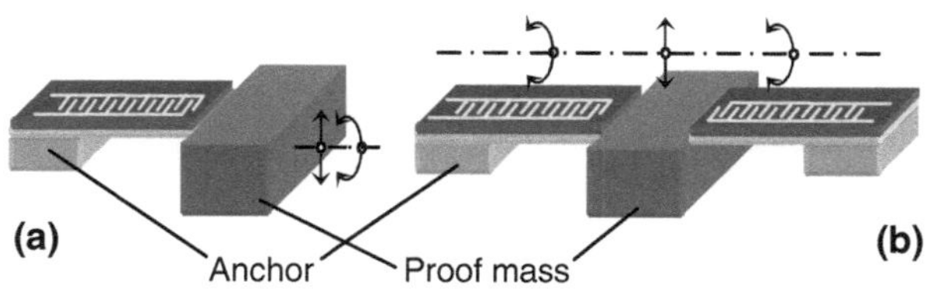

Fig. 20.6 Two simple structures used in MEMS accelerometers: (**a**) cantilever with tip mass and (**b**) bridge (two cantilevers) with intermediate mass

Brownian noise, which is inversely proportional to the mass. To create bigger proof mass (lower noise), many have utilized bulk micromachining techniques, while surface micromachining has also been demonstrated in creation of integrated discrete proof mass devices (DeVoe and Pisano 2001).

Gyroscopes are accelerometers meant to measure the angular velocity of the body with respect to a reference system. This is usually performed by a resonant/vibrating body moving in a defined plane with respect to a reference frame, in absence of any external forces. Any angular acceleration is translated through the Coriolis force into an out-of-plane acceleration of the vibrating structure. Although piezoelectric materials are widely used in bulk gyroscope both for creating the base vibration and for sensing the out-of-plane motion, to date electrostatic sensing has been prevalent among commercial microdevice gyroscopes. Piezoelectric MEMS sensors offer several advantages over electrostatic devices, including use of a single material to create vibration and to sense the out-of-plane velocity, and the insensitivity to the small gap accuracy (which is required in electrostatic devices).

20.3.2.2 Resonant Microsensors

Mass sensing (bio- and nanosensors) can be achieved by functionalizing the top surface of a piezoelectric cantilever, bridge, or membrane with required receptors, as shown in Fig. 20.7. Very small amounts of material absorbed on the functionalized site can be detected by the change in the resonant frequency of the structure. The reported sensing methods range from bulk optical detection systems to the use of integrated active materials within the cantilever MEMS.

The integrated detection by use of piezoelectric materials in resonant structures offers the possibility of creation of an effectively minimized lab-on-chip system, where the same active material is used for oscillation and signal detection. In such a configuration, femtogram mass variations have been reported (Sanghun and Jaichan 2006). One of the drawbacks of this method is the effective damping in viscous fluids and reduction of the Q factor.

Piezoelectric thin films can also be used for creation of standing waves, excited at resonance, on the surface of microbridges or micromembranes. These resonators

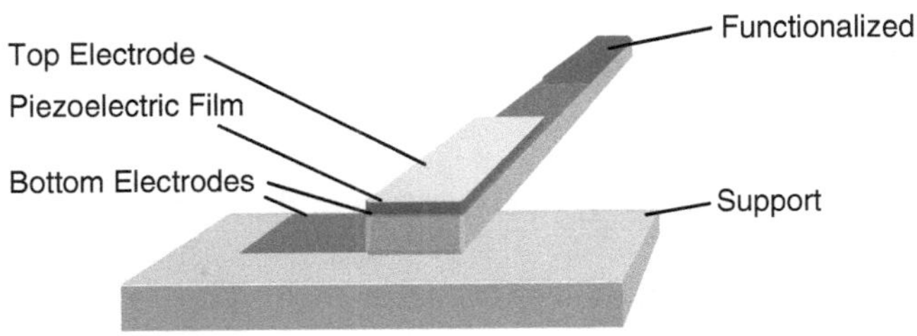

Fig. 20.7 A simple microbalance cantilever resonant biosensor. The functionalized area can cover the whole top surface of the cantilever if contact to the top and bottom electrode is provided by separate vias

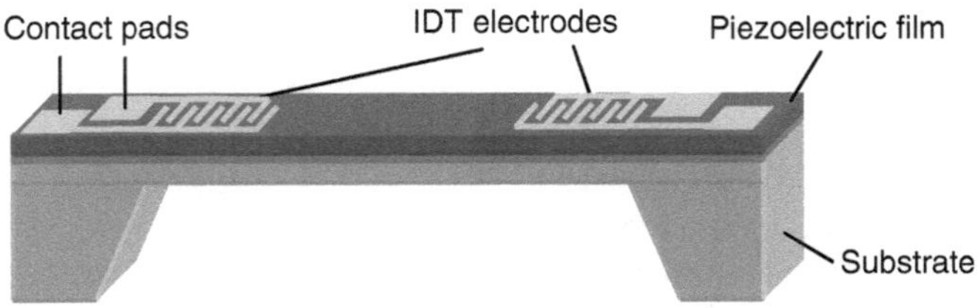

Fig. 20.8 Schematic representation of a piezoelectric microresonator

can then be used for sensing variations in the physical variable by variations in the propagation time of the wave between the two electrodes. Piezoelectric microresonators combine the high quality factor of the Si passive layer with the possibility of exciting and sensing waves in the same material. Two sets of IDT electrodes, positioned on sites of maximum deflection, are used to excite a deflection wave and to detect the response, respectively. The detected response from the second set of electrodes is then amplified and used to drive the first electrode, creating a standing wave at resonance. The frequency variation upon alteration of the physical variables can be used as a high-sensitivity sensing method.

While the microresonators' need for an amplifier, and therefore required electrical power supply for operation, constitutes a major drawback, the possibility of measuring static forces (not otherwise possible with piezoelectric materials) constitutes a major advantage. A schematic representation of a piezoelectric microsensor is shown in Fig. 20.8.

An applied force on the membrane (e.g., a differential pressure between the top and bottom sides of the membrane) will result into strain within the membrane and therefore a variation in the traveling speed and time of the standing wave and ultimately in its resonance frequency. It has to be noted that the quality factor of these devices can be reduced by vibrations and emissions transmitted to the bulk reference. Therefore, it is imperative to reduce transmission of moments and shear forces from the membrane (or bridge) from a design standpoint by choice of appropriate structures. Additionally, the quality factor can be reduced by damping of the resonator by the fluid (air or liquid) in which the structure is immersed, imposing a need for special packaging.

20.3.3 Other Devices

Wireless communications have proven to be one of the highest growing industries over the last decade and one of the major drivers for integrated, low cost, miniaturized filters and resonators. Bulk acoustic resonators (BAW) have proven to be one of the best solutions for filters, specifically in duplexer for mobile phones, both extending the principle of quartz crystals to higher frequencies (Aigner 2005). In its simplest form, a BAW consists of piezoelectric films sandwiched between two metal layers (electrodes). The metal layers function as acoustically reflecting surfaces, trapping energy in the piezoelectric layer and producing a characteristic resonance. The created resonance frequency is directly proportional to the acoustic velocity and inversely proportional to the thickness of the piezoelectric layer. One of the biggest challenges for nonsingle crystalline SMR and FBAR devices is the quality of the polycrystalline piezoelectric material, as high piezoelectric coupling coefficients accompanied by very low loss values are required.

20.4 Future Considerations

The MEMS industry has substantially proliferated, so much so that it has become consumer oriented more than ever. While piezoelectric MEMS have been initially hindered by lack of processing know-how and silicon technology integration issues, the advantages offered with regard to low power consumption and combined sensing and actuating capabilities have pushed the limits of their applications, and the growth of the piezoelectric MEMS sector seems to be extremely promising.

Processing in the context of lowering of the deposition temperature for the ferroelectric films and highly oriented AlN and ZnO films, and patterning techniques for both electrodes and ferroelectric films are some of the key challenges in the path to commercialization. Another important challenge to be addressed in future piezoelectric MEMS, especially in actuator applications, is device control.

The current control methods include complicated algorithms and multiple memory cells for a prediction of the device behavior, particularly at field levels (below coercive) where the extrinsic contributions are predominant, causing nonlinearities and deviation of piezoelectric coefficient from the low field values.

Separate groups have proposed quantitative models based on modification of the Rayleigh law to describe the ac field amplitude dependence of the dielectric permittivity and piezoelectric coefficients for thin films, but none has been attempted for actual devices (Bharadwaja et al. 2004; Gharb and Trolier-McKinstry 2005a, b). If successful quantitative models of the extrinsic contributions and nonlinearities are obtained, prediction of device behavior and the required control systems can be considerably simplified further promoting commercialization of piezoelectric MEMS.

References

Aigner, R., in MEMS in RF-filter applications: Thin film bulk-acoustic-wave technology, Seoul, South Korea, 2005 (Institute of Electrical and Electronics Engineers Computer Society, Piscataway, NJ 08855–1331, United States), 5.

Baborowski, J. (2004) Microfabrication of piezoelectric MEMS. J. Electroceram. 12: 33

Baborowski, J., Muralt, P., Ledermann, N. (2000), Mechanisms of $Pb(Zr_{0.53}Ti_{0.47})O_3$ thin film etching with ECR/RF reactor, 577.

Baborowski, J., Ledermann, N., Gentil, S., in Micromachining of piezoelectric MEMS, Munich, Germany, 2001 (Springer-Verlag), 596

Bassiri-Gharb, N., Fuji, I., Hong, E. (2007) Domain Wall Contributions to the Properties of Piezoelectric Thin Films, J. Electroceram.

Bharadwaja, S.S.N., Damjanovic, D., Setter, N. (2004) Analysis of the non linear domain wall response in ferroelectric thin films. Ferroelectrics 303: 59

Calame, F., Muralt, P. (2007) Growth and properties of gradient free sol–gel lead zirconate titanate thin films. Appl. Phys. Lett. 90: 62907

Campbell, G.A., Mutharasan, R. (2005) Detection of pathogen *Escherichia coli* O157:H7 using self-excited PZT-glass microcantilever. Biosens. Bioelectr. 21: 462

Carlotti, G., Socino, G., Petri, A. (1987) Elastic constants of ZnO sputtered films. IEEE Ultrason. Symp. Proc. 295

Chang-Il, K., Gwan-Ha, K., Dong-Pyo, K. (2006) Dry etching of $LaNiO_3$ thin films using inductively coupled plasma. Thin Solid Films 506–507: 217

Chen, Q., Qin, L., Wang, Q.-M. (2007) Property characterization of AlN thin films in composite resonator structure. J. Appl. Phys. 101: 084103

Coudevylle, J.-R., Basrour, S., De Labachelerie, M. (2001) Fabrication and characterization of high Q microresonators using thin plate mechanical mode. SPIE Proc. 5343: 273

DeVoe, D.L., Pisano, A.P. (2001) Surface micromachined piezoelectric accelerometers (PiXLs). J. MEMS 10: 180

Eun Sun, L., Hyun Woo, C., Sung Hoon, L. (2005) Comparison of forming gas effects on the ferroelectric properties between more-oriented and less-oriented $Pb(Zr_{0.52}Ti_{0.48})O_3$ thin films. J. Vac. Sci. Tech. 23: 773

Fabula, T., Wagner, H.J., Schmidt, B. (1994) Triple-beam resonant silicon force sensor based on piezoelectric thin films, Sen. Actuat. A 42: 375

Gharb, N.B., Trolier-McKinstry, S. (2005a) Dielectric nonlinearity of (100) oriented PYbN-PT thin films, IEEE ISAF, 95.

Gharb, N.B., Trolier-Mckinstry, S. (2005b) Dielectric nonlinearity of $Pb(Yb_{1/2}Nb_{1/2})O_3$–$PbTiO_3$ thin films with $\{100\}$ and $\{111\}$ crystallographic orientation. J. Appl. Phys. 97: 064106

Gross, S.J. (2004) Micromachined switches and cantilever actuators based on piezoelectric lead zirconate titanate (PZT). Ph.D. Thesis, The Pennsylvania State University, University Park, PA

Gross, S.J., Tadigadapa, S., Jackson, T.N. (2003) Lead-zirconate-titanate-based piezoelectric micromachined switch. Appl. Phys. Lett. 83: 174

Gualtieri, J.G., Kosinski, J.A., Ballato, A. (1994) Piezoelectric materials for acoustic wave applications. IEEE Trans. Ultrason. Ferroelectr. Freq. Contr. 41: 53

Hall, D.A. (2001) Nonlinearity in piezoelectric ceramics J. Mater. Sci. 36: 4575

Hee-Chul, L., Jae-Hyoung, P., Yong-Hee, P. (2007) Development of shunt type ohmic RF MEMS switches actuated by piezoelectric cantilever. Sens. Actuat. A 136: 282

Hickernell, F.J., Yue, R.X., Hickernell, F.S. (1997) Statistical modeling for the optimal deposition of sputtered piezoelectric films. IEEE Trans. Ultrason. Ferroelectr. Freq. Contr. 44: 615

Hong, E. (2004) Surface micromachined peristaltic pumps using lead zirconate titanatefilms. Ph.D. Thesis, The Pennsylvania State University, University Park, PA

Hong, E., Krishnaswamy, S.V., Freidhoff, C.B. (2005) Micromachined piezoelectric diaphragms actuated by ring shaped interdigitated transducer electrodes. Sens. Actuat. A 119: 520

Hong, E., Smith, R., Krishnaswamy, S.V. (2006) Residual stress development in $Pb(Zr,Ti)O_3/ZrO_2/SiO_2$ stacks for piezoelectric microactuators. Thin Solid Films 510: 213

Hongxia, Q., Jinsong, Z., Zhiqiang, J. (2000) Oriented growth in PZT thin films. Integr. Ferroelectr. 30: 175

IEEE (2003) Draft 16 of a working document for a proposed standard to be entitled: IEEE standard definitions of terms associated with ferroelectric and related materials. IEEE Trans. Ultrason. Ferroelectr. Freq. Contr. 50: 1613

Jiang, S.W., Zhang, Q.Y., Huang, W. (2006) Texture control of $Pb(Zr, Ti)O_3$ thin films with different post-annealing processes. Appl. Surf. Sci. 252: 8756

Judy, J.W., Polla, D.L., Robbins, W.P. (1990) A linear piezoelectric stepper motor with submicrometer step size and centimeter travel range. IEEE Trans. Ultrason. Ferroelectr. Freq. Contr. 37: 428

Kanno, I., Tazawa, Y., Suzuki, T. (2007a) Piezoelectric unimorph microactuators with X-shaped structure composed of PZT thin films. Microsyst. Tech. 13: 825

Kanno, I., Kunisawa, T., Suzuki, T. (2007b) Development of deformable mirror composed of piezoelectric thin films for adaptive optics. IEEE J. Select. Topics Quantum Electron. 13: 155

Kennedy, M.S., Zosel, M., Richards, C.D. (2004) Optimization of film stresses utilized in composite piezoelectric membrane microgenerators. MRS Proc. 795: 503.

Kholkin, A.L., Calzada, M.L., Ramos, P. (1996) Piezoelectric properties of Ca-modified $PbTiO_3$ thin films. Appl. Phys. Lett. 69: 3602

Kotera, H., Kanno, I., Inoue, T., in Deformable micro-mirror for adaptive optics composed of piezoelectric PZT thin films, Oulu, Finland, 2005 (Institute of Electrical and Electronics Engineers Computer Society, Piscataway, NJ 08855–1331, United States), 31

Kugeler, C., Tappe, S., Bottger, U. (2005) A novel design for integrated RF-MEM switches using ferroelectric thin films. Integr. Ferroelectr. 76: 59

Kugeler, C., Tappe, S., Bottger, U. (2006) Piezoelectric actuated MEMS for integrated RF switches based on PZT thin film bridges. Ferroelectrics 338: 89

Ledermann, N., Muralt, P., Baborowski, J. (2003) {100}-textured, piezoelectric $Pb(Zr_x, Ti_{1-x})O_3$ thin films for MEMS: integration, deposition and properties. Sens. Actuat. A A105: 162

Lee, K.P., Jung, K.B., Srivastava, A. (1999) Dry etching of $BaSrTiO_3$ and $LaNiO_3$thin films in inductively coupled plasmas. J. Electrochem. Soc. 146: 3778

Loebl, H.P., Metzmacher, C., Milsom, R.F. (2004) RF bulk acoustic wave resonators and filters. J. Electroceram. 12: 109

Maeder, T., Muralt, P., Sagalowicz, L. (1996) $Pb(Zr, Ti)O_3$ thin films on zirconium membranes for micromechanical applications. Appl. Phys. Lett. 68: 776

Miller, D.C., Herrmann, C.F., Maier, H.J. (2007) Thermo-mechanical evolution of multilayer thin films. I. Mechanical behavior of Au/Cr/Si microcantilevers. Thin Solid Films 515: 3208

Miyahara, Y., II, Fujii, T., Watanabe, S. (1999) Lead zirconate titanate cantilever for noncontact atomic force microscopy. Appl. Surf. Sci. 140: 428.

Ouyang, J., Ramesh, R., Roytburd, A.L. (2005) Intrinsic effective piezoelectric coefficient $e_{31,f}$ for ferroelectric thin films. Appl. Phys. Lett. 86: 152901

Pan, W., Thio, C.L., Desu, S.B. (1998) Reactive ion etching damage to the electrical properties of ferroelectric thin films. J. Mater. Res. 13: 362

Park, S.E., Shrout, T.R. (1997) Relaxor based ferroelectric single crystals for electro-mechanical actuators. Mater. Res. Innovat. 1: 20

Polla, D.L., Erdman, A.G., Robbins, W.P. (2000) Microdevices in medicine. Ann. Rev. Biomed. Eng. 2: 551–576

Rogers, B., Manning, L., Sulchek, T. (2004) Improving tapping mode atomic force microscopy with piezoelectric cantilevers. Ultramicroscopy 100: 267–276

Sang-Gook, K., Kyu-Ho, H. (2000) Thin-film micromirror array (TMA) for large information-display systems. J. Soc. Inf. Display 8: 177

Sanghun, S., Jaichan, L. (2006) Fabrication and sensing behavior of ultrasensitive piezoelectric microcantilever-based precision mass sensor K. J. Phys. Soc. 49: 608

Schwartz, R.W., Boyle, T.J., Lockwood, S.J. (1995) Sol-gel processing of PZT thin films. A review of the state-of-the-art and process optimization strategies. Integr. Ferroelectr. 7: 259

Subasinghe, S.S., Goyal, A., Tadigadapa, S.A. (2006) High aspect ratio plasma etching of bulk lead zirconate titanate. SPIE Proc. 6409: 61090

Taylor, D.V., Damjanovic, D. (2000) Piezoelectric properties of rhombohedral $Pb(Zr, Ti)O_3$ thin films with (100), (111), and "random" crystallographic orientation. Appl. Phys. Lett. 76: 1615

Tonisch, K., Cimalla, V., Foerster, C. (2006) Piezoelectric properties of polycrystalline AlN thin films for MEMS application. Sens. Actuat. A 132: 658

Trolier, S.E. (1987) Use of photolithography and chemical etching in the preparation of miniature piezoelectric devices from lead zirconate titanate (PZT) ceramics. M.S. Thesis, The Pennsylvania State University, University Park, PA

Trolier-Mckinstry, S., Muralt, P. (2004) Thin film piezoelectrics for MEMS. J. Electroceram. 12: 7

Tsaur, J., Zhang, L., Maeda, R. (2002) 2D micro scanner actuated by sol–gel derived double layered PZT. IEEE MEMS Proc. 548

Tuttle, B.A., Voigt, J.A., Garino, T.J. (1992) Chemically prepared $Pb(Zr,Ti)O_3$ thin films: the effects of orientation and stress. IEEE ISAF, 344

Varadan, V.K., Varadan, V.V. (2000) Micro pump and venous valve by micro stereo lithography. SPIE Proc. 3990: 246

Wang, G.S., Remiens, D., Soyer, C. (2006) Combined annealing temperature and thickness effects on properties of $PbZr_{0.53}Ti_{0.47}O_3$ films on $LaNiO_3/Si$ substrate by sol–gel process. J. Cryst. Growth 293: 370

Wang, L.P., Wolf, R., Zhou, Q. (2001) Wet-etch patterning of lead zirconate titanate (PZT) thick films for microelectromechanical systems (MEMS) applications, MRS Proc. 657: 5

Xu, F., Trolier-McKinstry, S., Ren, W. (2001) Domain wall motion and its contribution to the dielectric and piezoelectric properties of lead zirconate titanate films. J. Appl. Phys. 89: 1336

Yao, Z.J., Chen, S., Eshelman, S. (1999) Micromachined low-loss microwave switches. J. MEMS 8: 129–134

Zeto, R.J., Rod, B.J., Dubey, M. (1998) High-resolution dry etch patterning of PZT for piezoelectric MEMS devices. ISAF 11: 89

Zhang, Z., Park, J.H., Trolier-McKinstry, S. (2000) (001)-oriented $LaNiO_3$ bottom electrodes and (001)-textured ferroelectric thin films on $LaNiO_3$. MRS Proc. 596: 73

Zhou, Q., Cannata, J.M., Meyer, R.J., Jr. (2005) Fabrication and characterization of micromachined high-frequency tonpilz transducers derived by PZT thick films. IEEE Trans. Ultrason. Ferroelectr. Freq. Contr. 52: 350

Chapter 21
High-Frequency Ultrasonic Transducers and Arrays

K. Kirk Shung,* Jonathan M. Cannata, and Qifa Zhou

21.1 Introduction

Ultrasonic imaging is one of the most important and still growing diagnostic tools today. Ultrasound is more appealing as a clinical imaging modality compared with such modalities as magnetic resonance imaging (MRI), nuclear imaging, and x-ray computed tomography (CT) in that it is more cost-effective, noninvasive, capable of real-time operation, and portable while providing images of comparable quality and resolution. State-of-the-art ultrasonic scanners offer real-time gray scale images of anatomical detail with millimeter spatial resolution, which are superimposed on a map of Doppler blood flow, displaying the information in color thereby (Shung 2006). Clinical applications of these devices are still expanding, and the operating frequencies of these devices seem to inch higher and higher. High-frequency (HF) imaging (higher than 20 MHz) yields improved spatial resolution at the expense of a shallower depth of penetration. There are a number of clinical problems that may benefit from high-frequency ultrasonic imaging (Lockwood et al. 1996). Intravascular imaging with probes mounted on catheter tips at frequencies higher than 20 MHz with the highest frequency being 60 MHz has been used to characterize atherosclerotic plaque and to guide stent placement and angioplasty procedures (Saijo and van der Steen 2003). Endoscopic imaging with probes mounted on the tip of an endoscope or a catheter at frequencies from 10 to 20 MHz has been proven clinically beneficial in diagnosing esophageal, gastrointestinal, and urinary lesions (Liu and Goldberg 1995). Medical efficacy of ultrasonic imaging of anterior segments of the eye at frequencies higher than 50 MHz in diagnosing glaucoma and ocular tumors and in assisting refractive surgery has been demonstrated (Pavlin and Foster 1995). The availability of a noninvasive imaging tool for dermatological

K.K. Shung
NIH Resource on Medical Ultrasonic Transducer Technology, Department of Biomedical Engineering, University of Southern California, Los Angeles, CA 90089-1111, USA

A. Safari, E.K. Akdoğan (eds.) *Piezoelectric and Acoustic Materials for Transducer Applications.*

applications could reduce the number of biopsies that are associated with patient discomfort and could better demarcate tumor involvement. An additional benefit is that the results are known immediately or shortly after the examination unlike biopsy where a substantial time lag is likely to occur between the examination time and the report of histological analysis. Small-animal imaging is another frontier of HF ultrasound. Small-animal imaging is of intense interest recently because of the utilization of small animals in drug and gene therapy research. MicroMR, microCT, and microPET have all been developed to meet this need. Ultrasound thus far has only played a very limited role.

The current major competing technology for HF ultrasound is optical coherent tomography (OCT), which has been used to image the anterior segment of the eye and the retina (Bouma and Tearney 2002). The advantages of OCT are that it requires no contact and has a better resolution. The wavelengths used are between 800 and 1,300 nm. The disadvantages are the smaller depth of penetration and its inability to penetrate light-opaque structures. Real-time OCT systems operating at 1,300 nm are being developed at several laboratories around the world. Applications in dermatology for defining tumor involvement and for monitoring treatment, in vascular surgery for characterizing atherosclerotic plaques, and in small-animal imaging are also being investigated (Bouma and Tearney 2002).

To further expand the role of HF ultrasound, probe performance must be improved since current HF ultrasound imaging devices, dubbed "Ultrasound Biomicroscopes," or UBMs by a number of investigators (Pavlin and Foster 1995) use mechanically scanned single element transducers. Many commercial HF single element transducers use the piezoelectric polymer poly(vinylidene fluoride) PVDF or the copolymer P(VDF-TrFE) because of its ease in fabrication and high receiving coefficient (Sherar and Foster 1989). Compared to electronic scanning with linear arrays, mechanical scanning of a single element transducer has three major drawbacks: poorer resolution, slower frame rate, and motion of the probe.

Single element transducers can only produce beams with a fixed focus, which means the spatial resolution of the device is best only within the depth of focus, i.e., in a very tight zone and degrades rapidly beyond the focal point. Mechanical motion of the transducer limits the frame rate, is unreliable, and may cause discomfort and, at worst, may be hazardous to the patient. Current commercial UBMs have been reported to be capable of producing a frame rate as high as 30–40 frames/s. However, this is still inadequate for imaging the hearts of small animals like rats or mice, which beat at a rate of 4–6 beats/s and is much faster than the typical 1–2-beats/s human heart rate.

These problems may be overcome by adopting annular array or linear array technology. HF Annular arrays that allow dynamic 2D focusing are a partial solution to these problems (Brown et al. 2004; Ketterling et al. 2005; Snook et al. 2006; Gottlieb et al. 2006). Since mechanical motion is still involved for annular arrays, a better solution would be to use linear arrays, which generate an image by electronically sweeping a beam (Shung 2006). There are three major challenges in HF linear array fabrication: small kerf (the gap between two elements) width, large electrical impedance mismatch, and increased crosstalk. To minimize grating lobes, typically in linear array and linear phased array designs, the pitch that is the distance between the centers of two elements must be less than 1λ and $^1/_2\lambda$, respectively (Shung and Zipparo 1996). This means that the pitch must be less than 50 and 30 μm at 30 and 50 MHz, respectively. To insure good sensitivity in linear arrays, the element size should be made as large as possible. Therefore, the kerf width should be made as small as possible, making it very difficult if not impossible to dice with current dicing machines. A confounding factor is that unless suitable kerf filler is found, the crosstalk level will increase as a result of the smaller kerf width. It should be noted here that in phased arrays the element size should also be judiciously chosen to assure a broad steering angle.

Efforts are now underway to develop 20–40-MHz linear arrays (Ritter et al. 2002; Michau et al. 2004; Oralkan et al. 2004; Yeh et al. 2005; Cannata et al. 2005). Piezoelectric composites have been used by Ritter et al. (2002), Michau et al. (2004), and Cannata et al. (2005) in their designs. A new technology that utilizes MEMS called cMUT (capacitive micromachined ultrasonic transducer) appears to hold great promise (Oralkan et al. 2004; Yeh et al. 2005). This technology is still under development. A drop in bandwidth in high-frequency operation, which is only 32% at 41 MHz, caused by the *fill effect* appears to be a major problem that needs to be overcome.

For cMUTs, the fill factor is defined as the ratio of the active area of a transducer to its total area. When it is low, each membrane acts like an individual element pushing the fluid forward as well as sideways, representing an increase in hydrodynamic loading of the transducer by the fluid (Oralkan et al. 2004). Currently the pitch in HF linear array designs is greater than 1λ, and their performance is not yet up to par with commercial linear arrays at lower frequencies. This means that much improvement in linear arrays in this frequency range can still be made. Developing linear arrays to 50 MHz or even higher frequencies and phased arrays at 30 MHz remains quite a challenge technically.

The performance of both the HF single element transducers and arrays can be improved by developing improved active and passive transducer materials. Low dielectric constant and high electromechanical coupling materials are desired for single element transducers whereas high dielectric constant and high electromechanical coupling materials are desired for array transducers. HF 1–3 composites, for which the dielectric constant may be adjusted and the electromechanical coupling coefficient increased, and the single crystal PMN-PT poled in [001] plane (Han et al. 2005), which has been shown to have extraordinary high electromechanical coupling coefficient, a velocity of 4,000–5,000m/s, and a relative free dielectric constant of 5,000–7,000, are some of the promising fronts that are worth investigating. In this chapter, efforts that have been undertaken in the development of HF single element transducers and arrays will be reviewed. The new piezoelectric materials are addressed in other chapters of this book.

21.2 High-Frequency Single Element Transducers

An ultrasonic transducer can be treated as a three-port network (Kino 1987; Shung and Zipparo 1996). It is necessary to condition the transfer of energy between the mechanical and electrical ports to optimize the transducer performance. The most important component of the transducer is the active piezoelectric element. There are many different active materials that can be used in a single element transducer design, all with their advantages and disadvantages. Frequently used materials include piezoceramics, piezoelectric polymers, and single crystals (Shung and Zipparo 1996; Oakley and Zipparo 2000). Piezoelectric polymers such as PVDF and copolymer P(VDF-TrFE) have a low acoustic impedance (~4 MRayl), which is ideal for acoustic matching of the transducer to human tissue (~1.5 MRayl). These materials also display a low relative dielectric constant (<10), ideal for electrical impedance matching, but unfortunately suffer from a poor thickness-mode coupling coefficient (<0.3). Alternately, PZT ceramics have higher electromechanical coupling coefficients (0.5 for PZT-5H) but display higher acoustic impedances and large dielectric constants (1,350 for PZT-5H), which can cause problems when acoustically and electrically matching a transducer, respectively. Current bulk piezoelectric ceramics also have finite grain sizes ranging from 1 to 10 μm, which may be a problem when fabricating very thin active elements.

Of all piezoelectric materials that have been used to make 40–100 MHz single element transducers, lithium niobate ($LiNbO_3$) and PVDF polymer have been proven to be the most desirable. Either press-focusing or the application of an acoustic lens has been used to focus the sound beam generated by $LiNbO_3$ transducers whereas PVDF devices are predominantly self-focused.

21.2.1 $LiNbO_3$ Single Crystal Transducers

Lithium niobate ($LiNbO_3$) possesses several material properties that are ideal for fabricating high-frequency transducers (Table 21.1). $LiNbO_3$ is a single crystal material and displays good electromechanical coupling, a low dielectric constant, and

Table 21.1 Relevant material properties of 36° rotated *Y*-cut $LiNbO_3$

Property	$LiNbO_3$
Electromechanical coupling coefficient (k_t)	0.49
Relative clamped dielectric constant ($\varepsilon^S/\varepsilon_0$)	39
Mechanical quality factor (Q_m)	10,000
Electrical quality factor (Q_E)	1,000
Density (g/cm^3)	4.64
Longitudinal wave velocity (m/s)	7,340
Acoustic impedance (MRayl)	34.0
Curie temperature (°C)	1,150

high longitudinal sound speed, ideal for designing sensitive large-aperture single-element transducers. Also this material is ideal for making high-frequency transducers, which require piezoelectric thicknesses in the order of 10–50 μm due to its lack of a finite grain size.

The basic transducer designs for $LiNbO_3$ transducers are shown in Fig. 21.2 (Snook et al. 2002; Cannata et al. 2003). The first matching layer consists of a 0–3 silver epoxy composite. Silver particles of 2–3.5 μm in size are centrifuged in an epoxy matrix to achieve an acoustic impedance of 7.3 MRayls. This 0–3 composite layer was cast and lapped on electroplated pieces of piezocrystal with the appropriate dimensions. A conductive epoxy backing was also cast onto the piezocrystal substrates. Once cured, the backed/matched piezocrystal was turned down to the appropriate diameter with a lathe and potted in the center of a brass tube/SMA connector housing for interconnect and shielding. The second matching layer or lens material was either vapor-deposited Parylene (Specialty Coating Systems, Indianapolis, IN), with an acoustic impedance of 2.6 MRayls, or Epotek 301 (Epoxy Tech., Billerica, MA) with an impedance of 3.05 MRayls. All transducers fabricated were focused by either casting an epoxy lens or by press-focusing the piezoelectric disk in the brass housing using a stainless steel spherical mold, heat, and force.

Numerous transducers operating in the 10–100 MHz range were constructed. The performance of these devices varied because of fabrication inconsistencies with −6-dB bandwidths ranging from 45–75% and insertion loss values ranging from 9 to 22 dB. Typical pulse-echo waveforms for press-focused $LiNbO_3$ transducers are shown in Fig. 21.1. The −6-dB bandwidth was 60% for the 50 MHz device and 73% for the 80 MHz device. The measured insertion loss values were 9.6 and 13.5 dB for the 50 and 80 MHz devices, respectively (Cannata et al. 2003).

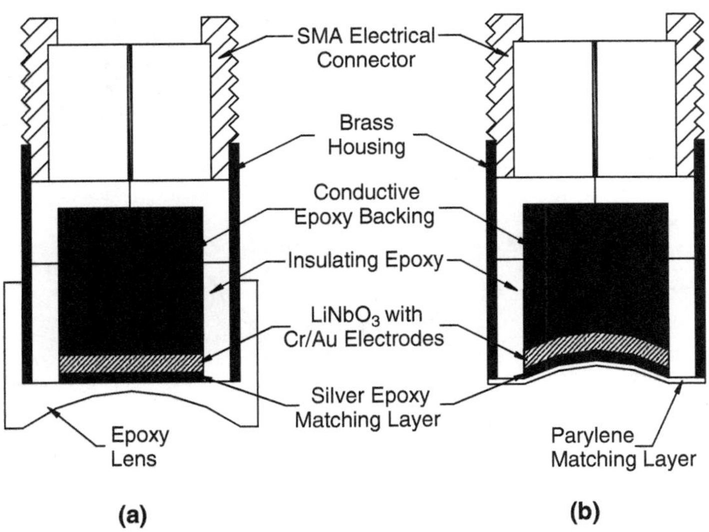

Fig. 21.1 Time-domain pulse/echo response (**a**) and normalized frequency spectrum (**b**) for a 50 MHz (*above*) and 80 MHz (*below*) press-focused $LiNbO_3$ transducer

21.2.2 *Single Crystal PMN-PT Needle Transducers for Doppler Flow Measurements*

PMN-33%PT needle transducers have been designed and fabricated to detect blood flow in the retinal region of the eye (Fig. 21.2) using a pulsed Doppler system (Zhou et al. 2007).

The construction of the needle transducers was similar to the pressed-focused transducer design shown in Fig. 21.3b. A $0.4\,\text{mm} \times 0.4\,\text{mm} \times 0.05\,\text{mm}$ acoustically backed, matched piece of PMN-PT material was housed in a 20-gauge needle (<1-mm diameter) to produce a 45-MHz device center frequency. The entire finished transducer was coated in Parylene. This layer served as the second acoustic matching layer for the device, as well as protected it from the environment. The small-aperture device required no mechanical focusing to produce a natural focus at a range of $\sim$1.2 mm. The sensitivity of the devices produced was acceptable despite their small size. The measured bandwidth and insertion loss for a representative needle transducer was 45% and 15 dB, respectively. The device insertion loss was quite low considering that the transducer was unfocused and that diffractive losses were not compensated for in the measurement.

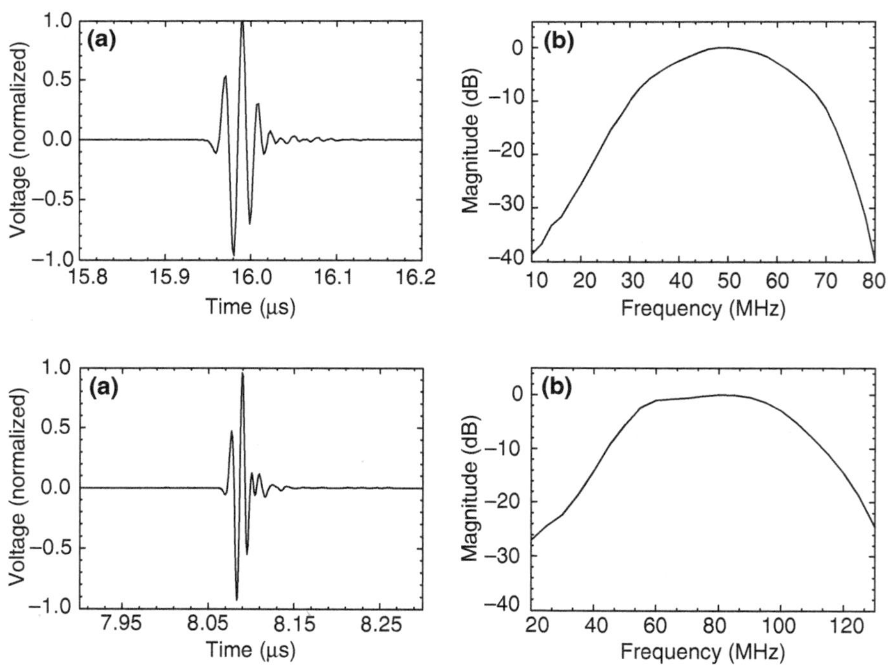

Fig. 21.2 Overview of eye anatomy and orientation of transducer for Doppler measurements (*left*), and a photo of the finished needle transducer (*right*)

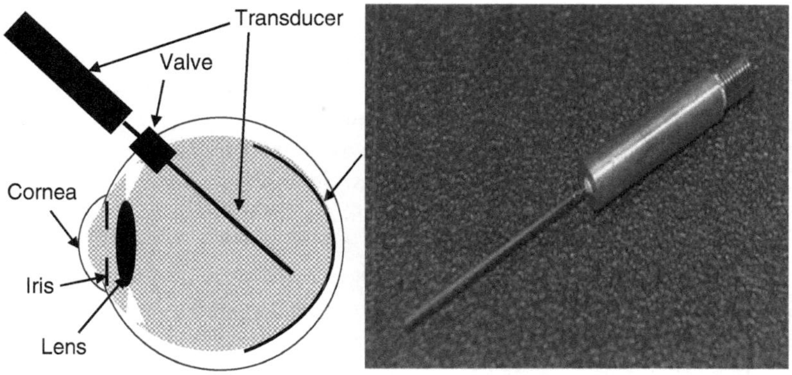

(a) (b)

Fig. 21.3 Design cross sections of the lens-focused (**a**) and press-focused (**b**) transducers. The center conductor of the SMA connector was electrically connected to the negative electrode of the piezoelectric crystal disk via the conductive epoxy backing. A thin layer of sputtered chrome/gold was used to connect and shield the positive side of the piezoelectric disk to the brass housing and outer SMA conductor

21.2.3 Polymer Transducers

The design of PVDF high-frequency single element transducers has been described extensively (Sherer and Foster 1989; Brown 2000). A thin polymer transducer can resonate at $1/2$ wavelength (λ) or $1/4\lambda$ mode depending on if a light- or a heavy-backing is used. For instance, it was shown by Brown (2000) that a 100-μm P(VDF-TrFE) copolymer film backed by a material with an acoustic impedance of 2 and 12 MRayl would resonate at 12 and 6 MHz, respectively. Typically, polymer transducers have one or no matching layer. Because the polymers have a very low melting temperature and low Curie temperature, depositing the electrode and attaching the polymer onto a backing have to be carefully done. For the former, sputtering Cr/Au or applying a conductive epoxy appears to be the best approach rather than soldering. For the latter direct deposition via spin coating onto a substrate such as aluminum eliminating the glue layer yields the best results.

21.2.4 PZT Thick Film Transducers

For a thickness mode transducer, the center frequency is inversely proportional to the thickness of the piezoelectric element. Typically, a thickness less than 20 μm is required for a ceramic transducer to operate at a frequency over 100 MHz. It is very difficult and time consuming to lap down the ceramic to a very thin layer and to handle it in a traditional transducer fabrication process. One of the solutions to this problem is to use piezoelectric thick-film technology.

It was shown that the fabrication of thick films could be improved by utilizing a ceramic powder/sol–gel solution composite approach (Zhou et al. 2005). In this process, sol–gel PZT precursor derived from 2-methoxyethanol was used as the matrix. The composition of the precursor was $Pb(Zr_{0.52}Ti_{0.48})O_3$ near the morphotropic phase boundary where the piezoelectric properties were most desirable. The composition of the PZT powder, which had an average particle size of 5 μm, was similar to PZT 5H. After ball-milling the powder at 200 rpm for 15 h, the average particle size of the powder was reduced to 300 nm. The PZT composite solution was prepared by mixing the PZT powder with the PZT precursor, followed by ultrasonic dispersion for 30 min before depositing the films. A platinum-coated silicon wafer (Pt/Si) was used as the substrate.

PZT thick film was deposited by spin-coating the composite solution onto Pt/Si substrate at 2,000 rpm for 30 s, followed by a two-stage pyrolysis sequence to drive out the solvent and decompose the organic compounds (Zhou et al. 2005). This sequence consisted of a 2-min heat treatment at 150 °C followed by another at 400 °C (2 min). The amorphous layer was then crystallized at 750 °C for 1 min in a rapid thermal annealer (RTA). The thickness of each layer was about 2 μm for the composite with a PZT powder weight fraction of 50%. The spin-coating procedure was repeated to produce the desired thickness. Generally, thick films derived from the modified sol–gel method were porous, which degraded their dielectric and piezoelectric properties. Vacuum infiltration was introduced to fill the pores. After placing a crystallized film into a vacuum chamber, a PZT sol–gel solution was dipped onto the film surface. Allowing a few minutes for infiltration, the film was taken out from the chamber and immediately put on a spin-coating stage to remove residual solution from the surface. The film was then dried and annealed again at 750 °C in the RTA. The filling process was performed to each layer to obtain a dense film shown by SEM (scanning electron microscope cross-sectional image in Fig. 21.4).

Fig. 21.4 SEM cross-sectional image of PZT thick film fabricated with the described filling process

The thick film single element transducer was fabricated (Zhang et al. 2006). An acoustically lossy conductive epoxy (E-solder 3022, Von Roll Isola Inc., New Haven, CT) was used as the backing material. This backing layer was cast onto the Cr/Au electrode following by removing Si with XeF_2 etching. Si is isotropically etched in XeF_2 gas, and PZT was not attacked by XeF_2. The backed sample was then mounted in an SMA coaxial connector, and a thin layer of Cr/Au was sputtered onto the SMA surface to form a ground electrode. A press-focus method was used to focus the transducer. A spherical ball bearing was gently pressed into the PZT film surface at 60 °C to form a curved surface. The PZT thick film had a lower risk of cracking during the press-focusing process than $LiNbO_3$ single crystal or bulk PZT because it was far less brittle. When a PZT thick film with thickness of 13 μm and diameter of 1.2 mm was used, a spherical ball with a diameter of 3 mm was chosen for pressing, which made a transducer with an f-number of 1.25.

The Panametrics 5910 PR (Waltham, MA) was used as both a pulser and receiver for the pulse-echo measurement. Distilled water at room temperature was used as the propagation medium. A polished quartz flat was used as the target placed at the focal plane of the transducer. The transducer had a center frequency of 103 MHz and −6-dB bandwidth of 70% (Fig. 21.5). Using the equation: $\delta_{\text{axial}} = 2C_{\text{water}}\Delta f \ln 2/\pi$ (Foster et al. 2000), a 9.2-μm axial resolution was calculated for this transducer.

An 8-μm diameter wire was used as the target and mounted to the three-axis positioning system for a two-way lateral beam profile test. Both the transducer and the wire were placed in a water tank filled with distilled water. Using the same setup as in the pulse-echo measurement, the receive signals obtained from scanning the wire displayed on a digital oscilloscope (Lecroy LC534, Chestnut Ridge, NY), which was connected to the computer via an IEEE-488 interface. The results of this test showed that the transducer had a lateral resolution of 33 μm at −6 dB.

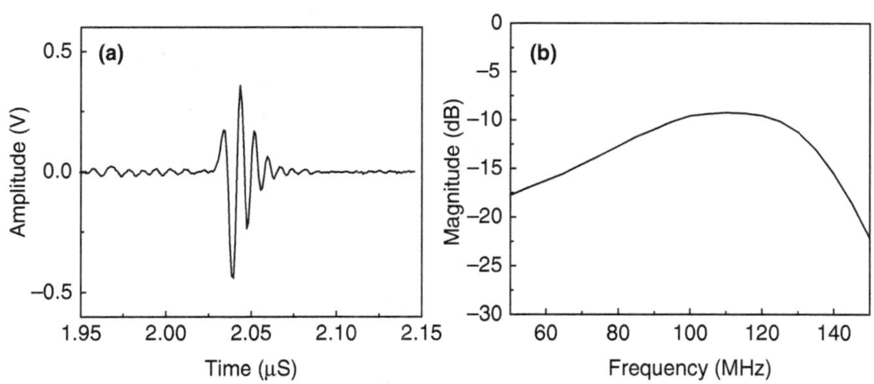

Fig. 21.5 Pulse echo of PZT thick film transducer (**a**) time domain response and (**b**) Fourier transform spectrum

21.3 High-Frequency Annular Arrays

Single element ultrasound transducers have been exclusively used in high-frequency ultrasonic imaging for a number of years. They have been able to provide an adequate solution in a number of clinical applications including imaging the anterior segments of the eye in ophthalmology and imaging small animals. These devices however are less than ideal due to their single geometrical focus, and must be mechanically scanned to form an image.

Dynamic focusing is a distinct advantage that array transducers possess over single element transducers. Though linear arrays, and more recently multidimensional arrays, are the most popular of these transducer types, annular arrays also have significant advantages. Annular arrays provide significant benefits in relation to linear arrays at high frequencies (>20 MHz), allowing for 2D focusing, and less complex beamforming and fabrication methods due to fewer number of elements. Their drawback is that mechanical scanning is still required. This means that annular array transducers are an intermediate solution for improving the performance of current single element transducer-based UBM imaging systems.

HF annular arrays have been investigated by several groups (Brown et al. 2004; Ketterling et al. 2005; Snook et al. 2006; Gottlieb et al. 2006). They can be divided into a kerfless design and a conventional design. The kerfless design is a compromise in which performance is sacrificed for its simplicity for fabricating high-frequency (>50 MHz) annular arrays using a thin layer of piezopolymer or ceramic (Demore et al. 2006). The conventional design was achieved with laser dicing or chemical etching, since it is impossible to create concentric annuli with a mechanical dicing saw.

21.3.1 Conventional Annular Array

The level of interactivity, or crosstalk, between elements is of particular concern for array transducer designs. Mechanical and electrical coupling between elements can degrade the performance of arrays by introducing extra ringing or magnifying the effective aperture of the array elements. To reduce crosstalk, a six-element, 3-mm diameter, 45-MHz annular array was developed using a fine-grain lead titanate ($PbTiO_3$) ceramic, which exhibits much lower planar coupling than traditional PZT ceramics. The array incorporated six elements, which were fully diced to further reduce mechanical coupling. Since the ceramic could not be press-focused without cracking and disconnecting the active electrodes, an acoustic lens was utilized to provide a coarse focus before dynamic focusing, and to allow for imaging closer to the array face.

Individual $PbTiO_3$ array elements were created by a commercial laser micromachining system (MicroConnex, Snoqualmie, WA), which used a tripled Nd:YAG laser operating at 355 nm. The full width at half-maximum power was kept smaller than 35 μm. The kerfs were cut over 100 μm into the plate, fully separating the

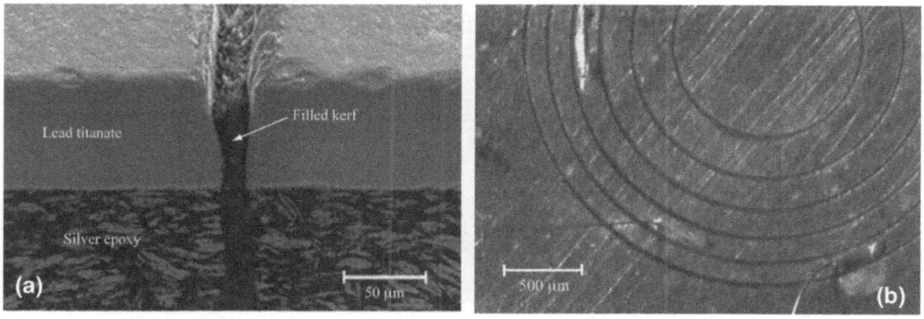

Fig. 21.6 An SEM image of 25-μm kerf after laser micromachining and filling with epoxy (**a**). The silver epoxy substrate served as a rigid support and heat sink during the laser ablation process. A photo showing solder connections between 50-μm wires and PT during construction (**b**)

elements and insuring that the sides of the annuli were as perpendicular as possible, given the Gaussian energy profile of the laser. An SEM image of the kerf is shown in Fig. 21.6. The kerf widths were ~25 μm. The width of the outermost element was 113 μm. The kerf was filled with degassed epoxy before the final lapping to 40 μm and removal of the silver epoxy substrate used as a heat sink during the laser ablation process.

Interconnection between the coaxial cables and piezoelectric was completed with 50-μm wires soldered directly between the elements and small Cr/Au pads on a machinable ceramic disk, as shown in Fig. 21.6b. The indium-based solder Indalloy #2 (Indium Corp., Utica, NY) was used to keep the temperature as low as possible during connection. The center of the transducer was filled with a mixture of Insulcast 502 epoxy and 40 wt% of 3-μm Al_2O_3 to ensure that the backing was acoustically lossy and remained nonconductive. Heavier, conductive silver epoxy used in single element transducers would be an ideal backing to improve bandwidth, though they would short the active electrodes of the elements together. The center wire of 38-gage, 50-Ω coaxial cables (NEEWC, Lisbon, NH) was then soldered to the Cr/Au pads, and the ground connections were soldered to a brass ring with posts cut out. This ring provided tension support as well as the ground connection to the front face of the transducer. Electrical shielding was completed with a brass housing. An epoxy lens was fabricated apart from the transducer and glued to the front face. The lens was aligned over the transducer and glued into place with additional degassed epoxy to minimize bond line effects.

The pulse-echo impulse response, crosstalk, and insertion loss were determined for each element in the array. The pulse-echo response was measured in a water tank with a flat, steel target 7 mm from the array. A Panametrics 5900PR was used to excite the transducer elements, and the echoes were digitized at 500 MHz. Combined electrical and acoustical crosstalk was characterized over the pass band using sinusoid bursts from a function generator (Sony/Tetronix model AFG2020, Beaverton, OR). The insertion loss was measured using a 40-MHz gated toneburst reflected from the steel target placed at the focus, with corrections applied for diffractive effects. The completed array attained an average impulse response that exhibited a

43-MHz center frequency, 30% relative bandwidth, and an average insertion loss of 31 dB at 45 MHz. Maximum next-element crosstalk was −27 dB in water (Snook et al. 2006).

21.3.2 Kerfless Annular Array

Kerfless annular arrays have been designed and fabricated with polymers and ceramics (Brown et al. 2004; Ketterling et al. 2005; Gottlieb et al. 2006). If a piezoceramic material is used for fabrication, considerable care must be taken to mechanically lap it to the $\lambda/2$ resonance thickness. Since most piezoceramics have high acoustic impedances in the order of 30 MRayls, $\lambda/4$ acoustic matching layers in the order of 10 μm need to be added to the substrate for broad bandwidth operation. Again, lapping these matching layers can be very time consuming and meticulous. The advantage of using a copolymer film is that neither lapping nor matching is needed. Interconnect methods typically used in transducer manufacturing include ultrasonic wire bonding and hot-temperature soldering. These methods can be problematic when trying to interconnect to a small element in the order of two wavelengths at high frequencies. In one approach (Gottlieb et al. 2006) that followed the work by Ketterling et al. (2004), the copolymer substrate was ohmically bonded to a flexible circuit consisting of a two-layer gold electroded polyimide surface instead of one side which was gold plated. Individual elements were defined by the flex circuit traces. The annular array flexible circuit is illustrated in Fig. 21.7. The array had eight elements with a total annular array aperture of 3.12 mm. The area of each element was 0.79 mm^2. The flexible circuit was fabricated at Microconnex (Snoqualmie, WA). The advantage of using a double-sided flexible circuit for array interconnect was in having the positive annuli electrode pattern bonded to the copolymer substrate top layer and solder connections on the other side. Microvias then could be used to connect the positive traces to the cable interconnect pads on the bottom side of the circuit. Thus, if properly housed the array could be fully shielded from the outside RF noise by sputtering an additional ground plane

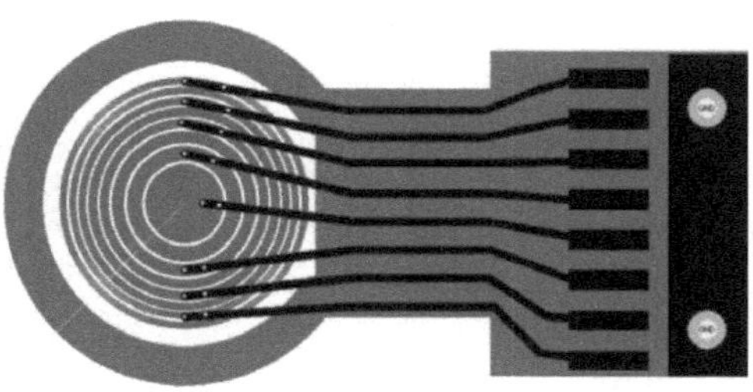

Fig. 21.7 A drawing of the annular array flexible circuit. The gray traces are the active annuli and ground plane on the top side of the flex while the black traces are on the bottom side connected to the top through microvias. The bottom traces terminate to solder pads for coaxial cable interconnects. A ground plane surrounds the entire circuit on the top side and is connected to the bottom side through vias

over the entire assembly. The top layer electrode pattern consisted of eight annuli of equal area with a 30-μm separation; the electrode thickness consisted of 1-μm copper, 0.05-μm chrome, and 0.45-μm gold. The center electrode had an aperture of 1-mm diameter. Each 30-μm via hole was electroplated with copper.

As was discussed earlier, the advantage of a kerfless array is its simplicity. A flex circuit was first bonded on the end of a brass tube with positive annuli traces facing outward. The flex was bent so that the coaxial cable solder connections were inside the brass tube and not covered with epoxy. Care was taken to ensure that the center of annuli was concentric with the housing. The brass tube was filled with degassed nonconductive epoxy. After the flex circuit and housing were left to dry overnight, the assembly was cleaned and prepped for bonding of the copolymer sheet. A poled piece of 9-μm P(VDF-TrFE) piezopolymer material was applied to the top side of the flexible circuit interconnect using the nonconductive epoxy with the aid of an adhesion promoter. A jig was used to apply pressure to the copolymer material and secure it in place as well as to create a bond line of 1 μm or less. An AA ground electrode was then sputtered over the P(VDF-TrFE) film to complete the circuit and shield the array.

The high electrical impedance of the array elements was matched to the send/receive electronics using an inductor and coaxial cable. Inductors (1.2 μH) were connected to the flex circuit solder pads then using short wires. One-meter-length coaxial cables, 75 Ω, 40 AWG, coax no. 171-1019-XX (Precision Interconnect, Portland, OR), were then soldered directly to the inductors. The combination of the high-impedance coax and the inductors impedance-matched each element to 50 Ω at resonance by canceling the negative phase of the transducer and transforming the real component, respectively.

The pulse-echo response, insertion loss, and crosstalk of a single flat, nonmechanically focused array were measured in the same way as reported for the $PbTiO_3$ annular array. For the echo responses, the average center frequency and bandwidth measured were 55 MHz and 47%, respectively. The average insertion loss was 33.5 dB after compensating for attenuation and diffractive losses in the water bath. Finally, the measured crosstalk between center and adjacent elements remained below −25 dB over the device passband.

To assess the performance of the array, a vertically spaced wire target phantom was imaged with a single transmit beamformer and dynamic focusing receive beamformer. The phantom consisted of vertically spaced 0.02-mm diameter tungsten wires (California Fine Wire Co., Grover Beach, CA) separated by 1.5 mm. The transmit beamformer was made out of a Panametrics 5900 pulser, and eight-way power splitter (MiniCircuits, ZCSC-8–1, Brooklyn, NY) and carefully trimmed 50-Ω coaxial cables to deliver an equal amplitude pulse to each annulus with the appropriate delays to produce a 10-mm focus. The coaxial delay lines were placed between the power splitter and the annuli. For receive, a 1-m-length coaxial probe connected a single annulus to the receiver portion of the Panametrics 5900. Twenty-five scan lines were acquired per element over a distance of 1 mm corresponding to a 40-μm scan line width. The averaged (32:1) received echo waveform was displayed on an oscilloscope (50-Ω coupling) and downloaded to file for postprocessing.

A total of eight RF scan lines were recorded at each position to collect the received waveform from each annulus. Dynamic receive focusing time delays were calculated based on every sample corresponding to a 4-ns resolution and applied to each received waveform. Once all the waveforms were delayed, they were coherently summed to form one corresponding scan line. The phantom image before and after dynamic focusing is shown in Fig. 21.8. Figure 21.9 shows a comparison between the measured and Field II (Jensen 1996) simulated −6 dB lateral resolutions throughout the depth of field before and after dynamic focusing. This plot shows a large improvement in image resolution over a 9-mm range after dynamic receive focusing was applied.

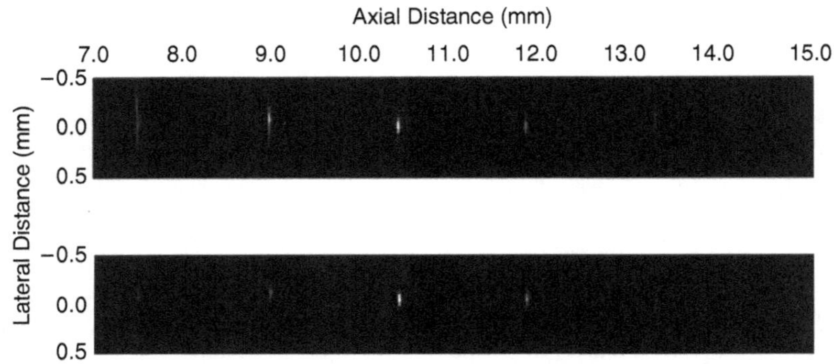

Fig. 21.8 Measured wire target phantom images (48-dB dynamic range): before dynamic focusing (*top*) and after (*bottom*) dynamic focusing

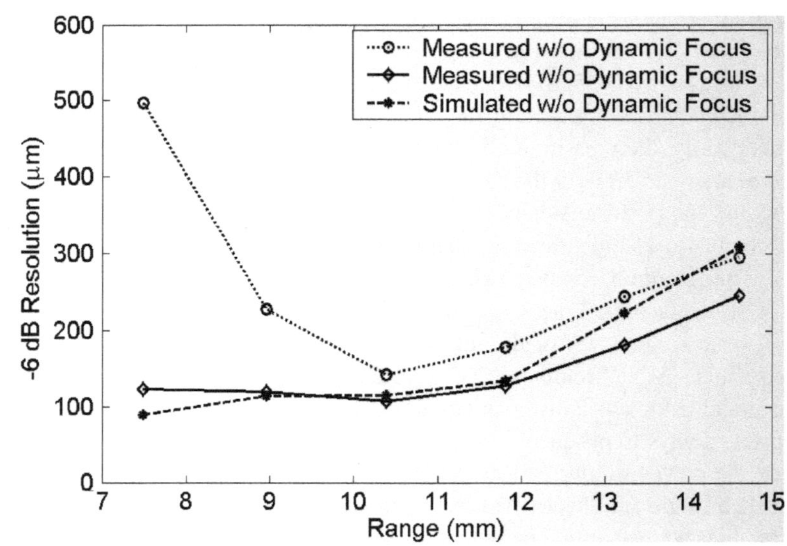

Fig. 21.9 Measured and simulated lateral resolution (−6 dB) from wire targets before and after dynamic receive focusing

21.4 High-Frequency Linear Arrays

Linear and multidimensional ultrasonic array systems have significant advantages over signal element and annular array imaging systems, which require mechanical scanning to form an image slice. Array systems use electronic scanning to form an image slice and therefore can achieve higher image frame rates. Also, the emitted sound beam can be steered and dynamically focused in the image plane. Finally, the lack of movable parts make these arrays less likely to be hazardous to patients. Unfortunately at the present time, commercial array systems are not yet available at frequencies above 20 MHz due mainly to limitations in fabrication technology and equipment, as well as a lack of quality high-frequency materials and electronics. Current efforts in HF linear arrays have relied on conventional dice-and-fill technology and laser dicing (Ritter et al. 2002; Michau et al. 2004; Lukacs et al. 2005; Cannata et al. 2005; Brown et al. 2006). A new technology that appears to be very promising in HF linear arrays is MEMS-based cMUTs (Yeh et al. 2005). Two different types of linear arrays have been under development at the authors' laboratory.

21.4.1 The 48-Element 30-MHz Linear Array

The 48-element, 30-MHz piezocomposite array (Ritter et al. 2002) incorporated a patented process for making fine-scale piezoelectric composites that involved carefully stacking and bonding layers of piezoceramic and polymer. A cutaway view of the major components in this array's design is shown in Fig. 21.10. The composite elements consisted of 28-μm-wide ceramic slabs separated by a 5.3-μm polymer kerf. Brass posts (40 μm × 40 μm × 2.5-mm long) provided interconnect to each

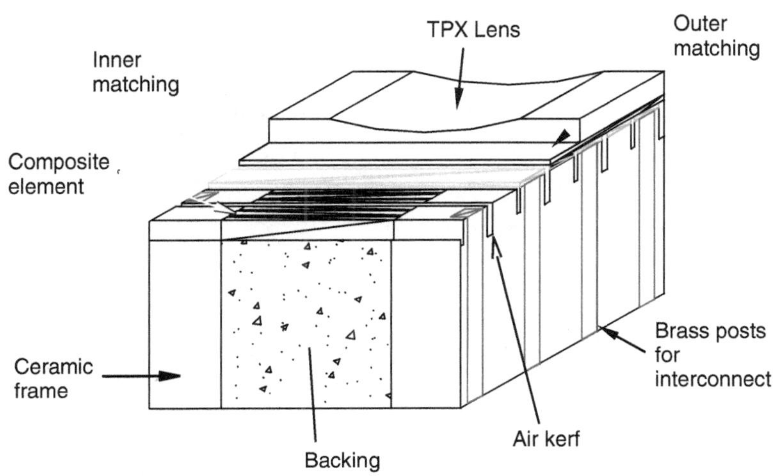

Fig. 21.10 A cutaway view of the 30-MHz array, not to scale, listing each of the components

element of the array. These posts extend along the side of the array and are designed for attachment of a flex circuit. This flex circuit (not shown in the figure) then connected the brass posts to the individual coaxes assigned to each element. The first matching layer consisted of a 0–3 composite engineered specifically for this project. Silver particles 2–3.5 μm in size were centrifuged in an epoxy matrix to achieve an acoustic impedance of 7.3 MRayls and an attenuation of 14 dB/mm. The layer was then lapped to the final thickness of 14 μm. The second matching layer is a 20-μm thick unfilled epoxy, and it was bonded to the array after the kerfs were diced. The elements were spaced on 0.100 mm centers, with a 0.018-mm air kerf separating each element. The air kerf was achieved using a fast curing, viscous epoxy (Araldite 2012, Ciba Specialty Chemicals, East Lansing, MI) that is applied in a very thin film and then B-staged before bonding to limit infiltration into the kerfs. The TPX lens, used for elevation focusing, was bonded to the first matching layer and machined using a precision process to ensure alignment of the lens with the elevation aperture of the array. To match the high-impedance array elements to a 50-Ω load, a coaxial cable was selected with the proper electrical length and impedance. RF shielding in the form of a conductive shell (not shown in the figure) was also incorporated for immunity to EMI.

The completed array was tested for center frequency, bandwidth, sensitivity, electrical input impedance, insertion loss, and element-to-element crosstalk. The results of these tests are listed in Table 21.2. Electrical impedance was measured using an HP4194a vector impedance meter. Pulse-echo testing, used to evaluate center frequency, bandwidth, focal length, and sensitivity, was performed using a polished flat plate target placed at the focal point of the lens. A 1-μJ pulse from a Panametrics model 5900 pulser was used to excite each element, while a 50-Ω receiver with +20 dB of gain was used. Combined electrical and acoustical crosstalk was characterized over the passband using a sinusoid burst from a Wavetek model 81 function generator. Insertion loss was measured using a 30-MHz gated toneburst reflected from a flat plate target placed at the elevation focus, with corrections applied for diffraction in the azimuthal direction and attenuation due to water. The effective

Table 21.2 Measured properties for the 48-element 30-MHz array

Property	Value
Number of elements	48
Number of open/shorted elements	5/0
Average center frequency	30.6 MHz
Average bandwidth (−6 dB)	59%
Average sensitivity	1.76 V
−20 dB pulse length[a]	110 ns
Electrical impedance magnitude (at 30 MHz)[a]	52 Ω
Electrical impedance phase angle (at 35 MHz)[a]	42°
Focal point[a]	6.2 mm
Acceptance angle, −6-dB transmit[a]	40°
Insertion loss[a]	15 dB

[a]Measured on a representative array element

element width, as determined from a measurement of element directivity, was used for the diffraction correction in the insertion loss measurement reported.

The array performed quite well considering the issues associated with fabricating such a complicated structure on such a small scale. Wire phantom images generated with the array using a monostatic synthetic aperture algorithm showed the potential of the array to resolve structures in the order of 100 μm both laterally and axially. Although this array was designed for optimum performance, the design is complicated and costly to fabricate. In addition, the 0.1-mm element-to-element pitch, although it provides a large aperture with a small number of elements, is too coarse for some imaging applications. Therefore, a more simplified approach was developed to make the second-generation high-frequency linear arrays. These newer arrays include a 64-element, 35-MHz array with a 0.05-mm pitch and an additional 64-element, 30-MHz array with a 0.100-mm pitch built to provide a direct design/performance comparison with the previously reported 30-MHz composite array.

21.4.2 The 64-Element 35-MHz Array

This 35-MHz array was developed to simplify the fabrication process without allowing major compromises in imaging performance (Cannata et al. 2005). Individual composite array elements were spaced at a 50-μm pitch (1.2λ in water) and were 3-mm long in the elevation direction. Mechanical dicing was used to form the composite elements out of a bulk piece of fine-grain high-density Navy Type VI (PZT-5H) ceramic at a 25-μm pitch (two composite posts per element). A picture of the diced composite is shown in Fig. 21.11. Elevation focusing was achieved using

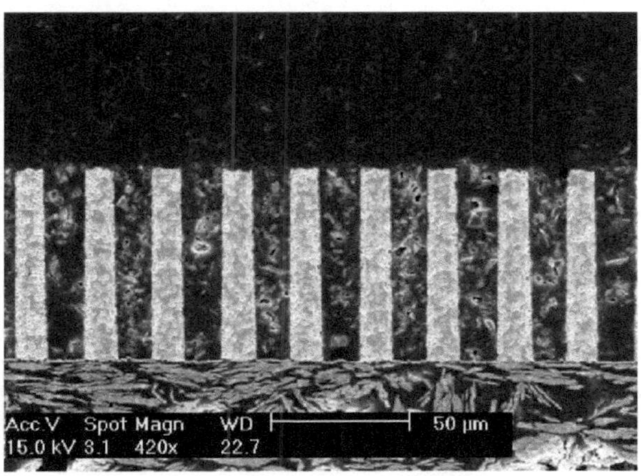

Fig. 21.11 An SEM image of a lapped and backed 2–2 composite. The lighter portions of this image are the ceramic posts whereas the darker portions are the kerf filler. The irregular textured section under the composite is the silver epoxy backing layer

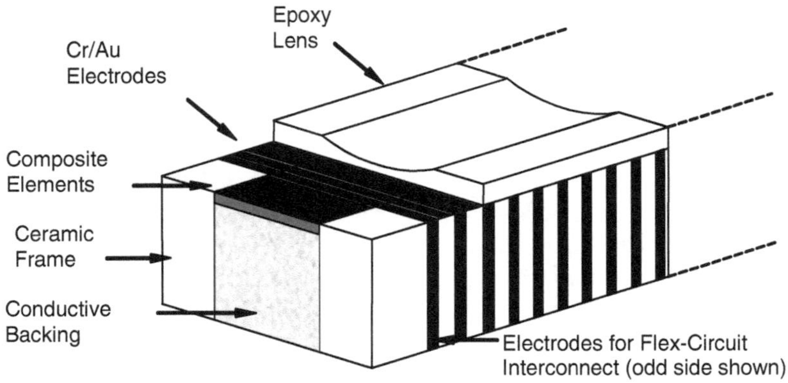

Fig. 21.12 A cutaway drawing of the 35-MHz array, not to scale, showing all of the major design components

a cylindrically cast-molded epoxy lens. Element interconnection was made with a custom-designed flexible circuit, and a $^1/_4$-wavelength, 75-Ω coaxial cable.

Figure 21.12 displays a cutaway view of the array and all major design components. Each array element incorporated two piezoceramic slabs separated by an Al_2O_3-loaded epoxy and with Cr/Au electroplating on both surfaces. The array composite was backed by a lossy conductive epoxy, which also served as the ground connection for the elements. The backed array was housed in a nonconductive ceramic frame, which enabled the electrical connection of individual array elements to a flexible circuit via sputtered Cr/Au electrodes. A cylindrical acoustic lens was molded onto the surface of the array using an unloaded epoxy. This lens provides a fixed elevation focus and acoustical impedance matching between the composite and the load medium, as well as protects the fragile array elements from the environment. To provide RF shielding, the array was encased in a brass tube capped off with a stainless steel end piece. Only the active area of the array was exposed through a small opening in the end piece. The metal housing assembly surrounding the array was also connected to the coaxial cable bundle shield and therefore provided continuous RF shielding from the distal array face to the proximal connector box.

The major differences between this array design and the 48-element 30-MHz array are as follows: (1) A more repeatable mechanical dice-and-fill technology was used to create the composite for this array as opposed to the laborious "stack-and-bond" method used in the 30-MHz array. (2) Separation of elements electrically was simplified for the 35-MHz array design by removing the brass interconnect posts and utilizing the mechanical dicing saw to scratch-dice the sputtered electrode on the surface of the composite elements and ceramic frame. Connection of the flex circuit to the array was done by ohmic bonding using a low-viscosity epoxy as opposed to indium soldering, which was used to interconnect the 30-MHz array. (3) A low volume fraction composite (44%) was produced by mechanical dicing. This resulted in a low effective composite impedance of 17.5 MRayls. Because of this, no acoustic matching layers were needed in the design to achieve a relative

Table 21.3 Measured properties for the 64-element 35-MHz array

Property	Value
Number of elements	64
Number of open/shorted elements	1/0
Average center frequency	35.3 MHz
Average bandwidth (−6 dB)	55%
Highest/lowest bandwidth	62%/49%
Average sensitivity	403 mV
−20 dB pulse length[a]	94 ns
Electrical impedance magnitude (at 35 MHz)[a]	31 Ω
Electrical impedance phase angle (at 35 MHz)[a]	−48°
Focal point[a]	9.5 mm
Acceptance angle, −6-dB transmit[a]	44°
Insertion loss[a]	22.8 dB

[a]Measured on a representative array element

bandwidth of higher than 50% as opposed to the two matching layers needed for the 30-MHz array. Only a 3.05-MRayls epoxy lens (Epotek 301) was used for acoustic matching. (4) Finally, because the composite elements were not separated by air kerfs, elevated interelement crosstalk was expected from this design.

The array was characterized in the same way as described for the 30-MHz array. The results of these performance tests are shown in Table 21.3. Bandwidths averaging 55%, 23-dB insertion loss, and crosstalk less than −24 dB were measured. This array was capable of resolving structures smaller than 50 μm axially and 95 μm laterally.

As expected, the 35-MHz array suffered from a higher element-to-element crosstalk (−24 dB max) than the previously built 30-MHz array (−32 dB max). The additional crosstalk levels contributed to a reduction in the azimuthal acceptance angle, and therefore limit the ability of the array to resolve structures in the near field. The measured insertion loss for this array was 7.8 dB higher than that of the 30-MHz array and was likely caused by attenuation in the epoxy lens as well as electrical and acoustical impedance mismatches. However, these performance tradeoffs are outweighed by the simplicity of the design.

21.5 Conclusion

This chapter discusses the development of high-performance HF single element transducers at frequencies as high as 100 MHz and arrays at 20–50 MHz to date. New piezoelectric materials and methodologies will be needed to further improve the performance and increase the frequency of these devices. There is no reason not to believe that in the next 5 years linear arrays at 50 MHz and HF imaging systems that are capable of acquiring images of similar quality to those acquired by conventional ultrasonic scanners will be developed.

References

Bouma BE, Tearney GJ (2002) Handbook of optical coherent tomography. Marcel Dekker, New York.

Brown LF (2000) Design considerations for piezoelectric polymer ultrasound transducers. IEEE Trans Ultrason Ferroelectr Freq Contr 47: 1377–1396.

Brown JA, Demore CEM, Lockwood GR (2004) Design and fabrication of annular arrays for high frequency ultrasound. IEEE Trans Ultrason Ferroelectr Freq Contr 51: 1010–1017.

Brown JA, Foster FS, Needles A, Lockwood GR (2006) A 40 MHz linear array based on a 1–3 composite with geometric elevational focusing. 2006 IEEE Ultrason Symp Proc, Vancouver, BC, pp 256–260.

Cannata JM, Ritter TA, Chen WH, Silverman RH, Shung KK (2003) Design of efficient, broadband single-element (20–80 MHz) ultrasonic transducers for medical imaging application. IEEE Trans Ultrason Ferroelectr Freq Contr 50: 1548–1557.

Cannata JM, Williams JA, Zhou QF, Ritter TA, Shung KK (2005) Development of a 35 MHz piezo-composite ultrasound array for medical imaging. IEEE Trans Ultrason Ferroelectr Freq Contr 52: 224–236.

Demore CEM, Brown JA, Lockwood (2006) Investigation of cross-talk in kerfless annular array for high frequency imaging. IEEE Trans Ultrason Ferroelectr Freq Contr 53: 1046–1056.

Gottlieb EJ, Cannata JM, Hu CH, Shung KK (2006) Development of a high frequency (>50 MHz) copolymer annular array ultrasound transducer. IEEE Trans Ultrason Ferroelectr Freq Contr 53: 1037–1045.

Foster FS, Pavlin CJ, Harasiewicz KA, Christopher DA, Turnbull DH (2000) Advances in ultrasound biomicroscopy. Ultrasound Med Biol 26: 1–27.

Han P, Yan W, Tian J, Huang X, Pan H (2005) Cut directions for the optimization of piezoelectric coefficients of lead magnesium niobate-lead titanate ferroelectric crystals. Appl Phys Lett 86: 052902-1-3.

Jensen J (1996) Field: A program for simulating ultrasound systems. Med Biol Eng Comput 34: 351–353.

Ketterling JA, Aristizabal O, Turnbull DH, Lizzi FL (2005) Design and fabrication of a 40 MHz annular array transducer. IEEE Trans Ultrason Ferroelectr Freq Contr 52: 672–681.

Kino GS (1987) Acoustic waves: Devices, imaging, and analog signal processing. Prentice Hall, Upper Saddle River, NJ.

Liu B, Goldberg B (1995) Endoluminal vascular and nonvascular sonography: Past, present and future. Am J Roent 165: 765–775.

Lockwood GR, Turnbull DH, Christopher DA, Foster FS (1996) Beyond 30 MHz: Application of high frequency ultrasound imaging. Eng Med Biol Mag 15: 60–71.

Lukacs M, Yin J, Pang G, Garcia R, Cherin E, Williams R, Foster FS (2005) Performance and characterization of high frequency linear arrays. 2005 IEEE Ultrason Symp Proc, Rotterdam, The Netherlands, pp 105–109.

Michau S, Maucharmp P, Dufait R (2004) Piezocomposite 30 MHz linear array for medical imaging: Design challenges and performances evaluation of 128 element array. 2004 IEEE Ultrason Symp Proc, Montreal, Canada, pp 898–902.

Oakley CG, Zipparo MJ (2000) Single crystal piezoelectrics: A revolutionary development for transducers. 2000 IEEE Ultrason Symp Proc, Puerto Rico, pp 1157–1167.

Oralkan O, Hansen ST, Bayram B, Ergan S, Khuri-Yacob BT (2004) High-frequency CMUT arrays for high resolution medical imaging. 2004 IEEE Ultrason Symp Proc, Montreal, Canada, pp 399–402.

Pavlin CJ, Foster FS (1995) Ultrasound biomicroscopy of the eye. Springer-Verlag, New York.

Ritter TA, Shrout TR, Tutwiler R, Shung KK (2002) A 30-MHz composite ultrasound array for medical imaging applications. IEEE Trans Ultrason Ferroelectr Freq Contr 49: 217–230.

Saijo Y, van der Steen AFW (2003) Vascular ultrasound. Springer, Tokyo.

Sherar MD, Foster FS (1989) The design and fabrication of high frequency poly(vinylidene fluoride) transducers. Ultrason Imaging 11: 75–94.

Shung KK (2006) Diagnostic ultrasound: Imaging and blood flow measurements. CRC Press, Boca Raton, FL.

Shung KK, Zipparo MJ (1996) Ultrasonic transducers and arrays. IEEE Eng Med Biol 15: 20–30.

Snook KA, Zhao JZ, Alves CH, Cannata JM, Chen WH, Meyer RJ Jr, Ritter TA, Shung KK (2002) Design, fabrication, and evaluation of high frequency, single-element transducers incorporating different materials. IEEE Trans Ultrason Ferroelectr Freq Contr 49: 169–176.

Snook KA, Shrout TR, Hu CH, Shung KK (2006) Development of a high frequency annular array imaging system. I. Annular array design. IEEE Trans Ultrason Ferroelectr Freq Contr 53: 300–308.

Yeh DT, Oralkan O, Wygant IO, Wong JH, Khuri-Yakub BT (2005) High resolution imaging with high frequency 1-D linear CMUT arrays. 2005 IEEE Ultras Symp Proc, Rotterdam, The Netherlands, pp 665–669.

Zhang QQ, Djuth FT, Zhou QF, Hu CH, Cha JH, Shung KK (2006) High frequency broadband PZT thick film ultrasonic transducers for medical imaging applications. Ultrasonics 44: e711–e715.

Zhou QF, Cannata JM, Meyer RJ, Van Tol JD, Hughes WJ, Shung KK, Trolier-McKinstry S (2005) Fabrication and characterization of micro tonpilz high frequency transducer derived by PZT thick films. IEEE Trans Ultrason Ferroelectr Freq Cont 52: 350–357.

Zhou QF, Xu X, Gottlieb E, Sun L, Cannata JM, Ameri H, Humayun M, Han PD, Shung KK (2007) PMN-PT single crystal high frequency ultrasonic needle transducers for pulsed wave Doppler application. IEEE Trans Ultrason Ferroelectr Freq Contr 54: 668–675.

Chapter 22
Micromachined Ultrasonic Transducers

Massimo Pappalardo,* Giosue Caliano, Alessandro S. Savoia, and Alessandro Caronti

In this chapter the basic principles, the fabrication process, and some modeling approaches of the novel micromachined ultrasonic transducers (MUTs) are described. These transducers utilize the flextensional vibration of an array of micromembranes. They are usually called cMUT (capacitive MUT) or pMUT (piezoelectric MUT) depending on the actuation principle, electrostatic or piezoelectric. For water coupling applications, both these kinds of transducers offer a better matching to the load compared with the typical piezoelectric transducers and therefore they have a larger intrinsic bandwidth. Here emphasis is given to the cMUTs because they have shown good electroacoustic characteristics, which parallel, or even exceed, those of conventional piezoelectric transducers. Good echographic images of internal organs of the human body have been obtained demonstrating the possibilities of this technology to be utilized in commercial 1D and 2D probes for medical applications. At present pMUTs are in a very early stage of development and the potential advantages over the cMUTs are still to be demonstrated.

22.1 Introduction

Piezoelectric transducer fabrication is nowadays a mature technology, which gives a bandwidth as wide as 80% and good sensitivity, but it is expensive and it suffers from some restrictions. The electrostatic principle is an alternative to the piezoelectric effect for the manufacturing of ultrasonic transducers.

Electrostatic transducers made of a thin, metallized, dielectric membrane stretched over a rough conducting backplate have been reported in the early 1950s (Kuhl et al. 1954) for air-coupled applications, whereas the first device for applications in liquids was developed in 1979 (Cantrell et al. 1979). These devices

---

M. Pappalardo
Aculab – Department of Electronics, University "Roma Tre", Roma, Italy

A. Safari, E.K. Akdoğan (eds.) *Piezoelectric and Acoustic Materials for Transducer Applications.*

are based on the electrostatic attraction between the backplate and the metallized membrane. In transmission, the membrane is forced into flexural vibration by a time-varying voltage; in reception, when the membrane is put into vibration by an incident pressure wave, the capacitance modulation caused by the membrane displacement is used to detect the signal.

Transducers with a rough backplate are able to generate ultrasounds in the very low megahertz range. To obtain a more repeatable structure and a better control of the transducer performance, silicon micromachined backplates have been employed (Suzuki et al. 1989; Scheeper et al. 1991; Hietanen et al. 1992; Carr and Wykes 1993; Anderson et al. 1995; Schindel et al. 1995). In 1994, Haller and Khuri-Yakub presented an ultrasonic transducer based on the electrostatic effect, operating in air and made of silicon by means of a micromachining technique (Haller and Khuri-Yakub 1994). They used the planar silicon technology, currently employed for microelectromechanical systems (MEMS), to manufacture electrostatic cells of very small dimensions.

The transducer consists of an array of these miniaturized electrostatic cells, electrically connected in parallel. For operation in the megahertz range, the lateral dimension of each cell is in the range of tens of microns, and for sufficient sensitivity, the number of cells must be in the order of thousands. Because of the working principle and the technology used, this transducer was called cMUT (capacitive micromachined ultrasonic transducer). In the following years Eccardt et al. (1996) and Ladabaum et al. (1998) reported cMUT transducers working in air and water illustrating the manufacturing process and providing the first attempts to a mathematical modeling.

The promising results together with the high flexibility offered by the planar silicon technology have stimulated many authors to develop new manufacturing processes and adequate models to obtain a deeper physical insight and new design tools (Caliano et al. 2000; Caronti et al. 2000, 2002; Eccardt and Niederer 2000; Foglietti et al. 2000, 2001; Ge 1999; Jin et al. 1998, 1999; Ladabaum et al. 1998; Ronnekleiv et al. 1997).

The major efforts have been addressed toward the realization of multielement probes for echographic medical applications. Echographic images of internal organs of the human body have been obtained showing a quality comparable or even better than those currently obtained with standard commercial piezoelectric transducers (Caliano et al. 2005a; Mills and Smith 2003).

The good results obtained in the field of echographic probes with flextensional microcells obtained by means of micromachining technology have stimulated the experimentation of piezoelectric, rather than capacitive actuation (Akasheh et al. 2004; Baborowsky et al. 2002; Bernstein et al. 1999; Eccardt et al. 1996; Muralt et al. 2001; Wang et al. 2006; Yamaschita et al. 2002). The piezoelectric micromachined ultrasonic transducer (pMUT) has basically the same structure as the cMUT but the actuation of the micromembrane is given by the radial forces generated by a thin piezoelectric film deposited over the membrane. The difficulty of producing good quality, crack-free piezoelectric films has so far retained the development of these new devices. At present pMUTs are in a very early stage of development; they seem to have some potential advantages over the cMUTs that, however, are still to be demonstrated.

22.2 Capacitive Micromachined Ultrasonic Transducer

22.2.1 Basic Principle of cMUT

A cMUT transducer includes a plurality of electrostatically driven micromembranes connected in parallel that are fabricated over a substrate using surface micromachining techniques. Each electrostatic cell consists of a vibrating membrane with built-in top electrode, separated from the fixed bottom electrode by a vacuum cavity, as shown schematically in Fig. 22.1a. The micromachining technology allows the electrode separation to be very small, in the submicron range, so as to achieve high electric fields inside the cavity and efficient transduction, required to enable medical imaging applications.

The basic operation of the elementary cell is the well-known electrostatic transduction mechanism. In transmission, the membrane is driven by the electrostatic force produced by a voltage signal applied between the electrodes, with consequent generation of ultrasounds. Conversely, when the membrane is set into vibration by an incident ultrasonic wave, the resulting capacitance change can be detected as a voltage or current signal, provided that a charge is present on the electrodes. Therefore, a DC bias voltage must be used in reception for proper signal detection, whereas it is required in transmission for linear operation. In fact, because the electrostatic force depends on the square of the voltage, a sinusoidal voltage applied to the transducer without polarization would drive the membrane to oscillate at double frequency. The alternating voltage V_{ac} can be superimposed to the bias voltage V_{DC} through the simple polarization circuit shown in Fig. 22.1b, where R_B is a high-value bias resistance, and the coupling capacitance C_B must be much higher than the transducer capacitance C_o.

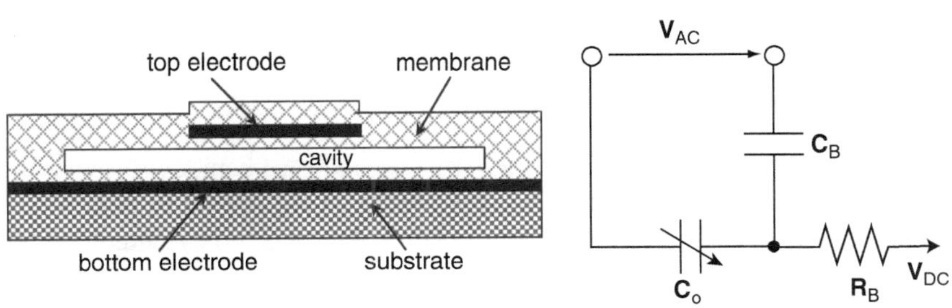

Fig. 22.1 Schematic showing a cMUT electrostatic cell (*left*), polarization circuit for capacitive transducers (*right*)

22.2.2 Membrane Collapse and Snapback

If a DC voltage V_{DC} is applied between the top and bottom electrodes, an attractive electrostatic force arises, which displaces the membrane from its equilibrium position until the electrostatic force is balanced by the elastic restoring force.

Assuming a parallel plate capacitor model for the electrostatic cell, and a piston-like motion of the membrane (one-degree-of-freedom system), the electrostatic force can be expressed as

$$F_E = \frac{\varepsilon_0 A V_{DC}^2}{2\left(d_{eq} - x\right)^2}, \tag{22.1}$$

where ε_0 is the dielectric constant of vacuum, A is the electrode area, d_{eq} is the effective electrode spacing, accounting for all the dielectric layers placed between the electrodes, and x is the membrane displacement. The equilibrium displacement x_0 is obtained by solving the static equilibrium equation

$$\frac{\varepsilon_0 A V_{DC}^2}{2\left(d_{eq} - x_0\right)^2} = kx_0, \tag{22.2}$$

where k is the equivalent spring constant of the membrane (in N/m). The three roots of this cubic equation for an arbitrary applied voltage (below a critical value) are real and correspond to one unphysical solution ($x_0 > d_{eq}$), one stable position, and one unstable point. These points get closer when increasing the voltage, and finally combine at the position $x_0 = d_{eq}/3$. Beyond this point, the electrostatic force is no more balanced by the elastic restoring force and the membrane collapses onto the bottom electrode. The voltage that causes this collapse is called the pull-in (or collapse) voltage and is

$$V_{col} = \sqrt{\frac{8kd_{eq}^3}{27\varepsilon_0 A}}. \tag{22.3}$$

Once the membrane has been pulled down, the snapback does not occur until the applied voltage is decreased below a certain value. This is called the snapback voltage, and it results in a hysteretic behavior (Ladabaum et al. 1998).

22.2.3 FEM Computation of the Collapse Voltage

An accurate calculation of the collapse voltage, and the membrane static deformation at the operating bias point, is needed to predict the overall performances of the cMUT transducer. This requires a coupled electrostatic structural analysis, taking into account the coupling between the electrical and mechanical fields. The commercial software program ANSYS (Ansys Inc., Canonsburg, PA), based on the finite element method (FEM), makes use of either sequential or direct coupling approaches to perform this task. In the sequentially coupled method, the electrostatic

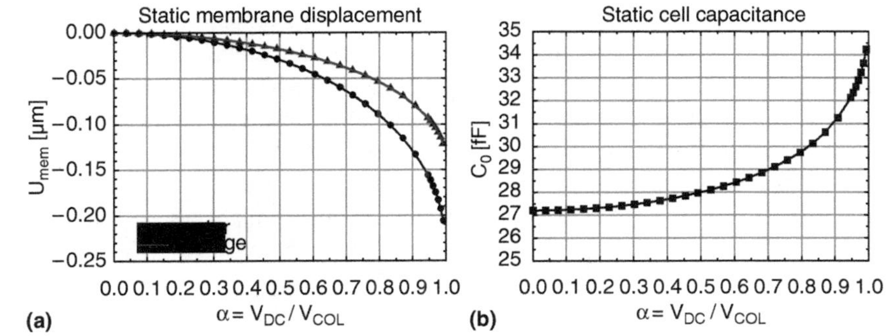

Fig. 22.2 (**a**) Center and average membrane displacement, (**b**) single cell capacitance vs. the applied bias voltage, normalized to the collapse voltage

field and displacement field are coupled by applying the results from one analysis as loads in the other analysis: the iterations between the two field solutions are performed until a desired level of convergence is achieved. Hence, the collapse voltage computation is based on a nonconvergence criterion: when the applied voltage is higher than the collapse voltage, the successive iterations do not converge to a stable deformed shape.

Figure 22.2 shows the membrane displacement and the static capacitance of a representative cMUT cell, as a function of the applied bias voltage, normalized with respect to the collapse voltage (before membrane pull-down). The simulation results were obtained using the ESSOLV sequential macrosolver, where PLANE121 electrostatic elements and PLANE82 solid elements were used in the electrostatic and mechanical finite element domains, respectively. The capacitance was calculated using the CMATRIX command macro, relating the charges to the voltage drop across the electrodes.

The top metallization of the membranes can affect significantly both the static and the dynamic performance of the cMUT cell. In particular, because the electrode thickness is often nonnegligible in comparison with the membrane thickness, the mass loading of the electrode affects the mechanical resonance frequency, and the collapse voltage also changes as a function of the electrode size, depending on its thickness. Therefore, the electromechanical effects of the metal layer must be taken into account in cMUT design and optimization. A FEM study of these effects is reported in Caronti et al. (Caronti et al. 2004).

22.2.4 Spring Softening

The fundamental resonant frequency of an electrostatically driven micromembrane is lowered by the applied bias voltage, by a considerable amount as the voltage approaches the collapse voltage of the structure. This effect, which is referred to as electrostatic spring softening, has a simple physical explanation: for increasing values of the bias voltage, as the membrane is more attracted toward the bottom electrode, the elastic spring force is opposed by the increased DC electrostatic force (which acts in opposite direction), so that the apparent spring constant decreases in dynamic conditions.

By a first order-analysis, based on a parallel plate capacitor model, the small signal softened spring constant of the membrane results (Ladabaum et al. 1998)

$$k_{\text{soft}} = k - \frac{\varepsilon A V_{\text{DC}}^2}{d_{\text{eq}}^3}. \tag{22.4}$$

The spring softening has also important consequences in immersed capacitive transducers such as cMUT arrays for medical imaging applications because cMUTs are usually operated near the collapse voltage; both the center frequency and fractional bandwidth are significantly affected by this phenomenon.

22.2.5 Equivalent Circuit Modeling of cMUTs

Electromechanical transducers, converting electrical energy into mechanical energy and vice versa, are mostly nonlinear in nature. Nonetheless, their operation can be linearized around a bias point, so that they can be described by means of a two-port network with a single electrical and a single mechanical port, as illustrated in Fig. 22.3. Using the phasor notation, $V = Ve^{j\omega t}$ is the sinusoidal voltage with pulsation ω, I is the current, F is the reaction force of the surrounding medium on the vibrating surface, and $\langle v \rangle$ is the spatial average velocity of the surface.

The canonical equations for a single electrostatic cell may be written in a matrix form as follows (Kinsler et al. 2000)

$$\begin{cases} V = Z_{\mathbf{Eb}} I + \phi_{\mathbf{em}} \langle v \rangle \\ F = \phi_{\mathbf{me}} I + Z_{\mathbf{mo}} \langle v \rangle \end{cases}, \tag{22.5}$$

where Z_{Eb} is the blocked electrical impedance, Z_{mo} is the open-circuit mechanical impedance, φ_{em} and φ_{me} are the complex transduction coefficients for the mechanical to electrical conversion and the electrical to mechanical conversion, respectively.

In the parallel plate or piston-like approximation, i.e., assuming a uniform membrane motion and electrostatic pressure, the transduction coefficients are real and equal, reducing to the simplified form of the transformation factor

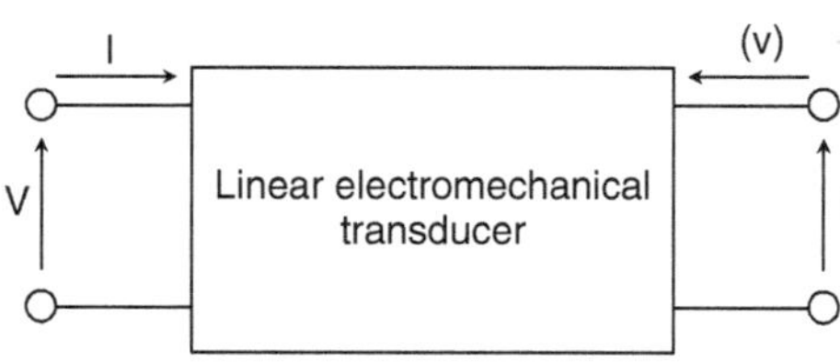

Fig. 22.3 Electromechanical transducer represented as a two-port network

$$\phi = \frac{C_0 V_{DC}}{d_{eq}}, \tag{22.6}$$

whereas the mechanical impedance is given by $\mathbf{Z_{mo}} = \mathbf{Z_{ms}} + \phi^2 \mathbf{Z_{Eb}}$, $\mathbf{Z_{ms}}$ being the short-circuit mechanical impedance of the membrane. All the parameters in the earlier equations must be calculated at a given bias point V_{DC}. More generally, the membrane is partially metallized and clamped along its perimeter, which results in a static deformation $U_{DC}(r)$ function of the radial position, and the transduction coefficients depend on the deflection profile and the distribution of electrostatic pressure over the membrane (Caronti et al. 2004, Certon et al. 2005).

The mechanical impedance $\mathbf{Z_{ms}}$ can be derived by solving the equation governing the vibration of the membrane. Assuming that the membrane is circular with a radius R_m and a uniform thickness d_m, and that the elastic restoring force has contributions from both the inherent stiffness of the material and an applied tension, the equation of motion for symmetric vibrations can be written

$$D_m \nabla^2 \left(\nabla^2 x\right) - d_m T \nabla^2 x + p_C + p_E + p_R = \rho_m d_m \frac{\partial^2 x}{\partial t^2}, \tag{22.7}$$

where ∇^2 is the Laplacian operator in cylindrical coordinates, T is the tensile stress, D_m is the flexural rigidity, and ρ_m is the density of the membrane material. In addition, p_E is the electrostatic pressure actuating the membrane, p_C is the pressure on its lower surface due to the presence of a gas inside the cavity, and p_R is the radiation pressure on the upper surface of the membrane in contact with the surrounding medium. In general, all the terms in the earlier equation are functions of the radial position. For a cMUT with vacuum-sealed cavities, designed for immersion operation, the term p_C in (22.7) can be ignored.

A closed-form harmonic solution for the dynamic displacement of the membrane can be obtained under simplifying assumptions. Following Mason's approach (Mason 1948), the response of the transducer to a driving pressure is derived in vacuum, i.e., taking $p_R = 0$; the fluid loading is then described by means of a radiation impedance, which adds to the mechanical impedance of the transducer. The basic hypothesis underlying this approach is that the vibration shape of the radiator is not affected by the fluid. However, because of the mass loading of the fluid, the mechanical antiresonance of a single membrane in flexural vibration shifts to a lower frequency (Yaralioglu et al. 2003); in addition, for a cMUT including a plurality of cells, because of the acoustic interaction effects between the cMUT micromembranes, the Mason's approach of separately lumped mechanical and radiation impedances does not result in accurate calculation (Caronti et al. 2005).

If all tension, stiffness, and cavity effects are considered simultaneously, while assuming a uniform electrostatic pressure and a uniform static displacement for the DC solution, an analytical solution for the mechanical impedance of a circular membrane can be obtained (Caronti et al. 2002).

To extend the description of one single cell to a plurality of cells, it is necessary to rewrite the canonical equations in such a way as to include the number of membranes and their coupling through the surrounding medium and the substrate.

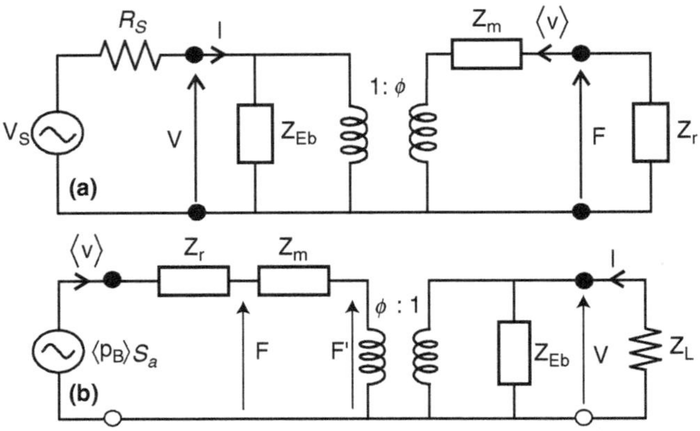

Fig. 22.4 Equivalent circuit models of a cMUT: (**a**) in transmission, (**b**) in reception

A common way to take acoustic coupling into account is to define a global radiation impedance $Z_\mathbf{r} = F/\langle\tilde{v}\rangle$, where $\langle\tilde{v}\rangle$ is the average velocity of the collection of membranes. In general, the reaction force of the fluid on one single micromembrane is influenced by the vibration of all the adjacent membranes, depending on their mutual interactions. As a rough approximation, the collection of membranes driven in parallel vibrates all in phase, similar to a piston transducer with a radiating area equivalent to that of the membranes. Therefore, as long as the wavelength of the ultrasonic wave in the fluid is much lower than the linear dimensions of the cMUT, and the membrane motion approximates that of a rigid piston, the radiation impedance reduces to a pure resistive load and plane waves are radiated. With the aforementioned assumptions and restrictions, yet the linear operation of the cMUT can be described by the canonical (22.5), where appropriate parameters must be used in the transformation matrix.

On the basis of (22.5), an equivalent circuit can be derived for a cMUT working in transmission, as shown in Fig. 22.4a. The transducer is driven by a voltage source V_S whose internal resistance is R_S. The blocked electrical impedance Z_Eb includes the total capacitance (with a parasitic contribution), the bias resistor providing the transducer with the bias voltage, and a parallel loss resistance. Figure 22.4b shows the equivalent circuit of the transducer working as a receiver. A force generator $\langle p_\mathbf{B}\rangle S_\mathrm{a}$, where $\langle p_\mathbf{B}\rangle$ is the average pressure over the surface of the cMUT and S_a is the active area, is used to represent the incident pressure field. The load impedance Z_L is the input impedance of the receiving amplifier.

22.2.6 *Acoustic Coupling*

In all respects, acoustic coupling between the micromembranes and the surrounding fluid medium is one of the most important aspects of cMUT operation, because it affects the performances of the transducer in a fundamental way. This subject has received much attention and several investigations by researchers, by means of experimental measurements, and analytical and finite element modeling. The first experimental study of the cross-coupling in cMUT arrays based on laser interferometry measurements was reported by Caronti et al. (2003); a comparison between finite element calculations and optical displacement measurements in a 1D cMUT array, for both conventional and collapsed modes of operation was presented by Bayram et al. (2007), where the effects of a surface coating on crosstalk were also investigated. An analytical description of the crosstalk between cMUT membranes, based on a theory of evanescent surface waves, was presented by Eccardt et al. (2005). Finally, a transfer function matrix approach based on programmable waveforms to reduce interelement crosstalk in transmission was proposed by Zhou and Hossack (2007).

Unlike conventional piezoelectric transducer arrays, where each mechanism such that the response of the single element is influenced by the adjacent elements is regarded as an unwanted source of crosstalk, the pulse-echo operational bandwidth of a cMUT array is formed significantly by the acoustic coupling between one element and the adjacent elements through the fluid medium. Therefore, acoustic coupling in cMUTs cannot be simply regarded as a source of cross-coupling, although undesirable acoustic interaction effects between the membranes may also occur, resulting in dips and ripples at some frequencies within the bandwidth, and thus degrading the response of the array.

22.2.6.1 FEM Modeling of Acoustic Coupling

Because of the complexity of the problem, the FEM is an essential tool for the analysis and understanding of the acoustic coupling in cMUTs, including the effect of the supporting substrate. The simplest FE model for the analysis of acoustic coupling is the model of an infinity of identical membranes, all driven in phase with the same excitation (Caronti et al. 2005; Yaralioglu et al. 2005). This unbounded transducer model can provide a first-order approximation of the actual response of a cMUT array, although more complex interaction effects originating from the surface waves propagating in the fluid cannot be taken into account.

Figure 22.5 shows the average displacement and transmit pressure vs. frequency, as computed over the surface of the membranes, for different values of the ratio between the membrane center-to-center distance (or pitch) p_{mem}, and the membrane diameter D_{mem}. The dip around 22.5 MHz occurs between the first and second symmetrical modes of the membrane, where its average displacement is minimum: this can be called a mechanical antiresonance. As can be seen, the immersion bandwidth and center frequency of the transducer decrease as the membrane pitch increases, because of the change of radiation impedance. Therefore, it is important to closely pack the membranes to achieve a large bandwidth. For a similar reason, because the reactive part of the radiation impedance changes with the radiating area, in a finite-aperture transducer the transmit bandwidth is affected by the number of elements connected in parallel (Caronti et al. 2006; Daft et al. 2006). If a cMUT element with

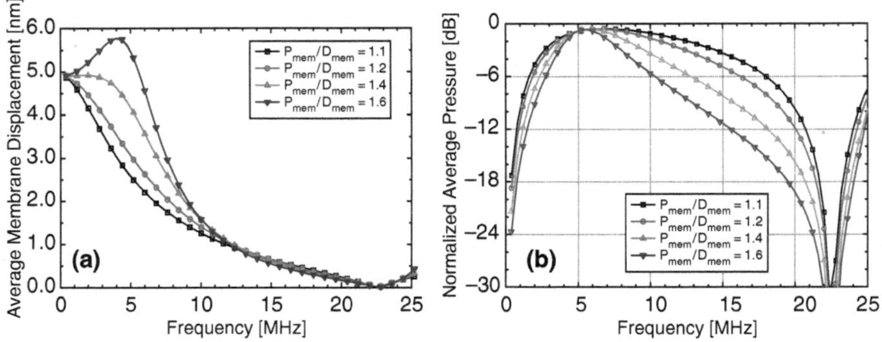

Fig. 22.5 (**a**) Average membrane displacement, (**b**) normalized transmit pressure of an unbounded cMUT transducer in water, with increasing separation between the membranes

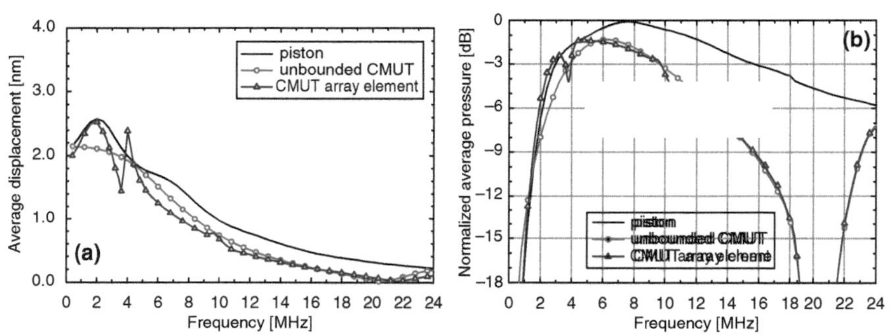

Fig. 22.6 (**a**) Average displacement, (**b**) normalized transmit pressure of a cMUT array element in water. A comparison with the corresponding unbounded cMUT and its equivalent piston transducer is also reported

finite dimensions is considered, acoustic interactions between the membranes arise, which are caused by the reaction pressure over the membranes of the propagating acoustic waves in the fluid (Caronti et al. 2005). The simulated displacement and pressure responses of a representative cMUT array element in water are shown in Fig. 22.6.

Although the membranes in the element are driven in phase, they are subjected to a different acoustic loading and may exhibit a different behavior, depending on their position in the element. The dips in the average displacement and transmit pressure at some frequencies are caused by the excitation of an additional mode through the fluid coupling, such that the inner column of membranes in the element vibrate out of phase with respect to the outermost ones. The remarkable pressure bandwidth of the equivalent piston transducer, having the same mechanical impedance as the cMUT membranes, is due to the absence of the higher order modes and the reactive effects of the fluid.

22.3 Technology of cMUT Transducers

In this paragraph, we point out our attention on the technology used to fabricate an ultrasonic transducer on a silicon substrate. We start with the basic technological process, which introduces the methods. We will analyze the PECVD (plasma-enhanced chemical vapor deposition), and finally a different approach to cavity formation called reverse technology (RT) will be discussed in the last section.

22.3.1 Basic Technology of the cMUT

To manufacture an array of electrostatic cells with dimensions in the micron range and with accurately defined geometry, silicon micromachining technology is employed. The conceptual steps of the fabrication process are the following: a sacrificial layer is grown onto a silicon substrate; its thickness H will define the gap height of the cells (Fig. 22.7a). A second structural layer (e.g., silicon nitride) of thickness h' is deposited over the sacrificial layer (Fig. 22.7b), and a small hole is made to create a via to remove the underlying sacrificial layer (Fig. 22.7c). A selective etchant solution is used to remove only the sacrificial layer; depending on the etching time, a wide circular cavity of radius a is realized under the structural layer, which remains suspended above the cavity (Fig. 22.7d). Finally, the etchant hole is sealed using a second thin layer h'' and the actual thickness of the membrane becomes the total thickness $h = h' + h''$ of the two layers (Fig. 22.7e). In the final device, the backplate of each cell is the heavily doped silicon substrate, whereas the second electrode is simply created by evaporating a thin metallic layer over the membrane, which is then patterned by means of a lithographic step (Fig. 22.7f).

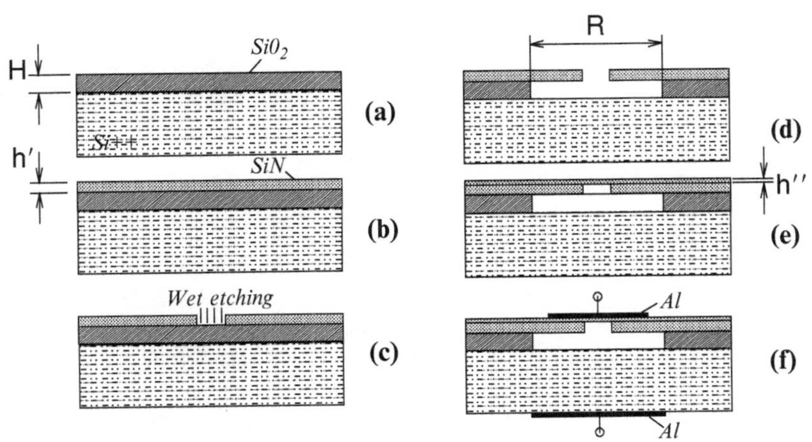

Fig. 22.7 Basic technological process of a single cMUT cell

22.3.2 Sacrificial Release Process: PECVD Technology

Silicon nitride is usually used as structural layer of the micromembranes that vibrate during operation (Cianci et al. 2006a, 2006b); therefore, the film must be characterized by a low tensile stress to obtain flat membranes suspended over the substrate (Foglietti et al. 2002; Ronnekleiv et al. 1997). Furthermore the silicon nitride structural layer must be a good dielectric to sustain the biasing voltage, and an effective barrier layer to protect the device against water when it works in contact with liquids (Lin et al. 1998; Schmid et al. 1998). The requirements for this material are high, and they become even more stringent due to the need of temperature compatibility to allow the use of metal layers and eventually the integration with electronics on the same chip. Furthermore, this material must be characterized by a good step coverage to conformably cover patterned substrates. PECVD technique allows growing films using a low temperature, but the properties of the film largely depend on the deposition process, which, as a consequence, must be optimized for the required application. The effects on the film properties of the deposition parameters, such as temperature, gas pressure, and gas flow rates, have been widely investigated (Cianci et al. 2005; Lin and Lee 1992; Parsons et al. 1991). Also, the effect of the plasma excitation frequency has been studied (Claasen et al. 1985), showing that the addition of a low-frequency excitation, to the standard one at 13.56 MHz, provides a mean to control the internal stress in the silicon nitride film, enhancing the film barrier properties and the deposition step coverage (Pearce et al. 1992; Tarraf et al. 2004; Van de Ven et al. 1990). Using this technique, silicon nitride can be used for the fabrication of cMUTs, and this results in an important improvement of the transducer's performances.

If the cMUTs have to be built on the same silicon substrate with the CMOS electronic circuitry, the highest temperature throughout the process cannot exceed 400 °C. Such low-temperature process is often called *CMOS compatible*. The PECVD technology is a CMOS-compatible process; also the reverse technology, that we shall see in the next paragraph, is CMOS compatible; all other types are not (i.e., LPCVD technology).

22.3.3 Reverse Fabrication Process (RT)

All sacrificial release processes have in common the presence of microholes on the surface of the transducer, necessary to evacuate the cavities under the membranes or, in the case of a two-dimensional array, to electrically connect upper pads to lower pads and to allow the electrical connection to external circuits. The presence of etching holes for emptying the cavities has a bad effect for both the uniformity of the membranes and the sealing of the cavities underneath. Moreover, the holes strongly affect the filling factor, especially in the case of high-frequency transducers, limiting the possibility to achieve large bandwidth and good sensitivity. The use of deep trenches for the electrical connection on the backside of the device can introduce problems not easily solvable, and remarkable complications in the fabrication process (Cheng et al. 2000).

In view of these considerations, a new fabrication concept was proposed by Caliano et al. (2005). The fabrication of the capacitive cell is not performed, such as in the standard process, growing successive layers onto the silicon wafer, up to the silicon nitride structural layer of the micromembranes, but inverting the sequence of the layers. The device is built on the top of a LPCVD silicon nitride film grown over a silicon wafer, using a standard process and finally, once the cMUT has been realized, the bulk silicon wafer is completely removed by KOH process, allowing the silicon nitride membranes to freely vibrate. By working on the back of the device, there is no need to make holes in the structural silicon nitride layer to evacuate the cavities. The device is fabricated using a commercial silicon wafer with a low-stress LPCVD silicon nitride layer and the temperatures of the whole process are not higher than 350 °C, so that a wide range of materials can be used. Electrical connection pads are already available on the backside of the die, ready to be soldered, avoiding deep trenches or interface circuits.

Generally, cMUT devices, realized with the conventional technique, require the use of a large number of lithographic masks. The reverse technique allows to reduce the number of masks to four, and it is even possible (for a particular configuration) to further reduce the lithographic steps to three. A further innovation of the RT process concerns the acoustic backing. Because of the high acoustic impedance of silicon, the use of a silicon device for acoustical emission is tricky: in fact, it is difficult to find a good material as backing layer to prevent the ringing of the pulse-echo signal (Ladabaum et al. 2000). In all the reported technologies, a relatively thick silicon bulk ($>350\,\mu$m) is always supporting the active cMUT membranes. In the reverse process, there is only a few-micron-thick silicon nitride layer under the cavity, and a classical backing material is used, also acting as a mechanical support for the device. Therefore, the acoustic device overcomes the unwanted effects of the bulk silicon.

The fabrication process uses, as starting substrate, a commercial silicon wafer (380-μm thick, $\langle 100\rangle$ orientation), coated on both faces by a LPCVD silicon nitride layer (Fig. 22.8a). The thickness of the silicon nitride layer depends on the particular design of the cMUT array. The first operation is the evaporation of the ground electrode (aluminum) followed by passivation with PECVD silicon nitride. A chromium sacrificial layer is then deposited (Fig. 22.8b, c). Using the first mask, the sacrificial chromium is patterned into small circular regions by optical lithography and wet etching (Fig. 22.8d): they will form the cavities of the transducer. The process foresees a PECVD silicon nitride layer, with a thickness greater than 400 nm that represents the backplate layer (Fig. 22.8e). At this point, a metal layer (Al) is sputtered and patterned using the second mask (Fig. 22.8f). A layer of PECVD silicon nitride is then deposited, to protect the electrode surface (Fig. 22.8g). Using the next mask, the etching holes (4 μm in diameter) are opened to access the sacrificial layer and to evacuate the cavities under the membrane (Fig. 22.8h); the cavities are then sealed using a PECVD nitride layer (Fig. 22.8i). The fourth patterning step assures the opening of the contact pads. The last layer of silicon nitride makes the structure robust, as it will be in touch with the backing layer to support the device after the

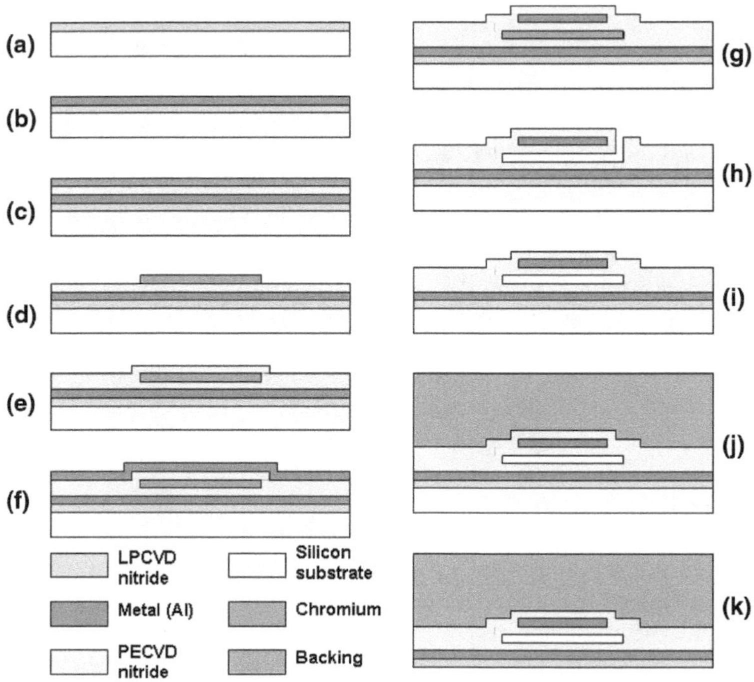

Fig. 22.8 Steps of the fabrication of a cMUT, using the *reverse fabrication process*

membrane release and the removal of the bulk silicon. The device is now diced and soldered on a flexible circuit board (e.g., using a flip-chip bonding technique) or a rigid PCB. At this step the transducer is electrically ready but not acoustically, being the membrane structural layer still attached to the silicon wafer. Mounted in a special case, an acoustic backing material is coated over the transducer (Fig. 22.8j), which works not only as an acoustic layer, but also as a mechanical support. The last step is the removal of the LPCVD layer and the bulk silicon (by KOH process), to release the membrane (Fig. 22.8l).

22.4 cMUT Characterization and Performances

The intrinsic differences between the operation principles and the fabrication technology of cMUTs in respect to classical piezoelectric transducers lead to the necessity to define a new and adequate characterization procedure able to extract the main functional parameters. Many measurement setups and procedures are shown in this paragraph, some of them typical and necessary for performance evaluation and benchmarking and some others useful for understanding the behavior of cMUTs.

A first characterization of cMUTs is made to determine the yield of the micromachining fabrication process. This level typically involves electrical impedance as well as laser-interferometry measurements. A second characterization is carried out to test the radiation and detection performances of cMUTs. In this case transmission and pulse/echo measurements are performed.

22.4.1 Electrical and Mechanical Characterization

22.4.1.1 Impedance

Impedance measurements are used to extract electrical and mechanical parameters of cMUTs. The assessment of these parameters among a set of transducers built on the same silicon wafer can be used to evaluate the fabrication process uniformity and yield. In fact, the electrical parameters can be related to some electrical and mechanical properties of the materials used in the transducer fabrication.

As explained in Sect. 22.2.5, the cMUT can be seen as a two-port (electrical and mechanical) network. The electrical impedance is thus affected by the acoustic load exerted by the propagating medium. To estimate the mechanical resonance frequency by finding the peaks of the impedance curves it is necessary to perform all impedance measurements while operating in air.

Figure 22.9 shows the impedance measurement results (Caliano et al. 2005a) performed on a 64-element array using an impedance analyzer (HP4194A, HP, Inc., California, USA). The impedance curves are superimposed to estimate the variation across the array elements. Much information about the electrical and mechanical properties of the materials used in the fabrication can be extracted by properly processing the electrical impedance data. It is possible to evaluate the parallel capacitance, the series resistance, and the mechanical resonance frequency of each array element. The distribution of these quantities across the transducers can be related

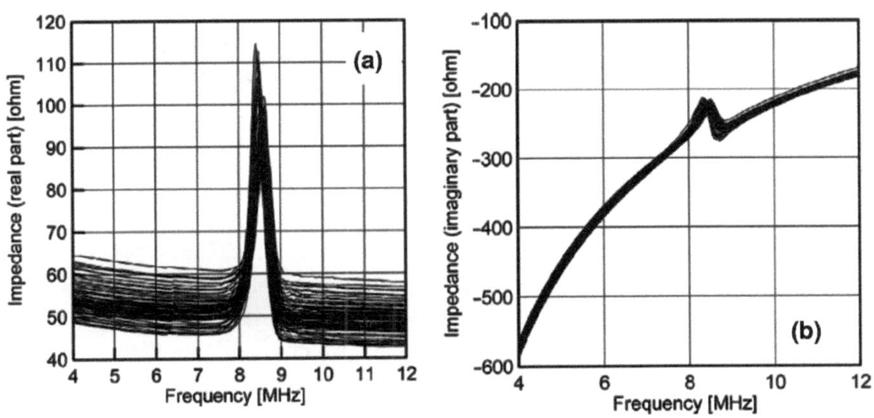

Fig. 22.9 Electrical impedance measurements of a 64-element cMUT array (**a**) real part, (**b**) imaginary part

to the uniformity of the process over the silicon wafer area. The series resistance, which can be extracted by the impedance real-part curves of Fig. 22.9, is proportional to the resistivity of the metal films used for electrode fabrication.

The parallel capacitance, obtainable by the impedance imaginary-part curves of Fig. 22.9, is related to the geometrical dimensions of the capacitive cells and to the dielectric properties of the materials used as structural and isolation layers. The average resonance frequencies of the array elements are extracted by calculating the frequency values of the maximum of the impedance real-part curves.

In Fig. 22.10 the resonance frequencies are plotted as a function of the array element number. This kind of representation is useful to evaluate the structural film thickness uniformity over the array and over the whole silicon wafer. The impedance imaginary part of a cMUT is substantially capacitive except near the mechanical resonance frequency where a modulation occurs. Thus, it is possible to extract the static capacitance of a cMUT by fitting the impedance curve with a pure capacitive impedance curve neglecting the resonance modulation.

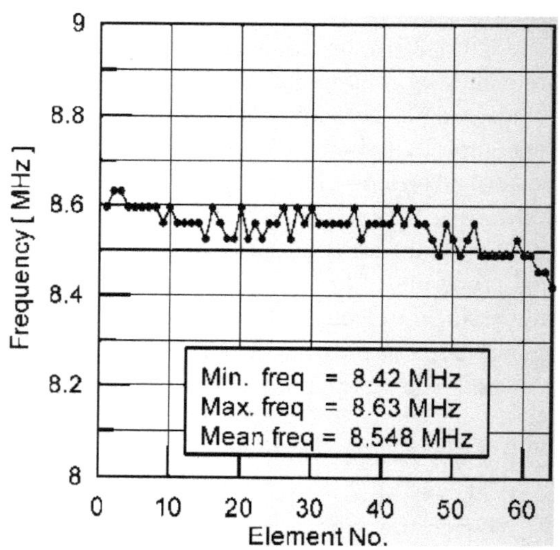

Fig. 22.10 Resonance frequency of the 64-array elements

22.4.1.2 Vibrometry

A typical laser Doppler vibrometry (LDV) experimental setup (Caronti et al. 2006b), shown in Fig. 22.11a, makes it possible to measure out-of-plane displacements of vibrating surfaces in the subnanometer range. The cMUT is placed under the interferometer on a holder that may contain a liquid (water or oil) in order to possibly investigate the mechanical behavior in immersion operation. An AC driving voltage, superimposed to a DC bias voltage, is applied to the cMUT by means of a function generator. The vibrometer and the function generator are connected to an oscilloscope. The cMUT and the holder can be placed on a micrometric motorized

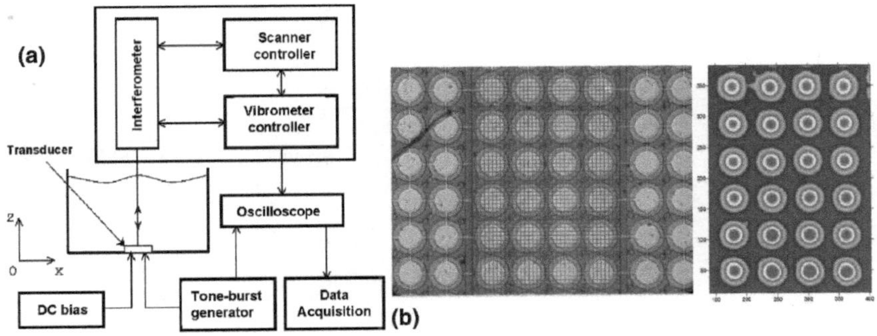

Fig. 22.11 (**a**) Laser Doppler vibrometer setup, (**b**) vibration map of a cMUT array element

stepper stage able to shift the position along the x and y directions, making it possible to measure the displacement of large vibrating surfaces. Figure 22.11b shows a vibration map of a portion of a cMUT array element. The scanned area is highlighted on the left side of the figure. A single displacement acquisition is performed on each point of the grid while driving the cMUT with a sinusoidal signal. The resulting displacement signal is acquired, and a vibration map is reconstructed by plotting the displacement amplitude as a function of the point coordinates. Vibration maps give important information about the mechanical behavior of cMUTs. The deformed shape of the membrane can be investigated in depth, and structural imperfections can be easily recognized. These local imperfections are related to the quality of the materials used for the device fabrication.

22.4.2 Acoustical Characterization

22.4.2.1 Transmission and Pulse/Echo

To characterize the performances of cMUTs, acoustical measurements must be performed. Typical acoustical characterization procedures involve transmission and pulse-echo measurements. The aim of these measurements is to obtain information about the radiation and detection capabilities of cMUT devices. For both kinds of measurements, the cMUT radiating area must be coupled to a propagating medium (typically water). The biased cMUT is driven using a function or a pulse generator. In the transmission setup, an hydrophone is placed in front of the radiating area and aligned along the axial direction (perpendicular to the transducer surface). In the pulse-echo setup, a flat reflector is placed in front of the transducer radiating area and aligned perpendicularly to the axial direction. In Fig. 22.12 transmission and pulse-echo setups are shown.

If the cMUT array is focused in the elevation direction (e.g., using an acoustic lens), the measurement distance may be conveniently chosen as the focal distance in order to minimize the effects of diffraction loss. In both transmission and pulse-echo

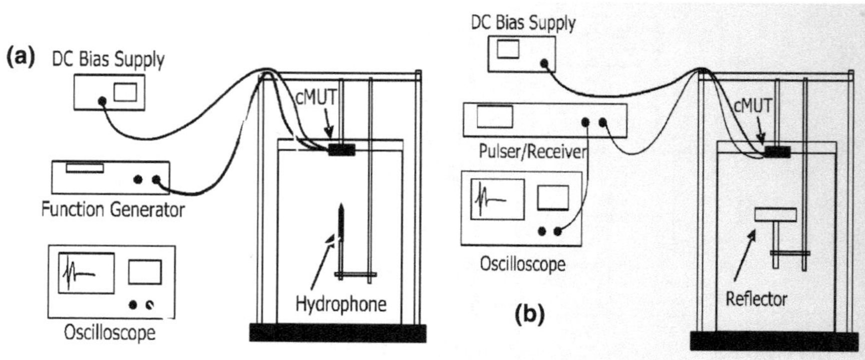

Fig. 22.12 (**a**) Transmission experimental setup, (**b**) pulse-echo experimental setup

measurements, a high-voltage broadband pulse is typically used as the excitation signal. The advantage of using such an excitation signal is related to the possibility of extracting the transmission and the two-ways transfer functions over a wide frequency range performing only a single measurement.

In the transmission measurement, the hydrophone detects the time-variant pressure signal and converts it to a voltage signal. In the pulse-echo measurement, the cMUT detects the round-trip pressure signal from the aligned flat target. In both measurements, the electric signals are acquired and digitized using an oscilloscope. The digitized data is then processed using FFT and Hilbert transforms to obtain the frequency spectrum and the envelope, respectively. Transmission and pulse-echo sensitivities, central frequency, fractional bandwidth, and pulse length at various decibel levels are the typical parameters used for cMUT performance characterization.

In Fig. 22.13 transmission and pulse-echo measurement results for a cMUT array are represented in both time and frequency domains. The main parameters, such as central frequency, −6-dB fractional bandwidth, and −20-dB pulse length are reported for both measurements. As can be seen, immersion-operating cMUTs show a very broadband response.

22.4.2.2 Medical Imaging

Figure 22.14 shows a linear scan image of a human carotid artery obtained by connecting a 128-element cMUT array to a commercial ultrasound system (Caliano et al. 2007). The scan was performed at 6.5 MHz using a single-cycle excitation burst. The scanning depth is 31 mm, and the transmission focus is set at about 17 mm. A good improvement in contrast and definition compared with conventional PZT arrays, due to the higher bandwidth, can be clearly observed in the image obtained with cMUT, especially in the part of the sternocleidomastoid muscle, where it is possible to distinguish the striations of the muscular tissue.

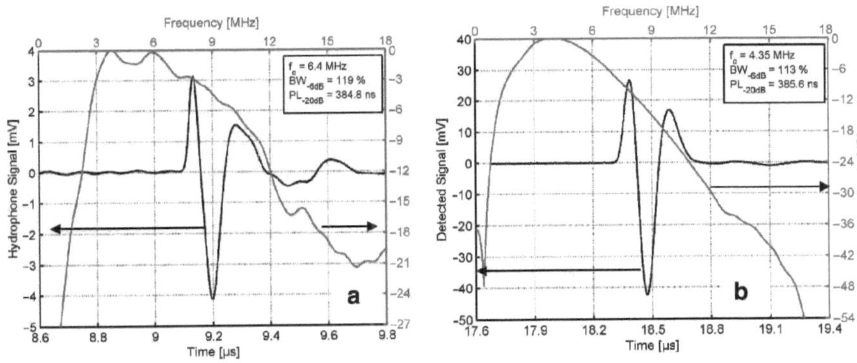

Fig. 22.13 (**a**) Transmission (**b**) pulse-echo measurements of a cMUT array

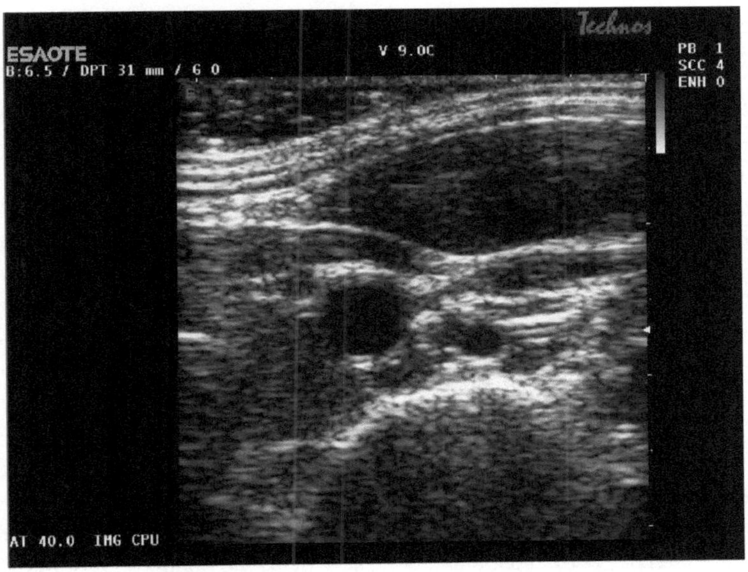

Fig. 22.14 In-vivo ultrasound imaging obtained with a 128-element cMUT linear array

22.5 Piezoelectric Micromachined Ultrasonic Transducer

22.5.1 Basic Principles of the pMUT

The pMUT basically has the same structure as the cMUT; it consists of a grid of micromembranes vibrating in a flextensional mode with the main difference that they are piezoelectrically, and not electrostatically, actuated.

Figure 22.15a shows the structure of the typical ultrasonic transducer – a thin piezoceramic plate, metallized on both surfaces, poled and vibrating in thickness direction, bonded to a base (backing), which is able to absorb the back radiation of

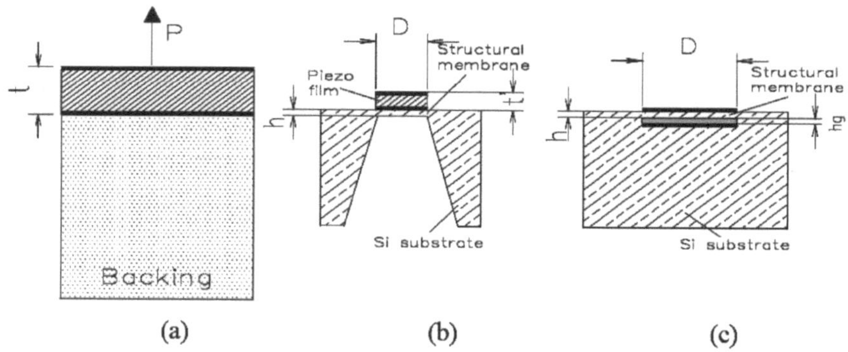

Fig. 22.15 Basic structures of a typical (**a**) piezoelectric transducer, (**b**) pMUT, and (**c**) cMUT

the plate. Figure 22.15b shows the typical cell of a pMUT structure; generally, the membrane is of silicon or silicon oxide and it is covered by a thin piezoelectric film. Finally, for comparison, Fig. 22.15c shows a cell of the cMUT structure. The working frequency of the structure of Fig. 22.15a is mainly governed by the thickness of the piezoceramic plate while, for pMUT and cMUT structures, the resonance frequency is mainly governed by the lateral dimension D of the membranes.

The fundamental difference between a pMUT and a cMUT cell is that the first can be open, i.e., without bottom, while the second must be closed in order to allocate the bottom electrode. For this reason there is not a physical limit to the displacement amplitude of the pMUT membrane while, as mentioned in the previous sections, for the cMUT, gap height, hg, and maximum membrane displacement are conflicting parameters.

The pMUT membrane works like the composite membrane of a commercial buzzer operating at audible frequencies (see Fig. 22.16). A metallized, transversally poled PZT plate of typical thickness 0.2–0.1 mm is bonded on a metal plate clamped on the border. When a varying electric voltage is applied to the electrodes the piezoelectric plate attempts to expand and contract due to the coupling, described by the piezoelectric constant e_{31} between the transverse electric field E_3 and the planar stress σ_1. Because the PZT plate is bonded to the metal plate, a flexural vibration takes place, governed by the diameters of the metal and PZT plates and by their thickness. The pMUT cell operates in the same way as a buzzer plate with the following differences: (a) the lateral dimension of the pMUT membrane is in the order

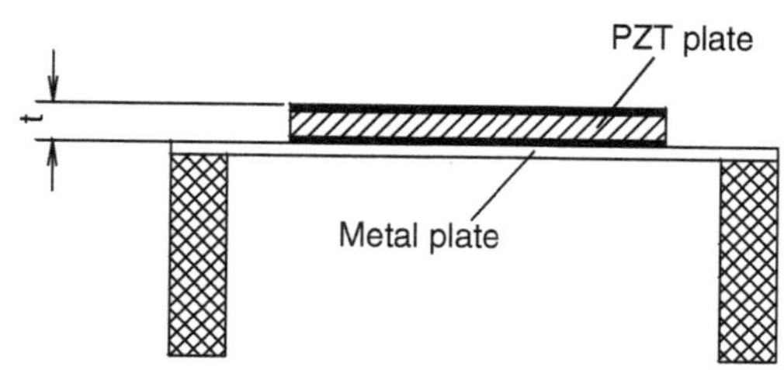

Fig. 22.16 Basic structure of a commercial buzzer

of tens of microns to allow vibrations in the mega-Hertz range, (b) the pMUT membrane is made of silicon, silicon oxide, or silicon nitride by means of micromachining techniques and has a thickness of few microns, (c) the PZT plate of a buzzer is made with a tape-casting technology while for the pMUT the PZT film is generally made with a chemical solution deposition (CSD), as we will see later.

The major advantage of the pMUT structure, compared with the typical ultrasonic transducer and the cMUT operating in the same frequency range, is evident in Fig. 22.15, keeping in mind that the distance between the electrodes dictates the electrical capacitance of the device, i.e., the electrical impedance: the smaller is this distance the higher is the capacitance and the smaller the impedance for the unit emitting surface. In fact, the distance between the electrodes in the typical transducer is equal to the thickness t of the PZT plate, which must be in the order of hundreds of microns to have resonance, in the thickness mode in the mega-Hertz range, while the PZT film of a pMUT has a thickness of few microns, which scarcely influence the resonance frequency of the membrane. Compared with the cMUT, the distance between the electrodes is of the same order (fractions of microns for the cMUT and few microns for the pMUT) but with completely different dielectric constants. Assuming for the gap of the cMUT cell the vacuum dielectric constant ε_0, the ratio between the dielectric constant of the PZT and that of the vacuum is the relative dielectric constant ε_r, which, as it is well known, for this material is one of the highest.

High capacitance not only means low electrical impedance and better matching with the front-end electronics but also implies that unavoidable parasitic capacitances are less significant for the sensitivity of the transducer. A square pMUT element with lateral dimensions of 100 μm has a capacitance of 121 pF for 1 μm and 76 pF for 2-μm PZT film thickness, while the capacitance of a bulk PZT element as large as 300 μm is $<$10 pF and only 0.1 pF for a cMUT of the same dimensions and gap height of 0.5 μm (Dausch et al. 2006).

22.5.2 pMUT Fabrication Process

Basically, pMUT fabrication involves building a grid of structural membranes suspended over the board of open cells created on the rear of a silicon wafer and covered on the top with a metallized piezoelectric layer (see Fig. 22.15b). A typical fabrication process is reported by Yamaschita et al. (2004). Perçin and Khuri-Yakub (2002) reported a fabrication process more similar to that usually used for cMUT transducers; it makes use of a sacrificial layer, deposited over a silicon wafer, to create a vacuum chamber under the suspended silicon nitride membranes. This process appears more convenient for dense 1D and 2D arrays because the silicon wafer, as in the classical cMUT, works like a base for all the vibrating membranes.

22.5.3 Piezoelectric Film

The fabrication of a good quality, crack-free, uniform piezoelectric film is at present the major obstacle for the development of high-density, high-frequency pMUT arrays. Thick, single layer films of PZT up to 15–30 μm can easily be obtained by a screen-printing technique, but generally they have low density and they must be cured at a temperature of 800 °C or higher to assure adequate sintering of the PZT powders. Unfortunately, this temperature is not compatible with the silicon micromachining. Thin piezoelectric films of zinc oxide (ZnO) or PZT can be obtained with different deposition techniques. Sputtered ZnO is largely used for high-frequency (above 100 MHz) devices, such as delay lines and saw filters, because of the intrinsic low mechanical losses, but because of its poor electromechanical coupling, PZT is usually preferred for power applications, such as ultrasonic transducers.

Most of the existing deposition techniques have been investigated to make PZT films. Physical techniques, such as ion beam sputtering, rf planar magnetron sputtering, pulsed laser deposition, etc., have given poor results. Chemical methods such as metal organic-chemical vapor deposition and CSD are now established methods to make PZT films. Both these methods require much lower temperature, about 600 °C, than the typical sintering temperature (1,000 °C) used for bulk PZT ceramics. At present, the CSD technique appears more promising for MEM applications but it requires postannealing treatments for the crystallization of the film. A typical fabrication process able to fabricate crack-free films of PZT up to some micrometers is the spin-coating of the PZT sol–gel, followed by two drying cycles at 150 and 350 °C (Akasheh et al. 2004). A total of 15–32 layers have to be deposited using crystallization annealing at 600–700 °C every four to five layers. Another problem arises from the interdiffusion and oxidation reactions between silicon and PZT. To prevent these phenomena a chemical barrier between these materials must be used. Usually the barrier function is provided by the sequence of layers PZT/Pt/Ti/SiO2/Si, which include the Pt electrode, where Ti is needed as adhesion layer. Patterning of these electrode layers, such as the PZT layer, is difficult. Wet etching cannot be used, and dry etching is required with a quite low etching rate. Finally, after the electrode metallization and patterning, the PZT film must be poled applying an electric field in the order of 15 MV/m.

The piezoelectric constants of films are quite difficult to measure; for this reason, only few data are reported in papers on pMUT, where usually only characterization of the device is reported omitting the material characterization. Baborowsky et al. (2002) reported a sol–gel film with good piezoelectric parameters. The transverse parameter d_{33} improves increasing layer thickness and reaches a value very similar to that of bulk PZT for a thickness of 4 μm. The appropriate piezoelectric constant for planar deformation e_{31} is reported to be independent of the film thickness and equal to half of that of a bulk PZT.

At present only very few acoustic characterizations of pMUT devices can be found in literature. Bandwidth and insertion loss are not yet reported. Dausch et al. (2006) were the first to demonstrate the possibility of obtaining pulse B mode echographic images with a 5 × 5 element, 6.25-MHz pMUT array. They reported a pressure up to 5.5 MPa generated with a single pMUT cell with lateral dimensions ranging from 50 to 200 μm.

22.6 Conclusion

cMUT and pMUT are both flextensional micromachined transducers but with different actuation principles. cMUT has received greater attention from researchers than pMUT; it has shown good electroacoustic characteristics, which parallel, and even exceed, those of conventional transducers. Very good 2D images have been obtained, demonstrating the possibilities of this new technology to be usefully utilized in commercial 1D and 2D probes for medical applications. At present pMUT is in a very early stage of development. There is not a clear evidence of its potentialities in the field of ultracoustic imaging. Large bandwidths are not experimentally demonstrated, dense arrays are not yet fabricated, and acceptable uniformity of the piezoelectric activity of the different membranes is still to be proven. The deposition process of crack-free PZT layers is quite difficult, time consuming, and not fully compatible with the standard micromachining techniques. As a final consideration, we can say that pMUTs have some potential advantages over the cMUT cells but they are much more difficult to fabricate, and the advantages are still to be demonstrated.

References

Akasheh F, Myers T, Fraser JD, Bose S, Bandyopadhyay A (2004) Development of piezoelectric micromachined ultrasonic transducers. Sens. Actuat. A, vol. 111, pp. 275–287

Anderson MJ, Hill JA, Fortunko CM, Doğan NS, Moore RD (1995) Broadband electrostatic transducers: modeling and experiments. J. Acoust. Soc. Am., vol. 97, no. 1, pp. 262–272

Baborowsky J, Ledermann N, Muralt P (2002) Piezoelectric micromachined transducers (pMUT) based on PZT films. IEEE Ultrason. Symp., pp. 1051–1054

Bayram B, Kupnik M, Yaralioglu GG, Oralkan O, Ergun AS, Lin DS, Wong SH, Khuri-Yakub BT (2007) Finite element modeling and experimental characterization of crosstalk in 1-D CMUT arrays. IEEE Trans. UFFC, vol. 54, no. 2, pp. 418–430

Bernstein JJ, Bottari J, Huston K, Kirkos G, Miller R (1999) Advanced MEMS ferroelectric ultrasound 2D arrays. IEEE Ultrason. Symp., pp. 1145–1153

Caliano G, Foglietti V, Galanello F, Caronti A, Lamberti N, Carotenuto R, Pappalardo M (2000) Micromachined ultrasonics transducers using silicon nitride membrane fabricated in PECVD technology. IEEE Ultrason. Symp., pp. 963–967

Caliano G, Carotenuto R, Cianci E, Foglietti V, Caronti A, Iula A, Pappalardo M (2005a) Design, fabrication and characterization of a capacitive micromachined ultrasonic probe for medical imaging. IEEE Trans. UFFC, vol. 52, pp. 2259–2269

Caliano G, Pappalardo M, Caronti A, Minotti A, Foglietti V, Cianci E, Nencioni A (2005b) Surface micromechanical process for manufacturing micromachined capacitive ultracoustic transducers and relevant micromachined ultracoustic transducer. Patent no. RM2005A93, PCT/IT2006/000126

Caliano G, Savoia A, Caronti A, Longo C, Gatta P, Pappalardo M (2007) Echographic images improvements with a cMUT probe. 19th International Congress on Acoustics (ICA), Madrid, p. 375

Cantrell JH, Heyman J, Yost W, Torbett M, Breazeale M (1979) Broadband electrostatic acoustic transducer for ultrasonic measurements in liquids. Rev. Sci. Instrum., vol. 50, pp. 31–33

Caronti A, Caliano G, Iula A, Pappalardo M (2000) Electrical impedance mismatch in capacitive micromachined ultrasonic transducers. IEEE Ultrason. Symp., pp. 925–930

Caronti A, Iula A, Caliano G, Pappalardo M (2002) An accurate model for capacitive micromachined ultrasonic transducers. IEEE Trans. UFFC, vol. 49, no. 2, pp. 159–168

Caronti A, Fiasca D, Caliano G, Pappalardo M (2003) Experimental study of acoustic coupling in CMUT arrays by optical interferometry. IEEE Ultrason. Symp., pp. 1960–1964

Caronti A, Carotenuto R, Caliano G, Pappalardo M (2004) The effects of membrane metallization in capacitive microfabricated ultrasonic transducers. J. Acoust. Soc. Am., vol. 115, no. 2, pp. 651–657

Caronti A, Savoia AS, Caliano G, Pappalardo M (2005) Acoustic coupling in capacitive microfabricated ultrasonic transducers: modeling and experiments. IEEE Trans. UFFC, vol. 52, no. 12, pp. 2220–2234

Caronti A, Caliano G, Carotenuto R, Savoia AS, Pappalardo M, Cianci E, Foglietti V (2006a) Capacitive micromachined ultrasonic transducer (CMUT) arrays for medical imaging. Microelectron. J., vol. 37, pp. 770–777

Caronti A, Longo C, Savoia AS, Gatta P, Caliano G, Pappalardo M (2006b) Analysis of acoustic interaction effects and crosstalk in CMUT linear arrays for medical imaging. IEEE Ultrason. Symp., pp. 582–585

Carr H, Wykes C (1993) Diagnostic measurements in capacitive transducers. Ultrasonics, vol. 31, no. 1, pp. 13–20

Certon D, Teston F, Patat F (2005) A finite difference model for cMUT devices. IEEE Trans. UFFC, vol. 52, no. 12, pp. 2199–2210

Cheng CH, Chow EM, Jin X, Ergun AS, Khuri-Yakub BT (2000) An efficient electrical addressing method using through-wafer vias for two-dimensional ultrasonic arrays. IEEE Ultrason. Symp., pp. 1179–1182

Cianci E, Pirola F, Foglietti V (2005) Analysis of stress and composition of silicon nitride thin films deposited by ECR-PECVD for microfabrication processes. J. Vac. Sci. Technol. B, vol. 23, no. 1, p. 168

Cianci E, Foglietti V, Minotti A, Caronti A, Caliano G, Pappalardo M (2006a) Fabrication techniques in micromachined capacitive ultrasonic transducers and their applications. In: Leondes CT (ed.) MEMS/NEMS Handbook: Techniques and Applications, vol. 2, Springer, New York

Cianci E, Schina A, Minotti A, Quaresima S, Foglietti V (2006b) Dual frequency PECVD silicon nitride for fabrication of CMUTs' membranes. Sens. Actuat. A, vol. 127, pp. 80–87

Claasen WAP, Valkenburg WGYN, Willemsen MFC, van de Wijgert WM (1985) Influence of deposition temperature, gas pressure, gas phase composition, and RF frequency on composition and mechanical stress of plasma silicon nitride layers. J. Electrochem. Soc., vol. 132, no. 4, p. 893

Daft C, Panda S, Wagner P, Ladabaum I (2006) Two approaches to electronically scanned 3D imaging using cMUTs. IEEE Ultrason. Symp., pp. 685–688

Dausch D, Castellucci JB, Chou DR, Von Ramm OT (2006) Piezoelectric micromachined ultrasound transducer (pMUT) arrays for 3D imaging probe. IEEE Ultrason. Symp., pp. 930–933

Eccardt P, Niederer K (2000) Micromachined ultrasound transducers with improved coupling factors from a CMOS compatible process. Ultrasonics, vol. 38, pp. 774–780

Eccardt P, Niederer K, Scheiter T, Hierold C (1996) Surface micromachined ultrasound transducers in CMOS technology. IEEE Ultrason. Symp., pp. 959–962

Eccardt PC, Lohfink A, von Garssen HG (2005) Analysis of crosstalk between fluid coupled membranes. IEEE Ultrason. Symp., pp. 593–596
Foglietti V, Caliano G, Cianci E, Galanello F, Pappalardo M (2000) Fabrication of micromechanical capacitive ultrasonic transducers by surface micromachining. Proc. MICROtec 2000, VDE World Micro-Technologies Congress, Berlin, p. 79
Foglietti V, Memmi D, Cianci E, Caliano G, Pappalardo M (2001) Capacitive ultrasonic transducers fabricated by a low temperature surface-micromachined process. Sixth National Conference on Sensors and Microsystems, AISEM 2001, Pisa
Foglietti V, Cianci E, Memmi D, Caliano G, Caronti A, Pappalardo M (2002) Fabrication of capacitive ultrasonic transducers by a low temperature and fully surface-micromachined process. Precision Eng., vol. 26, p. 347
Ge LF (1999) Electrostatic airborne ultrasonic transducers: modeling and characterization. IEEE Trans. UFFC, vol. 46, no. 5, pp. 1120–1127
Haller M, Khuri-Yakub BT (1994) A surface micromachined electrostatic ultrasonic air transducer. IEEE Ultrason. Symp., pp. 1241–1244
Hietanen J, Stor-Pellinen J, Luukkala M (1992) A model for an electrostatic ultrasonic transducer with a grooved backplate. IEEE Meas. Sci. Technol., vol. 3, pp. 1095–1097
Jin X, Ladabaum I, Khuri-Yakub BT (1998) The microfabrication of capacitive ultrasonic transducers. IEEE J. Microelectromech. Syst., vol. 7, no. 3, pp. 295–302
Jin X, Ladabaum I, Degertekin FL, Calmes S, Khuri-Yakub BT (1999) Fabrication and characterization of surface micromachined capacitive ultrasonic immersion transducers. IEEE J. Microelectromech. Syst., vol. 8, no. 1, pp. 100–114
Kinsler LE, Frey AR, Coppens AB, Sanders JV (2000) Fundamentals of Acoustics. Wiley, New York
Kuhl W, Schodder G, Schroder F (1954) Condenser transmitters and microphones with solid dielectric diaphragms for airborne ultrasonics. Acustica, vol. 4, pp. 520–532
Ladabaum I, Jin X, Soh HT, Atalar A, Khuri-Yakub BT (1998) Surface micromachined capacitive ultrasonic transducers. IEEE Trans. UFFC, vol. 45, no. 3, pp. 678–690
Ladabaum I, Wagner P, Zanelli C, Mould J, Reynolds P, Wojcik G (2000) Silicon substrate ringing in microfabricated ultrasonic transducers. IEEE Ultrason. Symp., pp. 943–946
Lin H, Xu L, Chen X, Wang X, Sheng M, Stubhan F, Merkel K, Wilde J (1998) Moisture-resistant properties of SiNx films prepared by PECVD. Thin Solid Films, vol. 333, p. 71
Lin KC, Lee SC (1992) The structural and optical properties of a-SiNx:H prepared by plasma-enhanced chemical-vapor deposition. J. Appl. Phys., vol. 72, no. 11, p. 5474
Mason WP (1948) Electromechanical transducers and wave filters. Van Nostrand Company, New York
Mills DM, Smith LS (2003) Real time in-vivo imaging with capacitive micromachined ultrasound transducer (cMUT) linear arrays. IEEE Ultrason. Symp., pp. 568–571
Muralt P, Schmitt D, Ledermann N, Baborowski J, Weber PK, Steichen W, Petigrand S, Bosseboeuf A, Setter N, Gaucher P (2001) Study of PZT coated membrane structures for micromachined ultrasonic transducers. IEEE Ultrason. Symp., pp. 907–911
Parsons GN, Souk JH, Batey J (1991) Low hydrogen content stoichiometric silicon nitride films deposited by plasma-enhanced chemical vapor deposition. J. Appl. Phys., vol. 70, p. 1553
Pearce CW, Fetcho RF, Gross MD, Koefer RF, Pudliner RA (1992) Characteristics of silicon nitride deposited by plasma-enhanced chemical vapor deposition using dual frequency radio-frequency source. J. Appl. Phys., vol. 71, no. 4, p. 1838
Perçin G, Khuri-Yakub BT (2002) Piezoelectrically actuated flextensional micromachined ultrasound transducers. Ultrasonics, vol. 40, pp. 441–448
Ronnekleiv A, Ladabaum I, Jin X, Khuri-Yakub BT (1997) An improved circuit model of MUTs. IEEE Ultrason. Symp., pp. 395–399
Scheeper PR, Voorthuzen JA, Bergveld P (1991) PECVD silicon nitride diaphragm for condenser microphones. Sens. Actuat. A, vol. 4, p. 79
Schindel D, Hutchins D, Zou L, Sayer M (1995) The design and characterization of micromachined air-coupled capacitance transducers. IEEE Trans. UFFC, vol. 42, no. 1, pp. 42–50

Schmid P, Orfert M, Vogt M (1998) Plasma deposition of Si N and Si O passivation layers on three-dimensional sensor devices. Surf. Coat. Technol., vol. 98, p. 1510

Suzuki K, Higuchi K, Tanigawa H (1989) A silicon electrostatic ultrasonic transducer. IEEE Trans. UFFC, vol. 36, no. 6, pp. 620–627

Tarraf A, Daleiden J, Irmer S, Prasai D, Hillmer H (2004) Stress investigation of PECVD dielectric layers for advanced optical MEMS. J. Micromech. Microeng., vol. 14, p. 317

Van de Ven EP, Connick IW, Harrus AS (1990) Advantages of dual frequency PECVD for deposition of ILD and passivation films. Proc. IEEE VMIC Conf., p. 194

Wang Z, Zhu W, Miao J, Zhu H, Chao C, Tau OK (2006) Micromachined thick film piezoelectric transducer array. Sens. Actuat. A, vol. 130–131, pp. 485–490

Yamaschita K, Katata H, Okuyama M, Miyoshi H, Koto G, Aoyagi S, Suzuki Y (2002) Arrayed ultrasonic microsensors with high directivity for in air use using PZT thin film on silicon diaphrams. Sens. Actuat. A, vol. 97–98, pp. 302–307

Yamashita K, Chansamphou L, Miracami H, Okuyama M (2004) Ultrasonic micro array sensors using piezoelectric thin film and resonant frequency tuning. Sens. Actuat. A, vol. 114, pp. 147–153

Yaralioglu GG, Badi MH, Ergun AS, Khuri-Yakub BT (2003) Improved equivalent circuit and finite element method modeling of capacitive micromachined ultrasonic transducers. IEEE Ultrason. Symp., pp. 469–472

Yaralioglu GG, Ergun AS, Khuri-Yakub BT (2005) Finite-element analysis of capacitive micromachined ultrasonic transducers. IEEE Trans. UFFC, vol. 52, no. 12, pp. 2185–2198

Zhou S, Hossack JA (2007) Reducing inter-element acoustic crosstalk in capacitive micromachined ultrasound transducers. IEEE Trans. UFFC, vol. 54, no. 6, pp. 1217–1228

Index

A
Acceptor, 68, 69, 71
Acoustic axis, 222
Acoustic coupling, 263, 283, 460–462
Acoustic impedance, 161, 169, 179, 180, 184–186, 196, 201–206, 261, 276, 284, 434, 435, 442, 446, 465
Acoustic lens, 161, 163–172, 174, 175, 434, 440, 448, 469
Acoustic matching, 161, 203, 204, 434, 436, 442, 449
Acoustic silicon, 161–175
Admittance, 218, 219, 221, 283
AFM probe tips, 421
Alkaline niobates, 81
AlN, 40, 41, 415, 416
$A_{m-1}Bi_2B_mO_{3m+2}$, 105
Annular arrays, 198, 309, 310, 312, 313, 440–444
Antiferroelectric, 43, 63, 64, 81, 103, 309, 347
Arbitrary coordinate system, 23
a-texture, 28
Aurivillius phases, 50

B
Backing layer, 179, 194, 196, 211, 283, 439, 465
Bandwidth, 174, 175, 186, 193, 196, 204, 205, 207, 211, 224, 232, 283, 424, 433, 436, 449, 453, 462, 474
Bender bars, 230
Bi-laminar, 230
Bimorph, 323, 350, 351, 353, 363, 366, 380, 381, 383, 419, 420, 422
Bismuth layer-structured ferroelectrics, 105, 106
Boracite, 34, 35

C
Carbon fiber composite, 179–187
Cavitation, 223
Cell doubling, 45
Centrosymmetric, 7, 20–22, 39, 40, 72, 346
Ceramic-polymer composites, 72, 196, 234, 236–239, 269
Characteristic impedance, 192
Charge control, 332
Closed-loop methods, 332–335
cMUT, 433, 453–455, 458–460, 463, 466–471, 473, 475
Coercive field, 12, 14, 62, 68, 108, 138, 151, 225, 289, 290, 295, 300, 346, 417
Collapse voltage, 456, 457
Composition gradient actuators, 289–303
Connectivity, 199, 204, 205, 211, 234, 236, 264, 305, 311
Converse piezoelectric effect, 18, 19, 33, 132, 349
Copolymers, 135, 137, 0139, 143–146, 151, 152, 204, 442, 443
Correlation length, 4
Crystallographic classes, 19, 20
c-texture, 28
Curie point, 21, 81, 82, 106, 124
Curie-Weiss law, 7
Curie-Weiss temperature, 7
Cymbal transducer, 255–257, 354–357

D
Deformable mirrors, 364, 365, 422
Depth of penetration, 192, 431, 432
Dielectric anisotropy, 24, 25, 61
Dielectric loss, 68, 98, 203, 206, 219, 222, 255, 415, 416
Diesel injection valves, 365, 366

Direct extensional actuators, 322, 325
Directivity, 62, 222, 223, 257, 447
Discretization, 245, 246, 253, 295
Displacement transducers, 342, 343
Domain concentrations, 26
Domain switching, 12, 14
Domain wall motion, 14, 47, 140, 346, 417
Domain walls, 21, 25, 30, 32, 46–50, 65–68, 346, 417
Donor, 68, 71, 110
Doppler, 196, 197, 436, 437, 468, 469
Drive techniques, 361–363
d^0 transition metal cations, 42

E
Elastic compliance, 10, 28, 132, 135
Elastic Gibbs function, 10, 11
Elastography, 191
Electric displacement, 19, 243, 293, 294, 414
Electroacoustic efficiency, 222, 376
Electroconductive, 289, 296
Electromechanical anisotropy, 62
Electropolymers, 131–155
Electrostatic energy density, 10
Electrostriction, 6, 11, 28, 142–144, 146, 150, 343, 346, 347
Electrostrictive graft elastomers, 148–150
Electrostrictive polymers, 131, 133, 134, 142–152
Energy harvesting, 373–386

F
Ferroelastic domain walls, 25, 32, 417
Fiber composite, 179–187, 261–284, 377
Finite element method, 241–255
Flextensional transducers, 231–233, 351–360
Fourier transform, 207, 221, 253, 408
Frequency bandwidth, 196, 446
Functionally graded, 289, 292
Fused deposition of multimaterials, 291

G
Galerkin's method, 245
Gauss's equation, 243
Gibbs function, 10, 13
Grain orientation factor, 107, 110, 115, 117, 118

H
Harmonic analysis, 252, 256
Helmholtz free energy, 3, 7, 10
Homopolymer, 136, 139
Huygen's principle, 208
Hybridization, 61
Hydrophone designs, 234–239
Hysteresis loop, 13, 71, 108, 120, 151, 154, 294, 297, 346, 417

I
Imaging modalities, 191
Impedance matching, 217, 434, 448
Induced polarization, 39, 134, 299
Input impedance, 192, 207, 208, 446, 460
Internal bias, 14
Internal stresses, 26, 27, 67, 322, 464

K
Kerf, 432, 433, 440, 441, 445, 446

L
Landau-Ginzburg-Devonshire, 4–11, 61, 66
Langevin-type, 390, 404
Lattice composites, 310, 311
Linear array, 198, 432, 433, 440, 445–449, 471
Lone electron pairs, 42
Lotgering method, 107

M
Magnetic resonance imaging, 191, 431
Magnetoelectric coupling, 33
Magnetostriction, 343–345
Mechatronics, 131–155
Medical ultrasonic array, 161–175
MEMS, 290, 413–427
MicroCT, 432
MicroMR, 432
MicroPET, 432
Micropositioning, 319–337
Micropumps, 350, 419, 423–424
Modal analysis, 252–253
Monolithic bending actuators, 290
Monomorph, 296, 323, 349, 350, 420
Moonie, 351, 353–357, 361, 363, 377
Morphotropic phase boundary, 22, 45, 64–68, 81, 96, 265, 291, 297, 416, 438
Multiferroics, 33, 34

N
Neumann's law, 39
Nonlinear dielectric behavior, 133
Nonlinear dielectric media, 5

O
Octahedral coordination, 42, 52
Order parameter, 3, 4, 6, 7, 11, 33
Oxygen octahedra, 43, 45, 51, 53, 64, 69, 71

P
Pad-printing, 209
PECVD technology, 464
Perovskite structure, 5, 33, 43, 46, 48, 52, 60, 290, 296, 416
Phase transformation, 3, 4, 7, 48–50, 343, 344
Piezoceramic fibers, 261–263, 277, 281, 284
Piezomagnetism, 33
pMUT, 453–455, 471–475
Point groups, 6, 20, 39
Polyamides, 152–155
Poly(electrolyte), 307
Polymorphic, 48, 65, 136
Poly(vinylidene fluoride), 131, 432
Projector, 217, 222–234
Pulse-echo, 161, 192–193, 208, 362, 435, 441, 469–471
Pyroelectric, 5, 17, 18, 20, 39, 44, 103, 104, 117, 151, 299

R
Radiation impedance, 219, 220, 459–461
Radiation pattern, 192, 193, 196, 197, 200, 208, 209
Radius of curvature, 164, 165, 197
RAINBOW, 291, 293, 323, 351, 352, 357, 358, 360
Rapid prototyping, 305, 316, 381
Relaxor, 17, 19, 59, 64, 72–75, 145, 161, 374, 416
Reorientation of dipoles, 69
Reversal of polarization, 11, 14, 35, 153, 343
Ring projectors, 229, 230
Robocasting, 305–316
Roto-flextensional, 354–357

S
Screen-printing, 183, 474
Shape memory effect, 343, 344, 347
Signal to noise ratio, 192, 415
Single element transducers, 193–197, 211, 432, 434–439
Sol-gel, 84, 262, 264–271, 438, 474
Solid freeform fabrication, 305
Sonar, 19, 217–239, 241
Spontaneous polarization, 3, 5–11, 13, 14, 21, 24, 28, 29, 34, 39, 42, 44–48, 104, 119, 294, 300
Spontaneous strain, 6, 8, 10, 47, 62
Static analysis, 251, 252
Symmetry hierarchy, 39, 40

T
Temperature coefficient of resonance frequency, 43, 108, 121, 122, 126
Terpolymers, 146–148, 151
THUNDER, 359, 360, 378
Tilt transition, 43–45, 50
Tolerance factor, 46
Tonpilz projectors, 233, 234
Transducer arrays, 198–201, 461
Transient analysis, 253
Transmission efficiency, 161, 163
Transmitting power response, 224
Transmitting voltage response, 223, 257, 284
Tungsten bronzes, 53, 106, 234
2D arrays, 191, 213, 473

U
Unidirectional carbon-fiber composite, 180–185
Unimorph, 296, 323, 328, 349, 350, 357, 368, 377, 378, 420, 422

V
Variational principle, 241, 242, 244, 245, 248–250
Vibrometry, 468, 469

W
Wavenumber, 207, 220
Wind energy harvester, 383–386
Wire-bonding technology, 389–411
Woven carbon cloth, 182

Z
Zinc oxide, 415, 416

SPRINGER NATURE

GPSR Compliance

The European Union's (EU) General Product Safety Regulation (GPSR) is a set of rules that requires consumer products to be safe and our obligations to ensure this.

If you have any concerns about our products, you can contact us on ProductSafety@springernature.com

In case Publisher is established outside the EU, the EU authorized representative is:

Springer Nature Customer Service Center GmbH
Europaplatz 3
69115 Heidelberg, Germany

FSC
www.fsc.org
MIX
Papier | Fördert
gute Waldnutzung
FSC® C083411

Kolibri 360 GmbH
Ferdinand-Jühlke-Straße 7
99095 Erfurt, Deutschland
produktsicherheit@kolibri360.de